GROUP THEORY

GROUP THEORY

W. R. SCOTT

Professor of Mathematics
The University of Utah

DOVER PUBLICATIONS, INC.
NEW YORK

To Bobbie

Published in Canada by General Publishing Company, Ltd., 30 Lesmill Road, Don Mills, Toronto, Ontario.
Published in the United Kingdom by Constable and Company, Ltd., 10 Orange Street, London WC2H 7EG.

This Dover edition, first published in 1987, is an unabridged, corrected republication of the work first published by Prentice-Hall, Englewood Cliffs, N.J., 1964.

Manufactured in the United States of America
Dover Publications, Inc., 31 East 2nd Street, Mineola, N.Y. 11501

Library of Congress Cataloging-in-Publication Data

Scott, William R. (William Raymond), 1919–
  Group theory.

  Reprint. Originally published: Englewood Cliffs, N.J. : Prentice-Hall, 1964.
  Bibliography: p.
  Includes indexes.
  1. Groups, Theory of.   I. Title.
QA171.S42   1987   512′.22   87-562
ISBN 0-486-65377-3 (pbk.)

PREFACE

This book contains most of the standard basic theorems in group theory, as well as some topics which have not heretofore appeared in book form. The organization is as follows. Chapters are numbered 1 through 15; sections in Chapter 5, for example, are numbered 5.1, 5.2, etc., and theorems within sections, 5.2.1, 5.2.2, etc. Exercises follow nearly all of the sections and are numbered like theorems. Material other than theorems is not numbered (occasionally, Equations (1), (2), etc., or Examples 1, 2, etc., appear within a section). In addition to an index and bibliography, there is an index of notation. Chapters 4 through 15 are closed by a brief list of references.

The exercises are of a quite varied nature and are not graded in difficulty. Some are mere remarks of the "of course" variety. A number are counter-examples to various possible modifications of theorems in the next. A further type asks the reader to supply details omitted in proofs of theorems in the text. Many are provided with hints (in a few cases, simplifying published proofs). The reader should at least read the exercises since frequent reference is made to them in the body of the text.

An attempt has been made to keep the book as self-contained as possible. For example, in Chapter 13, in order to avoid the use of Dirichlet's theorem about the infinitude of primes in an arithmetic progression, a special case sufficient for the problem at hand is proved. A few theorems from elementary number theory are presupposed. Of a less elementary nature, the following background is needed at various places: (a) The easier properties of cardinal and ordinal numbers, though lack of this background will not hamper the reader very much; (b) Some facts about vector spaces and matrices are assumed, although others are proved; (c) The theorem that the algebraic integers form an integral domain is used in Chapter 12; (d) Various facts about polynomials are required; (e) In Chapter 14, some facts from Galois theory and the theory of division rings are assumed; (f) In Section 15.4, the solvability of groups of odd order (Feit and Thompson [1]) is assumed.

I wish to thank the following organizations and people: The University of Kansas for granting me sabbatical leave and Cornell University for providing office space during 1962–63. W. Feit for several conversations, a seminar, several theorems in Chapter 12, and for bringing to my attention a number of items, including the theorem in Section 15.4. R. Bercov for conversations about S-rings, including unpublished material of his (these results do not appear in the book because of space requirements). L. Sonneborn for reading a portion of the manuscript and for a number of suggestions. T. Head for calling my attention to a theorem. P. Hall, E. Gaughan, and C. Stuth for permission to include unpublished results of theirs. A number of former students for suggestions which have had an influence on the book. A. Kruse for a point of view, even though it has not always been adhered to rigidly. Finally, the members of my family, Roberta, Dorothy, Martha, and Janet Scott, for help in the preparation of the manuscript.

W. R. SCOTT

TABLE OF CONTENTS

1

2

3

4

Direct Sums 64

5

Abelian Groups 89

6

p-Groups and p-Subgroups 131

7

Supersolvable Groups 150

8

Free Groups and Free Products 174

9

Extensions 209

10

Permutation Groups   255

11

Symmetric and Alternating Groups   298

12

Representations   319

13

14

15

GROUP
THEORY

ONE

INTRODUCTION

1.1 Preliminaries

We shall begin by reviewing some notions from intuitive set theory.

Throughout the book, equality will mean identity. A set is determined by its members; that is, if S and T are sets, then $S = T$ iff $(*)$ $x \in S$ iff $x \in T$. The notation $\{1, 2, 3\}$, for example, means the set whose members are 1, 2, and 3. The notation $\{x \mid P\}$ means the set of x such that P is true. Various corruptions of this latter notation will be used. A few sets occur frequently enough to deserve a permanent symbol. Among these are the empty set $\varnothing$, the set $\mathcal{N}$ of natural numbers, the set $\mathcal{P}$ of primes, the set $\mathcal{P}^{\infty} = \{p^n \mid p \in \mathcal{P},\ n \in \mathcal{N}\}$ of powers of primes, the set (later group or ring) J of integers, and the set (or group or field) $\mathcal{R}$ of rationals. Other permanent notation will be introduced from time to time (see the index of notation).

The number of elements of a set S will be called the *order* of S and denoted by $o(S)$. A set S is a *singleton* iff $o(S) = 1$, and an *n-ton* iff $o(S) = n \in \mathcal{N}$. A set S is a *subset* of a set T, written $S \subset T$,* iff $(*)$ if $x \in S$, then $x \in T$. If S is a set and T_i is a set for each $i \in S$, then

$$\cup\,\{T_i \mid i \in S\} = \{x \mid \exists\ i \in S \text{ such that } x \in T_i\},$$
$$\cap\,\{T_i \mid i \in S\} = \{x \mid \text{if } i \in S \text{ then } x \in T_i\}.$$

* Beginning with Section 1.6, the notation $\subset$ will be used to mean "is a subgroup of," while the phrase "is a subset of" will be written out.

1

If $S = \{1, 2\}$, then $T_1 \cup T_2$ is usually used instead of $\cup \{T_i \mid i \in S\}$ and $T_1 \cap T_2$ instead of $\cap \{T_i \mid i \in S\}$, although there is a conceptual difference. The notation $\overset{\cdot}{\cup} \{T_i \mid i \in S\}$ will be used for $\cup \{T_i \mid i \in S\}$ if $T_i \cap T_j = \varnothing$ whenever $i \neq j$, and will be called the *disjoint union* of the family $\{T_i \mid i \in S\}$. If each element of S is a set, then $\cup \{i \mid i \in S\}$ will sometimes be written simply $\cup S$. The notations $\overset{\cdot}{\cup} S$ and $\cap S$ have similar meanings. If S and T are sets, then $S \backslash T$ will denote the set $\{x \mid x \in S \text{ and } x \notin T\}$.

Two ordered pairs (a, b) and (c, d) are equal iff $a = c$ and $b = d$. A *relation* R is a set* of ordered pairs. Its *inverse*

$$R^{-1} = \{(b, a) \mid (a, b) \in R\}.$$

The relation R is *symmetric* iff $R = R^{-1}$. The *domain* of R is the set

$$\text{Dom}(R) = \{a \mid \exists \, b \text{ with } (a, b) \in R\}$$

(that is, the domain of R is the set of first coordinates of elements of R). The *range* of R is given by $\text{Rng}(R) = \text{Dom}(R^{-1})$ (that is, the range of R is the set of second coordinates of elements of R). A relation R is said to be *from A into B* iff $\text{Dom}(R) = A$ and $\text{Rng}(R) \subset B$, and *onto B* iff $B = \text{Rng}(R)$. If $S \subset \text{Dom}(R)$, then the *restriction* of R to S is the set

$$R \mid S = \{(a, b) \mid (a, b) \in R \quad \text{and} \quad a \in S\}.$$

If $S \subset \text{Dom}(R)$, then we set $SR = \text{Rng}(R \mid S)$. If R and R' are relations, then their *product* is the relation

$$RR' = \{(a, c) \mid \exists \, b \quad \text{such that} \quad (a, b) \in R \quad \text{and} \quad (b, c) \in R'\}.$$

A relation R is *transitive* iff $RR \subset R$. The product of relations is associative, that is

1.1.1. *If R_1, R_2, and R_3 are relations, then $R_1(R_2 R_3) = (R_1 R_2) R_3$.*

Proof. Let $(a, d) \in R_1(R_2 R_3)$. Then $\exists \, b$ such that $(a, b) \in R_1$ and $(b, d) \in R_2 R_3$. Hence $\exists \, c$ such that $(b, c) \in R_2$ and $(c, d) \in R_3$. Therefore $(a, c) \in R_1 R_2$ (because of the properties of b). But then $(a, d) \in (R_1 R_2) R_3$ (because of the properties of c). Hence $R_1(R_2 R_3) \subset (R_1 R_2) R_3$. Similarly, one shows that $(R_1 R_2) R_3 \subset R_1(R_2 R_3)$, and the equality follows. ‖

The *Cartesian product* of two sets S and T is the set

$$S \times T = \{(a, b) \mid a \in S \text{ and } b \in T\}.$$

A relation R is *on S* iff $R \subset S \times S$. A relation R on S is *reflexive* iff $\{(a, a) \mid a \in S\} \subset R$. An *equivalence* relation on S is a relation on S which is reflexive, symmetric, and transitive. If R is an equivalence relation on S,

---

* Or class. The same remark applies to functions.

then a subset B of S is an *equivalence class* with respect to R iff $\exists\, a \in S$ such that $B = \{b \mid (a, b) \in R\}$. It turns out (Exercise 1.1.6) that S is then the disjoint union of its equivalence classes.

While the preceding concepts are useful, the notion which will be used most frequently is that of function. A *function* is a relation f such that if $(a, b) \in f$ and $(a, c) \in f$, then $b = c$. If f is a function and $(a, b) \in f$, then we shall often write $af = b$ or $f(a) = b$ (usually the former). It is clear that two functions f and g are equal iff $\mathrm{Dom}(f) = \mathrm{Dom}(g)$, and $af = ag$ for all $a \in \mathrm{Dom}(f)$. A function f is 1–1 iff the relation f^{-1} is a function. If S is a set, the *identity* function on S is $I_S = \{(s, s) \mid s \in S\}$. I_S will often be denoted by I. It is 1–1 from S onto S, and such that $sI_S = s$ for all $s \in S$.

1.1.2. *The product of functions is a function.*

Proof. Let f and g be functions. Let $(a, d) \in fg$ and $(a, e) \in fg$. Then $\exists\, b$ such that $(a, b) \in f$ and $(b, d) \in g$, and $\exists\, c$ such that $(a, c) \in f$ and $(c, e) \in g$. Since f is a function, $b = c$. Therefore, since g is a function, $d = e$. Hence fg is a function. $\parallel$

A *binary operation* on S is a function $\circ$ from $S \times S$ into S. The fact that $(a, b)\circ = c$ will be written $a \circ b = c$. A binary operation $\circ$ on S is *commutative* iff

  (i)   if $a \in S$ and $b \in S$, then $a \circ b = b \circ a$.

A binary operation $\circ$ on S is *associative* iff

  (ii) if $a \in S$, $b \in S$, and $c \in S$, then $(a \circ b) \circ c = a \circ (b \circ c)$.

If $\circ$ is a binary operation on S, $T \subset S$, and $U \subset S$, then $T \circ U$ is defined to be the set $\{x \circ y \mid x \in T \text{ and } y \in U\}$.

It will now be shown that associativity implies general associativity. Toward this end, let $\circ$ be a binary operation on a set S. Let f_1 be the function with domain S such that $f_1(a_1) = \{a_1\}$ for all $a_1 \in S$. Recursively for $n \in \mathcal{N}$, define f_n to be the function whose domain is the set of n-tuples $(a_1, \ldots, a_n)$ with $a_i \in S$, such that

$$f_n(a_1, \ldots, a_n) = \cup \{f_r(a_1, \ldots, a_r) \circ f_{n-r}(a_{r+1}, \ldots, a_n) \mid 0 < r < n\}.^*$$

It is inductively evident that $f_n(a_1, \ldots, a_n)$ is not empty. Moreover, the operation $\circ$ is associative iff $f_3(a_1, a_2, a_3)$ is a singleton for all $a_1 \in S$, $a_2 \in S$, and $a_3 \in S$.

1.1.3. (*General associative law.*) *If $\circ$ is an associative binary operation on S and $a_1 \in S, \ldots, a_n \in S$, then $f_n(a_1, \ldots, a_n)$ is a singleton for all $n \in \mathcal{N}$.*

---

* The notation $f_n(a_1, \ldots, a_n)$ is used instead of $f_n((a_1, \ldots, a_n))$.

Proof. This is obvious for $n = 1$ and $n = 2$. Now use induction, and let $n > 2$. Let $z \in f_n(a_1, \ldots, a_n)$ and $z' \in f_n(a_1, \ldots, a_n)$. Then $\exists \; x, y, x', y'$, and natural numbers r and t such that

$$z = x \circ y, z' = x' \circ y', x \in f_r(a_1, \ldots, a_r),$$

$$y \in f_{n-r}(a_{r+1}, \ldots, a_n), x' \in f_t(a_1, \ldots, a_t),$$

$$y' \in f_{n-t}(a_{t+1}, \ldots, a_n).$$

If $r = t$, then by the induction hypothesis, $x = x'$ and $y = y'$, so $z = x \circ y = x' \circ y' = z'$. If $r < t$, say, then by the induction hypothesis, $f_{t-r}(a_{r+1}, \ldots, a_t)$ is a singleton $\{v\}$. Hence, again by the induction hypothesis, $x' = x \circ v$ and $y = v \circ y'$. Therefore, by associativity,

$$z = x \circ y = x \circ (v \circ y') = (x \circ v) \circ y' = x' \circ y' = z'.$$

Hence, by induction, $f_n(a_1, \ldots, a_n)$ is a singleton. ‖

The sole element of $f_n(a_1, \ldots, a_n)$ is denoted by $a_1 \circ \ldots \circ a_n$ or by $\pi \{a_i \mid 1 \leq i \leq n\}$. It is intuitively clear that Theorem 1.1.3 states that all legal insertions of parentheses into the product $a_1 \circ a_2 \ldots \circ a_n$ which make the result computable lead to the same result (see Exercise 1.1.19).

In a similar manner, it may be shown that if $\circ$ is a commutative and associative binary operation on S, then the product of any n elements of S is defined and is independent of the order of factors and positions of parentheses. The proof of this fact will be omitted.

EXERCISES

1.1.4. Show that $\cup$ and $\cap$ are commutative and associative (as binary operations).

1.1.5. If A and B are sets, then $A = B$ iff $A \subset B$ and $B \subset A$.

1.1.6. Let R be an equivalence relation on a set S. Prove that S is the disjoint union of its equivalence classes with respect to R.

1.1.7. If R is a symmetric and transitive relation on S, then $R \cup \{(a, a) \mid a \in S\}$ is an equivalence relation on S.

1.1.8. Show how a plane may be considered as the Cartesian product of the set of real numbers with itself.

1.1.9. In the sense of Exercise 1.1.8, find the geometric condition under which a relation in the plane is symmetric; under which it is reflexive; under which it is transitive.

1.1.10. If $o(S) = n$ is finite, then

    (a) $o(S \times S) = n^2$,

    (b)   there are 2^{n^2} relations on S,

    (c)   there are 2^{n^2-n} reflexive relations on S,

    (d)   there are $2^{(n^2+n)/2}$ symmetric relations on S,

    (e)   there are $2^{(n^2-n)/2}$ reflexive and symmetric relations on S.

1.1.11. Let R be a relation on S, and let

$$R^* = \{(a, b) \mid \exists\, n \in \mathcal{N} \quad \text{and} \quad x_0 = a, x_1, \ldots, x_n = b$$

such that all $(x_i, x_{i+1}) \in R\}$. Prove that

    (a) R^* is the smallest transitive relation on S which contains R;

    (b)   If R is symmetric, so is R^*;

    (c)   If R is reflexive and symmetric, then R^* is an equivalence relation.

1.1.12. Let S be a set, T the set of relations on S, and let

$$A \circ B = AB \text{ if } A \in T \text{ and } B \in T.$$

Then $\circ$ is an associative binary operation on T. Hence the product of relations satisfies the general associative law.

1.1.13. Let S be a set and let F be the set of relations on S which are functions. Prove that if $\circ$ is the product as in Exercise 1.1.12, then $\circ \mid F$ is an (associative binary) operation on F.

1.1.14. Prove that if R_1 and R_2 are relations, then

$$(R_1 R_2)^{-1} = R_2^{-1} R_1^{-1}.$$

1.1.15. Which of the four processes of arithmetic—addition, subtraction, multiplication, and division—are binary operations on J? Which of those that are operations are commutative? Which are associative?

1.1.16. If S and T are sets, then $o(S) = o(T)$ iff $\exists\, a$ 1–1 function from S onto T. Prove that if R is a relation, then $o(R) = o(R^{-1})$.

1.1.17. If $o(S) = n$ is finite, then

    (a)   there are n^{n^2} binary operations on S;

    (b)   there are $n^{(n^2+n)/2}$ commutative binary operations on S.

1.1.18. Define unary operation, ternary operation, and n-ary operation on S. How many of each are there if $o(S) = m$?

1.1.19. (a)   Show that the general associative law implies that if $\circ$ is associative, then

$$((a \circ b) \circ c) \circ d = (a \circ (b \circ c)) \circ d = (a \circ b) \circ (c \circ d)$$
$$= a \circ ((b \circ c) \circ d) = a \circ (b \circ (c \circ d)).$$

    (b)   Give a similar interpretation of Theorem 1.1.3 for the case of a product of five elements.

1.1.20. (a) Show that if $\circ$ is a binary operation on a set S, then, in the notation of 1.1.3, $f_n(a_1, \ldots, a_n)$ is a set containing at most $g(n)$ elements, where $g(1) = 1$, $g(2) = 1$, and recursively,

$$g(n) = \Sigma\{g(r)g(n - r) \mid 0 < r < n\}.$$

   (b) Compute $g(4)$ and compare with Exercise 1.1.19. (a).

1.1.21. Show that the bound given in Exercise 1.1.20.(a) is the best possible, i.e., that for each natural number n there are a set S, a binary operation $\circ$ on S, and $a_1 \in S, \ldots, a_n \in S$ such that $f_n(a_1, \ldots, a_n)$ is a $g(n)$-ton.

1.1.22. Prove that if f and g are functions, $x \in \mathrm{Dom}(f)$, and $xf \in \mathrm{Dom}(g)$, then $x(fg) = (xf)g$.

1.1.23. If f is a function, then

$$f^{-1}f = \{(x, x) \mid x \in \mathrm{Rng}(f)\}.$$

1.1.24. Prove that if $S \times T = U \times V$, then $S = U$ and $T = V$. In particular, if $S \times S = T \times T$, then $S = T$.

1.1.25. Prove that if f is a binary operation on S and also a binary operation on T, then $S = T$.

1.2 Definitions and first properties

A *group* is an ordered pair $(G, \circ)$ such that G is a set, $\circ$ is an associative binary operation on G, and $\exists\, e \in G$ such that

(i)   if $a \in G$, then $a \circ e = a$,

(ii) if $a \in G$, then $\exists\, a^{-1} \in G$ such that $a \circ a^{-1} = e$.

When there is no danger of misunderstanding, the group $(G, \circ)$ will be denoted by G, and the element $a \circ b$ by ab (see Exercises 1.2.11 and 1.2.16). Sometimes the operation $\circ$ will be denoted by $+$ and called addition. Since general associativity holds (Theorem 1.1.3), parentheses will usually be omitted from products of several factors. The element e is called the *identity* of G. (See Theorem 1.2.5 for justification of the article "the".) The symbol e will always denote the identity of whatever group G is under consideration (e_G will be used if necessary for the sake of clarity). The element a^{-1} is called the *inverse* of a (see Theorem 1.2.6).

1.2.1. $a^{-1}a = e$.

Proof. Compute $a^{-1}aa^{-1}(a^{-1})^{-1}$. (This means that one expands

$$a^{-1}aa^{-1}(a^{-1})^{-1}$$

in two directions to obtain the proof:

$$a^{-1}a = a^{-1}ae = a^{-1}a(a^{-1}(a^{-1})^{-1})$$
$$= a^{-1}e(a^{-1})^{-1} = a^{-1}(a^{-1})^{-1} = e.)$$

1.2.2. $ea = a$.

Proof. Compute $aa^{-1}a$.

1.2.3. *If* $a \in G$ *and* $b \in G$, *then* $\exists \mid x \in G$ *such that* $ax = b$.

Proof. If $ax = b$, then $x = ex = a^{-1}ax = a^{-1}b$ by 1.2.2 and 1.2.1. Conversely, if $x = a^{-1}b$, then $ax = aa^{-1}b = eb = b$.

1.2.4. *If* $a \in G$ *and* $b \in G$, *then* $\exists \mid x \in G$ *such that* $xa = b$.

1.2.5. *The element* e *is unique.*

Proof. In Theorem 1.2.3 let $b = a$.

1.2.6. *If* $ab = e$, *then* $b = a^{-1}$.

Proof. In Theorem 1.2.3 let $b = e$.

1.2.7. $(a_1 \ldots a_n)^{-1} = a_n^{-1} \ldots a_1^{-1}$.

1.2.8. $(a^{-1})^{-1} = a$.

Proof. By 1.2.1. ‖

If $a \in G$, define $a^0 = e$. If $n \in \mathcal{N}$, define a^n to be the product of n a's, and a^{-n} as $(a^{-1})^n$.

1.2.9. (*Exponents.*) *If* $a \in G$ *and* r *and* s *are integers, then* (i) $a^r a^s = a^{r+s}$, *and* (ii) $(a^r)^s = a^{rs}$.

Outline of Proof. If either r or s is 0, the theorem is obvious. If r and s are positive, then (i) follows from the definition of the sum of finite cardinals (and the general associative law) and (ii) from the definition of product of finite cardinals. If one of r and s is negative and the other positive, then (i) follows from successive cancellations. In particular, it follows that $a^{-r} = (a^r)^{-1}$. As to (ii), letting both $r > 0$ and $s > 0$,

$$(a^{-r})^s = ((a^{-1})^r)^s = (a^{-1})^{rs} = a^{-(rs)} = a^{(-r)s}.$$

Again

$$(a^r)^{-s} = ((a^r)^{-1})^s = (a^{-r})^s = a^{(-r)s} = a^{r(-s)}.$$

by an earlier remark and the last sentence. Finally, if $r > 0$ and $s > 0$, then

$$a^{-r}a^{-s} = (a^{-1})^r(a^{-1})^s = (a^{-1})^{r+s} = a^{-(r+s)} = a^{(-r)+(-s)},$$

and

$$(a^{-r})^{-s} = ((a^{-r})^{-1})^s = (a^r)^s = a^{rs} = a^{(-r)(-s)}$$

by an earlier remark and earlier cases.

EXERCISES

1.2.10. State the theorems of this section in the additive notation.

1.2.11. Let $G = \{a, b\}$. Prove that there are exactly two binary operations $\circ$ on G such that $(G, \circ)$ is a group. (This proves that the convention of denoting a group $(G, \circ)$ by G is actually incorrect. However, we shall continue to use this abbreviation.)

1.2.12. State (ii) in the definition of group in terms of a unary operation on G.

1.2.13. If $(G, \circ)$ is a group, then G is not empty.

1.2.14. Let $S = \{a, b\}$.

    (a) Define a binary operation on S which is commutative but not associative.

    (b) Define a binary operation on S which is associative but not commutative.

1.2.15. Generalize Exercise 1.2.14 to any set S with more than one element.

1.2.16. Prove that if $(G, \circ)$ and $(H, \circ)$ are groups, then $G = H$. This means that the set G of a group is determined by the operation $\circ$ of the group. This fact permits one to define a group as an operation $\circ$ with certain properties. Make this definition.*

1.3 Permutation groups

    Examples of groups will be given in the next section. In this section, a special type of group, permutation groups, will be introduced, both to facilitate the construction of examples and for its own sake.

    A *permutation* of a set M is a 1–1 function from M onto M.

---

    * If one defines a group as an operation $\circ$ with certain properties, one gains precision of language at the cost of simplicity, although the loss is not as great as one would imagine. However, the terminology would be nonstandard in any case, and we prefer to use the incorrect but standard method of referring to the group G.

1.3.1. *If M is a set and* Sym(M) *is the set of permutations of M, then* Sym(M) *is a group.*

Proof. If $f \in$ Sym(M) and $g \in$ Sym(M), then by Theorem 1.1.2, fg is a function from M into M. If $x \in M$, then, since g is onto M, $\exists\ y \in M$ such that $yg = x$. Since f is onto M, $\exists\ z \in M$ such that $zf = y$. Hence $z(fg) = x$. Therefore fg is onto M. By Exercise 1.1.14, $(fg)^{-1} = g^{-1}f^{-1}$. Since f and g are 1–1, f^{-1} and g^{-1} are functions. By Theorem 1.1.2, $(fg)^{-1}$ is a function. Hence fg is 1–1. This proves that fg is a permutation of M, that is that $fg \in$ Sym(M).

By Exercise 1.1.13, the associative law is satisfied in Sym(M). Clearly, $I \in$ Sym(M). If $x \in M$ and $g \in$ Sym(M), then $x(gI) = (xg)I = xg$. It follows that $gI = g$.

If $f \in$ Sym(M), then f^{-1} is a function since f is 1–1. Since f is onto M, f^{-1} is from M onto M. To prove that f^{-1} is 1–1, suppose that $x \in M$ and $y \in M$ are such that $xf^{-1} = yf^{-1}$. Then $\exists\ a \in M$ such that $(x, a) \in f^{-1}$ and $(y, a) \in f^{-1}$, hence $(a, x) \in f$ and $(a, y) \in F$. Since f is a function, $x = y$. This proves that f^{-1} is 1–1 and therefore that $f^{-1} \in$ Sym(M). It remains to be shown that the inverse function f^{-1} is an inverse with respect to the operation. If $x \in M$ and $f \in$ Sym(M), then $(x, xf) \in f$ so $(xf, x) \in f^{-1}$, and $x(ff^{-1}) = (xf)f^{-1} = x = xI$. Hence $ff^{-1} = I$. Therefore Sym(M) is a group. ∥

Definition. The group Sym(M) is called the *symmetric* group on the set M.

Definition. A *permutation group* is an ordered pair (M, G), where M is a set and G is a group of permutations of M (and where the operation in G is the usual multiplication of functions described earlier). The *degree* of (M, G) is $o(M)$. The elements of M are called *letters*.

If V and U are subsets of M and G, respectively, then

$$VU = \{vu \mid v \in V, u \in U\}.$$

If V or U is a singleton, say $U = \{u\}$, the notation is simplified to Vu (instead of $V\{u\}$).

If $a_i \in M$ for $i = 1, \ldots, n$ and $a_i \neq a_j$ if $i \neq j$, then $(a_1, \ldots, a_n)$ means the permutation $g \in$ Sym(M) such that $a_i g = a_{i+1}$ for $i = 1, \ldots, n-1$, $a_n g = a_1$, and $bg = b$ for all other $b \in M$. Such a permutation is called an n-*cycle*. Similarly, $(\ldots, a_{-1}, a_0, a_1, \ldots)$ means the $g \in$ Sym(M) such that $a_i g = a_{i+1}$ for all integers i, and $bg = b$ for all other $b \in M$. This latter permutation is called an ∞-*cycle*. Any 1-cycle equals e, of course. Two cycles with no letters in common are called *disjoint*.

Note that

$$(1, 2, \ldots, n) = (2, \ldots, n, 1) = \ldots = (n, 1, \ldots, n - 1).$$

Let (M, G) be a permutation group of finite degree, and let $g \in G$. There are two standard ways of writing g, a single example of which will suffice for purposes of illustration. If $M = \{1, 2, 3, 4, 5, 6\}$, $1g = 3$, $2g = 5$, $3g = 4$, $4g = 1$, $5g = 2$, and $6g = 6$, then g may be written in the forms:

$$g = \begin{pmatrix} 1\ 2\ 3\ 4\ 5\ 6 \\ 3\ 5\ 4\ 1\ 2\ 6 \end{pmatrix} = (1, 3, 4)(2, 5)(6) = (1, 3, 4)(2, 5).$$

The first form will be called a *two-row form* of g. Either the second or the third form is called a *cyclic decomposition* of g (or a *one-row form*).

1.3.2. *If (M, G) is a permutation group of finite degree and $g \in G$, then g is the product of pairwise disjoint cycles. This cyclic decomposition is unique except for order of the cycles and inclusion or omission of 1-cycles.*

Proof. The theorem is obvious if $o(M) = 1$. Induct on $o(M)$. Let $g \in G$ and $x_1 \in M$. Then $\exists\ i \in \mathcal{N}$ and distinct letters $x_1, \ldots, x_i$ of M such that

$$x_1 g = x_2,\ x_2 g = x_3, \ldots, x_{i-1} g = x_i,\ x_i g = x_1,$$

by finiteness of the set M and the fact that g is 1-1. Now $g \mid M \backslash \{x_1, \ldots, x_i\}$ is a permutation (if $M \backslash \{x_1, \ldots, x_i\}$ is nonempty), hence by the induction hypothesis $g \mid M \backslash \{x_1, \ldots, x_i\}$ is the product $c_2 \ldots c_m$ of pairwise disjoint cycles, and $g = (x_1, \ldots, x_i) c_2 \ldots c_m$. Therefore a cyclic decomposition always exists. In any cyclic decomposition of g, the cycle containing x_1 must be $(x_1, \ldots, x_i)$ so that the uniqueness follows readily. ∥

If M is infinite and $g \in \mathrm{Sym}(M)$, then g need not be the product of a finite number of cycles. Nevertheless, Theorem 1.3.2 is still true in a certain sense.

If $S \subset \mathrm{Sym}(M)$ and for each $a \in M$, there is at most one $y \in S$ such that $ay \neq a$, then we define a function $\pi \{y \mid y \in S\}$ by, for $a \in M$,

$$a(\pi \{y \mid y \in S\}) = \begin{cases} ay & \text{if } ay \neq a \text{ and } y \in S, \\ a & \text{if } ay = a \text{ for all } y \in S. \end{cases}$$

We call $\pi \{y \mid y \in S\}$ the *formal product* of the members of S, and will abbreviate it πy when no confusion is possible.

1.3.3. *If M is a set and $\pi \{y \mid y \in S\}$ is a formal product of elements of $\mathrm{Sym}(M)$, then $\pi \{y \mid y \in S\} \in \mathrm{Sym}(M)$.*

Proof. It is clear from the definition of formal product and $a(\pi y)$ that $\pi \{y \mid y \in S\}$ is a function from M into M.

Suppose that $a(\pi y) = b(\pi y)$ with $a \neq b$. If $b(\pi y) = b$, then $\exists\, x \in S$ such that $ax = b$, while $by = b$ for all $y \in S$. Therefore $ax = b = bx$ and x is not 1-1, a contradiction. Hence WLOG,* $a(\pi y) = b(\pi y) = c$, $c \neq a$, $c \neq b$. Thus $\exists\, x \in S$ and $y \in S$ such that $ax = c = by$. Since $a \neq b$, $x \neq y$. But then $cx \neq c$ and $cy \neq c$, contradicting the definition of formal product. It follows that if $a \neq b$, then $a(\pi y) \neq b(\pi y)$, so that πy is 1-1.

Let $b \in M$. If $by = b$ for all $y \in S$, then $b(\pi y) = b$. Suppose $\exists\, x \in S$ such that $bx \neq b$. Since $x \in \mathrm{Sym}(M)$, $\exists\, a \in M$ such that $ax = b$. Then $a \neq b$, so that $a(\pi y) = b$. Therefore πy maps M onto M, so that $\pi y \in \mathrm{Sym}(M)$.

1.3.4. *If $g \neq e$ is a permutation, then g is a formal product of pairwise disjoint cycles, $g = \pi \{c \mid c \in S\}$, with no $c \in S$ a 1-cycle. If $g = \pi \{c \mid c \in S'\}$ is a second such product, then $S = S'$.*

Proof. Let $g \in \mathrm{Sym}(M)$, $a \in M$, and $ag \neq a$. Let

$$c = \begin{cases} (a, ag, \ldots, ag^{n-1}) \text{ if } ag^n = a, n > 1, n \text{ minimal}, \\ (\ldots, ag^{-1}, a, ag, \ldots) \text{ if } ag^n \neq a \text{ for all } n \in \mathcal{N}. \end{cases} \tag{1}$$

It is obvious that c defined in (1) is a cycle and $ag = ac$. It follows that if $g = \pi \{c \mid c \in S\}$ is a formal product of cycles of length at least 2, then S is the set of cycles of the form (1).

Conversely, let S be the set of cycles of the form (1). Suppose $b \in M$, $c_1 \in S$, $c_2 \in S$, $bc_1 \neq b$, and $bc_2 \neq b$. Then, if

$$c_1 = (a, ag, \ldots, ag^i = b, ag^{i+1}, \ldots, ag^{n-1}),$$

then

$$c_1 = (b, bg, \ldots, bg^{n-1}),$$

while if

$$c_1 = (\ldots, ag^{-1}, a, ag, \ldots, ag^i = b, \ldots),$$

then

$$c_1 = (\ldots, bg^{-1}, b, bg, \ldots).$$

In either case, it follows that

$$c_1 = c_2.$$

Therefore $\pi \{c \mid c \in S\}$ is a formal product. Moreover, from (1), $ag = a(\pi c)$ in all cases. Hence $g = \pi c$. In a formal product of a set S of cycles of length at least 2, the cycles are automatically disjoint. ‖

If $a \in M$, $x \in \mathrm{Sym}(M)$, and $y \in \mathrm{Sym}(M)$, then $a(x(x^{-1}yx)) = a(yx)$. From this simple fact, there results the following useful rule.

---

* Without loss of generality.

1.3.5. *If x and y are permutations of M, then a cyclic decomposition of $x^{-1}yx$ is formed from one of y by replacing each letter by the letter standing below it in a two-row form of x.*

Proof. In a cyclic decomposition of y, the letter a is followed (cyclically) by the letter ay. The letters standing under a and ay in a two-row form of x are ax and ayx, respectively. By the remark preceding the theorem, ayx is the letter following ax in a cyclic decomposition of $x^{-1}yx$. ‖

Example. Let

$$x = \begin{pmatrix} 1\ 2\ 3\ 4\ 5\ 6 \\ 5\ 1\ 4\ 3\ 6\ 2 \end{pmatrix}, \quad y = (1, 3, 4)(2, 5)(6).$$

Then $x^{-1}yx = (5, 4, 3)(1, 6)(2)$.

1.3.6. *If x and y are permutations of M, then y and $x^{-1}yx$ have the same number of cycles of each length in their cyclic decompositions.*

Proof. By Theorem 1.3.5.

EXERCISES

1.3.7. $(1, 2, \ldots, n) = (1, 2)(1, 3) \ldots (1, n)$.

1.3.8. If g is a permutation of a finite set, then g is a finite product of (not necessarily disjoint) 2-cycles.

1.3.9. The *character* of $g \in \mathrm{Sym}(M)$ is the number $\mathrm{Ch}(g)$ of letters in M fixed by g. Thus, if $M = \{1, 2, 3, 4\}$ and $g = (1, 2, 3)$, then $\mathrm{Ch}(g) = 1$. Prove the following statements:

(a) $\mathrm{Ch}(e) = o(M)$;

(b) $\mathrm{Ch}(x^{-1}yx) = \mathrm{Ch}(y)$.

1.3.10. Compute the following products:

(a) $(1, 2, 3)(2, 3, 5)(1, 4, 5)$;

(b) $x^{-1}yx$ if $x = \begin{pmatrix} 1\ 2\ 3\ 4\ 5 \\ 2\ 1\ 4\ 3\ 5 \end{pmatrix}$ and $y = (1, 3, 5)(2, 4)$;

(c) $x^{-1}yx$ if $x = (1, 2, 3)$ and $y = (1, 3)$.

1.3.11. Prove the following converse of Theorem 1.3.6. If y and z are permutations of M such that if $1 < n \in \mathcal{N}$ or $n = \infty$ then y and z have the same number of n-cycles and $\mathrm{Ch}(y) = \mathrm{Ch}(z)$, then $\exists x \in \mathrm{Sym}(M)$ such that $z = x^{-1}yx$. (M may be infinite.)

1.3.12. Show that $(1, 2)(1, 3) = (1, 2, 3)$, whereas $(1, 3)(1, 2) = (1, 3, 2)$. Hence if $o(M) \geqq 3$, then $\text{Sym}(M)$ has noncommutative operation.

1.4 Examples of groups

Although little machinery of group theory has been set up, it seems worth while to give some examples of groups at this stage.

Definition. A group G is *Abelian* iff (*) if $a \in G$ and $b \in G$, then $ab = ba$.

Example 1. If M is a set, then $G = \text{Sym}(M)$ is a group (Theorem 1.3.1). If $o(M) < 3$, then G is Abelian; if $o(M) \geqq 3$, then G is non-Abelian (see Exercise 1.3.12). If M is finite, then $o(\text{Sym}(M)) = (o(M))!$ If M is infinite, then $o(\text{Sym}(M)) = 2^{o(M)}$.

Example 2. Let $M = \{1, 2, 3, 4\}$ and let

$$G = \{(1), (1, 2)(3, 4), (1, 3)(2, 4), (1, 4)(2, 3)\}.$$

It is readily verified that (M, G) is a permutation group with identity element (1). Each element is its own inverse, and the product of any two of the three nonidentity elements is the third one. This group is often called the *four-group*. It is Abelian.

Example 3. The integers J form an Abelian group under the operation of addition.

Example 4. The rationals $\mathscr{R}$ form an Abelian group under addition.

The following permanent notation will be used occasionally. If G is a group with identity e, let $G^{\#} = G \backslash \{e\}$.

Example 5. $\mathscr{R}^{\#}$ is an Abelian group under multiplication.

Example 6. Let P be a plane and $Q \in P$. Let G be the set of rotations of P about Q. Then (P, G) is an Abelian permutation group.

Example 7. Let $n \in \mathscr{N}$, and let G be the set of rotations of a plane P about a point $Q \in P$ through $2\pi k/n$ radians, $k = 0, 1, \ldots, n-1$. Then (P, G) is an Abelian permutation group, and $o(G) = n$.

Definition. An *isometry* of Euclidean n-space E_n is a function T from E_n into E_n such that if $x \in E_n$ and $y \in E_n$, then $d(x, y) = d(xT, yT)$, where $d(x, y)$ is the distance from x to y.

Example 8. If G is the set of all isometries of E_n, then (E_n, G) is a permutation group.

Proof. It is clear that the product of two isometries is an isometry and that I_{E_n} is an isometry. Associativity follows from the fact that the operation of multiplying functions is associative. If $T \in G$ and x and y are distinct points of E_n, then $xT \neq yT$. Hence T is 1–1. If T is onto M, then T^{-1} is a function from E_n onto E_n which preserves distance, and is therefore an isometry. Hence, it only remains to be shown that an isometry T from E_n into E_n is onto. A proof of this fact will now be sketched.

If x, y, and z are points such that

$$d(x, z) = d(x, y) + d(y, z),$$

then

$$d(xT, zT) = d(xT, yT) + d(yT, zT).$$

It follows readily that if L is a line in E_n, then LT is also a line. An *affine subspace* of E_n is a nonempty subset S such that if a and b are distinct points of S then the line through a and b is in S. It follows from the preceding remarks that, if S is an affine subspace, so is ST. If $S_0, S_1, \ldots, S_n = E_n$ is a strictly increasing sequence of affine subspaces (which clearly exists), so is $S_0 T, S_1 T, \ldots, S_n T$. Therefore, $\mathrm{Dim}(S_{i+1} T) > \mathrm{Dim}(S_i T)$ (where Dim denotes dimension). Hence $\mathrm{Dim}(S_n T) \geqq n$. Therefore, $S_n T = E_n$, and T is onto E_n. ‖

Example 9. Let S be a nonempty subset of E_n. The set G of isometries T of E_n such that $ST = S$ is a group.

Example 10. In Example 9, let $n = 2$, and let S be a regular polygon with k sides. The resulting group G of isometries is called a dihedral group. If $k \geqq 3$, then G is non-Abelian. It has $2k$ elements: k rotations about the center of S, and k products of a rotation and a fixed reflection about a line through the center and one vertex. ‖

Let $\{M_s \mid s \in S\}$ be a family of sets. Then the *Cartesian product* $\times \{M_s \mid s \in S\}$ is the set of functions f from S such that $sf \in M_s$ for all $s \in S$. (In case $o(S) = 2$, this differs slightly from the previous definition in that the present product is unordered, while the previous one was ordered.)

Example 11. Let $\{(G_s, \circ_s) \mid s \in S\}$ be a family of groups. Then the *direct product* $\pi \{(G_s, \circ_s) \mid s \in S\}$ has $\times \{G_s \mid s \in S\}$ as set, and addition defined by the rule

$$s(f + g) = (sf) \circ_s (sg)$$

(that is, addition is made componentwise). The direct product of groups is again a group. If S is finite, say $S = \{1, \ldots, n\}$, it is more customary to call

$$G_1 \times \ldots \times G_n = \{(x_1, \ldots, x_n) \mid x_s \in G_s\},$$

the direct product. However, there is a formal distinction between the two concepts in this case (see Exercise 2.1.32 and earlier remarks). The latter notation will also be used.

Example 12. The *external direct sum* $\Sigma_E \{G_s \mid s \in S\}$ of a family $\{G_s \mid s \in S\}$ of groups is the set of $f \in \pi \{G_s \mid s \in S\}$ such that sf is the identity of G_s for all but a finite number (perhaps 0) of $s \in S$. The external direct sum is a group under the operation given in Example 11.

Example 13. As a special case of Example 11, let S be a set and G a group. Then the sum of two functions f and g from S into G is given by

$$s(f + g) = (sf)(sg), s \in S.$$

The set of these functions forms a group $\pi \{G \mid s \in S\}$ under addition. The identity elements of this group will be denoted by O_{SG} or O, even in situations where the group itself is not under consideration. Thus

$$sO = e \text{ for all } s \in S.$$

The inverse of $T \in \pi \{G \mid s \in S\}$ will be denoted by $-T$. Thus

$$s(-T) = (sT)^{-1}.$$

EXERCISES

1.4.1 Prove that the external direct sum and direct product of Abelian groups is Abelian.

1.4.2. In Example 9, let $n = 3$ and let S be a cube. Show that $o(G) = 48$.

1.4.3. In Example 9, if S is an n-dimensional cube, what is $o(G)$? Describe the elements of G in case S is the unit cube.

1.5 Operations with subsets

Definition. If $A_1, \ldots, A_n$ are subsets of a group G, then the product $A_1 A_2 \ldots A_n$ is the set of elements of G of the form $a_1 a_2 \ldots a_n$, where $a_i \in A_i$. The following convention will be used: if a factor A of a product is a singleton

$\{x\}$, then x will be used instead of $\{x\}$. For example, if $x \in G$, then SxT means $S\{x\}T$.

1.5.1. $S(TU) = (ST)U = STU$ for all subsets S, T, and U of G.

1.5.2. If G is Abelian, then $AB = BA$ for all subsets A and B of G.

1.5.3. $A(\cup \{B_i \,|\, i \in S\}) = \cup \{AB_i \,|\, i \in S\}$,
$(\cup \{B_i \,|\, i \in S\})A = \cup \{B_iA \,|\, i \in S\}$.

1.5.4. $A(\cap \{B_i \,|\, i \in S\}) \subset \cap \{AB_i \,|\, i \in S\}$.

EXERCISES

1.5.5. If "$\subset$" is replaced by "$=$" in Theorem 1.5.4, the resulting statement is false.

1.5.6. What is a product of subsets if one of the subsets is empty?

1.5.7. If A is a subset of a group G, then $Ae = eA = A$.

1.5.8. If $x \in G$, then $xG = Gx = G$.

1.6 Subgroups

A subgroup of a group G is a subset which is a group under the operation in G. More precisely, a *subgroup* of a group $(G, \circ)$ is a group $(H, *)$ such that H is a subset of G and $*$ is the restriction of $\circ$ to $H \times H$. This means that multiplication in H is the same as in G.

The fact that $(H, *)$ is a subgroup of $(G, \circ)$ will usually be written $H \subset G$. If G is a group, then $G \subset G$ and $\{e\} \subset G$. The subgroup $\{e\}$ will henceforth be denoted by E. A subgroup H of G is *proper* iff $H \neq G$, and is *nontrivial* iff $H \neq G$ and $H \neq E$. The fact that H is a proper subgroup of G will be written $H < G$.

1.6.1. *If $H \subset G$, then (i) $e_H = e_G$, and (ii) if $a \in H$, then the inverse of a in H is the inverse of a in G.*

Proof. $H \neq \varnothing$ (by the definition of group). Let $a \in H$. Then $ae_G = a = ae_H$. By Theorem 1.2.3, $e_G = e_H$. If b is the inverse of a in H and c the inverse of a in G, then $ab = e_H = e_G = ac$, so that $b = c$ by Theorem 1.2.3.

1.6.2. *A subset of a group G is a subgroup iff* (i) *S is not empty,* (ii) *if $a \in S$ and $b \in S$ then $ab \in S$, and* (iii) *if $a \in S$ then $a^{-1} \in S$.*

Proof. First suppose that $(S, *)$ is a subgroup of $(G, \circ)$. Since S contains an identity element, it is not empty. Statement (ii) must hold since $*$ is an operation on S, and $*$ and $\circ$ agree on S. As to (iii), if $a \in S$, then the inverse b of a in S is an element of S, hence by Theorem 1.6.1, $a^{-1} = b \in S$.

Conversely, suppose that (i), (ii), and (iii) are true, and let $\circ$ be the operation of G. It follows from (ii) that the restriction $*$ of $\circ$ to $S \times S$ is an operation on S. The associativity of $*$ follows immediately from that of $\circ$. Since $S \neq \varnothing$, $\exists\, a \in S$. By (iii), $a^{-1} \in S$ and, by (ii), $e = aa^{-1} \in S$. If $b \in S$ then $be = b$ and, by (iii), $\exists\, b^{-1} \in S$ such that $bb^{-1} = e$. Hence S is a subgroup of G.

1.6.3. *A subset S of a group G is a subgroup iff* (i) *S is not empty, and* (ii) *if $a \in S$ and $b \in S$, then $ab^{-1} \in S$.*

Proof. If $a \in S$ and $b \in S$, then (ii) implies that $e = aa^{-1} \in S$, hence by (ii) again $b^{-1} = eb^{-1} \in S$, and $ab = a(b^{-1})^{-1} \in S$. The theorem now follows from Theorem 1.6.2.

1.6.4. *A finite subset S of a group G is a subgroup iff* (i) *S is not empty, and* (ii) *if $a \in S$ and $b \in S$, then $ab \in S$.*

Proof. If $a \in S$ then, by (ii) and the finiteness of S, there are natural numbers r and s such that $a^r = a^s$ and $r > s + 1$. Hence $a^{-1} = a^{r-s-1} \in S$. The theorem now follows from Theorem 1.6.2. ‖

The following theorem furnishes examples of subgroups of a group which are usually nontrivial.

1.6.5. (*The powers of an element form a subgroup.*) *If G is a group and $g \in G$, then the set $\langle g \rangle = \{g^n \mid n \in J\} \subset G$.*

Proof. Use Theorem 1.6.3 and the laws of exponents (Theorem 1.2.9).

1.6.6. (*The intersection of subgroups is a subgroup.*) *If $\{H_i \mid i \in S\}$ is a family of subgroups of a group G, then $\cap \{H_i \mid i \in S\} \subset G$.* ‖

If $\{H_i \mid i \in S\}$ is a family of subgroups of a group G, then $\langle H_i \mid i \in S \rangle$ is defined to be the smallest subgroup of G containing all the H_i. If H and K are subgroups of G, $\langle H, K \rangle$ has an analogous meaning.

The existence of $\langle H_i \mid i \in S \rangle$ follows almost immediately from Theorem 1.6.6. However, a more explicit theorem can be stated as follows.

1.6.7. *If $\{H_i \mid i \in S\}$ is a family of subgroups of a group G, then $\langle H_i \mid i \in S \rangle$ is the set of all finite products $x_1 \ldots x_n$ such that each $x_j \in H_{i_j}$ for some $i_j \in S$.*

Proof. This follows readily from Theorems 1.6.3 and 1.2.7. ‖

Remark. It has been implicitly assumed in the above discussion that the set S was nonempty in each case. If S is empty, then $\cap \{H_i \mid i \in S\}$ is defined to be G, while $\langle H_i \mid i \in S \rangle = E$ (from the definition already given). However, Theorem 1.6.7 is false if S is empty.

1.6.8. *If H and K are subgroups of a group G, then HK is a subgroup iff $HK = KH$.*

Proof. Suppose that $HK \subset G$. Let $h \in H$ and $k \in K$. Then $kh = (h^{-1}k^{-1})^{-1}$ is the inverse of an element of HK and is, therefore, itself in HK. Thus KH is a subset of HK. If $x \in HK$, then $x^{-1} \in HK$ and $\exists\, h' \in H$ and $k' \in K$ such that $x^{-1} = h'k'$. Therefore $x = k'^{-1}h'^{-1} \in KH$. Hence HK is a subset of KH. Therefore $HK = KH$.

Conversely, suppose that $HK = KH$. Clearly $HK \neq \varnothing$. Let $h \in H$, $h_1 \in H$, $k \in K$, and $k_1 \in K$. Then

$$(hk)(h_1 k_1)^{-1} = h(kk_1^{-1}h_1^{-1}).$$

Since

$$(kk_1^{-1})h_1^{-1} \in KH = HK,$$

$\exists\, h_2 \in H$ and $k_2 \in K$ such that $(kk_1^{-1})h_1^{-1} = h_2 k_2$. Therefore

$$(hk)(h_1 k_1)^{-1} = hh_2 k_2 \in HK.$$

By Theorem 1.6.3, HK is a subgroup of G.

EXERCISES

1.6.9. If $K \subset H$ and $H \subset G$, then $K \subset G$.

1.6.10. If G is a group then $E = \{e\}$ is a subgroup of G.

1.6.11. Use Theorem 1.6.5 to find four nontrivial subgroups of $\mathrm{Sym}(\{1, 2, 3\})$.

1.6.12. Find all subgroups of the group J of integers.

1.6.13. (Hard) Find all subgroups of $\mathscr{R}$. *Hint:* It is slightly easier to find those subgroups H such that $1 \in H$.

1.6.14. True or false? $\mathrm{Sym}(\{1, 2, 3\}) \subset \mathrm{Sym}(\{1, 2, 3, 4\})$. If true, prove it; if false, fix the statement up so that a true statement results.

1.6.15. If G is a group, H is a subset of $L \subset G$, and K is a subset of G, then

  (a) $L \cap (HK) = H(L \cap K)$.

  (b) If $G = HK$, then $L = H(L \cap K)$.

1.6.16. If H and K are proper subgroups of G, then there is an element x in G which is not in H or K.

1.7 Cosets and index

Definition. If H is a subgroup of a group G, then a *right coset* of H is a subset S of G such that $\exists\, x \in G$ for which $S = Hx$. *Left coset* is defined similarly. If K is also a subgroup of G, then a *double coset* of H and K is a subset S of G such that $\exists\, x \in G$ for which $S = HxK$.

1.7.1. (*Disjointness of double cosets.*) *If* $H \subset G$, $K \subset G$, $x \in G$, *and* $y \in G$, *then* HxK *and* HyK *are equal or disjoint.*

Proof. If $HxK \cap HyK \neq \varnothing$, then $\exists\, h_1 \in H$, $h_2 \in H$, $k_1 \in K$, and $k_2 \in K$ such that $h_1 x k_1 = h_2 y k_2$. It follows (Exercise 1.5.8) that

$$HxK = Hh_1 x k_1 K = Hh_2 y k_2 K = HyK.$$

1.7.2. (*Disjointness of cosets.*) *If* $H \subset G$, $x \in G$, *and* $y \in G$, *then* Hx *and* Hy *are equal or disjoint, and* xH *and* yH *are equal or disjoint.*

Proof. In Theorem 1.7.1, let $K = E$ or $H = E$.

1.7.3. $Hx = Hy$ *iff* $xy^{-1} \in H$. $xH = yH$ *iff* $y^{-1}x \in H$.

1.7.4. $x \in HxK$, $x \in Hx$, *and* $x \in xH$. ‖

Let us make the following temporary definition. If H is a subgroup of a group G, then the *right index* of H in G is the number of right cosets of H.

1.7.5. (*Right index = left index.*) *If* H *is a subgroup of a group* G, *then the right index of* H *in* G *equals the left index of* H *in* G.

Proof. Consider the relation $T = \{(Hx, x^{-1}H) \mid x \in G\}$. Now $Hx = Hy$ iff $xy^{-1} \in H$ (Theorem 1.7.3), which is true iff $(xy^{-1})^{-1} \in H$, i.e., iff $yx^{-1} \in H$, i.e., iff $(y^{-1})^{-1}x^{-1} \in H$, which by Theorem 1.7.3 is true iff $x^{-1}H = y^{-1}H$. It follows that T is a 1–1 function from the set of right cosets of H onto the set of left cosets of H. The theorem follows. ‖

Because of Theorem 1.7.5 there is no distinction between right and left index. We therefore define the *index* $[G:H]$ of H in G to be the number of right cosets of H.

1.7.6. *If S is a subset of a group G and $x \in G$, then $o(Sx) = o(S) = o(xS)$.*

Proof. The function T defined by $sT = sx$ is 1–1 from S onto Sx by Theorem 1.2.4. Hence $o(Sx) = o(S)$. Similarly $o(xS) = o(S)$.

1.7.7. (*Lagrange's Theorem.*) *If H is a subgroup of a group G, then $o(G) = o(H)[G:H]$.*

Proof. By Theorems 1.7.4 and 1.7.2, there is a subset S of G such that $G = \dot\cup \{Hg \mid g \in S\}$. Hence (Theorem 1.7.6)

$$o(G) = \Sigma \{o(Hg) \mid g \in S\} = o(H)[G:H].$$

1.7.8. (*Lagrange's Theorem.*) *If H is a subgroup of a finite group G, then $o(H) \mid o(G)$.*

1.7.9. *If $\{H_i \mid i \in S\}$ is a family of subgroups of a group G, then*

$$[G: \cap \{H_i \mid i \in S\}] \leqq \pi\{[G:H_i] \mid i \in S\}.$$

Proof. Let $K = \cap \{H_i \mid i \in S\}$. Let R be the set of right cosets of K, and R_i the set of right cosets of H_i. Define the relation

$$T = \{(Kx, U) \mid x \in G, U \in \times \{R_i \mid i \in S\} \quad \text{and} \quad iU = H_ix\}.$$

It may then be verified that T is a 1–1 function from R into $\times \{R_i \mid i \in S\}$. The theorem follows.

1.7.10. (*Poincaré.*) *The intersection of a finite number of subgroups of finite index is of finite index.*

1.7.11. *If $K \subset H \subset G$, then $[G:K] = [G:H][H:K]$.*

Proof. There are subsets S of G and T of H such that

$$G = \dot\cup \{Hx \mid x \in S\} \quad \text{and} \quad H = \dot\cup \{Ky \mid y \in T\}.$$

It follows (see Theorem 1.5.3) that

$$G = \cup \{Kyx \mid y \in T \quad \text{and} \quad x \in S\}.$$

Moreover, if $Kyx = Ky'x'$, then both $Hx \supset Kyx$ and $Hx' \supset Ky'x' = Kyx$. Since G is the *disjoint* union $\dot\cup \{Hx \mid x \in S\}$, $x = x'$. Hence $Kyx = Ky'x$, and therefore (multiplying on the right by x^{-1}), $Ky = Ky'$. Thus $y = y'$.

But this shows that

$$G = \overset{\cup}{} \{Kyx \mid y \in T \text{ and } x \in S\}.$$

The theorem follows.

EXERCISES

1.7.12. $o(G) = [G:E]$.

1.7.13. If $[G:H] = m$ and $[G:K] = n$ are finite, then $[G:H \cap K]$ is at least as large as the least common multiple of m and n.

1.7.14. If $H \subset G$ and $K \subset G$, then $[G:K] \geq [H:H \cap K]$.

1.7.15. If $[G:H] = m$ and $[G:K] = n$ are finite and relatively prime, then $[G:H \cap K] = mn$.

1.7.16. Let $M = \{1, \ldots, n\}$. Let

$$H = \{x \in \text{Sym}(M) \mid nx = n\}.$$

    (a)   Prove that $H \subset \text{Sym}(M)$.

    (b)   Find $[\text{Sym}(M):H]$.

    (c)   Find a set T such that $\text{Sym}(M) = \overset{\cup}{} \{Hx \mid x \in T\}$.

1.7.17. Let H be the subgroup of J consisting of all multiples of 4.

    (a)   Find the cosets of H.

    (b)   Find $[J:H]$.

1.7.18. Let $G = \text{Sym}(M)$ where $M = \{1, 2, 3\}$, and let $H = \{(1), (1, 2)\}$.

    (a)   List the right cosets of H in G.

    (b)   List the left cosets of H.

    (c)   Are the lists the same?

    (d)   Is the product of two right cosets of H again a right coset of H?

    (e)   Answer parts (a) through (d) for

$$H = \{(1), (1, 2, 3), (1, 3, 2)\}.$$

1.7.19. (a)   If $H_0 = G$, $H_{n+1} \subset H_n \subset G$ for $n \in \mathcal{N}$, and $H = \cap H_n$, then $[G:H] \leq (\pi[H_n:H_{n+1}])$.

    (b)   Unsolved problem. Does (a) hold for well-ordered descending sequences of subgroups?

1.7.20. A subset S of group G cannot be a right coset of two distinct subgroups of G.

1.8   Partially ordered sets

We now append a few definitions and facts about partially ordered sets. This material could have been included in the first section, but was delayed in order that group theory itself could be started sooner.

A *partially ordered* set is an ordered pair (S, R) where S is a set and R is a transitive and reflexive relation on S such that if $(a, b) \in R$ and $(b, a) \in R$, then $a = b$. One usually refers to S itself rather than to (S, R) as a partially ordered set. An *ordered* set or *chain* is a partially ordered set such that if $a \in S$ and $b \in S$ then $(a, b) \in R$ or $(b, a) \in R$.

Example. Let S be the set of nonnegative integers and let R be the relation on S such that $(a, b) \in R$ iff $a \mid b$. Then (S, R) is a partially ordered set but not an ordered set.

Example. Let M be a set, S the set of subsets of M, and R the relation on S such that $(a, b) \in R$ iff a is a subset of b. Then (S, R) is a partially ordered, but (provided M has at least two elements) not an ordered, set. S is said to be partially ordered by inclusion.

Example. Let G be a group, S the set of subgroups of G, partially ordered by inclusion. Then S is partially ordered.

Example. Let S be any subset of the reals ordered by $\leq$. Then S is a chain.

If (S, R) is a partially ordered set and $a \in S$, then a is *maximal* iff $(a, b) \in R$ implies $a = b$. There may be infinitely many maximal elements of S. In fact, if $R = \{(a, a) \mid a \in S\}$, then (S, R) is a partially ordered set and every element of S is maximal. An element $a \in S$ is *maximum* iff (∗) if $b \in S$ then $(b, a) \in R$. There is at most one maximum element in a partially ordered set. For if a and b are maximum elements, then $(a, b) \in R$ (since b is maximum) and $(b, a) \in R$ (since a is maximum), and therefore $a = b$ (from the definition of partially ordered set). *Minimal* and *minimum* are defined similarly. A finite chain has a minimum element and a maximum element. A maximal element of an ordered set is a maximum element. Note that an ordered set need not have a maximal (or maximum) element. For example, J under the ordering $\leq$ has no maximal element.

Example. If S is the set of subgroups of a group G, partially ordered by inclusion, then S has a maximum element, namely G, and a minimum element, namely E.

If (S, R) is a partially ordered set and T is a nonempty subset of S, then an *upper bound* of T is an $x \in S$ such that if $a \in T$, then $(a, x) \in R$. A *least upper bound* of T is an upper bound x of T such that if y is an upper bound of T, then $(x, y) \in R$. It follows easily that there is at most one least upper bound of T. However, there may not be any (Exercise 1.8.5). If $T = \{a, b\}$ and the least upper bound of T exists, then it is denoted by $a \cup b$. In general it will be denoted by $\cup \{x \mid x \in T\}$ or simply $\cup T$. *Greatest lower bound* is defined similarly, and is denoted by $\cap$ when it exists.

A lattice is a partially ordered set S such that for any doubleton T, the greatest lower bound and least upper bound of T exist. (That is, for all $a \in S$ and $b \in S$, $a \cup b$ and $a \cap b$ exist.)

A lattice L is *complete* iff for every nonempty subset T of L the greatest lower bound and least upper bound of T exist. This implies that a complete lattice has a maximum element and a minimum element.

1.8.1. *If G is a group, then the set* Lat(G) *of subgroups of G, partially ordered by inclusion, is a complete lattice.*

Proof. This is the content of Theorems 1.6.6 and 1.6.7.

EXERCISES

1.8.2. The definition of partially ordered set (S, R) could have been given in terms of R alone.

1.8.3. Let M be an infinite set, S the set of finite subsets of M, and R the inclusion relation on S. Prove that S is a lattice with a minimum element but no maximum element.

1.8.4. Prove that the partially ordered set S of all subsets of a set M is a complete lattice.

1.8.5. Prove that if (S, R) is a partially ordered set and T is a nonempty subset of S, then there is at most one least upper bound of T. Give an example where there is no least upper bound of T.

1.8.6. If G is a group, $H \subset G$, and Lat(H, G) the set of subgroups of G containing H, then Lat(H, G) is a complete lattice under inclusion.

1.8.7. Let (S, R) be a nonempty ordered set of groups such that if $A \in S$, $B \in S$, and $(A, B) \in R$ then $A \subset B$. Prove that an operation can be defined in $G = \cup \{A \mid A \in S\}$ in one and only one way so that G is a group containing all $A \in S$ as subgroups.

1.8.8. A lattice which has a maximum element and which is such that every nonempty subset has a greatest lower bound is complete.

ISOMORPHISM THEOREMS

2.1 Homomorphisms

Two groups may be essentially alike even though they are not equal. For example, consider the groups Eng $= \{\ldots,$ one, two, three, $\ldots\}$ and Ger $= \{\ldots,$ eins, zwei, drei, $\ldots\}$ under addition. There is an obvious 1–1 function T from Eng onto Ger (one might call T the translation function). Furthermore, if, for example, one adds one and two and translates the result into German the result is the same as when one first translates and then adds (in German). The above facts will be expressed by saying that Eng is isomorphic to Ger and that the translation function T is an isomorphism of Eng onto Ger (these terms are defined below). A more general type of function, a homomorphism, is of still greater importance. This chapter is devoted to the study of these functions.

A *homomorphism* of a group $(G, \circ)$ into a group $(H, *)$ is a function T of G into H such that if $x \in G$ and $y \in G$, then $(x \circ y)T = (xT) * (yT)$. (Compare with the example above.) An *endomorphism* of G is a homomorphism of G into G. An *isomorphism* is a 1–1 homomorphism. An *automorphism* of G is an isomorphism of G onto G. A group G is *isomorphic* to a group H, written $G \cong H$, iff there is an isomorphism of G onto H.

Examples. The function T such that $xT = 2^x$ is an isomorphism of the

24

additive group of real numbers onto the multiplicative group of positive real numbers. The function U such that $xU = -x$ is an automorphism of J. The function V such that $zV = z^2$ is an endomorphism (but not an automorphism) of the multiplicative group of nonzero complex numbers.

If G and H are groups, let $\mathrm{Hom}(G, H)$, $\mathrm{End}(G)$, and $\mathrm{Aut}(G)$ denote the set of homomorphisms of G into H, the set of endomorphisms of G, and the set of automorphisms of G, respectively. Later, $\mathrm{Aut}(G)$ and, in some cases, $\mathrm{Hom}(G, H)$ and $\mathrm{End}(G)$ will be given an algebraic structure. Let $\mathrm{Iso}(G)$ and $\mathrm{Hom}(G)$ be the class of isomorphisms and homomorphisms of G, respectively.

2.1.1. *If G and H are groups and $T \in \mathrm{Hom}(G, H)$, then* (i) $e_G T = e_H$ *and* (ii) $(g^{-1})T = (gT)^{-1}$.

Proof. (i) Compute $(e_G e_G)T$. (ii) Compute $(g^{-1}g)T$. ‖

Definition. A *word* f in the letters $x_1, \ldots, x_n$ is an expression of the form $x_{i_1}^{r_1} \ldots x_{i_k}^{r_k}$, where all $r_j \in J$. If $g_1, \ldots, g_n$ are elements of a group G, then for the above word f, $f(g_1, \ldots, g_n)$ will mean $g_{i_1}^{r_1} \ldots g_{i_k}^{r_k}$.

2.1.2. *If G is a group, $T \in \mathrm{Hom}(G)$, f is a word in $x_1, \ldots, x_n$, and $g_i \in G$ for $1 \leqq i \leqq n$, then*

$$(f(g_1, \ldots, g_n))T = f(g_1 T, \ldots, g_n T).$$

Proof. This follows from the definition of homomorphism, Theorem 2.1.1, and induction.

2.1.3. (*Product of homomorphisms is a homomorphism.*) *If G, H, and K are groups, $T \in \mathrm{Hom}(G, H)$, and $U \in \mathrm{Hom}(H, K)$, then $TU \in \mathrm{Hom}(G, K)$.*

2.1.4. *If $T \in \mathrm{Hom}(G, H)$, then $GT \subset H$.*

Proof. Use Theorems 1.6.2 and 2.1.1 (or 1.6.3 and 2.1.2).

2.1.5. *If $T \in \mathrm{Hom}(G, H)$ and $K \subset G$, then $(T \mid K) \in \mathrm{Hom}(K, H)$.* ‖

The homomorphism $T \mid K$ will often be denoted simply by T. If G and H are groups and $T \in \mathrm{Hom}(G, H)$, then the *kernel* of T is the set $\mathrm{Ker}(T) = \{x \in G \mid xT = e_H\}$.

2.1.6. *If G is a group and $T \in \mathrm{Hom}(G)$, then* (i) $\mathrm{Ker}(T) \subset G$, *and* (ii) *if $x \in G$ then $x^{-1}(\mathrm{Ker}(T))x \subset \mathrm{Ker}(T)$.* ‖

A subgroup H of G is *normal* in G, written $H \lhd G$, iff $x^{-1}Hx \subset H$ for all $x \in G$.

Theorem 2.1.6 may be restated as follows.

2.1.7. *A kernel of a homomorphism is a normal subgroup.*

2.1.8. *If G is a group and $H \subset G$, then $H \lhd G$ iff $x^{-1}Hx = H$ for all $x \in G$, hence iff $Hx = xH$ for all $x \in G$.*

Proof. The last equivalence is obvious, as is the fact that if $x^{-1}Hx = H$ for all x then $H \lhd G$. Conversely, suppose $H \lhd G$ and let $x \in G$. If $x^{-1}Hx \neq H$, then $\exists\, y \in H \backslash x^{-1}Hx$, so $xyx^{-1} \notin H$ and

$$xyx^{-1} \in xHx^{-1} = (x^{-1})^{-1}H(x^{-1}),$$

contradicting the normality of H.

2.1.9. *If G is a group and $T \in \mathrm{Hom}(G)$, then $T \in \mathrm{Iso}(G)$ iff $\mathrm{Ker}(T) = E$.*

Proof. If T is an isomorphism, then $x \in \mathrm{Ker}(T)$ implies $xT = e = eT$ by Theorem 2.1.1, hence $x = e$ since T is 1–1. Since $e \in \mathrm{Ker}(T)$ in any case, it is true that $\mathrm{Ker}(T) = E$. Conversely, suppose that $\mathrm{Ker}(T) = E$. Then if $xT = yT$, we have

$$(y^{-1}x)T = (yT)^{-1}(xT) = e \quad \text{(by Theorem 2.1.2)},$$

so that $y^{-1}x = e$ and $y = x$. Hence T is 1–1 and therefore an isomorphism.

2.1.10. *If G and H are groups, $T \in \mathrm{Hom}(G, H)$, and $K \subset G$, then $KT \subset H$. If, furthermore, $K \lhd G$, then $KT \lhd GT$.* ‖

If $T \in \mathrm{Hom}(G, H)$ and S is a subset of H, then T^{-1} is a relation, so that (see Section 1.1)

$$ST^{-1} = \{x \in G \mid xT \in S\}.$$

2.1.11. *If G and H are groups, $T \in \mathrm{Hom}(G, H)$, and $y \in GT$, then $\{y\}T^{-1}$ is a coset of $\mathrm{Ker}(T)$.*

Proof. Since $y \in GT$, $\exists\, x \in G$ such that $xT = y$. If $g \in G$, then $gT = y$ iff $(xT)^{-1}(gT) = e$, which holds iff $(x^{-1}g)T = e$, i.e., iff $x^{-1}g \in \mathrm{Ker}(T)$, which is equivalent to $g \in x(\mathrm{Ker}(T))$.

2.1.12. *If G and H are groups, $K \subset H$, and $T \in \mathrm{Hom}(G, H)$, then $KT^{-1} \subset G$; if $K \lhd H$, then $KT^{-1} \lhd G$.* ‖

An isomorphism of a lattice (S, R) onto a lattice (S', R') is a 1–1 function T from S onto S' such that $(x, y) \in R$ iff $(xT, yT) \in R'$ (see Exercise 2.1.38).

2.1.13. (*Lattice theorem.*) *If G and H are groups,* $T \in \mathrm{Hom}(G, H)$, *and* $GT = H$, *then T induces an isomorphism of the lattice of subgroups between* $\mathrm{Ker}(T)$ *and G onto the lattice* $\mathrm{Lat}(H)$ *of subgroups of H. T also induces an isomorphism of the lattice of normal subgroups of G between* $\mathrm{Ker}(T)$ *and G onto the lattice of normal subgroups of H.* (*See Exercises 1.8.6 and 2.1.21.*)

Proof. By Theorem 2.1.10, T is a function from $\mathrm{Lat}(G)$ into $\mathrm{Lat}(H)$. By Theorem 2.1.12 and the fact that $T^{-1}T$ is the identity function on $\mathrm{Lat}(H)$ (see Exercise 1.1.23), T is onto $\mathrm{Lat}(H)$. In fact, since T^{-1} is into the lattice (Exercise 1.8.6) L of subgroups of G which contain $\mathrm{Ker}(T)$, T is a function from L onto $\mathrm{Lat}(H)$. If $M \supset \mathrm{Ker}(T)$, then M is a union of cosets of $\mathrm{Ker}(T)$. If $x \notin M$, then x is in none of these cosets, so that by Theorem 2.1.11, $xT \notin MT$. Therefore, if $K \neq M$, $K \supset \mathrm{Ker}(T)$, and $M \supset \mathrm{Ker}(T)$, then $KT \neq MT$. Thus T is 1–1. That T preserves inclusion is clear, hence T induces an isomorphism of L onto $\mathrm{Lat}(H)$.

The portion of the theorem dealing with normal subgroups follows from Theorems 2.1.10 and 2.1.12, and Exercises 2.1.20 and 2.1.21. ‖

We now consider isomorphisms of permutation groups briefly.

Definition. If (L, G) and (M, H) are permutation groups, then an *isomorphism* of (L, G) onto (M, H) is an ordered pair (T, U) such that T is a 1–1 function from L onto M, U is an isomorphism of G onto H, and if $a \in L$ and $g \in G$ then $(ag)T = (aT)(gU)$. If there is an isomorphism of (L, G) onto (M, H), then (L, G) is *isomorphic* to (M, H) (written $(L, G) \cong (M, H)$). (The words *similarity* and *similar* are usually used in place of isomorphism and isomorphic in this context.)

The definition guarantees that isomorphic permutation groups are the same except for notation. In particular, isomorphic permutation groups have the same degree. However, if (L, G) and (M, H) are permutation groups such that G and H are isomorphic, it does not follow that (L, G) and (M, H) are isomorphic, even if their degrees are equal.

2.1.14. *The relation "is isomorphic to" is an equivalence relation on the class of permutation groups.*

2.1.15. (*Symmetric groups of degree n.*) *If L and M are sets such that* $o(L) = o(M)$, *then* $(L, \mathrm{Sym}(L)) \cong (M, \mathrm{Sym}(M))$.

Proof. Since $o(L) = o(M)$, ∃ a 1–1 function T from L onto M. Define U as follows: if $g \in \mathrm{Sym}(L)$, then $gU = T^{-1}gT$. Since T^{-1} is 1–1 from M onto L, g is 1–1 from L onto L, and T is 1–1 from L onto M, it follows that $gU \in \mathrm{Sym}(M)$. If $g \in \mathrm{Sym}(L)$, $h \in \mathrm{Sym}(L)$, and $gU = hU$, then

$$g = T(gU)T^{-1} = T(hU)T^{-1} = h.$$

Hence U is 1–1. If $k \in \mathrm{Sym}(M)$, then $TkT^{-1} \in \mathrm{Sym}(L)$ as above, and

$$(TkT^{-1})U = T^{-1}(TkT^{-1})T = k,$$

so U is onto $\mathrm{Sym}(M)$. It is easy to verify that if $g \in \mathrm{Sym}(L)$ and $h \in \mathrm{Sym}(L)$, then $(gh)U = (gU)(hU)$. Hence U is an isomorphism of $\mathrm{Sym}(L)$ onto $\mathrm{Sym}(M)$. Finally, if $a \in L$ and $g \in \mathrm{Sym}(L)$, then

$$(aT)(gU) = aTT^{-1}gT = (ag)T,$$

so (T, U) is an isomorphism of $(L, \mathrm{Sym}(L))$ onto $(M, \mathrm{Sym}(M))$. ‖

For each positive cardinal number n, let M_n be the set consisting of the first n nonzero ordinals. Let $\mathrm{Sym}(n) = \mathrm{Sym}(M_n)$. Then if M is a nonempty set, the permutation group $(M, \mathrm{Sym}(M))$ is isomorphic to one and only one $(M_n, \mathrm{Sym}(M_n))$. In fact, $\mathrm{Sym}(M)$ is isomorphic to exactly one $\mathrm{Sym}(n)$. This is obvious for finite sets M; for infinite sets it will be proved later (Theorem 11.3.7).

The following theorem will be useful on several occasions.

2.1.16. *If S is a set, G a group, and T a 1–1 function from S into G, then there is a group H containing S as a subset, and an isomorphism U of H onto G such that $U \mid S = T$.*

Proof. There is an infinite set M such that $o(M) > o(G)$ (if G is finite, let $M = J$; if G is infinite, let M be the set of subsets of G). Since $o(S) \leq o(G)$, $o(M \backslash S) > o(G)$. Hence there is a 1–1 function V from $G \backslash (ST)$ into $M \backslash S$. Let

$$H = S \cup (G \backslash (ST))V,$$

$$hU = \begin{cases} hT & \text{if } h \in S, \\ hV^{-1} & \text{if } h \in H/S. \end{cases}$$

Then U is a 1–1 function from the set H onto the set

$$(G \backslash ST) \cup (ST) = G \quad \text{and} \quad U \mid S = T.$$

Define multiplication in H as follows:

$$h_1 * h_2 = ((h_1 U)(h_2 U))U^{-1}. \tag{1}$$

Then $*$ is an operation on H. We have

$$\begin{aligned}
(h_1 * h_2) * h_3 &= (((((h_1 U)(h_2 U))U^{-1})U)(h_3 U))U^{-1} \\
&= (((h_1 U)(h_2 U))(h_3 U))U^{-1} = ((h_1 U)((h_2 U)(h_3 U)))U^{-1} \\
&= ((h_1 U)((((h_2 U)(h_3 U))U^{-1})U)))U^{-1} = h_1 * (h_2 * h_3).
\end{aligned}$$

Hence $*$ is associative. Again,

$$h * (eU^{-1}) = ((hU)((eU^{-1})U))U^{-1}$$
$$= (hU)U^{-1} = h,$$

and eU^{-1} is a right identity. Moreover

$$h * ((hU)^{-1}U^{-1}) = ((hU)(((hU)^{-1}U^{-1})U))U^{-1}$$
$$= ((hU)(hU)^{-1})U^{-1} = eU^{-1}.$$

Hence H is a group. Operating on (1) by U, one has

$$(h_1 * h_2)U = (h_1U)(h_2U)$$

so that U is a homomorphism. Since U was 1–1 from H onto G, U is an isomorphism of H onto G.

EXERCISES

2.1.17. If G is an Abelian group, H a group, and $G \cong H$, then H is Abelian.

2.1.18. Generalize Exercise 2.1.17 to a similar statement about homomorphisms.

2.1.19. If G is a group, $H \subset G$, and $T \in \mathrm{Iso}(G)$, then $[GT:HT] = [G:H]$. If $T \in \mathrm{Hom}(G)$, then $[GT:HT] \leqq [G:H]$.

2.1.20. If $\{H_i \mid i \in S\}$ is a family of normal subgroups of a group G, then $\cap \{H_i \mid i \in S\}$ and $\langle H_i \mid i \in S \rangle$ are normal subgroups of G.

2.1.21. If G is a group and $H \subset G$, then the set of A such that $H \subset A \lhd G$ is a complete lattice (under inclusion).

2.1.22. If G is a group and $xT = x^{-1}$ for all $x \in G$, then
  (a) T is 1–1 from G onto G,
  (b) $T \in \mathrm{Aut}(G)$ iff G is Abelian.

2.1.23. If G is an Abelian group, $n \in J$, and $xT = x^n$ for all $x \in G$, then $T \in \mathrm{End}(G)$.

2.1.24. (a) $G \lhd G$.
  (b) $E \lhd G$.
  (c)   If $H \lhd G$ and $H \subset K \subset G$, then $H \lhd K$.

2.1.25. If G is the set of complex numbers and $(a + bi)T = a - bi$ for all real a and b, then
  (a) T is an automorphism of $(G, +)$,
  (b) T is an automorphism of $(G^{\#}, \cdot)$.

2.1.26. If $x \in G$ and $yT_x = x^{-1}yx$ for all $y \in G$, then $T_x \in \mathrm{Aut}(G)$.

2.1.27. A subgroup H of an Abelian group G is normal in G.

2.1.28. If $H \lhd G$, it does *not* follow that $x^{-1}hx = h$ for all $x \in G$ and $h \in H$ (see Exercise 1.7.18).

2.1.29. Show that if a group H is obtainable from a group G by a finite number of applications of the processes

    (a)   taking a subgroup, and

    (b)   taking a homomorphic image,

           then H is a homomorphic image of a subgroup of G (i.e., only two applications are necessary).

2.1.30. (a)   The identity function I_G is an automorphism of the group G (the *identity* automorphism).

    (b)   If G and H are groups then O_{GH} is a homomorphism (the *zero* homomorphism) from G into H.

2.1.31. Show that there are two permutation groups (L, G) and (M, H) of degree 4 such that $G \cong H$ but $(L, G) \not\cong (M, H)$.

2.1.32. Let $S = \{1, \ldots, n\}$ and let G_i be a group for $i \in S$.

    (a) $\pi\{G_i \mid i \in S\} \cong G_1 \times \ldots \times G_n$.

    (b)   If $T \in \mathrm{Sym}(n)$, then $G_1 \times \ldots \times G_n \cong G_{1T} \times \ldots \times G_{nT}$.

2.1.33. (a)   (Associativity of the direct product.) If S is a set of groups and $S = \dot{\cup} \{S_i \mid i \in T\}$, then

$$\pi\{G \mid G \in S\} \cong \pi\{\pi\{G \mid G \in S_i\} \mid i \in T\}.$$

    (b)   Formulate and prove the corresponding property for external direct sums.

2.1.34. Show that $\mathrm{Sym}(3)$ has nonnormal subgroups.

2.1.35. Isomorphism is an equivalence relation.

2.1.36. If G and H are groups and T is an isomorphism of G into H, then there is a group $K \supset G$, and an isomorphism U of K onto H such that $U \mid G = T$.

2.1.37. If T is a homomorphism of a group G, $x \in G$, and $n \in J$, then $x^n T = (xT)^n$.

2.1.38. If T is an isomorphism of a lattice (S, R) onto a lattice (S', R'), $x \in S$ and $y \in S$, then

$$(x \cup y)T = xT \cup yT \quad \text{and} \quad (x \cap y)T = xT \cap yT.$$

2.2   Factor groups

    2.2.1. *If $H \lhd G$, then the set of cosets of H in G forms a group under multiplication (see Section 1.5).* ‖

    The group in Theorem 2.2.1 is called the *factor group* G/H. (It is also called the quotient group, or sometimes, if the operation is addition, the difference group $G - H$.) In the group G/H, H is the identity, and $(Hx)^{-1} = Hx^{-1}$.

2.2.2. *If $H \lhd G$ and $gT = Hg$ for $g \in G$, then T is a homomorphism of G onto G/H with kernel H.* ‖

The above function T is called the *natural* homomorphism of G onto G/H.

It often happens that when one tries to define a homomorphism from one group into another, it is not even clear that the defined object is a function. For this reason it is convenient to have available the following technical lemma on relations.

2.2.3. *Let G and H be groups and R a relation from G into H such that* (i) *if $(x, a) \in R$ and $(y, b) \in R$, then $(xy, ab) \in R$, and* (ii) *if $(e_G, a) \in R$ then $a = e_H$. Then $R \in \text{Hom}(G, H)$.*

Proof. Let $(x, a) \in R$ and $(x, a') \in R$. $\exists\, b \in H$ such that $(x^{-1}, b) \in R$. Hence by (i), $(xx^{-1}, ab) = (e, ab) \in R$, so by (ii), $a = b^{-1}$. Similarly, $a' = b^{-1}$, hence $a' = a$. Therefore R is a function. It is now obvious from (i) that R is a homomorphism.

2.2.4. *If $T \in \text{Hom}(G, H)$, $K \lhd G$, $K \subset \text{Ker}(T)$, and $T^* = \{(gK, gT) \,|\, g \in G\}$, then $T^* \in \text{Hom}(G/K, H)$.*

Proof. Since $K \lhd G$, G/K is defined. If $(gK, gT) \in T^*$ and $(hK, hT) \in T^*$, then

$$((gK)(hK), (gT)(hT)) = (ghK, (gh)T) \in T^*.$$

If $gK = K$ then $g \in K$, so $gT = e$. Hence, by Theorem 2.2.3, $T^* \in \text{Hom}(G/K, H)$.

2.2.5. (*Homomorphism theorem.*) *If $T \in \text{Hom}(G, H)$, $GT = H$, and $U = \{(g\text{Ker}(T), gT) \,|\, g \in G\}$, then U is an isomorphism of $G/\text{Ker}(T)$ onto H. Thus $G/\text{Ker}(T) \cong H$.*

Proof. By Theorem 2.2.4, U is a homomorphism of $G/\text{Ker}(T)$ into H. Since $GT = H$, it is clear that U is onto H. If $g \in G$ and $(g\text{Ker}(T))U = e$, then $gT = e$, and $g\text{Ker}(T) = \text{Ker}(T)$, the identity element of $G/\text{Ker}(T)$. By Theorem 2.1.9, U is an isomorphism.

EXERCISES

2.2.6. Give examples of factor groups. (See, for example, Exercises 1.7.18, 1.6.10, and 1.6.12.)

2.2.7. Let H be a nonnormal subgroup of G. Prove that the set of right cosets of H do not form a group under multiplication.

2.2.8. Show that the homomorphism theorem 2.2.5 may be stated more fully as follows. If V is a homomorphism of a group G onto a group H, then $V = TU$ where T is the natural homomorphism of G onto $G/\mathrm{Ker}(V)$ (Theorem 2.2.2), and U is the isomorphism of $G/\mathrm{Ker}(V)$ onto H given in Theorem 2.2.5.

2.2.9. If $H_i \lhd G_i$ for all $i \in S$, then

(a) $\dfrac{\pi G_i}{\pi H_i} \cong \pi \dfrac{G_i}{H_i}$

(b) $\dfrac{\sum_E G_i}{\sum_E H_i} \cong \sum_E \dfrac{G_i}{H_i}$

2.2.10. If A, B, G, and H are (possibly infinite) groups such that $A/B \cong G/H$, then $o(A)o(H) = o(B)o(G)$.

2.3   Isomomorphism theorems

2.3.1. *If G is a group, $H \subset G$, and $K \lhd G$, then $H \cap K \lhd H$.*

2.3.2. *If G is a group, $H \subset G$, and $K \lhd G$, then $HK = KH = \langle H, K \rangle$.*

2.3.3. *(Isomorphism theorem.) If G is a group, $H \subset G$, $K \lhd G$, and $T = \{(hK, h(H \cap K)) \mid h \in H\}$, then T is an isomorphism of HK/K onto $H/H \cap K$. Thus*

$$\frac{HK}{K} \cong \frac{H}{H \cap K}.$$

Proof. The factor group HK/K is defined since $K \lhd HK$, and $H/H \cap K$ is defined by Theorem 2.3.1. Let

$$(h_i K, h_i(H \cap K)) \in T, \qquad i = 1, 2.$$

Then

$$((h_1 K)(h_2 K), h_1(H \cap K)h_2(H \cap K)) = (h_1 h_2 K, h_1 h_2(H \cap K)) \in T.$$

If $hK = K$ and $h \in H$, then $h \in H \cap K$ and $h(H \cap K) = H \cap K$. Hence, by Theorem 2.2.3, T is a homomorphism of HK/K into $H/H \cap K$. It is clear that T is onto $H/H \cap K$. If $hK \in \mathrm{Ker}(T)$, then $h \in H \cap K$, so $hK = K$, and T is an isomorphism by Theorem 2.1.9.

2.3.4. *If G is a group, $H \triangleleft G$, and $K \triangleleft G$, then*

$$\frac{HK}{K} \cong \frac{H}{H \cap K} \quad \text{and} \quad \frac{HK}{H} \cong \frac{K}{H \cap K}.$$

2.3.5. *If $T \in \mathrm{Hom}(G, H)$, $GT = H$, $K \triangleleft H$, $M = KT^{-1}$, and*

$$U = \{(gM, (gT)K) \mid g \in G\},$$

then U is an isomorphism of G/M onto H/K. Thus $G/M \cong H/K$.

Proof. Let V be the natural homomorphism of H onto H/K. Then TV is a homomorphism of G onto H/K with kernel M. The theorem now follows from the homomorphism theorem 2.2.5.

2.3.6. *(Freshman theorem.)** *If $K \triangleleft H \triangleleft G$, $K \triangleleft G$, and*

$$U = \{(gH, (gK)(H/K)) \mid g \in G\},$$

then U is an isomorphism of G/H onto $(G/K)/(H/K)$. Thus

$$\frac{G}{H} \cong \frac{G/K}{H/K}.$$

Proof. Let T be the natural homomorphism of G onto G/K. Then $(H/K)T^{-1} = H$ and $H/K \triangleleft G/K$ by Theorem 2.1.10. Now apply Theorem 2.3.5.

EXERCISES

2.3.7. (a) If $H \subset G$, $K \subset G$, and $HK \subset G$, then $[HK:H] = [K:H \cap K]$.

  (b) If $H \subset G$ and $K \triangleleft G$, then $[HK:H] = [K:H \cap K]$.

2.3.8. If $H \subset G$, $K \triangleleft G$, and $H \cap K = E$, then $HK/K \cong H$.

2.3.9. Let $G = \mathrm{Sym}(4)$ and let H be the four-group

  $$H = \{e, (1, 2)(3, 4)\ (1, 3)(2, 4), (1, 4)(2, 3)\}.$$

  (a) Prove that H is an Abelian normal subgroup of G.

  (b) Use Exercise 2.3.8 to show that $G/H \cong \mathrm{Sym}(3)$ (see Exercise 1.7.16).

  (c) Find a normal subgroup K of H which is not normal in G.

  (d) Conclude that normality is not transitive.

2.3.10. Show that if H and K are subgroups of G, then it does not follow that HK is a subgroup of G. (Compare with Theorem 2.3.2.)

2.3.11. In $G = \mathrm{Sym}(4)$, let H be the 4-subgroup in Exercise 2.3.9, $K = \langle (1, 2, 3) \rangle$.

  (a) $HK \subset G$. Henceforth this group HK will be denoted by Alt(4).

  (b) Normality is not transitive in Alt(4).

* So-called because the isomorphism follows by "cancellation" of K.

2.4   Cyclic groups

If S is a subset of a group G, then $\langle S \rangle$ will denote the smallest (see Theorem 1.6.6) subgroup containing S, and will be called the subgroup *generated* by S. S is a *generating subset* of G iff $G = \langle S \rangle$. A group G is *cyclic* iff there is a generating subset which is a singleton $\{x\}$. Some corruptions of the notation $\langle S \rangle$ will be used. For example, if $S = \{x\}$, $\langle x \rangle$ will be used instead of $\langle \{x\} \rangle$.

It is clear (Theorem 1.6.5) that G is cyclic iff $\exists\, x \in G$ such that G is the set of powers of x.

2.4.1. *If G is cyclic and T is a homomorphism of G, then GT is cyclic.*

Proof. $\exists\, x$ such that $G = \langle x \rangle$. Then $GT = \langle xT \rangle$. ‖

Let us determine all cyclic groups. If $n \in J$, $n \geq 0$, then $nJ = \{ni \mid i \in J\}$ is a subgroup of J. Conversely (Exercise 1.6.12) if $H \subset J$, then $\exists\, n \in J, n \geq 0$, such that $H = nJ$. Moreover, if $m > n \geq 0$, then $mJ \neq nJ$. It is worth noting, however, that if $n \neq 0$, then the function $T\colon iT = ni$, is an isomorphism of J onto nJ. The factor group J/nJ will be denoted by J_n (if $n > 0$), and called the (additive) group of integers (mod n). Note that $o(J_n) = n$. Denote the elements of J_n by $[0]_n, \ldots, [n-1]_n$.

Now let G be a group and $x \in G$. Let $nT = x^n$ for $n \in J$. Then T is a homomorphism (Theorem 1.2.9) of J onto the group $\langle x \rangle$ of powers of x. By the homomorphism theorem 2.2.5, $J/\mathrm{Ker}(T) \cong \langle x \rangle$. Thus a cyclic group $\langle x \rangle \cong J$ if $x^n = e$ implies $n = 0$, and $\langle x \rangle \cong J_n$ if n is the smallest positive integer such that $x^n = e$. The *order* $o(x)$ is ∞ and n, respectively, in the situations just described. Recapitulating:

2.4.2. *There is a cyclic group of order n for each natural number n. There is an infinite cyclic group. Any two cyclic groups of the same order are isomorphic.*

2.4.3. *If G is a finite group and $x \in G$, then $o(x) \mid o(G)$.*

Proof. For $o(x) = o(\langle x \rangle)$, which divides $o(G)$ by Lagrange's theorem 1.7.8. ‖

We have already found all subgroups and factor groups of J, and therefore, by Theorem 2.4.2, of any infinite cyclic group. It remains to do the same for finite cyclic groups.

2.4.4. (*Subgroups of cyclic groups are cyclic.*) *If G is cyclic of finite order n, then G has exactly one cyclic subgroup of order m for each positive divisor m of n, and no other subgroups.*

Proof. Since $G \cong J_n$, it suffices to consider J_n. By the lattice theorem 2.1.13, since $J_n = J/nJ$, $\text{Lat}(J_n)$ is isomorphic to the lattice of subgroups of J containing nJ. Hence $\text{Lat}(J_n)$ is isomorphic to the lattice of subgroups mJ of J such that $m \mid n$. This isomorphism is induced by the natural homomorphism T of J onto J/nJ. Since mJ is cyclic, $(mJ)T$ is cyclic (Theorem 2.4.1). By Theorem 2.3.5, $[J_n : (mJ)T] = [J : mJ] = m$, so $o((mJ)T) = n/m$.

2.4.5. *If o(G) is prime, then G is cyclic and has no nontrivial subgroups.*

Proof. G has no nontrivial subgroups by Lagrange. $\exists\, x \in G^{\#}$. Since $\langle x \rangle \neq E$, $\langle x \rangle = G$, and G is cyclic.

EXERCISES

2.4.6. (Factor groups of cyclic groups are cyclic.) Prove that if G is cyclic of finite order n, then G has exactly one cyclic factor group of order m if $m \mid n$, $m > 0$, and has no other factor groups.

2.4.7. Find all automorphisms of J_{12}.

2.4.8. Find all automorphisms of J_n.

2.4.9. Find all homomorphisms of J_{12} into J_6.

2.4.10. (a)   If S is a nonempty subset of a group G, then $\langle S \rangle$ is the set of words in the letters of S.

(b)   Let $G = \langle S \rangle$, and let U be a function from S into a group H. Prove that there is at most one homomorphism T of G into H such that $T \mid S = U$.

2.4.11. If $G = \langle S \rangle$ and $T \in \text{Hom}(G)$, then $GT = \langle ST \rangle$.

2.4.12. If $T \in \text{Hom}(G)$ and x is an element of finite order in G, then $o(xT) \mid o(x)$.

2.4.13. If every element of a group has order 1 or 2, then the group is Abelian.

2.4.14. If H is a finite subgroup of G, K is a normal subgroup of G of finite index, and $(o(H), [G:K]) = 1$, then $H \subseteq K$.

2.5   Composition series

A group is *simple* iff it has no nontrivial normal subgroups.

2.5.1. *The only simple Abelian groups are the cyclic groups of order* 1 *or a prime.*

Proof. It is trivial (Theorem 2.4.5) that the groups mentioned are simple. Conversely, let G be a simple Abelian group, $o(G) > 1$. Then $\exists\, x \in G^{\#}$. Since any subgroup of an Abelian group is normal, $G = \langle x \rangle$. It follows from Theorem 2.4.4 and the comments after Theorem 2.4.1 that if G is not of prime order, then it has nontrivial subgroups. Hence G has prime order.

2.5.2. *If G is a group, $H \lhd G$, and $H < G$, then G/H if simple iff H is a maximal proper normal subgroup of G.* *

Proof. This follows immediately from the lattice theorem 2.1.13.

2.5.3. *If A and B are distinct maximal proper normal subgroups of G, then $A \cap B$ is a maximal proper normal subgroup of A and of B.*

Proof. $AB \lhd G$ by Theorems 2.3.2 and 2.1.20. Since $AB \supset A$, $AB = A$ or $AB = G$ by the maximality of A. But if $AB = A$, then $B \subset A$, hence $B < A$ (since $A \neq B$), contradicting the maximality of B. Thus $AB = G$. Therefore $G/B \cong A/A \cap B$ by the isomorphism theorem 2.3.3. Application of Theorem 2.5.2 shows that G/B is simple, hence (see Exercise 2.5.11) that $A \cap B$ is a maximal, proper, normal subgroup of A. Since the hypotheses are unchanged by interchange of A and B, the same conclusion follows for B. ‖

A *normal series* of G is a finite sequence $(A_0, \ldots, A_r)$ of subgroups such that $E = A_0 \lhd A_1 \lhd \ldots \lhd A_r = G$. An *invariant series* is a normal series such that each $A_i \lhd G$. The *factors* of a normal series are the factor groups A_{i+1}/A_i, $0 \leq i \leq r - 1$. Two normal series $(A_0, \ldots, A_r)$ and $(B_0, \ldots, B_s)$ of G are *equivalent* (denoted by $\sim$) iff $s = r$ and $\exists\, T \in \mathrm{Sym}(r)$ such that,

$$\frac{A_i}{A_{i-1}} \cong \frac{B_{iT}}{B_{iT-1}} \qquad \text{for } 1 \leq i \leq r.$$

2.5.4. *$\sim$ is an equivalence relation on the set of normal series of a group G.* ‖

A *composition series* of G is a normal series without repetition whose factors are all simple. Thus a composition series is a normal series $(A_0, \ldots, A_r)$ in which each A_i is a maximal, proper, normal subgroup of A_{i+1}. The factors, of a composition series are called *composition factors*. E has just one composition series, (E), and the series has no factors.

2.5.5. *A finite group has a composition series.*

Proof. This follows easily by induction. ‖

---

* This statement is an abbreviation of the statement "H is a maximal element in the set of proper normal subgroups of G partially ordered by inclusion." A similar interpretation is to be given later uses of maximal, minimal, maximum, and minimum.

In fact, a generalization of this theorem is true. A normal series $(B_0, \ldots, B_s)$ is a *refinement* of a normal series $(A_0, \ldots, A_r)$ iff $\exists$ a 1–1 function T from $\{0, \ldots, r\}$ into $\{0, \ldots, s\}$ such that $A_i = B_{iT}$ for all i. If the series $(A_0, \ldots, A_r)$ is without repetitions, this is simply the requirement that each A_i be some B_j. The generalization of Theorem 2.5.5 then reads as follows.

2.5.6. *If G is a finite group and S is a normal series of G without repetitions, then there is a composition series of G which is a refinement of S.* ‖

It should be remarked that Theorems 2.5.5 and 2.5.6 are not true for groups in general (see Exercise 2.5.12).

2.5.7. *If $H \lhd G$ and $(A_0, \ldots, H = B_0, B_1, \ldots, B_s = G)$ is a composition series of G, then $(B_0/H, B_1/H, \ldots, B_s/H)$ is a composition series of G/H.*

Proof. By the freshman theorem 2.3.6, for each i,

$$\frac{B_{i+1}/H}{B_i/H} \simeq \frac{B_{i+1}}{B_i},$$

which is simple. Hence $(B_0/H, \ldots, B_s/H)$ is a composition series of G/H. ‖

Easily the most important fact about composition series is the Jordan-Hölder theorem which follows.

2.5.8. (*Jordan-Hölder theorem.*) *If G is a finite group, then any two composition series of G are equivalent.*

Proof. Induct on $o(G)$.* Let $(A_0, \ldots, A_r)$ and $(B_0, \ldots, B_s)$ be composition series of G. If $A_{r-1} = B_{s-1}$, then the theorem follows from the inductive hypothesis. If $A_{r-1} \neq B_{s-1}$, let $(C_0, \ldots, A_{r-1} \cap B_{s-1})$ be a composition series of $A_{r-1} \cap B_{s-1}$ (Theorem 2.5.5). Then the four composition series (see Theorem 2.5.3) of G:

$$S_1 = (A_0, \ldots, A_{r-1}, G),$$

$$S_2 = (C_0, \ldots, A_{r-1} \cap B_{s-1}, A_{r-1}, G),$$

$$S_3 = (C_0, \ldots, A_{r-1} \cap B_{s-1}, B_{s-1}, G),$$

$$S_4 = (B_0, \ldots, B_{s-1}, G),$$

are such that $S_1 \sim S_2$ and $S_3 \sim S_4$ by induction, while $S_2 \sim S_3$ by Theorem 2.3.4. Hence (Theorem 2.5.4), $S_1 \sim S_4$.

---

* Groups of order 1 are so trivial that here and elsewhere the verification that the theorem is true if $o(G) = 1$ will usually be omitted.

2.5.9.   *If* $(A_0, \ldots, A_r)$ *is a normal series of* G, $H \subset G$, *and* $H_i = H \cap A_i$, *then* H_{i+1}/H_i *is isomorphic to a subgroup of* A_{i+1}/A_i.

Proof. $H_{i+1} \subset A_{i+1}$, $A_i \lhd A_{i+1}$, and $H_{i+1} \cap A_i = H_i$. Therefore, by the isomorphism theorem (Theorem 2.3.3),

$$\frac{H_{i+1}}{H_i} \cong \frac{H_{i+1}A_i}{A_i} \subset \frac{A_{i+1}}{A_i}.$$

EXERCISES

2.5.10. If G is a simple group and $T \in \mathrm{Hom}(G)$, then either $T \in \mathrm{Iso}(G)$ or $o(GT) = 1$ and $T = O$.

2.5.11. If G is a simple group and $T \in \mathrm{Iso}(G)$, then GT is simple.

2.5.12. The group J of integers has no composition series.

2.5.13. Find all composition series of J_{60} and verify the Jordan-Hölder theorem.

2.5.14. Use Exercise 2.3.9. to find a composition series of $\mathrm{Sym}(4)$.

2.5.15. Prove the following generalization of the Jordan-Hölder theorem. If G has a composition series, then any two composition series are equivalent. (This can be proved by nearly the same argument as for Theorem 2.5.8. For a different proof, see Theorem 2.10.2.)

2.6   Solvable groups

A group is *solvable* iff it has a normal series whose factors are all Abelian.*

2.6.1.   *If* G *is a solvable group and* $H \lhd G$, *then* G/H *is solvable.*

Proof. There is a normal series $(A_0, \ldots, A_r)$ of G whose factors are Abelian. Then $H = HA_0 \lhd HA_1 \lhd \ldots \lhd HA_r = G$ (Exercise 2.6.7), and $HA_{i+1} = (HA_i)A_{i+1}$. By the isomorphism theorem and the freshman theorem

$$\frac{HA_{i+1}}{HA_i} \cong \frac{A_{i+1}}{HA_i \cap A_{i+1}} \cong \frac{A_{i+1}/A_i}{(HA_i \cap A_{i+1})/A_i},$$

which, being a factor group of an Abelian group, is Abelian. Hence $(HA_0/HA_0, \ldots, HA_r/HA_0)$ is a normal series of G/H with Abelian factors. Thus G/H is solvable.

---

* There are several inequivalent definitions of solvable groups in the literature. All are equivalent for finite groups.

2.6.2. *If G is a solvable group and H $\subset$ G, then H is solvable.*

Proof. There is a normal series $(A_0, \ldots, A_r)$ of G whose factors are Abelian. By Theorem 2.5.9, $(A_0 \cap H, \ldots, A_r \cap H)$ is a normal series of H whose factors are Abelian. Hence H is solvable.

2.6.3. *If H and G/H are solvable groups, then G is a solvable group.*

Proof. Let T be the natural homomorphism of G onto G/H, A be a normal series of H with Abelian factors, and B be a normal series of G/H with Abelian factors. Then the normal series of G formed by following A by BT^{-1} has Abelian factors.

2.6.4. *A finite solvable group G has a composition series whose factors are (cyclic) of prime order.*

Proof. If $o(G) = 1$ or a prime, the result is trivial (if $o(G) = 1$, then (G) is a composition series of G without factors). Now induct, and suppose that $o(G)$ is not 1 or a prime. G has a normal series with Abelian factors. Hence, if G is non-Abelian, then G has a nontrivial, normal subgroup H. If G is Abelian, the same is true by Theorem 2.5.1. By the induction assumption, H and G/H have composition series with factors of prime order. One then obtains a composition series of G as in the proof of Theorem 2.6.3 (after removing the second occurrence of H).

EXERCISES

2.6.5. If $G_1, \ldots, G_n$ are solvable groups (n finite), then their direct product is solvable.

2.6.6. A simple group $G \neq E$ is solvable iff it is of prime order.

2.6.7. If $H \lhd G$ and $A \lhd B \subset G$, then $HA \lhd HB$.

2.6.8. Exercise 2.6.7 is false if the hypothesis $H \lhd G$ is omitted, even if both HA and HB are subgroups of G.

2.6.9. If A and B are normal solvable subgroups of G, so is AB.

2.7   Operator groups. Homomorphisms.

Let S be a set, fixed throughout the next four sections. An S-group is an ordered pair $(G, *)$ such that G is a group and $*$ is a function from $G \times S$

into G such that if $a \in G$, $b \in G$, and $s \in S$, then $(ab) * s = (a * s)(b * s)$. An *operator group* is an S-group for some S.

From now on, $a * s$ will be denoted by as, and the S-group $(G, *)$ by G. Note that each $s \in S$ induces an endomorphism of G. However, S cannot be considered as a set of endomorphisms of G, since distinct elements s and s' of S may induce the same endomorphism. Examples of operator groups will be given in Section 2.11.

An *S-subgroup* of an S-group G is a subgroup H of G such that $Hs \subset H$ for all $s \in S$. An S-subgroup is made into an S-group in the natural way.

An *S-homomorphism* of an S-group G into an S-group H (same set S) is a homomorphism T of the group G into the group H such that if $g \in G$ and $s \in S$, then $(gs)T = (gT)s$. S-endomorphisms, S-isomorphisms, and S-automorphisms of S-groups are defined similarly.

The development of the theory of S-groups will be largely parallel to the earlier one beginning with Section 2.1. Many details of proofs are purely routine and will be omitted.

2.7.1. *If G, H, and K are S-groups, T an S-homomorphism of G into H, and U an S-homomorphism of H into K, then TU is an S-homomorphism of G into K.*

2.7.2. *If T is an S-homomorphism of an S-group G into an S-group H, then GT is an S-subgroup of H.*

2.7.3. *If T is an S-homomorphism of an S-group G into an S-group H and K is an S-subgroup of G, then $T \mid K$ is an S-homomorphism of K into H.*

2.7.4. *The kernel of an S-homomorphism of an S-group is a normal S-subgroup.*

2.7.5. *If G and H are S-groups, T is an S-homomorphism of G onto H, and K an S-subgroup of G, then KT is an S-subgroup of H. If K is normal, then KT is normal.*

2.7.6. *If G and H are S-groups, T is an S-homomorphism of G onto H, and M is an S-subgroup of H, then MT^{-1} is an S-subgroup of G. If M is normal, so is MT^{-1}.*

2.7.7. *If G and H are S-groups and T is an S-homomorphism of G onto H, then T maps the lattice of S-subgroups between $\mathrm{Ker}(T)$ and G isomorphically onto the lattice of S-subgroups of H. T maps the lattice of normal S-subgroups between $\mathrm{Ker}(T)$ and G isomorphically onto the lattice of normal S-subgroups of H.*

EXERCISES

2.7.8. The relation "is S-isomorphic to" is an equivalence relation on the class of S-groups.

2.7.9. Prove the theorems of this section.

2.7.10. The intersection or union of a set of S-subgroups of an S-group is again an S-subgroup.

2.7.11. If G is an S-group and $M \neq \varnothing$ a subset, what elements of G are in the smallest S-subgroup which contains M?

2.8   Operator groups. Factor groups.

If G is an S-group, H a normal S-subgroup, $g \in G$, $g' \in G$, $gH = g'H$, and $s \in S$, then $g^{-1}g' \in H$ so that $(g^{-1}g')s \in H$, $(gs)^{-1}(g's) \in H$, and $(gs)H = (g's)H$. This fact permits the following definition: If $g \in G$ and $s \in S$, then $(gH)s = (gs)H$.

2.8.1. *If G is an S-group and H a normal S-subgroup, then G/H is an S-group.*

2.8.2. *If G is an S-group and H a normal S-subgroup, then the natural homomorphism of the group G onto the group G/H is also an S-homomorphism.*

2.8.3. *If T is an S-homomorphism of an S-group G onto an S-group H, then the relation*

$$U = \{(g\mathrm{Ker}(T), gT) \mid g \in G\}$$

is an S-isomorphism of $G/\mathrm{Ker}(T)$ onto H.

EXERCISE

2.8.4. Supply proofs for the theorems of this section.

2.9   Operator groups. Isomorphism theorems.

2.9.1. *If $\{H_i \mid i \in M\}$ is a family of S-subgroups of an S-group, then $\cap \{H_i \mid i \in M\}$ and $\langle H_i \mid i \in M \rangle$ are also S-subgroups.*

2.9.2. *If G is an S-group, H an S-subgroup, and K a normal S-subgroup, then the relation*

$$T = \{(hK, h(H \cap K)) \mid h \in H\}$$

is an S-isomorphism of HK/K onto $H/(H \cap K)$.

2.9.3. *If H and K are normal S-subgroups of an S-group G, then*

$$\frac{HK}{K} \cong \frac{H}{H \cap K} \quad and \quad \frac{HK}{H} \cong \frac{K}{H \cap K}$$

as S-groups.

2.9.4. *If G and H are S-groups, T an S-homomorphism of G onto H, K a normal S-subgroup of H, and $M = KT^{-1}$, then the relation*

$$U = \{(gM, (gT)K) \mid g \in G\}$$

is an S-isomorphism of G/M onto H/K.

2.9.5. *If G is an S-group, H a normal S-subgroup, and K an S-subgroup of H which is normal in G, then the relation*

$$U = \{(gH, (gK)(H/K)) \mid g \in G\}$$

is an S-isomorphism of G/H onto $(G/K)/(H/K)$. ‖

In addition to the analogues of earlier theorems, another isomorphism theorem will be given and proved. This theorem could, of course, have been stated for (nonoperator) groups first, but wasn't needed in the proof of the Jordan-Hölder theorem.

2.9.6. (*Zassenhaus' lemma.*) *If G is an S-group, A, B, C, and D are S-subgroups of G, $A \lhd B$, and $C \lhd D$, then the relation*

$$T = \{(x(A(B \cap C)), x(C(D \cap A))) \mid x \in B \cap D\}$$

is an S-isomorphism of $A(B \cap D)/A(B \cap C)$ onto $C(D \cap B)/C(D \cap A)$.

Proof. The fact that $A(B \cap D)$, $A(B \cap C)$, $C(D \cap B)$, and $C(D \cap A)$ are S-subgroups of G follows from Theorems 2.9.1 and 2.3.2. Normality of $A(B \cap C)$ in $A(B \cap D)$ is left as an exercise, 2.9.7. Normality of $C(D \cap A)$ in $C(D \cap B)$ follows by symmetry.

Any element of $A(B \cap D)/A(B \cap C)$ is of the form $axA(B \cap C)$ where $a \in A$ and $x \in B \cap D$. But since $A \lhd B$, $ax = x(x^{-1}ax) = xa'$ with $a' \in A$, so

$$axA(B \cap C) = xa'A(B \cap C) = xA(B \cap C)$$

Similarly, any element of $C(D \cap B)/C(D \cap A)$ is of the form $x'C(D \cap A)$ with $x' \in B \cap D$.

If $u \in A(B \cap C) \cap B \cap D = A(B \cap C) \cap D$, then $u = ac$ with $a \in A$ and $c \in B \cap C$, so that $u = ca' \in D$ where $a' \in A$. Hence $a' \in A \cap D$ and $u \in C(D \cap A) \cap B \cap D$. By symmetry, $u \in A(B \cap C) \cap B \cap D$ iff $u \in C(D \cap A) \cap B \cap D$. It follows that if $x \in B \cap D$ and $y \in B \cap D$, then $xA(B \cap C) = yA(B \cap C)$ iff $xC(D \cap A) = yC(D \cap A)$. Therefore T is a 1–1 function from $A(B \cap D)/A(B \cap C)$ onto $C(D \cap B)/C(D \cap A)$. It is routine to prove that T is an isomorphism (Exercise 2.9.7). If $s \in S$ and $x \in B \cap D$, then

$$(xA(B \cap C)s)T = ((xs)A(B \cap C))T = (xs)C(D \cap A)$$

$$= (xC(D \cap A))s = ((xA(B \cap C))T)s.$$

Hence T is an S-isomorphism.

EXERCISE

2.9.7. In the notation of Zassenhaus' lemma, prove

    (a) $B \cap C \lhd B \cap D$,

    (b)   (see Exercise 2.6.7) $A(B \cap C) \lhd A(B \cap D)$,

    (c)   the function T preserves multiplication.

2.10   Operator groups. Composition series.

Use of the notion of S-group and Zassenhaus' lemma permits a two-way generalization of the Jordan-Hölder theorem. The terms "normal S-series of an S-group G," "equivalent", and "refinement" may be considered self-defining (see Section 2.5).

2.10.1. *Any two normal S-series of an S-group G have equivalent refinements.*

Proof. Let $(A_0, \ldots, A_r)$ and $(B_0, \ldots, B_s)$ be normal S-series of G. Let

$$A_{ij} = A_i(A_{i+1} \cap B_j), \; B_{ji} = B_j(B_{j+1} \cap A_i).$$

Then $A_{i0} = A_i$ and $B_{j0} = B_j$. It follows from Exercise 2.9.7 that

$$A_{00}, A_{01}, \ldots, A_{0s} = A_{10}, A_{11}, \ldots, A_{1s} = A_{20}, \ldots, A_{r-1,s}$$

is a normal S-series refining $(A_0, \ldots, A_r)$ and $(B_{00}, B_{01}, \ldots, B_{s-1,r})$ is a normal S-series refining $(B_0, \ldots, B_s)$. By Zassenhaus' lemma (Theorem 2.9.6), $A_{i,j+1}/A_{i,j}$ is S-isomorphic to

$$\frac{B_{j,i+1}}{B_{j,i}}, \quad 0 \le i \le r - 1, \quad 0 \le j \le s - 1.$$

Hence the constructed refinements are equivalent. ∥

A *composition S-series* of an S-group G is a series $(A_0 = E, \ldots, A_r = G)$ in which each A_i is a maximal proper normal S-subgroup of A_{i+1}.

2.10.2. *If an S-group has a composition S-series, then*

(i) *any normal S-series without repetitions can be refined to a composition S-series, and*

(ii) *any two composition S-series are equivalent.*

Proof. A normal S-series equivalent to a composition S-series is a composition S-series by the lattice theorem, 2.7.7. By Theorem 2.10.1, a normal S-series R without repetitions and a composition S-series T have equivalent refinements R' and T'. After removal of repetitions, one obtains a refinement R'' of R and T, with R'' equivalent to T. This proves (i) and, in fact, (ii) also, since if R is a composition S-series, then $R'' = R$.

EXERCISE

2.10.3. Verify Theorem 2.10.1 in the case where $G = J$, $S = \varnothing$, and the normal S-series are $(\langle 0 \rangle, \langle 12 \rangle, \langle 3 \rangle, \langle 1 \rangle)$ and $(\langle 0 \rangle, \langle 30 \rangle, \langle 5 \rangle, \langle 1 \rangle)$.

2.11   Operator groups. Examples.

Example 1. Let G be any group and let S be the empty set. Then G is an S-group in a trivial way. Moreover, an S-subgroup of G is just a subgroup, an S-homomorphism is just a homomorphism, and so on. Note that, even in this case, Theorems 2.10.1 and 2.10.2 are generalizations of the Jordan-Hölder theorem, 2.5.8. For, in the first place, an infinite group may have a composition series, in which case, by Theorem 2.10.2, the conclusion of the Jordan-Hölder theorem holds, and, in the second place, even if there is no composition series, the more general Theorem 2.10.1 holds.

2.11.1. *If G is a group, then the set* Aut(G) *of automorphisms of G is a group (under the product of functions defined earlier).* ∥

A subgroup H of a group G is *characteristic* in G iff $HT \subset H$ for all $T \in \text{Aut}(G)$. Let $\text{Char}(G)$ denote the lattice (2.11.16) of characteristic subgroups of G.

2.11.2. *If H is a subgroup of a group G, then the following statements are equivalent*:

(i) $H \in \text{Char}(G)$;

(ii) *If $T \in \text{Aut}(G)$, then $HT \subset H$*;

(iii) *If $T \in \text{Aut}(G)$, then HT is a subset of H*;

(iv) *If $T \in \text{Aut}(G)$, then $HT = H$.* ‖

Example 2. Let G be any group and let $S = \text{Aut}(G)$. Then G is an S-group. A subgroup H of G is an S-subgroup iff $H \in \text{Char}(G)$. A normal S-series of the S-group G will be called a *characteristic series* of the group G. Theorem 2.10.2 then says that any two maximal characteristic series without repetition of a group G are equivalent.

2.11.3. *If G is a group, $x \in G$, and T_x is the function from G into G defined by $yT_x = x^{-1}yx$, then $T_x \in \text{Aut}(G)$.* ‖

The function T_x given above is called the *inner automorphism* of G induced by x. The set of inner automorphisms of G will be denoted by $\text{Inn}(G)$.

2.11.4. *If G is a group, then $\text{Inn}(G) \lhd \text{Aut}(G)$.*

Proof. A computation shows that if $a \in G$ and $b \in G$, then $T_a T_b = T_{ab}$. Therefore, if F is the function on G such that $aF = T_a$, then F is a homomorphism of G into $\text{Aut}(G)$, and $GF = \text{Inn}(G)$. Hence (Theorem 2.1.4), $\text{Inn}(G)$ is a subgroup of $\text{Aut}(G)$. Let $U \in \text{Aut}(G)$, $x \in G$, and $y \in G$. Then

$$y(U^{-1}T_x U) = (yU^{-1})T_x U = (x^{-1}(yU^{-1})x)U$$

$$= (xU)^{-1}y(xU) = yT_{xU}.$$

Therefore, $U^{-1}T_x U = T_{xU} \in \text{Inn}(G)$. Hence $\text{Inn}(G) \lhd \text{Aut}(G)$. ‖

Example 3. Let G be any group, and let $S = \text{Inn}(G)$. Then G is an S-group. A subgroup H of G is an S-subgroup iff it is normal. A normal S-series of G is therefore an invariant series of the group G. A composition S-series of the S-group G is called a *principal series* of the group G. Thus a series $(A_0 = E, \ldots, A_r = G)$ is a principal series of G iff each A_i is a maximal proper subgroup of A_{i+1} which is normal in G. Theorem 2.10.2 then says that any two principal series of a group are equivalent.

Example 4. Let G be any group and let $S = \text{End}(G)$, the set of endo-morphisms of G. Then G is an S-group. A subgroup H of G is called *fully characteristic** iff H is an S-subgroup [for $S = \text{End}(G)$]. Theorem 2.10.2 says that any two maximal fully characteristic series (without repetitions) of a group are equivalent.

Example 5. Let G be an Abelian group, written additively. Let S be the set of integers. Then (writing operators on the left, as is customary in this case), G is an S-group where, for example, $3x = x + x + x$. In this case, S has an algebraic structure of its own since addition and multiplication are both defined. This situation will be investigated a little further in Section 5.6. Note that any subgroup of G is an S-subgroup of G.

EXERCISES

2.11.5. If $H \lhd G$ and $T \in \text{Aut}(G)$, then $HT \lhd G$.

2.11.6. If H is the only subgroup of its order in G, then H is characteristic in G.

2.11.7. If H is the only normal subgroup of its order in G, then $H \in \text{Char}(G)$.

2.11.8. If H is the only subgroup of its index in G, then $H \in \text{Char}(G)$.

2.11.9. Every subgroup of J is fully characteristic.

2.11.10. Every subgroup of a cyclic group is fully characteristic.

2.11.11. If a subgroup is characteristic, then it is normal.

2.11.12. If $H \in \text{Char}(K)$ and $K \in \text{Char}(G)$, then $H \in \text{Char}(G)$ (characteristicity is transitive).

2.11.13. If $H \in \text{Char}(K)$ and $K \lhd G$, then $H \lhd G$.

2.11.14. A fully characteristic subgroup is characteristic.

2.11.15. If H is a fully characteristic subgroup of K and K is a fully characteristic subgroup of G, then H is a fully characteristic subgroup of G.

2.11.16. $\text{Char}(G)$ is a complete lattice.

2.11.17. If G is the four-group (Section 1.4), then $\text{Aut}(G) \simeq \text{Sym}(3)$.

2.11.18. The four-group has subgroups which are normal but not characteristic.

2.11.19. There is only one principal series of $\text{Sym}(4)$. It is also a characteristic series.

2.11.20. Show that $\text{Sym}(2) \times \text{Sym}(3)$ has a subgroup which is characteristic but not fully characteristic.

---

* The term *fully invariant* is more common.

2.11.21. If $U \in \text{End}(G)$ and $x \in G$, then $T_x U = U T_{xU}$. It follows that the set $\text{Inn}(G) \cdot U$ is contained in the set $U \cdot \text{Inn}(G)$. Show that these sets are not always equal.

2.11.22. If $A \in \text{Char}(G)$ and $B/A \in \text{Char}(G/A)$, then $B \in \text{Char}(G)$.

2.11.23. If $A/K \in \text{Char}(H/K)$, $K \lhd G$, and $H \lhd G$, then $A \lhd G$.

2.11.24. If G is an infinite group, then $o(\text{Aut}(G)) \leq 2^{o(G)}$.

2.11.25. Any left coset of a subgroup H of a group G is a right coset of some subgroup of G.

TRANSFORMATIONS AND SUBGROUPS

3.1 Transformations

The following theorem of Cayley says that any group is isomorphic to a group of permutations.

3.1.1. *(Cayley.) Let G be a group, and, for each $x \in G$, let R_x be the function from G into G such that $yR_x = yx$ for all $y \in G$. If T is defined by $xT = R_x$ for $x \in G$, then T is an isomorphism of G into* $\mathrm{Sym}(G)$.

Proof. By Theorem 1.2.4, R_x is 1–1 from G onto G, hence $R_x \in \mathrm{Sym}(G)$. If $x \in G$, $y \in G$, and $x \neq y$, then $eR_x = x \neq y = eR_y$, so $R_x \neq R_y$. Hence T is 1–1. Finally, if x, y, and z are in G, then

$$z(R_x R_y) = (zx)R_y = zxy = zR_{xy},$$

so $R_x R_y = R_{xy}$. Hence

$$(xy)T = R_{xy} = R_x R_y = (xT)(yT).$$

Therefore T is an isomorphism. ∥

The isomorphism T in Cayley's theorem is called the *regular representation* of G.

48

Let G and H be groups. The sum of two elements of $\text{Hom}(G, H)$ was defined in Example 13 of Section 1.4.

3.1.2. *If G is a group and H an Abelian group, then* $\text{Hom}(G, H)$ *is an Abelian group under addition.* ∥

A *ring* is an ordered triple $(R, +, \cdot)$ such that $(R, +)$ is an Abelian group, $\cdot$ is an associative operation on R, and the distributive laws hold:

$$a \cdot (b + c) = a \cdot b + a \cdot c,$$
$$(b + c) \cdot a = b \cdot a + c \cdot a$$

for all a, b, c in R. If $\exists f \neq 0$ in R such that $a \cdot f = f \cdot a = a$ for all $a \in R$, then f is called the *identity* of R (there is at most one identity in a ring). The product $a \cdot b$ is normally written ab.

3.1.3. *If H is an Abelian subgroup of a group G, then* $\text{Hom}(G, H)$ *is a ring. If G is an Abelian group, then* $\text{End}(G)$ *is a ring with identity.*

EXERCISES

3.1.4. Using the notations of Cayley's theorem and Exercise 1.3.9, show that if $x \in G^{\#}$, then $\text{Ch}(R_x) = 0$.

3.1.5. If G is a group and $x \in G$, define a function L_x from G into G by the rule: $yL_x = xy$, if $y \in G$. State and prove some analogue of Cayley's theorem involving L instead of R.

3.1.6. (Distributive laws for homomorphisms.) If G is a group, H and K are Abelian groups, $T \in \text{Hom}(G, H)$, and $V \in \text{Hom}(H, K)$, then

(a)   if $U \in \text{Hom}(H, K)$, then $T(U + V) = TU + TV$;

(b)   if $U \in \text{Hom}(G, H)$, then $(T + U)V = TV + UV$.

3.1.7. If G, H, K, and L are Abelian groups, $T \in \text{Hom}(H, G)$, $V \in \text{Hom}(K, L)$, and for each $U \in \text{Hom}(G, K)$, $Us = TUV$, then $s \in \text{Hom}(\text{Hom}(G, K), \text{Hom}(H, L))$, (all $\text{Hom}(A, B)$ being considered as groups).

3.1.8. If, in Exercise 3.1.7, T and V are isomorphisms onto, so is s.

3.1.9. Compute $\text{Hom}(J, J)$.

3.2   Normalizer, centralizer, and center

If S is a subset of a group G, then the *centralizer* $C_G(S)$ of S in G is defined by

$$C_G(S) = \{x \in G \mid \text{if } s \in S \text{ then } xs = sx\}.$$

When there is no ambiguity, $C(S)$ will be used instead of $C_G(S)$. If $S = \{y\}$, $C(y)$ will be written instead of $C(\{y\})$. The centralizer of G in G is called the *center* of G and is denoted by $Z(G)$ or Z. The normalizer $N_G(S)$ of S in G is defined by

$$N_G(S) = \{x \in G \mid xS = Sx\}.$$

Again, $N(S)$ will be used usually instead of $N_G(S)$. Note that if $x \in G$, then $N(x) = C(x)$.

3.2.1. *If S is a subset of a group G, then $C(S)$, $N(S)$, and $Z(G)$ are subgroups of G.*

3.2.2. *If G is a group and $H \subset G$, then $H \subset N(H)$, and $N(H)$ is the largest subgroup of G in which H is normal.* ‖

The following almost trivial theorem is of great importance in the theory of groups.

3.2.3. *(N/C theorem.) If G is a group and $H \subset G$, then $N(H)/C(H)$ is isomorphic to a subgroup of* $\mathrm{Aut}(H)$ *(written $N(H)/C(H) \overset{\cong}{\cong} \mathrm{Aut}(H)$).*

Proof. For $x \in N(H)$, let T_x be the automorphism of G induced by x (see Theorem 2.11.3). The function U, defined by $xU = T_x \mid H$ for $x \in N(H)$, is a homomorphism of $N(H)$ into $\mathrm{Aut}(H)$ with kernel $C(H)$. The conclusion follows from the homomorphism theorem, 2.2.5.

3.2.4. $\mathrm{Inn}(G) \cong G/Z.$

Proof. Set $H = G$ in the proof of Theorem 3.2.3.

3.2.5. *If T is a homomorphism of a group G onto a group H, then $Z(G)T \subset Z(H)$. Hence $Z(G)$ is characteristic in G.*

3.2.6. *If G is a group and H an Abelian subgroup, then $HZ(G)$ is also an Abelian subgroup of G.*

Proof. Since $Z \in \mathrm{Char}(G)$, $Z \vartriangleleft G$, hence $HZ \subset G$ (Theorem 2.3.2). If $x \in HZ$ and $y \in HZ$, then $\exists\, h_i \in H$ and $z_i \in Z$ such that $x = h_1 z_1$ and $y = h_2 z_2$. A simple computation shows that $xy = yx$. Hence HZ is Abelian.

3.2.7. *If $x \in G \backslash Z$, then $\langle Z, x \rangle$ is Abelian.*

Proof. By the preceding theorem with H replaced by $\langle x \rangle$.

3.2.8. *If G is a non-Abelian group, then G/Z is not cyclic.*

Proof. Deny. Then $\exists\ x \in G$ such that $G/Z = \langle xZ \rangle$. Therefore $G = \langle Z, x \rangle$, which is Abelian by 3.2.7. ‖

Some generalizations of the last theorem will now be given.

3.2.9. *If G is a non-Abelian group, then G/Z is not the union of an increasing sequence of cyclic groups.*

Proof. Deny. Then $\exists\ \{H_n \mid n \in \mathcal{N}\}$ such that $H_n/Z(G)$ is cyclic, $Z(G) \subset H_1 \subset H_2 \subset \ldots$, and $\cup\ H_n = G$. Thus $\exists\ x_n \in G$ such that $H_n = \langle Z(G), x_n \rangle$. Hence H_n is Abelian. Let $a \in G$ and $b \in G$. Since G is clearly the point set union of the H_n, $\exists\ m$ such that $a \in H_m$ and $b \in H_m$. Therefore $ab = ba$. Hence G is Abelian, contrary to assumption.

3.2.10. (*Miller* [1].) *If G is a group, $x \in G^{\#}$, and S is a generating subset of G such that if $y \in S$ then $x \in \langle y \rangle$, then $\nexists$ a group H such that $H/Z(H) \cong G$.*

Proof. Deny. *WLOG* $G = H/Z(H)$. The elements of G are now cosets of $Z(H)$. Let $a_x \in x$, and for each $y \in S$, $a_y \in y$. Then $H = \langle Z(H), \{a_y \mid y \in S\} \rangle$. If $y \in S$ then $\exists\ n \in J$ and $z \in Z(H)$ such that $a_x = a_y^n z$. Now $a_x z^{-1}$ commutes with a_y as does z. Since $C_H(a_y)$ is a group, $a_x \in C_H(a_y)$. Therefore $a_y \in C_H(a_x)$ for all $y \in S$, and since $Z(H) \subset C_H(a_x)$, $C_H(a_x) = H$. Thus $a_x \in Z(H)$, contradicting the fact that $x \neq e$. ‖

The groups G prohibited by Theorem 3.2.10 form a fairly large class of groups. Some of these are indicated in the exercises below and others will be given later.

EXERCISES

3.2.11. (a) $Z(G) = \cap \{C(x) \mid x \in G\}$.

     (b) If $H \subset G$, then $C(H) = \cap \{C(x) \mid x \in H\}$.

3.2.12. $Z(\Sigma_E \{G_i \mid i \in S\}) = \Sigma_E \{Z(G_i) \mid i \in S\}$.

3.2.13. $Z(\pi\{G_i \mid i \in S\}) = \pi\{Z(G_i) \mid i \in S\}$.

3.2.14. Find formulas for $C(x)$ where x is an element of an external direct sum or direct product of groups.

3.2.15. If G is a group, then any subgroup of $Z(G)$ is normal in G.

3.2.16. (a) A subgroup H of G is *strictly characteristic* in G iff $HT \subset H$ for all endomorphisms T of G onto G. Prove that $Z(G)$ is strictly characteristic in G.

(b) If A is a strictly characteristic subgroup of G and B/A is a strictly characteristic subgroup of G/A, then B is a strictly characteristic subgroup of G.

3.2.17. (a) If $T \in \mathrm{Iso}(G)$ and S is a subset of G, then

$$N_{GT}(ST) = (N_G(S))T \quad \text{and} \quad C_{GT}(ST) = (C_G(S))T.$$

(b) Investigate the situation for homomorphisms T.

3.2.18. (a) If $H \triangleleft G$, then $C(H) \triangleleft G$.

(b) If $H \in \mathrm{Char}(G)$, then $C(H) \in \mathrm{Char}(G)$.

3.2.19. Let $p \in \mathscr{P}$, $n \in \mathscr{N}$, H a group such that $h \in H$ implies $o(h) = p^r m$ with $r < n$ and $p \nmid m$, and $G = J_{p^n} \times H$. Use Theorem 3.2.10 to prove that $\nexists K$ such that $K/Z(K) \cong G$.

3.2.20. Prove that there is no group H such that $H/Z(H)$ is isomorphic to any subgroup or factor group of $\mathscr{R}$ (other than E). (Use Theorem 3.2.9.)

3.2.21. If S is a subset of a subgroup H of a group G, then

$$N_H(S) = H \cap N_G(S) \quad \text{and} \quad C_H(S) = H \cap C_G(S).$$

3.2.22. Let S be a subset of T and T a subset of a group G. Prove

(a) $C(S) \supset C(T)$,

(b) $C(C(S))$ contains S.

(c) $C(C(C(S))) = C(S)$.

3.2.23. (See Scorza [1, Chapter 4].) Let x and y be elements of a group G. Then

(a) $x \in C(x)$,

(b) $C(C(x)) = Z(C(x))$,

(c) $C(x) \subset C(y)$ iff $y \in Z(C(x))$,

(d) $C(x) \subset C(y)$ iff $Z(C(x)) \supset Z(C(y))$,

(e) The relation $R = \{(C(x), Z(C(x))) \mid x \in G\}$ is a 1–1 function from a certain set of subgroups (so-called *fundamental* subgroups) onto a second set of subgroups (*normo-centers*).

3.2.24. (Baer [9].) If H is a finite maximal Abelian normal subgroup of G and K is a normal Abelian subgroup of G, then K is finite. (Use N/C.) (Compare with Theorem 9.2.17.)

3.3   Conjugate classes

If S and S' are subsets of a group G, then S is *conjugate* to S' iff $\exists\, x \in G$ such that $S' = x^{-1}Sx$. The notation $S^x = x^{-1}Sx$ will often be used. We have $(S^x)^y = S^{xy}$ for all x and y in G, and $S^x = S$ *iff* $x \in N(S)$. If $H \subset G$, then S^H will denote the subgroup of G generated by all S^h, $h \in H$.

3.3.1. *Conjugacy is an equivalence relation.* ‖

The *conjugate class* of a subset S of a group G is the set $\mathrm{Cl}(S)$ of subsets S' of G which are conjugate to S. Frequent use is made of the next two theorems.

3.3.2. *If S is a subset of a group G, then $[G:N(S)] = o(\mathrm{Cl}(S))$.*

Proof. If $x \in G$ and $y \in G$, then $S^x = S^y$ iff $xy^{-1} \in N(S)$, hence iff $N(S)x = N(S)y$. The assertion follows.

3.3.3. *If G is a group and $x \in G$, then $[G:C(x)] = o(\mathrm{Cl}(x))$.*

Proof. This follows from Theorem 3.3.2 when $S = \{x\}$. ‖

If $H \subset G$, then the *core* of H is the subgroup

$$\mathrm{Core}(H) = \cap \{K \mid K \in \mathrm{Cl}(H)\}.$$

If S is a subset of a group G, then the *normal closure* of S is S^G.

3.3.4. *If G is a group, $H \subset G$, and S is a subset of G, then $\mathrm{Core}(H)$ is the maximum normal subgroup of G contained in H, and S^G is the minimum normal subgroup of G containing S.*

Proof. Let $T \in \mathrm{Inn}(G)$. It follows easily from the fact that conjugacy is an equivalence relation that T induces permutations of $\mathrm{Cl}(H)$ and $\mathrm{Cl}(S)$. Hence both $\mathrm{Core}(H)$ and S^G are invariant under T. Therefore both are normal subgroups of G. If $A \subset H$ and $A \lhd G$, then for all $T \in \mathrm{Inn}(G)$, $A = AT \subset HT$, so that

$$A \subset \cap \{HU \mid U \in \mathrm{Inn}(G)\} = \mathrm{Core}(H).$$

Hence $\mathrm{Core}(H)$ is the maximum normal subgroup of G contained in H. The assertion about S^G is proved similarly.

3.3.5. *If G is a group and H a subgroup of finite index, then $\mathrm{Core}(H) \lhd G$, $\mathrm{Core}(H) \subset H$, and $G/\mathrm{Core}(H)$ is finite.*

Proof. Since $H \subset N(H)$, $[G:N(H)]$ is finite (Theorem 1.7.11). Hence $\mathrm{Cl}(H)$ is finite (Theorem 3.3.2). If $K \in \mathrm{Cl}(H)$, then $[G:K] = [G:H]$ by Exercise 2.1.19. By Poincaré's theorem, 1.7.10, and the preceding theorem, $\mathrm{Core}(H)$ is a normal subgroup of finite index in G contained in H.

3.3.6. *If G is an infinite group, $x \in G^{\#}$, and $\mathrm{Cl}(x)$ is finite, then G is not simple.*

Proof. If $o(\mathrm{Cl}(x)) > 1$, then, by Theorem 3.3.3, $C(x)$ is a proper subgroup of finite index in G, and by the preceding theorem, G is not simple. Now suppose $o(\mathrm{Cl}(x)) = 1$. Then $x \in Z$, so that $\langle x \rangle \lhd G$. Therefore, either G is not simple or it is cyclic. But in the latter case, G has nontrivial normal subgroups, hence is not simple. ‖

We next prove an interesting theorem, 3.3.8, about finite groups due to Landau [1]. A lemma is needed first.

3.3.7. *If* $r \in \mathcal{N}$ *and* $0 < x \in \mathcal{R}$, *then there are (at most) a finite number of r-tuples* $(i_1, \ldots, i_r)$ *of natural numbers such that* $\Sigma\, (1/i_j) = x$.

Proof. The statement is obvious if $r = 1$. Induct on r. Since $r!$ is finite, it need only be shown that the number of such r-tuples with $i_1 \leq i_2 \leq \ldots \leq i_r$ is finite. For such an r-tuple satisfying the equation (suitable r-tuple), $x \leq r/i_1$, so $i_1 \leq r/x$. But for each natural number $k \leq r/x$ there are only a finite number of suitable r-tuples $(k, i_2, \ldots, i_r)$ by the inductive hypothesis. Hence there are only a finite number of suitable r-tuples altogether.

3.3.8. *If* $r \in \mathcal{N}$, *then there are (at most) a finite number of isomorphism classes of finite groups G with exactly r conjugate classes of elements.*

Proof. It is sufficient to prove that there is a number M such that, if G is a finite group with exactly r conjugate classes of elements, then $o(G) < M$ (see Exercise 3.3.17). Let G have r conjugate classes $C_1, \ldots, C_r$, and let $o(G) = n$, $o(C_j) = c_j$. By Theorem 3.3.2, $i_j = n/c_j$ is a natural number. *WLOG* $C_r = \{e\}$, so that $i_r = n/1 = n$. Since the classes are disjoint, $c_1 + \ldots + c_r = n$. Division by n yields $\Sigma\, (1/i_j) = 1$. By the lemma, there are only a finite number of r-tuples $(i_1, \ldots, i_r)$ satisfying this equation and, in particular, only a finite number of choices of $i_r = n$. This proves the initial assertion, hence the theorem. ‖

The following theorem is due to Higman, Neumann, and Neumann [1].

3.3.9. *If G is a group, then there is a group $H \supset G$ such that any two elements of H of the same order are conjugate.*

Proof. First note that G is contained in an uncountable group. In fact, the direct product of G and an uncountable number of copies of J is uncountable and contains an isomorphic copy of G. The assertion then follows from Exercise 2.1.36. Hence *WLOG*, G is uncountable, say of order A.

Let $T: G \rightarrow G^*$ be the regular representation of G (Theorem 3.1.1). If two elements of G^* have the same order n, each is the formal product of A n-cycles (n may be infinite). Hence (Exercise 1.3.11), they are conjugate in

$\text{Sym}(G)$. By Cayley and Exercise 2.1.36, $\exists\ G_1 \supset G_0 = G$ such that $G_1 \cong \text{Sym}(G)$. Thus any two elements of G_0 of the same order are conjugate in G_1.

Inductively, if G_n is defined, then $\exists\ G_{n+1} \supset G_n$ such that any two elements of the same order in G_n are conjugate in G_{n+1}. Let $H = \cup\, G_n$ (see Exercise 1.8.7). Any two elements of H of the same order are contained in some G_n, hence are conjugate in G_{n+1}, and therefore are conjugate in H. ‖

Note that H has at most $\aleph_0$ conjugate classes of elements.

Finally, a theorem about double cosets will be proved. If $H \subset G$, $K \subset G$, and $x \in G$, then the double coset HxK is a union of right cosets of H and a union of left cosets of K. Let $[HxK:H]$ denote the number of right cosets of H in HxK and $[HxK:K]$ the number of left cosets of K. (In case $H = K$, this leads to an ambiguity which is of no consequence if H is finite. If H is infinite, these numbers may be different (see Exercise 9.2.12), and care must be taken to explain which index is meant.)

    3.3.10. *If H and K are subgroups of G and $x \in G$, then*

(i) $[HxK:H] = [K:H^x \cap K]$,

(ii) $[HxK:K] = [H:H \cap K^{x^{-1}}]$.

    Proof. (ii) Any left coset of K in HxK has the form hxK, $h \in H$. Now $h_1xK = h_2xK$ iff $x^{-1}h_2^{-1}h_1x \in K$, which is true iff $h_2^{-1}h_1 \in H \cap xKx^{-1}$. The assertion follows.

Statement (i) is proved similarly.

EXERCISES

3.3.11. Determine the conjugate classes of elements in $\text{Sym}(4)$. (See Theorem 1.3.6 and Exercise 1.3.11.)

3.3.12. If a group G has an element of infinite order, then it has $o(G)$ elements of infinite order.

3.3.13. If H is a proper subgroup of a finite group G, then $\exists\ x \in G$ such that x is not in any conjugate of H.

3.3.14. Define the terms characteristic closure and characteristic core and

    (a) prove the analogue of Theorem 3.3.4,

    (b) show that the analogue of Theorem 3.3.5 is false.

3.3.15. If A is an infinite cardinal and $[G:H] = A$, then $\text{Core}(H)$ is a normal subgroup of G of index at most 2^A. Hence if $2^A < o(G)$, then G is not simple.

3.3.16. Let G be an infinite group containing no proper subgroup of order $o(G)$.

    (a) The normal subgroups of G are G itself and subgroups of $Z(G)$.

    (b) If G is non-Abelian, then $G/Z(G)$ is simple, $o(G/Z(G)) = o(G)$, and G/Z has no proper subgroup of its own order.

[The only Abelian groups G satisfying these hypotheses are the p^∞-groups (see Exercise 5.2.28), all of which are countable. The existence of non-Abelian infinite groups with all proper subgroups of smaller order is an open question.]

3.3.17. If $n \in \mathcal{N}$, then there are only a finite number of isomorphism classes of groups of order n.

3.3.18. If G is a finite group and H is a subgroup of prime order such that $N(H) = H$, then $H^G = G$.

3.3.19. If G is a group, S a subset, and T a homomorphism of G onto a group K, then $S^G T = (ST)^K$.

3.3.20. If G is a group and S a nonempty subset, then $S^G = \{x \in G \mid$ if $T \in \mathrm{Hom}(G)$ and $ST = E$, then $xT = e\}$.

3.3.21. (a)   An infinite group has an infinite number of subgroups.

   (b)   In fact, if $o(G) = A$, then $o(\mathrm{Lat}(G)) \geqq A$.

3.3.22. (a)   If T is a homomorphism of G onto H, and S is a subset of G, then $(\mathrm{Cl}(S))T = \mathrm{Cl}(ST)$.

   (b)   If $H \lhd G$, $H \subset K \subset G$, and $H \subset L \subset G$, then K and L are conjugate in G iff K/H and L/H are conjugate in G/H.

3.3.23. If $H \subset G$ and if $x^2 \in H$ for all $x \in G$, then $H \lhd G$ and G/H is Abelian. (See Exercise 2.4.13).

3.4   Commutators

The *commutator* $[a, b]$ of two elements a and b of a group G is given by the equation $[a, b] = a^{-1}b^{-1}ab$. Let $[a, b, c] = [[a, b], c]$. The following facts may be verified directly.

3.4.1. *If a, b, and c are elements of a group G, then*

(i) $[a, b] = e$ iff $ab = ba$

(ii) $[a, b]^{-1} = [b, a]$,

(iii) $a^b = a[a, b]$,

(iv) $[a, bc] = [a, c][a, b]^c$,

(v) $[ab, c] = [a, c]^b[b, c]$,

(vi) $[a, b^{-1}, c]^b[b, c^{-1}, a]^c[c, a^{-1}, b]^a = e$.

3.4.2. *Let $x_1, \ldots, x_m, y_1, \ldots, y_n$ be elements of a group G, and let H be the subgroup generated by these elements. Then*

$$[x_1 \ldots x_m, y_1 \ldots y_n] = \pi[x_i, y_j]^{h_{ij}}, \qquad h_{ij} \in H,$$

where the order of the factors on the right is arbitrary but the h_{ij} depend on this order.

Proof. If there is one such product, then any desired order of the factors can be obtained by repeated use of the equation $ab = ba^b$. We shall prove the existence of one such product by induction.

First let $m = 1$. The statement is true for $n = 1$. Induct on n. Then, using the inductive hypothesis and Theorem 3.4.1, we get

$$[x_1, y_1 \ldots y_{n+1}] = [x_1, y_{n+1}][x_1, y_1 \ldots y_n]^{y_{n+1}}$$
$$= \pi[x_1, y_j]^{h_j}, \qquad h_j \in H.$$

Hence the result holds for all n if $m = 1$.

Now fix n, let $y = y_1 \ldots y_n$, and induct on m. Then

$$[x_1 \ldots x_{m+1}, y] = [x_1 \ldots x_m, y]^{x_{m+1}}[x_{m+1}, y]$$
$$= \pi[x_i, y]^{h_i'}, \qquad h_i' \in H,$$

by the inductive hypothesis. Therefore the assertion is true in general.

3.4.3. *If a, b, and c are elements of G and $c \in C([a, b])$, then $[a, bc] = [a, c][a, b]$ and $[ac, b] = [a, b][c, b]$.*

Proof. This follows from Theorem 3.4.1 (iv) and (v).

3.4.4. *If $x \in G$, $y \in G$, $[x, y] \in C(x) \cap C(y)$, $r \in J$, and $s \in J$, then*
(i) $[x^r, y^s] = [x, y]^{rs}$,
(ii) $(xy)^r = x^r y^r [y, x]^{r(r-1)/2}$.

Proof. (i) Let $H = \langle x, y \rangle$. Then $[x, y] \in Z(H)$. If $r > 0$ and $s > 0$, then (i) is true by Theorem 3.4.2. If $r = 0$ or $s = 0$, it clearly holds. By Theorem 3.4.3,

$$[x, y][x^{-1}, y] = [e, y] = e \quad \text{and} \quad [x, y^{-1}][x, y] = [x, e] = e,$$

so $[x^{-1}, y] = [x, y]^{-1} \in Z(H)$ and $[x, y^{-1}] = [x, y]^{-1} \in Z(H)$. Applying this result to $[x^{-1}, y]$ in place of $[x, y]$, one obtains

$$[x^{-1}, y^{-1}] = [x^{-1}, y]^{-1} = [x, y] \in Z(H).$$

Now let $r > 0$ and $s > 0$. By the first part of the proof,

$$[x^{-r}, y^s] = [(x^{-1})^r, y^s] = [x^{-1}, y]^{rs} = ([x, y]^{-1})^{rs}$$
$$= [x, y]^{(-r)s},$$
$$[x^r, y^{-s}] = [x^r, (y^{-1})^s] = [x, y^{-1}]^{rs} = ([x, y]^{-1})^{rs}$$
$$= [x, y]^{r(-s)},$$
$$[x^{-r}, y^{-s}] = [x^{-1}, y^{-1}]^{rs} = [x, y]^{rs} = [x, y]^{(-r)(-s)}.$$

Hence (i) holds in general

(ii) The statement is obvious for $r = 0$ or 1. Induct. Using (i) and the fact that powers of $[y, x] = [x, y]^{-1}$ are in $Z(H)$, we obtain

$$(xy)^{r+1} = (xy)(xy)^r = xyx^ry^r[y, x]^{r(r-1)/2}$$
$$= xx^r[x^r, y^{-1}]y^{r+1}[y, x]^{r(r-1)/2}$$
$$= x^{r+1}[x, y]^{-r}y^{r+1}[y, x]^{r(r-1)/2}$$
$$= x^{r+1}y^{r+1}[y, x]^r[y, x]^{r(r-1)/2}$$
$$= x^{r+1}y^{r+1}[y, x]^{(r+1)r/2}.$$

Hence (ii) holds for all $r > 0$. Again, with $r > 0$,

$$(xy)^{-r} = (y^{-1}x^{-1})^r = y^{-r}x^{-r}[x^{-1}, y^{-1}]^{r(r-1)/2}$$
$$= x^{-r}y^{-r}[y^{-r}, x^{-r}][x, y]^{r(r-1)/2}$$
$$= x^{-r}y^{-r}[y, x]^{r^2}[y, x]^{-r(r-1)/2}$$
$$= x^{-r}y^{-r}[y, x]^{(-r)(-r-1)/2}. \parallel$$

Now let H and K be subgroups of a group G. Then $[H, K]$ will denote the subgroup generated by the set of all commutators $[h, k]$ with $h \in H$ and $k \in K$. (The *set* of commutators $[h, k]$ with $h \in H$ and $k \in K$ is not always a subgroup. See Exercise 3.4.17.) By Theorem 3.4.1 (ii), $[K, H] = [H, K]$.

3.4.5. *Let H and K be subgroups of G. Then*
(i) $[H, K] \subset H^G \cap K^G$,
(ii)   *if $H \lhd G$, then $[H, K] \subset H$; if $K \lhd G$, then $[H, K] \subset K$.*
(iii) *If $H \lhd G$ and $K \lhd G$, then $[H, K] \lhd G$.*

Proof. If $a \in H$ and $b \in K$, then $[a, b] = a^{-1}a^b \in H^G$. Hence $[H, K] \subset H^G$. Similarly $[H, K] \subset K^G$, and (i) is true. Result (ii) follows from (i). Now let $a \in H$, $b \in K$, and $g \in G$. Then $[a, b]^g = [a^g, b^g]$, and (iii) follows readily.

3.4.6. *If H and K are subgroups of G such that $G = \langle H, K \rangle$, then $[H, K] \lhd G$.*

Proof. Let $a \in H$, $b \in H$, $c \in K$, $d \in K$. By Theorem 3.4.1 (iv) and (v),

$$[a, c]^b = [ab, c][b, c]^{-1} \in [H, K],$$
$$[a, c]^d = [a, d]^{-1}[a, cd] \in [H, K].$$

If $u \in [H, K]$, then $u = x_1 \ldots x_n$ where $x_i = [y_i, z_i]$ or $[y_i, z_i]^{-1}$ with $y_i \in H$ and $z_i \in K$. The above equations then show that $u^v \in [H, K]$ if $v \in H$ or $v \in K$. Since $G = \langle H, K \rangle$, it follows that $u^g \in [H, K]$ for all $g \in G$. Hence $[H, K] \lhd G$. $\parallel$

As for elements, if H, K, and L are subgroups of G, then $[H, K, L]$ will mean $[[H, K], L]$. Note that $[H, K, L] = [K, H, L]$.

3.4.7. (*Three subgroups theorem.*) *If H, K, and L are subgroups of G, $M \lhd G$, $[K, L, H] \subset M$, and $[L, H, K] \subset M$, then $[H, K, L] \subset M$.*

Proof. Let $x \in H$, $y \in K$, and $z \in L$. By hypothesis $[y, z^{-1}, x]^z \in M$ and $[z, x^{-1}, y]^x \in M$. By Theorem 3.4.1 (vi), $[x, y^{-1}, z]^y \in M$, hence $[x, y^{-1}, z] \in M$. Therefore, for all $a \in H$, $b \in K$, and $c \in L$, $[a, b, c] \in M$. Since $[K, L, H] = [L, K, H]$ and $[L, H, K] = [H, L, K]$, it is also true that $[[a, b]^{-1}, c] = [b, a, c] \in M$. It follows from 3.4.2 that $[H, K, L] = [[H, K], L] \subset M$.

3.4.8. (i) *If H, K, and L are subgroups of G such that $[K, L, H] = E$ and $[L, H, K] = E$, then $[H, K, L] = E$.*
(ii) *If H, K, and L are normal subgroups of G, then*

$$[H, K, L] \subset [K, L, H][L, H, K].$$

Proof. Statement (i) follows from the preceding theorem with $M = E$. If H, K, and L are normal, then by Theorem 3.4.5 (iii), $[K, L, H][L, H, K]$ is normal, so that (ii) follows from the three subgroups theorem also. ‖

The *commutator subgroup* G^1 of G is the subgroup $[G, G]$. The *derived series* of G is the sequence $(G^0 = G, G^1, G^2, \ldots)$ where, inductively, $G^{n+1} = (G^n)^1$.

3.4.9. *If G is a group, then G^1 is a fully characteristic subgroup of G. In fact, more generally, if H is a group and $T \in \mathrm{Hom}(G, H)$, then $G^1 T \subset H^1$.*

Proof. The second statement follows from the fact that a homomorphism preserves words and therefore sends a commutator into a commutator, and a product of commutators [see Theorem 3.4.1 (ii)] into a product of commutators.

3.4.10. *If G is a group, then G/G^1 is Abelian, and if G/H is Abelian, then $G^1 \subset H$.*

Proof. Let $x \in G$ and $y \in G$. Then, in the group G/G^1,

$$[xG^1, yG^1] = x^{-1}y^{-1}xyG^1 = G^1.$$

By Theorem 3.4.1 (i), G/G^1 is Abelian. In the group G/H, $H = [xH, yH] = [x, y]H$. Hence H contains all commutators, and therefore $H \supset G^1$.

3.4.11. *If G is a group and $G^1 \subset H \subset G$, then $H \lhd G$.*

Proof. H/G^1 is a subgroup of the Abelian group G/G^1, hence a normal subgroup thereof. By the lattice theorem, $H \lhd G$.

3.4.12. *A group G is solvable iff $G^n = E$ for some n.*

Proof. If $G^n = E$, then each G^i/G^{i+1} is Abelian, so G is solvable. Suppose, conversely, that G is solvable. Then there is a normal series $(E = A_n \lhd A_{n-1} \lhd \ldots \lhd A_0 = G)$ such that A_i/A_{i+1} is Abelian for all i. Now $G^0 \subset A_0$. If $G^i \subset A_i$, then $G^{i+1} \subset A_i^1 \subset A_{i+1}$ by Theorem 3.4.10. Hence, by induction, $G^n \subset A_n = E$ and $G^n = E$.

EXERCISES

3.4.13. $(\Sigma_E \{G_i \mid i \in S\})^1 = \Sigma_E \{G_i^1 \mid i \in S\}$.

3.4.14. If a, b, and c are elements of a group G, then

$[ab, c] = [a, c][a, c, b][b, c]$,

$[a, bc] = [a, c][a, b][a, b, c]$.

3.4.15. (P. Hall.) If A and B are subgroups of G such that $[A, B, B] = E$, then $[A, B]$ is Abelian. (Use Theorems 3.4.6 and 3.4.5.)

3.4.16. If H and K are both characteristic (strictly characteristic) (fully characteristic) subgroups of G, then $[H, K]$ is characteristic (strictly characteristic) (fully characteristic).

3.4.17. Let H and K be distinct subgroups of order 2 in Sym(3). Prove that $[H, K]$ is larger than the set of commutators $[h, k]$ with $h \in H$ and $k \in K$. (It is also true, but requires a more complex example, that G^1 is not necessarily the set of commutators. See Exercise 15.5.11.)

3.4.18. (a) $[a^{-1}, b] = ([a, b]^{-1})^{a^{-1}}$.

　　　(b) $[a, b^{-1}] = ([a, b]^{-1})^{b^{-1}}$.

　　　(c) If G is a group generated by S, then G^1 is the normal closure of the subset $\{[a, b] \mid a \in S, b \in S\}$. [Use (a), (b) and Theorem 3.4.2.]

3.4.19. If H is Abelian, then $\mathrm{Hom}(G, H) \cong \mathrm{Hom}(G/G^1, H)$.

3.4.20. If $G = G^1 H$ and $H \cap G^1 = E$, then $[G, G^1] = G^1$.

3.5   The transfer

The purpose of this section is to define a type of homomorphism of a group and to establish some of its properties. Applications will be deferred to later sections.

Let G be a group and H a subgroup of finite index n. Let $x_1, \ldots, x_n$ be such that $G = \cup \{x_i H \mid 1 \leq i \leq n\}$.

3.5.1. *If $G = \overset{\cdot}{\cup} \{x_iH \mid 1 \leq i \leq n\}$ and $y \in G$, then for each $i \; \exists \mid h \in H$ and $j \in J$ such that $yx_i = x_jh$.*

Proof. This follows from elementary facts about cosets. ‖

Let y be a fixed element of G. Because of Theorem 3.5.1,

$$G = \overset{\cdot}{\cup} x_iH, \qquad yx_i = x_{iA}h_i \tag{1}$$

where $h_i \in H$ and A is a function from $\{1, \ldots, n\}$ into itself.

3.5.2. *If (1) holds, then $A \in \mathrm{Sym}(n)$.*

Proof. If i and j are such that $iA = jA$, then

$$x_i^{-1}x_j = (yx_i)^{-1}(yx_j) = (x_{iA}h_i)^{-1}(x_{iA}h_j) = h_i^{-1}h_j \in H,$$

so $x_iH = x_jH$ and $i = j$.

3.5.3. *If (1) holds and also*

$$G = \overset{\cdot}{\cup} z_iH, \; yz_i = z_{iB}a_i, \; a_i \in H, \; B \in \mathrm{Sym}(n),$$

then $(\pi h_i)H^1 = (\pi a_i)H^1$.

Proof. $\exists \; c_i \in H$ and $D \in \mathrm{Sym}(n)$ such that $z_i = x_{iD}c_i$ for all i. Hence

$$yz_i = yx_{iD}c_i = x_{iDA}h_{iD}c_i$$

$$= z_{iDAD^{-1}}(c_{iDAD^{-1}})^{-1}h_{iD}c_i.$$

Therefore, by Theorem 3.5.1,

$$a_i = (c_{iDAD^{-1}})^{-1}h_{iD}c_i.$$

Since H/H^1 is Abelian, $D \in \mathrm{Sym}(n)$, and $DAD^{-1} \in \mathrm{Sym}(n)$, it follows that

$$(\pi a_i)H^1 = \pi((c_{iDAD^{-1}})^{-1}h_{iD}c_i)H^1$$

$$= (\pi c_i^{-1})(\pi h_i)(\pi c_i)H^1 = (\pi h_i)H^1. \; ‖$$

Theorem 3.5.3 allows the following definition. The *transfer* T of G into H is defined by the equation:

$$yT = (\pi h_i)H^1,$$

where h_i is given by (1).

Evidently T is a function from G into H/H^1. In case H is Abelian, it is customary to omit H^1 and thus make T a function from G into H.

3.5.4. *The transfer is a homomorphism.*

Proof. Let $y \in G$, $z \in G$, equations (1) hold, and $zx_i = x_{iB}c_i$, where $c_i \in H$ and $B \in \text{Sym}(n)$. Then

$$yzx_i = yx_{iB}c_i = x_{iBA}h_{iB}c_i.$$

Hence

$$(yz)T = (\pi h_{iB}c_i)H^1 = (\pi h_i)H^1(\pi c_i)H^1 = (yT)(zT). \quad \|$$

Since the image GT of G is Abelian, $\text{Ker}(T) \supset G^1$ (Theorem 3.4.10). Hence (Theorem 2.2.4) there is induced a homomorphism T^* of G/G^1 into H/H^1, where $(yG^1)T^* = yT$. The transfer is then transitive in the following sense.

3.5.5. (*Transitivity of transfer.*) *If G is a group, $K \subset H \subset G$, $[G:K]$ is finite, T is the transfer of G into H, U the transfer of H into K, and V the transfer of G into K, then $V^* = T^*U^*$.*

Proof. Let $G = \overset{.}{\cup} x_i H$ and $H = \overset{.}{\cup} z_j K$, so that $G = \overset{.}{\cup} x_i z_j K$. Let $y \in G$ and $yx_i = x_{iA}h_i$, $h_i \in H$, $A \in \text{Sym}([G:H])$. Then for each i and j $\exists A_i \in \text{Sym}([H:K])$ and $k_{i,j} \in K$ such that $h_i z_j = z_{jA_i}k_{i,j}$. Thus

$$yx_i z_j = x_{iA}z_{jA_i}k_{i,j}.$$

It follows that

$$(yG^1)V^* = yV = \left(\prod_{i,j} k_{i,j} \right) K^1;$$

$$(yG^1)T^*U^* = (yT)U^* = ((\prod h_i)H^1)U^* = (\prod h_i)U = \prod (h_i U)$$

$$= \prod_i \left(\prod_j k_{i,j} \right) K^1 = \left(\prod_{i,j} k_{i,j} \right) K^1,$$

where commutativity of K/K^1 has been used. Therefore $V^* = T^*U^*$. $\quad \|$

In practice, the transfer is often difficult to compute. The following theorem will be of help in some cases (see, for example, Exercise 3.5.8).

3.5.6. *If $G = \overset{.}{\cup} \{xH \mid x \in S\}$, S finite, $y \in G$, and T is the transfer of G into H, then there is a subset $S' = \{x_1, \ldots, x_r\}$ of S and $n_i \in \mathcal{N}$ such that*

$$yT = \pi(x_i^{-1}y^{n_i}x_i)H^1, \quad \sum n_i = [G:H], \quad \text{and}$$

n_i is minimal such that $x_i^{-1}y^{n_i}x_i \in H$.

Proof. We recall that $\exists A \in \text{Sym}(S)$ such that for all $x \in S$, $yx = (xA)h_x$, $h_x \in H$. Let $(x_{i_1}, \ldots, x_{i_m})$ be any cycle of A. Then

$$yx_{i_1} = x_{i_2}h_{i_1}, \ldots, yx_{i_{m-1}} = x_{i_m}h_{i_{m-1}}, yx_{i_m} = x_{i_1}h_{i_m};$$

$$x_{i_1}^{-1}y^m x_{i_1} = h_{i_m} \ldots h_{i_1} \in H.$$

Moreover, if $r < m$, then

$$x_{i_1}^{-1} y^r x_{i_1} = x_{i_1}^{-1} x_{i_{r+1}} h_{i_r} \ldots h_{i_1} \notin H,$$

so that m is minimal. By the definition of the transfer, yT is H^1 times the product, taken over all cycles of A, of elements of the above form, and the theorem follows.

EXERCISES

3.5.7. Give an example to show that if Equation (1) is replaced by the equations

$$G = \dot{\cup}\, x_i H, \qquad x_i y = x_{iU} h_i, h_i \in H, \tag{1*}$$

then U is not necessarily 1–1.

3.5.8. Let $[G:H] = n$ and let T be the transfer of G into H.

(a)   If $G = HK$, $K \subset G$, and $H \cap K = E$, then $K \subset \mathrm{Ker}(T)$.

(b)   If $G = HK$, $K \subset G$, $H \subset C(K)$, and $y \in G$, then $yT = y^n H^1$.

(c)   If $H \subset Z(G)$ and $y \in G$, then $yT = y^n$.

(d)   If $(n, o(H)) = 1$ and $H \subset Z(G)$, then $G = H \cdot \mathrm{Ker}(T)$ and $H \cap \mathrm{Ker}(T) = E$. (See Exercise 4.1.5 for a more precise result.)

FOUR

DIRECT SUMS

4.1 Direct sum of two subgroups

Direct sums and products will be studied in general in the next section. The simpler case of the direct sum of two subgroups is taken up in this section by way of introduction to the more complex general case.

A lemma is needed.

4.1.1. *If G is a group, A and B are normal subgroups, $A \cap B = E$, $x \in A$, and $y \in B$, then $xy = yx$.*

Proof. $x^{-1}(y^{-1}xy) \in A$ since $A \lhd G$, and $(x^{-1}y^{-1}x)y \in B$ since $B \lhd G$. Hence $x^{-1}y^{-1}xy = e$, and $xy = yx$. ‖

Let A_1 and A_2 be groups. Then the Cartesian product $A_1 \times A_2$ is a group under the operation: $(a, b)(c, d) = (ac, bd)$. If $x_1 \in A_1$ and $x_2 \in A_2$, define

$$x_1 j_1 = (x_1, e), \quad x_2 j_2 = (e, x_2), \quad (x_1, x_2)q_1 = x_1, \quad (x_1, x_2)q_2 = x_2.$$

Then it is easily verified (Exercise 4.1.6) that

(a) $\qquad j_r \in \mathrm{Hom}(A_r, A_1 \times A_2), \qquad q_r \in \mathrm{Hom}(A_1 \times A_2, A_r),$

$$j_r q_s = \begin{cases} I_{A_r} & \text{if } r = s, \\ O_{A_r A_s} & \text{if } r \neq s. \end{cases}$$

64

(b) $A_r j_r \lhd A_1 \times A_2$.

(c) $A_1 \times A_2 = (A_1 j_1)(A_2 j_2)$.

(d) $(A_1 j_1) \cap (A_2 j_2) = E$.

(e)   If $x_1 \in A_1$ and $x_2 \in A_2$, then $\exists \mid x \in A_1 \times A_2$ such that $x q_1 = x_1$ and $x q_2 = x_2$.

A converse is also true.

4.1.2. *If G, A_1, and A_2 are groups, and i_1, i_2, p_1, and p_2 are such that*

(f) $i_r \in \mathrm{Hom}(A_r, G)$, $p_r \in \mathrm{Hom}(G, A_r)$,

$$i_r p_s = \begin{cases} I & \text{if } r = s, \\ O & \text{if } r \neq s, \end{cases}$$

(g) $G = (A_1 i_1)(A_2 i_2)$,

then $\exists$ an isomorphism T of G onto $A_1 \times A_2$ such that

$$j_r = i_r T \quad \text{and} \quad p_r = T q_r.$$

Proof. If $g \in G$, define $gT = g(p_1 j_1 + p_2 j_2)$. By (f),

$$G p_r j_r \supset A_r i_r p_r j_r = A_r j_r \supset G p_r j_r,$$

hence $G p_r j_r = A_r j_r$. Now $p_r j_r \in \mathrm{Hom}(G, A_1 \times A_2)$. By (b) and (d), $G p_r j_r = A_r j_r \lhd A_1 \times A_2$ and $A_1 j_1 \cap A_2 j_2 = E$. Hence by Theorem 4.1.1, the elements of $G p_1 j_1$ and $G p_2 j_2$ commute. It follows (Exercise 4.1.11) that $T = p_1 j_1 + p_2 j_2$ is a homomorphism.

If $g \in \mathrm{Ker}(T)$, then $(g p_1 j_1)(g p_2 j_2) = e$, so

$$g p_1 j_1 \in A_1 j_1 \cap A_2 j_2 = E.$$

Therefore, $g p_1 j_1 = e$ and $g p_2 j_2 = e$. Hence, by (a), $e = g p_r j_r q_r = g p_r$, $r = 1, 2$. Now by (g), $\exists x_1 \in A_1$ and $x_2 \in A_2$ such that $(x_1 i_1)(x_2 i_2) = g$. Then by (f),

$$e = g p_1 = (x_1 i_1 p_1)(x_2 i_2 p_1) = x_1,$$

and similarly $x_2 = e$. Hence $g = (e i_1)(e i_2) = e$. Therefore $\mathrm{Ker}(T) = E$, and T is an isomorphism of G into $A_1 \times A_2$.

If $x_r \in A_r$, then

$$x_r i_r T = x_r (i_r p_1 j_1 + i_r p_2 j_2) = x_r j_r,$$

and $i_r T = j_r$. It follows from (c) that $GT = A_1 \times A_2$.

If $g \in G$, then by (a),

$$g T q_r = g(p_1 j_1 q_r + p_2 j_2 q_r) = g p_r,$$

so that $T q_r = p_r$. $\|$

The conclusion of Theorem 4.1.2 may be expressed in diagram language as follows. The diagrams

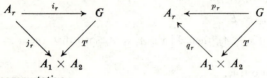

are commutative.

4.1.3. *If the hypotheses of Theorem* 4.1.2 *hold, then*

(h) $A_1 i_1 \cap A_2 i_2 = E,$

(i) $A_r i_r \lhd G, \quad r = 1, 2.$

Proof. Let T be the isomorphism guaranteed by Theorem 4.1.2. Then

$$(A_1 i_1 \cap A_2 i_2)T = A_1 j_1 \cap A_2 j_2 = E$$

by (d). Hence $A_1 i_1 \cap A_2 i_2 = E$. Also, by (b),

$$A_r i_r T = A_r j_r \lhd A_1 \times A_2,$$

so by the lattice theorem, $A_r i_r \lhd G$. ‖

Now consider the special case of Theorems 4.1.2 and 4.1.3 where $A_r \subseteq G$ and $i_r = I_{A_r}$. Then $A_1 \lhd G, \ A_2 \lhd G, \ G = A_1 A_2$, and $A_1 \cap A_2 = E$. The converse is also true.

4.1.4. *If* $A_1 \lhd G, \ A_2 \lhd G, \ G = A_1 A_2, \ A_1 \cap A_2 = E$, *and* $i_r = I_{A_r}$, *then* ∃ p_r *such that* (f) *holds. Hence* $G \cong A_1 \times A_2$.

Proof. If $x_1 x_2 = y_1 y_2$ with x_r and y_r in A_r, then

$$y_1^{-1} x_1 = y_2 x_2^{-1} \in A_1 \cap A_2 = E, \qquad x_1 = y_1, \ x_2 = y_2.$$

Hence, if $x \in G$, then ∃ $\mid x_r \in A_r$ such that $x = x_1 x_2$. Now define p_r by the equation $x p_r = (x_1 x_2) p_r = x_r$. By Theorem 4.1.1, $p_r \in \mathrm{Hom}(G, A_r)$. If $x_1 \in A_1$, then

$$x_1 i_1 p_1 = x_1 p_1 = (x_1 e) p_1 = x_1,$$
$$x_1 i_1 p_2 = x_1 p_2 = (x_1 e) p_2 = e.$$

Hence $i_1 p_1 = I_{A_1}$ and $i_1 p_2 = O$. Similarly, $i_2 p_2 = I_{A_2}$ and $i_2 p_1 = O$. Therefore (f) is true. By Theorem 4.1.2, $G \cong A_1 \times A_2$. ‖

Since the situation considered in Theorem 4.1.4 arises rather frequently, we make the following definition. A group G is the *direct sum** of its subgroups A_1 and A_2 iff each $A_r \lhd G, \ G = A_1 A_2$, and $A_1 \cap A_2 = E$. This fact

---

* In the literature, the term "direct product" is usually used for what we have called the direct sum.

will be written $G = A_1 + A_2$. A group G is *decomposable* iff $G = A + B$ with $A \neq E$ and $B \neq E$, and is indecomposable otherwise. A subgroup A of G is a *direct summand* of G iff $\exists\, B \subset G$ such that $G = A + B$. There are other equivalent ways of defining the direct sum of two subgroups (see Exercise 4.1.7).

EXERCISES

4.1.5. Exercise 3.5.8 (d) can be reworded as follows: if $[G:H] = n$, $(n, o(H)) = 1$, $H \subset Z(G)$, and T is the transfer of G into H, then $G = H + \mathrm{Ker}(T)$.

4.1.6. Verify Equations (a), (b), (c), (d), and (e).

4.1.7. Show that if A and B are subgroups of G, then $G = A + B$ iff (a) if $g \in G$, then $\exists\, | \, a \in A$ and $b \in B$ such that $g = ab$, and (b) if $a \in A$ and $b \in B$, then $ab = ba$.

4.1.8. Show that J_6 is decomposable, but Sym(3) is not.

4.1.9. Show that the four-group is decomposable and is the direct sum of two cyclic groups of order 2.

4.1.10. $(A + B)/A \cong B$.

4.1.11. If $T \in \mathrm{Hom}(G, H)$, $U \in \mathrm{Hom}(G, H)$, $GT \subset C(GU)$, and, as in Section 3.1, $g(T + U) = (gT)(gU)$, then $T + U \in \mathrm{Hom}(G, H)$.

4.1.12. An *antihomomorphism* of a group G into a group H is a function T from G into H such that if $a \in G$ and $b \in G$, then $(ab)T = (bT)(aT)$. *Antiautomorphism* is defined similarly. Prove

   (a)   If G or H is Abelian, then an antihomomorphism from G into H is a homomorphism.

   (b)   If T_i is a homomorphism or antihomomorphism from G_i into G_{i+1}, then $T_1 \cdots T_n$ is a homomorphism (antihomomorphism) if an even (odd) number of the T_i are antihomomorphisms.

   (c)   What sort of distributive laws hold for antihomomorphisms?

   (d)   The set Anti(G) of automorphisms and antiautomorphisms of G is a group.

   (e)   The function $T_{-1} : xT_{-1} = x^{-1}$ is in Anti(G) and is of order 2 or 1.

   (f)   If G is Abelian, then Anti(G) = Aut(G). If G is non-Abelian, then Anti(G) = Aut(G) + $\langle T_{-1} \rangle$.

4.1.13. If $G = A + B$ and $A \subset H \subset G$, then $H = A + (B \cap H)$.

4.1.14. Let $G = \text{Sym}(3) \times \langle(1, 2, 3)\rangle$, and let $(x, y)T = (y, e)$ for $(x, y) \in G$.

    (a) $T \in \text{End}(G)$.

    (b) $Z(G)T \not\subset Z(G)$, hence $Z(G)$ is not fully characteristic.

    (c)   There is a subgroup H of G such that $HT \subset H$, but $(N(H))T \not\subset N(H)$ and $(C(H))T \not\subset C(H)$.

    (d)   Use (c) to construct an S-group K with an S-subgroup L such that $C(L)$ and $N(L)$ are not S-subgroups.

4.2 Direct sum and product of a set of groups

Let S be a set, A_r a group for each $r \in S$, and πA_r and $\Sigma_E A_r$ their direct product and external direct sum respectively (Section 1.4). Define j_r and q_r as follows: if $x_r \in A_r$, then $x_r j_r$ is that element f of πA_s such that

$$sf = \begin{cases} x_r & \text{if } s = r, \\ e_s & \text{if } s \neq r; \end{cases}$$

and if $g \in \pi A_s$, then $gq_r = rg$. Then it is easily verified that

(a) $j_r \in \text{Hom}(A_r, \pi A_s), \qquad q_r \in \text{Hom}(\pi A_s, A_r)$,

$$j_r q_s = \begin{cases} I_{A_r} & \text{if } r = s, \\ O_{A_r A_s} & \text{if } r \neq s. \end{cases}$$

(b) $A_r j_r \lhd \pi A_s$.

(c) $\Sigma_E A_s = \langle A_s j_s \rangle$.

(d) $A_r j_r \cap \langle A_s j_s \mid s \neq r \rangle = E$.

(e)   If $x_r \in A_r$ for each $r \in S$, then $\exists \mid f \in \pi A_s$ such that $fq_r = x_r$ for each $r \in S$.

Two converses are also true.

4.2.1. *If G and A_r, $r \in S$, are groups and i_r and p_r are such that*

(f) $i_r \in \text{Hom}(A_r, G), \qquad p_r \in \text{Hom}(G, A_r)$,

$$i_r p_s = \begin{cases} I & \text{if } r = s, \\ O & \text{if } r \neq s, \end{cases}$$

(g) $G = \langle A_r i_r \rangle$,

then $\exists$ an isomorphism T of G onto $\Sigma_E A_s$ such that, for all r, $j_r = i_r T$ and $p_r = Tq_r$.

Proof. By (f), $Gp_rj_r \supset A_ri_rp_rj_r = A_rj_r \supset Gp_rj_r$, hence $Gp_rj_r = A_rj_r$. Now $p_rj_r \in \text{Hom}(G, \pi A_s)$. By (b) and (d), $Gp_rj_r = A_rj_r \triangleleft \pi A_s$, and $A_rj_r \cap A_sj_s = E$ if $s \neq r$. By Theorem 4.1.1, the elements of Gp_rj_r and Gp_sj_s commute if $r \neq s$. If $g \in G$, define

$$gT = g(\Sigma \, p_rj_r) = \Sigma \, (gp_rj_r).$$

By (g) and the commutativity already noted, $\exists \, x_1 \in A_{r_1}, \ldots, x_n \in A_{r_n}$ such that $g = \pi x_{k}i_{r_k}$. By (f) all but a finite number of the terms gp_rj_r equal O, so that T is well defined. Also

$$gT = \Sigma \, x_{k}i_{r_k}p_{r_k}j_{r_k} = \Sigma \, x_{k}j_{r_k}.$$

It follows readily that T is a homomorphism of G onto $\langle A_sj_s \rangle = \Sigma_E \, A_s$ [by (c)].

Let $g \in \text{Ker}(T)$. Then, with the above notation, $\Sigma \, x_kj_{r_k} = O$. Hence by (a), $e = Oq_{r_k} = x_k$. Thus $g = e$ and $\text{Ker}(T) = E$.

If $x_r \in A_r$, then

$$x_ri_rT = \Sigma \, \{x_ri_rp_sj_s \mid s \in S\} = x_rj_r,$$

so that $j_r = i_rT$. If $g \in G$, then

$$gTq_r = \pi\{gp_sj_sq_r \mid s \in S\} = gp_rj_rq_r = gp_r,$$

so that $Tq_r = p_r$. $\parallel$

The proof of the following theorem is left as an exercise, 4.2.5.

4.2.2. *If G and A_r, $r \in S$, are groups, and i_r and p_r are such that* (f) *and*

(h)   *if $x_r \in A_r$ for each $r \in S$ then $\exists \mid g \in G$ such that $gp_r = x_r$ for each $r \in S$,*

hold, then $\exists$ an isomorphism T of G onto πA_r such that

$$j_r = i_rT \quad \text{and} \quad p_r = Tq_r.$$

4.2.3. *If the hypotheses of Theorem* 4.2.1 *hold, then for all $r \in S$,*

(i) $A_ri_r \cap \langle A_si_s \mid s \neq r \rangle = E,$

(j) $A_ri_r \triangleleft G.$

Proof. Let T be the isomorphism guaranteed by Theorem 4.2.1. Then

$$(A_ri_r \cap \langle A_si_s \mid s \neq r \rangle)T = A_rj_r \cap \langle A_sj_s \mid s \neq r \rangle = E$$

by (d). Hence (i) holds. Also

$$A_ri_rT = A_rj_r \triangleleft \Sigma_E \, A_s$$

by (b), so $A_ri_r \triangleleft G$. $\parallel$

Now consider the special case where $A_r \subset G$ and each $i_r = I_{A_r}$. Conditions (g), (i), and (j) simplify to

$$G = \langle A_r \rangle, \quad A_r \cap \langle A_s \mid s \neq r \rangle = E \quad \text{and} \quad A_r \triangleleft G,$$

respectively. The converse is also true.

4.2.4. *If $A_r \triangleleft G$ for $r \in S$,*

$$G = \langle A_r \rangle, \quad A_r \cap \langle A_s \mid s \neq r \rangle = E,$$

and $i_r = I_{A_r}$, then $\exists\, p_r$ such that (f) holds. Hence $G \cong \Sigma_E\, A_r$.

Proof. Clearly $i_r \in \text{Hom}(A_r, G)$. The hypotheses imply that if $r \neq s$ then $A_r \cap A_s = E$, hence by Theorem 4.1.1, the elements of A_r and A_s commute. Therefore any element x of G can be written in the form $x_{r_1} \cdots x_{r_n}$ where $x_{r_u} \in A_{r_u}$ and $r_u \neq r_v$ if $u \neq v$. If x has another representation, then, inserting factors of e if necessary, $x_{r_1} \cdots x_{r_n} = y_{r_1} \cdots y_{r_n}$ with $y_{r_u} \in A_{r_u}$. Transposing, one gets $y_{r_u}^{-1} x_{r_u} \in \langle A_s \mid s \neq r_u \rangle$. Hence $y_{r_u}^{-1} x_{r_u} = e$ and $y_{r_u} = x_{r_u}$. Thus each $x \in G$ can be written in the form πx_r uniquely, where all but a finite number of factors are e, and πx_r is understood to be the product of the non-e factors (or e if all factors $x_r = e$). Now define p_r by $(\pi x_s)p_r = x_r$. Then $p_r \in \text{Hom}(G, A_r)$ by Theorem 4.1.1. If $x_r \in A_r$, then

$$x_r i_r p_s = x_r p_s = \begin{cases} x_r & \text{if } r = s, \\ e & \text{if } r \neq s. \end{cases}$$

This proves (f). $\parallel$

A group G is the *direct sum*, $\Sigma\,\{A_r \mid r \in S\}$, of subgroups A_r, $r \in S$, iff $A_r \triangleleft G$ for $r \in S$, $G = \langle A_r \rangle$, and $A_r \cap \langle A_s \mid s \neq r \rangle = E$ for $r \in S$ (i.e., iff the hypotheses of Theorem 4.2.4 are satisfied). A set $\{i_r, p_r \mid r \in S\}$ of homomorphisms satisfying (f) is called a *direct family* of homomorphisms. A direct family which also satisfies (g) gives a *representation of G as a direct sum* of the groups A_r, $r \in S$. A direct family which also satisfies (h) gives a *representation of G as a direct product* of the A_r, $r \in S$.

EXERCISES

4.2.5. Prove Theorem 4.2.2.

4.2.6. State and prove an analogue of Exercise 4.1.7 for direct sums of infinitely many groups.

4.2.7. If $\{i_r, p_r \mid r \in S\}$ gives a representation of G as a direct sum, then the set $\{i_r \mid r \in S\}$ determines the set $\{p_r \mid r \in S\}$.

4.2.8. (Associativity of direct sums.) If $G = \Sigma \{A_i \mid i \in S\}$ and each $A_i = \Sigma \{B_{ij} \mid j \in T_i\}$, then

$$G = \Sigma \{B_{ij} \mid i \in S, j \in T_i\}.$$

4.2.9. If $G = \Sigma \{A_r \mid r \in S\}$, $s \in S$, and $H \lhd A_s$, then $H \lhd G$.

4.3 Subdirect products

Let $\{H_i \mid i \in S\}$ be a family of groups, G a group, and T an isomorphism of G into the direct product πH_i. Let q_j be the canonical homomorphism of πH_i onto H_j (see Equation 4.2(a)). Then T is a *representation of G as a subdirect product* of $\{H_i \mid i \in S\}$ iff each homomorphism Tq_j is onto H_j. If $G \subset \pi H_i, T = I_G$, and Tq_j is onto H_j for all j, then G is a *subdirect product* of $\{H_i \mid i \in S\}$. If $K = \Sigma \{H_i \mid i \in S\}, G \subset K, T = I_G, p_j$ is the canonical homomorphism of K onto H_j, and Tp_j is onto H_j for all j, then G is called a *subdirect sum* of $\{H_i \mid i \in S\}$. Loosely speaking, G is a subdirect product or subdirect sum iff every element of every H_i is used as the ith coordinate of some element of G.

There may be nonisomorphic subdirect products of the same family $\{H_i \mid i \in S\}$ of groups. A given group may be represented as a subdirect product in an infinite number of ways.

There is a rather simple description of all subdirect sums of two groups H and K.

4.3.1. *If $G \subset H + K$, then G is a subdirect sum of H and K iff $\exists \, L \lhd H$, $M \lhd K$, and an isomorphism U of H/L onto K/M such that if $a \in H$ and $b \in K$, then $ab \in G$ iff $(aL)U = bM$.*

Proof. First suppose that G is a subdirect sum of H and K. Let $L = G \cap H$, $M = G \cap K$, $h \in H$, and $x \in L$. Since G is a subdirect sum of H and K, $\exists \, k \in K$ such that $hk \in G$. Also

$$h^{-1}xh = (hk)^{-1}x(hk) \in G \cap H = L,$$

hence $L \lhd H$. Similarly, $M \lhd K$. Let

$$U = \{(aL, bM) \mid a \in H, b \in K, ab \in G\}.$$

If $(a_iL, b_iM) \in U$, $i = 1, 2$, and $a_1L = a_2L$, then $a_1b_1 \in G$, $a_2b_2 \in G$, $a_1^{-1}a_2b_1^{-1}b_2 \in G$, $a_1^{-1}a_2 \in L \subset G$, so $b_1^{-1}b_2 \in M$ and $b_1M = b_2M$. Hence U is a function. A similar argument shows that U is 1–1. It is then immediate that U is an isomorphism from H/L onto K/M (from and onto, since G is a subdirect sum of H and K). It follows from the definition of U that $ab \in G$ iff $(aL)U = bM$.

Next suppose that L, M, and U exist with the stated properties. Let $a \in H$. Since $\mathrm{Dom}(U) = H/L$, $\exists\, b \in K$ such that $(aL)U = bM$. By the assumptions on U, $ab \in G$. Hence, in the notation of this section, $(ab)Tp_H = a$, and Tp_H is onto H. Similarly Tp_K is onto K. Therefore G is a subdirect sum of H and K. ‖

There appears to be no such simple description of subdirect sums of more than two groups (for the case of three groups, the brave reader may wish to consult Remak [1]).

EXERCISES

4.3.2. If $G \subset \pi\{H_i \mid i \in S\}$, then G is a subdirect product of a certain family $\{K_i \mid i \in S\}$, $K_i \subset H_i$.

4.3.3. A subdirect sum or product of Abelian groups is Abelian.

4.3.4. A subdirect sum of a finite number of solvable groups is solvable.

4.3.5. In the notation of Theorem 4.3.1,

    (a) $o(G) = o(H)o(M) = o(K)o(L)$.

    (b)   If $L = E$, then $G \cong K$; if $M = E$, then $G \cong H$.

    (c) $G \lhd H + K$ iff H/L (and K/M) are Abelian.

    (d)   (Gilbert [1]) If $G \lhd H + K$, then

$$\frac{H + K}{G} \cong \frac{G}{L + M} \cong \frac{H}{L}.$$

4.3.6. (a)   Find all subdirect products of J_2 and J_4, up to isomorphism.

    (b)   Find all subdirect products of two cyclic groups of order 4, up to isomorphism.

4.3.7. (a)   If G is a subdirect product of $\{H_i \mid i \in S\}$, then $Z(G) \subset \pi Z(H_i)$.

    (b)   In the above situation, $Z(G)$ need not be a subdirect product of $\{Z(H_i) \mid i \in S\}$. (Using the notation of Theorem 4.3.1, let $H = J_2$, $K = \mathrm{Sym}(3)$, $L = E$, and use Exercise 4.3.5).

4.4   Direct sums of simple groups

We assume

Zorn's lemma. If P is a nonempty, partially ordered set such that each chain C contained in P has an upper bound $u \in P$, then P has a maximal element.

4.4.1. *If U is a set of subgroups of G and $H \subset G$, then there is a subset V of U which is maximal with respect to the existence of $H + (\Sigma \{K \mid K \in V\})$.*

Proof. Denote the direct sum in the theorem by $X(V)$. Let P be the set of subsets V of U for which $X(V)$ exists. Since $X(\varnothing) = H$, $\varnothing \in P$, and P is not empty. Partially order P by inclusion. Let C be a chain in P. Let $W = \cup \{V \mid V \in C\}$. Then W is a subset of U, and is an upper bound of C. In order to apply Zorn's lemma, it remains to show that $W \in P$, i.e. that $X(W)$ exists. If $K \in W$, $K' \in W$, and $K \neq K'$, then there is a $V \in C$ such that $K \in V$ and $K' \in V$, since C is a chain. Since $X(V)$ exists, the elements of any two of the three subgroups H, K, and K' commute. It follows easily that H and K are normal in $\langle H, \{L \mid L \in W\}\rangle$. In order to show that $X(W)$ exists, it remains to prove that $H \cap \langle K \mid K \in W \rangle = E$ and if $L \in W$, then

$$L \cap \langle H, \{K \mid K \in W, K \neq L\}\rangle = E.$$

For simplicity, only the first statement will be proved, the other one being similar. If the first statement is false, then there is a finite subset $\{K_1, \ldots, K_n\}$ of W such that $H \cap \langle K_i \rangle \neq E$. But then $\exists\ Y \in C$ such that all $K_i \in Y$. Since $X(Y)$ exists, it follows that $H \cap \langle K \mid K \in Y \rangle = E$, and therefore $H \cap \langle K_i \rangle = E$, a contradiction. Therefore $X(W)$ exists. Thus $W \in P$, and W is an upper bound of C in P. By Zorn, P has a maximal element V. But this is just the conclusion of the theorem. ‖

Note that H may equal E in the above theorem. In this case the theorem states that there is a maximal subset V of U such that $\Sigma \{K \mid K \in V\}$ exists.

Let S be a set and G an S-group. An S-subgroup H of G is *S-characteristic* iff $HT \subset H$ for all S-automorphisms T of G. Since the S-automorphisms of G form a group, it follows that $HT = H$ if H is S-characteristic and T an S-automorphism of G. An S-group G is *S-simple* iff it has no nontrivial, normal S-subgroups. An S-group G is *characteristically S-simple* iff it has no nontrivial S-characteristic subgroups.

4.4.2. *If an S-group G is characteristically S-simple and H is a minimal, normal non-E S-group of G, then H is S-simple and $G = \Sigma\ H_i$, where each H_i is S-isomorphic to H.*

Proof. If T is an S-automorphism of G, then HT is S-isomorphic to H, and HT is a minimal, normal non-E S-subgroup of G (Theorem 2.7.7). It is easily verified that

$$\langle HT \mid T \text{ is an } S\text{-automorphism of } G\rangle$$

is an S-characteristic S-subgroup of G, hence it equals G.

By Theorem 4.4.1, there is a maximal set R of S-automorphisms of G such that the identity automorphism is in R and $K = \Sigma \{HT \mid T \in R\}$ exists.

If $K \neq G$, then $\exists$ an S-automorphism U of G such that $HU \not\subseteq K$. Since HU is a minimal normal non-E S-subgroup, while K is a normal S-subgroup, (Exercises 2.1.20 and 2.7.10), it follows (Exercises 2.1.20 and 2.7.10 again) that $HU \cap K = E$. Hence $HU + K$ exists, and R is not maximal (see Exercise 4.2.8). This contradiction shows that $G = \Sigma \{HT \mid T \in R\}$ and each HT is S-isomorphic to H.

If H is not S-simple, then $\exists$ an S-subgroup $M < H$ such that $M \neq E$ and $M \lhd H$. Since H is one of the summands in the decomposition $G = \Sigma \{HT \mid T \in R\}$, $M \lhd G$. (Exercise 4.2.9.). This contradicts the minimality of H. Hence H is S-simple. $\parallel$

In particular, of course, Theorem 4.4.2 is valid for ordinary groups.

4.4.3. *If H is a minimal, normal non-E subgroup of G and K is a minimal, normal non-E subgroup of H, then K is simple, and $H = \Sigma K_i$ where each K_i is conjugate to K in G.*

Proof. If H has a nontrivial characteristic subgroup L, then (Exercise 2.11.13) $L \lhd G$ a contradiction. Hence H is characteristically simple and, by Theorem 4.4.2, K is simple and $H = \Sigma K_i$, where each $K_i \cong K$. The fact that the K_i may be chosen as conjugates of K is proved in a manner similar to the proof of Theorem 4.4.2.

4.4.4. *If G is a finite group and H is a minimal, normal non-E subgroup of G, then H is the direct sum of isomorphic simple groups.*

Proof. In this case, a minimal, normal non-E subgroup K of H exists, and the theorem follows from Theorem 4.4.3.

4.4.5. *If G is a finite solvable group and H is a minimal, normal non-E subgroup of G, then H is the direct sum of cyclic groups of the same prime order.*

Proof. By Theorem 4.4.4, $H = \Sigma K_i$, where the K_i are simple and all isomorphic. Since K_1 is a subgroup of a solvable group, it is solvable (Theorem 2.6.2). Since K_1 is a simple solvable group, it is cyclic of prime order (Exercise 2.6.6).

4.4.6. *Let $G = \Sigma \{H \mid H \in S\}$ where each $H \neq E$. Then the following conditions are equivalent.*

(1) *If $M \lhd G$, then $\exists$ a subset T of S such that $M = \Sigma \{H \mid H \in T\}$.*

(2) (a) *If $H \in S$ then H is simple.*

(b) *If $H \in S$, $K \in S$, $H \neq K$, and $H \cong K$, then H is non-Abelian.*

Proof. (1) implies (2). (a) If $H \in S$ and $A \lhd H$, then $A \lhd G$ (Exercise 4.2.9). By definition of the direct sum,

$$H \cap \langle L \mid L \in S,\, L \neq H \rangle = E.$$

Hence

$$A \cap \langle L \mid L \in S,\, L \neq H \rangle = E.$$

By (1), A is a direct sum of some of the elements of S, hence $A = H$ or $A = E$. Therefore H is simple.

(b) Let T be an isomorphism of H onto K. Then the set $L = \{x(xT) \mid x \in H\}$ is a subgroup of G. If H is Abelian, then $N(L) \supset H$ and $N(L) \supset K$. Hence, by the fact that the elements of the terms in a direct sum commute, $L \lhd G$. It is clear, however, that L is not the direct sum of some of the elements of S, contradicting (1). Hence H is non-Abelian.

(2) implies (1). Let $T = \{H \in S \mid H \subset M\}$. Then $A = \Sigma \{H \mid H \in T\} \subset M$. Suppose that (1) is false. Then $A \neq M$ and $\exists\, x \in M \backslash A$. Now $x = x_1 \cdots x_n$ where $x_i \in H_i \in S$, $x_i \neq e$, and $H_i \neq H_j$ if $i \neq j$. WLOG, $H_i \not\subset M$ for all i.

CASE 1. One of the H_i is non-Abelian. Then, *WLOG*, H_1 is non-Abelian. Since $Z(H_1) \lhd H_1$ and H_1 is simple by (a), $Z(H_1) = E$. Hence $\exists\, y_1 \in H_1$ such that $y_1 x_1 \neq x_1 y_1$. By normality of M, $[y_1, x] \in M$. But

$$[y_1, x] = [y_1, x_1] \in H_1 \cap M.$$

Hence $H_1 \cap M \neq E$. Since $H_1 \cap M \lhd G$, $H_1 \subset M$ (by the simplicity of H_1), contradicting an earlier statement.

CASE 2. All of the H_i are Abelian. By (b), there are distinct primes $p_1, \ldots, p_n$ such that $H_i \cong J_{p_i}$. Then if $k = p_2 \cdots p_n$,

$$x^k = (x_1 \cdots x_n)^k = x_1^k \neq e$$

since $p_1 \nmid k$. Hence $x^k \in H_1 \cap M$, so that $H_1 \cap M \neq E$. Therefore, as in Case 1, $H_1 \subset M$, a contradiction. ∥

A group G is *completely reducible* iff every normal subgroup of G is a direct summand of G. Completely reducible groups are characterized in the following theorem.

4.4.7. *A group is completely reducible iff it is the direct sum of simple groups.*

Proof. (a) Let $G = \Sigma \{H \mid H \in S\}$ where each $H \in S$ is simple, and let $M \lhd G$. By Theorem 4.4.1, there is a maximal subset T of S such that $X(T) = M + (\Sigma \{H \mid H \in T\})$ exists. If $X(T) < G$, then there is an $H \in S$ such that $X(T) \cap H < H$. Therefore, by the simplicity of H, $X(T) \cap H = E$. But then $X(T')$ exists, where $T' = T \cup \{H\}$, a contradiction. Therefore $X(T) = G$. Hence G is completely reducible.

(b) Let G be completely reducible. By Zorn, there is a maximal subset S (possibly $\varnothing$) of simple normal subgroups of G such that $X(S) = \Sigma \{H \mid H \in S\}$ exists. By complete reducibility, $G = X(S) + K$. If $K = E$, we are done. Suppose that $K \neq E$. A normal subgroup M of K is normal in G, so $G = M + P$, $K = M + (P \cap K)$ (Exercise 4.1.13). Hence K is completely reducible and has no simple normal subgroups (by maximality of S).

Let $x \in K^{\#}$ and $M = x^K$. Then $K = M + P$, M is without simple normal subgroups, M is completely reducible, and $M = x^M$. Since M is not simple, $M = A_1 + B_1$, and by induction and the absence of simple normal subgroups of M,

$$M = A_1 + B_1 = \ldots = A_1 + \ldots + A_n + B_n = \ldots,$$

$$A_i \neq E, B_i \neq E.$$

Then ΣA_i exists and is normal in M, so $M = \Sigma A_i + D$. Therefore, letting $D = A_0$, $M = \Sigma A_i$. Now $x = a_0 \cdots a_r$, $a_i \in A_i$. Hence $x^M \subset A_0 + \ldots + A_r < M$, a contradiction.

EXERCISES

4.4.8. If $M_i \lhd H_i$ and $G = \Sigma H_i$, then $\Sigma M_i \lhd G$.

4.4.9. Let G be a finite group. Then each factor group A_{i+1}/A_i of a principal series $(A_0, \ldots, A_r)$ of G is the direct sum of isomorphic simple groups.

4.4.10. If G is a finite group, and

$$E = A_0 < A_1 < \ldots < A_r = G,$$

where each A_i is a maximal proper characteristic subgroup of A_{i+1}, then A_{i+1}/A_i is the direct sum of isomorphic simple groups.

4.4.11. If G is a finite group, and

$$E = A_0 < A_1 < \ldots < A_r = G$$

where each A_i is a maximal proper subgroup of A_{i+1} which is characteristic in G, then A_{i+1}/A_i is the direct sum of isomorphic simple groups.

4.4.12. (a) Define strictly S-characteristic S-subgroup, and strictly characteristically S-simple.

(b) Improve Theorem 4.4.2 by replacing "characteristically" by "strictly characteristically."

4.4.13. Let $G = \text{Sym}(3)$, T the inner automorphism induced by $(1, 2)$, and $S = \{T\}$, so that G is an S-group.

(a) The only S-automorphisms of G are T and the identity I.

(b) The subgroup $H = \{e, (1, 2)\}$ is S-characteristic.

(c) H is an S-characteristic S-subgroup which is not normal.

(d)   An S-characteristic S-subgroup X of a normal S-subgroup Y of an S-group A need not be normal in A (compare with Exercise 2.11.13).

4.4.14. A group contains a maximal Abelian subgroup and a maximal normal Abelian subgroup.

4.4.15. An Abelian group is completely reducible iff it is the direct sum of (cyclic) groups of prime orders.

4.4.16. If G is a group, then the following two properties are equivalent:

(a)   If $A \lhd G$, then $\exists \mid B$ such that $G = A + B$.

(b) $G = \Sigma \{H \mid H \in S\}$, where each H is simple and two different Abelian H's are not isomorphic.

4.4.17. If G is the direct sum of isomorphic simple subgroups, then G is characteristically simple.

4.5   Endomorphism algebra

Some elementary facts about endomorphisms required in the next section, as well as a few other theorems, will be derived or listed in this section.

Let G be a group and, for this section only, let H be the set of functions from G into G. Then

$$x(TU) = (xT)U, \; x(T + U) = (xT)(xU), \qquad x \in G, \; T \in H, \; U \in H.$$

4.5.1. (1) *H is a group with respect to addition with zero element $O_G = O$.*

(2)   *Multiplication is associative.*

(3)   *$TO = O$ for all $T \in H$.*

(4)   *$TI = IT = T$ for all $T \in H$.*

(5)   *If T, U, and V are in H, then $T(U + V) = TU + TV$.*

Note that it is not true that $OT = O$, or that the other distributive law is valid. For endomorphisms, somewhat more can be said.

4.5.2. (6) End(G) *is closed under multiplication.*

(7)   *$I \in$ Aut(G), which is contained in* End(G).

(8)   *$O \in$ End(G) and $OU = O$ if $U \in$ End(G).*

(9)   *If $V \in$ End(G), $T \in H$, and $U \in H$, then $(T + U)V = TV + UV$.*

Proof. Let us prove (9). If $x \in G$, then

$$x((T + U)V) = ((xT)(xU))V = ((xT)V)((xU)V) = (x(TV))(x(UV))$$
$$= x(TV + UV). \parallel$$

Let G be a group, H as before, and S the subset of H generated by End(G). Thus S is the intersection of all subsets of H which are closed under addition and multiplication and contain End(G). It should be noted that End(G) is not always closed under addition (see Exercise 4.5.11).

4.5.3. $S = \{T_1 + \ldots + T_n \mid T_i \in \text{End}(G)\}$.

Proof. Let

$$S' = \{T_1 + \ldots + T_n \mid T_i \in \text{End}(G)\}.$$

Then End(G) is contained in S', and since S is closed under addition, S' is contained in S. Moreover, S' is closed under addition. Let $T_i \in \text{End}(G)$ and $U_j \in \text{End}(G)$. Then by (5) and (9)

$$(T_1 + \ldots + T_m)(U_1 + \ldots + U_n)$$
$$= (T_1 + \ldots + T_m)U_1 + \ldots + (T_1 + \ldots + T_m)U_n$$
$$= T_1 U_1 + \ldots + T_m U_1 + \ldots + T_1 U_n + \ldots + T_m U_n,$$

and $T_i U_j \in \text{End}(G)$. Hence S' is closed under multiplication, so S is contained in S'. Therefore $S' = S$. $\parallel$

An element T of H is *nilpotent* iff $T^n = O$ for some $n \in \mathcal{N}$, and *idempotent* iff $T^2 = T$. Trivial examples are furnished by O which is both nilpotent and idempotent, and I which is idempotent. Less trivially, if K is a group, $G = K \times K$, and $(x, y)U = (e, x)$, then U is a nilpotent endomorphism of G, in fact $U^2 = O$. If $G = K_1 + \ldots + K_n$ and U_i is the canonical homomorphism of G onto K_i, then U_i is idempotent.

4.5.4. *If T and U are nilpotent endomorphisms of a group G such that $TU = UT$, then $T + U$ is nilpotent.*

Proof. For some $n \in \mathcal{N}$, $T^n = U^n = O$. By Exercise 4.5.10 $(T + U)^{2n-1}$ is a sum of terms of the form $V_1 \ldots V_{2n-1}$ where each $V_i = T$ or U. Since $TU = UT$, $V_1 \ldots V_{2n-1} = T^i U^j$, where either $i \geqq n$ or $j \geqq n$. By (3) and (8), $T^i U^j = O$, hence $(T + U)^{2n-1} = O$ also. $\parallel$

Let G be a finite group and $T \in \text{End}(G)$. Then

$$E \subset \text{Ker}(T) \subset \text{Ker}(T^2) \subset \ldots,$$

and for some smallest n, $\mathrm{Ker}(T^n) = \mathrm{Ker}(T^{n+1})$. If $x \in \mathrm{Ker}(T^{n+2})$, then

$$xT \in \mathrm{Ker}(T^{n+1}) = \mathrm{Ker}(T^n),$$

so that $xT^{n+1} = e$, and $x \in \mathrm{Ker}(T^{n+1})$. Hence

$$\mathrm{Ker}(T^{n+2}) = \mathrm{Ker}(T^n).$$

By induction

$$\mathrm{Ker}(T^n) = \mathrm{Ker}(T^{n+r})$$

for all $r \in \mathcal{N}$. Call $\mathrm{Ker}(T^n)$ the *final kernel* of T. It is a normal subgroup of G. Similarly,

$$G \supset GT \supset \ldots \supset GT^m = GT^{m+1}$$

(actually $m = n$). If m is minimal such that $GT^m = GT^{m+1}$, then $T \mid GT^m$ is an automorphism, hence $GT^m = GT^{m+r}$ for all $r \in \mathcal{N}$. Call GT^m the *final image* of T.

The final image of T need not be a normal subgroup. For example, let $G = \mathrm{Sym}(3)$ and let T be the (unique) endomorphism of G with $GT = \{e, (1, 2)\}$. Then T is idempotent, and GT is the nonnormal final image of T.

4.5.5. *If G is a finite group, $T \in \mathrm{End}(G)$, and A and B are the final kernel and final image of T, respectively, then $G = AB$ and $A \cap B = E$.*

Proof. For some k, $GT^k = B$ and $\mathrm{Ker}(T^k) = A$. Then $BT^k = B$, so that $\mathrm{Ker}(T^k) \cap B = E$, and therefore $A \cap B = E$. If $g \in G$, then $gT^k \in B = BT^k$, hence $\exists y \in B$ such that $yT^k = gT^k$. But $y \in B = GT^k$, hence $\exists x \in G$ such that $xT^k = y$, so that $(xT^k)T^k = gT^k$. Therefore

$$(g(x^{-1}T^k))T^k = e, \qquad g(x^{-1}T^k) \in A, \, g \in A(xT^k), \, g \in AB.$$

Hence $G = AB$. ∥

Let M be the normalizer of $\mathrm{Inn}(G)$ in $\mathrm{End}(G)$, i.e., let M be the set of endomorphisms T of G such that $T \cdot \mathrm{Inn}(G) = \mathrm{Inn}(G) \cdot T$. [It was pointed out in Exercise 2.11.21 that, for all endomorphisms T, $T \cdot \mathrm{Inn}(G)$ contains $\mathrm{Inn}(G) \cdot T$.] The set M is closed under multiplication. If $T \in M$ and $U \in \mathrm{Inn}(G)$, then $\exists V \in \mathrm{Inn}(G)$ such that $TU = VT$. Hence

$$(GT)U = G(TU) = G(VT) = (GV)T = GT,$$

so that $GT \lhd G$. If $T \in M$, then $T^n \in M$ and $GT^n \lhd G$ for all $n \in \mathcal{N}$.

4.5.6. (*Fitting's lemma.*) *If G is a finite group, $M = N_{\mathrm{End}(G)}(\mathrm{Inn}(G))$, $T \in M$, and A and B are the final kernel and final image, respectively, of T, then $G = A + B$.*

Proof. The theorem follows from Theorem 4.5.5, the above remarks, and the definition of the direct sum.

4.5.7. *If G is a finite indecomposable group and $T \in M = N_{\text{End}(G)}(\text{Inn}(G))$, then either $T \in \text{Aut}(G)$ or T is nilpotent.*

Proof. By Theorem 4.5.6, either $A = E$ or $B = E$. If $A = E$, then T is an automorphism; if $B = E$, then T is nilpotent.

4.5.8. *If G is a finite indecomposable group, $T \in M = N_{\text{End}(G)}(\text{Inn}(G))$, $U \in M$, and $T + U \in \text{Aut}(G)$, then T or U is an automorphism of G.*

Proof. Let $T + U = V$. Then $TV^{-1} + UV^{-1} = I$. Moreover, $\text{Aut}(G) \subset M$ since $\text{Inn}(G) \lhd \text{Aut}(G)$ (Theorem 2.11.4). Hence $TV^{-1} \in M$ and $UV^{-1} \in M$. Finally, if $TV^{-1} \in \text{Aut}(G)$, say, then $T \in \text{Aut}(G)$. Hence, *WLOG*, $T + U = I$.

If the theorem is false, then by Theorem 4.5.7, T and U are nilpotent. Moreover,

$$T^2 + TU = T(T + U) = TI = T = IT$$
$$= (T + U)T = T^2 + UT.$$

Hence by Theorem 4.5.1 (1), $TU = UT$. By Theorem 4.5.4, $I = T + U$ is nilpotent. This implies that $o(G) = 1$, hence T is an automorphism of G, a contradiction.

4.5.9. *If G is a finite indecomposable group, and $T_1, \ldots, T_n$ are endomorphisms in $M = N_{\text{End}(G)}(\text{Inn}(G))$ such that each $T_1 + \ldots + T_r \in M$ and $T_1 + \ldots + T_n \in \text{Aut}(G)$, then some $T_i \in \text{Aut}(G)$.*

Proof. This follows from Theorem 4.5.8 by induction.

EXERCISES

4.5.10. If $T_1, \ldots, T_n$ are endomorphisms of G and $r \in \mathcal{N}$, then

$$(T_1 + \ldots + T_n)^r = \Sigma\, T_{i_1} \ldots T_{i_r},$$

where the terms in the sum occur in the following order:
$T_{i_1} \ldots T_{i_r}$ precedes $T_{j_1} \ldots T_{j_r}$ iff $i_r = j_r, \ldots,$
$i_{k+1} = j_{k+1}, i_k < j_k$ (k may equal r).

4.5.11. Referring to Theorem 4.5.4, give an example of a group G and nilpotent endomorphisms T and U such that $TU = UT$, but $T + U$ is not an endomorphism.

4.5.12. An endomorphism T of G is *normal* iff $T \in C_{\text{End}(G)}(\text{Inn}(G))$.

    (a)   If T is normal, then $T \in N_{\text{End}(G)}(\text{Inn}(G))$.

    (b)   Restate Theorems 4.5.6 through 4.5.9 for normal endomorphisms.

(c)   The product of normal endomorphisms is a normal endomorphism.

(d)   If the sum of normal endomorphisms is an endomorphism, then it is normal.

(e)   If T, U, and $T + U$ are in End(G) with $T + U$ and one of T and U normal, then the other is also normal.

(f)   If $T \in$ End(G), then T is normal iff $g(I - T) \in C(GT)$ for all $g \in G$.

(g)   If T is a normal endomorphism of G, then $I - T$ is a normal endomorphism of G [use (f)].

(h)   If $G = A_1 + \ldots + A_n$ and p_i is the canonical endomorphism: $(\Sigma \, a_j)p_i = a_i$, then

   (i) p_i is a normal endomorphism of G,
   (ii) the sum of any (nonempty) set of distinct p_i's is a normal endomorphism of G.

4.5.13. An endomorphism T of G is *central* iff $g(I - T) \in Z(G)$ for all $g \in G$.

(a)   A central endomorphism is normal, but a normal endomorphism need not be central. [See Exercise 4.5.12 (h).]

(b)   Any endomorphism of an Abelian group is central.

(c)   If T is a central endomorphism of G and $U \in$ Aut(G), then $U^{-1}TU$ is central.

(d)   An automorphism T of G is normal iff it is central.

4.5.14. (Compare with Theorem 4.5.4.) Give an example of a group $G \neq E$ and central nilpotent endomorphisms T and U such that $T + U$ is an automorphism of G.

4.5.15. If T is a normal (central) endomorphism of G, $K \subset G$, and $KT \subset K$, then $T \mid K$ is a normal (central) endomorphism of K.

4.5.16. If $K = A_1 + \ldots + A_n$ and $T_i \in$ Hom(G, A_i), then

$$T_1 + \ldots + T_n \in \text{Hom}(G, K).$$

4.5.17. If G is a group and T an idempotent endomorphism of G, then $G = (GT)(\text{Ker}(T)), (GT) \cap (\text{Ker}(T)) = E.$

(See the proof of Theorem 4.5.5.)

4.6   Remak-Krull-Schmidt theorem

The analogue of the Jordan-Hölder theorem for direct decompositions would read: if

$$G = \Sigma \{A_i \mid i \in S\} = \Sigma \{B_j \mid j \in T\},$$

where the A_i and B_j are indecomposable and not E and where S and T are finite, then there is a 1–1 function U from S onto T such that $A_i \cong B_{iU}$. This theorem is false, even in the case where S and T both have order 2. (See Fuchs [1, p. 154].) It is true if G is finite, and a theorem implying this fact will be proved in this section.

In order to make the proof less unwieldy, a portion of it will be separated out in the form of a lemma.

4.6.1. *If a finite group $G = A_1 + \ldots + A_m = B_1 + \ldots + B_n$, A_i and B_j indecomposable and not E, $p_1, \ldots, p_m$, $q_1, \ldots, q_n$ the associated homomorphisms of G onto the A_i and B_j, then $\exists$ i such that p_1 induces an isomorphism of B_i onto A_1 and q_i an isomorphism of A_1 onto B_i, and such that*

$$G = B_i + A_2 + \ldots + A_m.$$

Proof. We have $q_1 + \ldots + q_n = I$, and any sum of distinct q_j's is a normal endomorphism of G [Exercise 4.5.12 (h)]. Since p_1 is normal, $q_j p_1$ is normal [Exercise 4.5.12 (c)], and any sum of terms $q_j p_1$ with distinct j's is a normal endomorphism [Exercise 4.5.12 (c) and Theorem 4.5.2 (9)]. Now $\Sigma q_j p_1 = p_1$, so that $\Sigma q_j p_1 = I$ on A_1. Since each $q_j p_1 \mid A_1$ and

$$(q_1 p_1 + \ldots + q_r p_1) \mid A_1$$

is normal (Exercise 4.5.15), it follows from Theorem 4.5.9 that $\exists$ i such that $q_i p_1 \mid A_1 \in \mathrm{Aut}(A_1)$. Hence q_i induces an isomorphism of A_1 into B_i and p_1 a homomorphism of B_i onto A_1. Now $p_1 q_i$ is a normal endomorphism of B_i. If it is not an automorphism of B_i, then (Theorem 4.5.7) it is nilpotent. Hence, for some n,

$$(q_i p_1)^{n+1} = q_i (p_1 q_i)^n p_1 = q_i O p_1 = O,$$

contradicting the fact that $q_i p_1 \in \mathrm{Aut}(A_1)$ and $A_1 \neq E$. Hence $p_1 q_i$ is an automorphism of B_i. Therefore p_1 is an isomorphism of B_i into A_1. Since it was already known to be onto A_1, p_1 is an isomorphism of B_i onto A_1. Similarly, q_i is an isomorphism of A_1 onto B_i.

If $x \in B_i \cap (A_2 + \ldots + A_m)$, then $x p_1 = e$, hence (since $p_1 \mid B_i$ is an isomorphism) $x = e$. Therefore

$$B_i \cap (A_2 + \ldots + A_m) = E.$$

Let $a \in A_1$. Since p_1 maps B_i onto A_1, $\exists$ $b \in B_i$ such that $b = ac$ with $c \in A_2 + \ldots + A_m$. Hence $a = bc^{-1} \in \langle B_i, A_2 + \ldots + A_m \rangle$. Thus

$$A_1 \subset \langle B_i, A_2 + \ldots + A_m \rangle.$$

Therefore

$$G = A_1 + (A_2 + \ldots + A_m) \subset \langle B_i, A_2 + \ldots + A_m \rangle,$$

so that $G = B_i + A_2 + \ldots + A_m$.

4.6.2. (*Remak-Krull-Schmidt.*) *If a finite group G has decompositions*

$$G = A_1 + \ldots + A_m = B_1 + \ldots + B_n,$$

where the A_i and B_j are indecomposable and not E, and if $p_1, \ldots, p_m, q_1, \ldots, q_n$ are the associated endomorphisms of G onto the A_i and B_j, then $m = n$ and $\exists \, T \in \mathrm{Sym}(m)$ such that

(1) $p_1 q_{1T} + \ldots + p_m q_{mT}$ *is a central automorphism of G which maps A_i onto B_{iT} for all i,*

(2) $G = B_{1T} + \ldots + B_{rT} + A_{r+1} + \ldots + A_m$ *for $1 \le r \le m$.*

Proof. Assume inductively that T is a 1–1 function from $\{1, \ldots, r\}$ into $\{1, \ldots, n\}$ such that

$$U_{r-1} = p_1 q_{1T} + \ldots + p_{r-1} q_{(r-1)T} + p_r + \ldots + p_m$$

is an automorphism of G inducing an isomorphism of A_i onto B_{iT} for $i \le r - 1$. This assumption is certainly valid for $r = 1$. Under U_{r-1}, the decomposition $G = \Sigma A_i$ yields

$$G = B_{1T} + \ldots + B_{(r-1)T} + A_r + \ldots + A_m.$$

Let $Q = B_{1T} + \ldots + B_{(r-1)T}$. Then

$$G = Q + A_r + \ldots + A_m = Q + B_{j_1} + \ldots + B_{j_s}.$$

Let V be the natural homomorphism of G onto G/Q. Then

$$G/Q = A_r V + \ldots + A_m V = B_{j_1} V + \ldots + B_{j_s} V.$$

Let $p_r^*, \ldots, p_m^*, q_{j_1}^*, \ldots, q_{j_s}^*$ be the associated endomorphisms of G/Q. By Theorem 4.6.1, $\exists \, j$, which will be denoted by rT, such that p_r^* is an isomorphism of $B_{rT} V$ onto $A_r V$, q_{rT}^* is an isomorphism of $A_r V$ onto $B_{rT} V$, $rT \ne iT$ for $i < r$, and

$$G/Q = B_{rT} V + A_{r+1} V + \ldots + A_m V.$$

We have

$$\langle B_{1T} + \ldots + B_{rT}, A_{r+1} + \ldots + A_m \rangle \supset Q,$$
$$\langle B_{1T} + \ldots + B_{rT}, A_{r+1} + \ldots + A_m \rangle V = G/Q,$$

hence

$$\langle B_{1T} + \ldots + B_{rT}, A_{r+1} + \ldots + A_m \rangle = G.$$
If
$$x \in (B_{1T} + \ldots + B_{rT}) \cap (A_{r+1} + \ldots + A_m),$$

then

$$xV \in B_{rT}V \cap (A_{r+1}V + \ldots + A_mV),$$

so $xV = Q$, $x \in Q$, hence $x = e$ since

$$Q \cap (A_r + \ldots + A_m) = E.$$

Therefore

(3) $G = B_{1T} + \ldots + B_{rT} + A_{r+1} + \ldots + A_m.$

Hence (Exercise 4.5.16)

$$U_r = p_1 q_{1T} + \ldots + p_r q_{rT} + p_{r+1} + \ldots + p_m \in \text{End}(G).$$

One verifies that $Vq_{rT}^* = q_{rT}V$. Now $V \mid A_r$ is an isomorphism since $G = Q + A_r + \ldots + A_m$, q_{rT}^* is an isomorphism of $A_r V$ onto $B_{rT}V$, and $V \mid B_{rT}$ is an isomorphism. Hence q_{rT} is an isomorphism of A_r onto B_{rT}. Therefore, it follows from the induction assumption and (3) that U_r has range G. Since $p_i q_{iT}$ is an isomorphism of A_i onto B_{iT}, $i = 1, \ldots, r$, the decompositions (3) and $G = \Sigma A_i$ show that U_r is an automorphism of G.

It follows by induction that there is a sequence $T_1, \ldots, T_m$ with T_r a 1–1 function from $\{1, \ldots, r\}$ into $\{1, \ldots, n\}$ such that

(a)   if $j > i$, then T_j is an extension of T_i,

(b) $U_r = p_1 q_{1T_r} + \ldots + p_r q_{rT_r} + p_{r+1} + \ldots + p_m$
      is an automorphism of G mapping A_i onto B_{iT_r} for $i \leq r$.

Letting $T = T_m$ and applying T to $G = \Sigma A_i$, we get

$$G = B_{1T} + \ldots + B_{mT} = B_1 + \ldots + B_n, \qquad B_j \neq E.$$

Hence $m = n$ and $T \in \text{Sym}(m)$. Equation (2) now follows from (a) and (b) above. Since each $p_i q_{iT}$ is a normal endomorphism of G, so is U_m. By Exercise 4.5.13 (d), U_m is central, so that (1) is satisfied. ‖

As an immediate corollary, we have

4.6.3.   *If G is a finite group, $Z(G) = E$, and*

$$G = \Sigma A_i = \Sigma B_j,$$

where the A_i and B_j are indecomposable and not E, then each B_j is equal to some A_i; i.e., G has only one decomposition into indecomposable groups up to order of factors.

Proof. The only central automorphism of G is I. Hence, using the notation of Theorem 4.6.2, $A_i = B_{iT}$.

EXERCISES

4.6.4. Show that the four-group has three indecomposable decompositions.

4.6.5. Show that Sym(3) × Sym(2) has two indecomposable decompositions. Find the central automorphism of G connecting these decompositions.

4.6.6. In the notation of the proof of the Remak-Krull-Schmidt theorem,

$$p_i U_m = U_m q_{iT} = p_i q_{iT}, \qquad U_m^{-1} p_i U_m = q_{iT}.$$

4.7 Generalizations

For the generalization of the Remak theorem, which will be sketched later in this section, we will introduce some new concepts here.

A partially ordered set Q satisfies the *maximal condition* iff every nonempty subset R of Q has a maximal element. This is equivalent to requiring that every ascending chain in Q be finite. *Minimal condition* is defined in an analogous manner. This terminology can be carried over to various classes of subgroups of a group in an almost self-explanatory fashion. For example, a group G satisfies the *maximal condition for subgroups* iff every nonempty set R of subgroups of G contains a subgroup maximal (in R).

The group J of integers satisfies the maximal condition for subgroups, but not the minimal condition for subgroups. In Section 5.2, we shall see a group, the p^∞-group, which satisfies the minimal condition for subgroups, but not the maximal condition for subgroups.

It has already (Exercises 2.3.9 and 2.3.11) been pointed out that normality is not transitive. Nonetheless, the notion "normal subgroup of normal subgroup of . . . normal subgroup of G" is an important one and will be formalized. A subgroup H of G is *subnormal* in G, written $H \lhd \lhd G$, iff ∃ subgroups $A_0 = H, A_1, \ldots, A_n = G$ such that $A_i \lhd A_{i+1}$ for all i. Thus H is subnormal iff it is a member of some normal series of G.

Let S be a set and G an S-group (see Section 2.7). G satisfies the *descending chain condition* iff every descending normal S-series without repetitions, $G \rhd A_0 \rhd A_1 \rhd \ldots$, is finite. G satisfies the *ascending chain condition* iff every subnormal S-subgroup H satisfies the maximal condition for normal S-subgroups.

4.7.1. *If a group $G \neq E$ satisfies the descending chain condition, then G contains a minimal normal non-E subgroup H, and every such subgroup $H = \Sigma K_i$, where the subgroups K_i are simple and all conjugate in G.*

Proof. If G contained no minimal normal non-E subgroup, then any normal subgroup $A \neq E$ would contain a subgroup $B < A$ with $E \neq B \lhd G$.

An infinite descending chain of normal subgroups would then exist. (For this part of the theorem, one need only assume that G satisfies the minimal condition for normal subgroups.)

H also has a minimal normal subgroup K by the same argument. The result now follows from Theorem 4.4.3.

4.7.2. *An S-group G has an S-composition series iff it satisfies the ascending and descending chain conditions.*

Proof. Let G satisfy both chain conditions. By the ascending chain condition and induction, there is a chain $C = (G = A_0, A_1, \ldots)$ such that each A_i is a maximal proper normal S-subgroup of A_{i-1}. By the descending chain condition, C is finite. It must terminate with E. By definition, C is an S-composition series (written in reverse order).

Now suppose that G has a composition S-series of length n, say. Then, by Theorem 2.10.2 any normal S-series of G can be refined to a composition S-series of length n. Hence no chain occurring in the definition of the chain conditions can have length greater than n. ∥

A decomposition $G = \Sigma\, A_i$ of an S-group G is an *S-decomposition* iff each A_i is an S-subgroup.

4.7.3. *If $\Sigma\, A_i$ is an S-decomposition of the S-group G, then each canonical homomorphism p_i of G onto A_i is an S-homomorphism.*

Proof. If $x = \pi a_j \in G$, $a_j \in A_j$, and $s \in S$, then

$$xp_i s = a_i s, \quad xsp_i = (\pi(a_j s))p_i = a_i s$$

since each $a_j s \in A_j$. Hence $p_i s = s p_i$ for all $s \in S$, and p_i is an S-homomorphism.

4.7.4. *If a sum of S-endomorphisms is an endomorphism, it is an S-endomorphism.*

Proof. Let $T_1, \ldots, T_n$ be S-endomorphisms of the S-group G such that $T_1 + \ldots + T_n$ is an endomorphism. If $s \in S$, then since s is essentially an endomorphism of G,

$$s(T_1 + \ldots + T_n) = sT_1 + \ldots + sT_n = T_1 s + \ldots + T_n s$$
$$= (T_1 + \ldots + T_n)s. \; \parallel$$

Now let S be a set of endomorphisms of G containing $\mathrm{Inn}(G)$, and let G be an S-group. If T is an S-endomorphism of G, then $GT \lhd G$. If, in addition, G satisfies both chain conditions, then the final kernel and final image of T exist and are normal in G. One may then write the analogues of Theorems 4.5.6, 4.5.7, and 4.5.9.

4.7.5. *If G is an S-group, S contains $\mathrm{Inn}(G)$, G satisfies both chain conditions, and T is an S-endomorphism with final kernel A and final image B, then $G = A + B$.*

4.7.6. *If G is an S-indecomposable S-group, S contains $\mathrm{Inn}(G)$, G satisfies both chain conditions, and T is an S-endomorphism of G, then T is an S-automorphism of G or T is nilpotent.*

4.7.7. *If G is an S-indecomposable S-group, S contains $\mathrm{Inn}(G)$, G satisfies both chain conditions, $T_1, \ldots, T_n$ are S-endomorphisms of G, each $T_1 + \ldots + T_r$ is an endomorphism, and $T_1 + \ldots + T_n$ is an automorphism of G, then some T_i is an S-automorphism of G.* $\parallel$

With these remarks, the proof of the Remak-Krull-Schmidt theorem can be carried over, almost word for word, to the case of operator groups satisfying both chain conditions. The statement of the theorem follows.

4.7.8. *If G is an S-group, S contains $\mathrm{Inn}(G)$, G satisfies both chain conditions, and*

$$G = A_1 + \ldots + A_m = B_1 + \ldots + B_n,$$

where the A_i and B_j are S-indecomposable and not E, and p_i and q_j are the associated S-endomorphisms of G onto the A_i and B_j, respectively, then $m = n$, and $\exists\, T \in \mathrm{Sym}(m)$ such that

(1) *$p_1 q_{1T} + \ldots + p_m q_{mT}$ is an S-automorphism of G which maps A_i onto B_{iT} for all i,*

(2) *$G = B_{1T} + \ldots + B_{rT} + A_{r+1} + \ldots + A_m$ for $1 \leqq r \leqq m$.*

4.7.9. *If S contains $\mathrm{Inn}(G)$, G is an S-group, G satisfies both chain conditions, and no non-E subgroup of $Z(G)$ is an S-subgroup of G, then G has just one S-indecomposable S-decomposition.*

Proof. If there are two such decompositions, then by Theorem 4.7.8, there is an S-automorphism T of G connecting them. Since S contains $\mathrm{Inn}(G)$, an S-automorphism is normal. By Exercise 4.5.13, T is central. Since T is central, it is easy to verify that $I - T$ is an endomorphism. If $x \in G$ and $s \in S$, then

$$x(I - T)s = (x(x^{-1}T))s = (xs)(x^{-1}Ts) = (xs)(x^{-1}sT)$$
$$= (xs)((xs)^{-1}T) = (xs)(I - T).$$

Hence $s(I - T) = (I - T)s$ for all $s \in S$, and $I - T$ is an S-endomorphism. By Theorem 2.7.2, $\text{Rng}(I - T)$ is an S-subgroup of G. Since T is central, $\text{Rng}(I - T) \subseteq Z(G)$. Since $T \neq I$, this contradicts the hypotheses of the theorem.

The fact that G has at least one S-indecomposable S-decomposition is left as the first exercise below.

EXERCISES

4.7.10. If G is an S-group and satisfies both chain conditions, then G has an S-indecomposable S-decomposition

$$G = A_1 + \ldots + A_n.$$

4.7.11. Carry out the proof of Theorem 4.7.8.

REFERENCES FOR CHAPTER 4

For Sections 4.1 and 4.2, Cartan and Eilenberg [1]; for 4.3, Remak [1]; Theorem 4.3.5, Gilbert [1]; Zorn's lemma, Zorn [1]; Theorem 4.4.7, B. Neumann in Wiegold [1]; Sections 4.5 to 4.7, Jacobson [1].

A B E L I A N   G R O U P S

5.1 Direct decompositions

The structure of all finite Abelian groups will be determined in this section. Additive notation will be used throughout this chapter. The group operation will be $+$, the identity 0, the inverse of a will be $-a$, na will be used instead of a^n, and $a + H$ instead of aH for cosets. However, factor groups will be denoted by G/H as before.

A group G is *periodic* iff every element of G is of finite order. G is *torsion-free* iff every element except 0 is of infinite order.

5.1.1. *Let G be a group and T the set of elements of finite order in G. If T is a subgroup of G, then T is fully characteristic and G/T is torsion-free.*

Proof. Let $x \in T$ and $U \in \text{End}(G)$. Then $\exists\, n \in \mathcal{N}$ such that $nx = 0$. Hence

$$n(xU) = (nx)U = 0U = 0,$$

and $xU \in T$. Therefore T is fully characteristic. Hence $T \lhd G$ and G/T exists. If $y \in G$ is such that $o(y + T) = m$ is finite, then $my \in T$, so $\exists\, r \in \mathcal{N}$ such that $(rm)y = r(my) = 0$. Hence $y \in T$. Therefore G/T is torsion-free.

5.1.2. *If G is an Abelian group and T is defined as in Theorem 5.1.1, then T is a fully characteristic subgroup of G and G/T is torsion-free.*

Proof. It need only be proved that T is a subgroup (by Theorem 5.1.1), and this is straightforward. ∥

The subgroup T is called the *torsion* subgroup of G.
If G is Abelian and $p \in \mathscr{P}$, let

$$G_p = \{x \in G \mid o(x) = p^n \text{ for some } n\}.$$

5.1.3. *If G is Abelian and $p \in \mathscr{P}$, then G_p is a fully characteristic subgroup of G.*

Proof. Let $x \in G_p$ and $y \in G_p$. Then $\exists\, m \geq 0$ and $n \geq 0$ such that $p^m x = 0 = p^n y$. Therefore $p^{m+n}(x + y) = 0$, and $o(x + y) \mid p^{m+n}$. Hence $x + y \in G_p$. Since $0 \in G_p$ and $o(-x) = o(x)$, it follows that G_p is a subgroup of G. It is easy to check that G_p is fully characteristic in G.

5.1.4. *If G is a group, x an element of finite order in G, and $n \in J$, then $o(nx) = o(x)/(n, o(x))$.*

Proof. Let $(n, o(x)) = d$, $n = dn'$, and $o(x) = dm$. Then $o(x)/(n, o(x)) = m$. Now $k(nx) = 0$ iff $dm \mid kn$, hence iff $m \mid kn'$. Since $(m, n') = 1$, this is true iff $m \mid k$. Therefore $o(nx) = m$.

5.1.5. *If G is a group, $x \in G$, $o(x) = n_1 \cdots n_r$, $n_i \in \mathscr{N}$, and $(n_i, n_j) = 1$ if $i \neq j$, then $\exists$ unique $a_1 \in G, \ldots, a_r \in G$ such that*

$$o(a_i) = n_i \text{ for } i = 1, \ldots, r, \tag{1}$$

$$x = a_1 + \ldots + a_r, \tag{2}$$

$$a_i + a_j = a_j + a_i \quad \text{for all } i \text{ and } j. \tag{3}$$

Proof. Let $q_i = \pi \{n_j \mid j \neq i\}$. Since the greatest common divisor of $q_1, \ldots, q_r$ is 1, $\exists\, c_i \in J$ such that $\Sigma\, c_i q_i = 1$.

Uniqueness. Assume that $a_1, \ldots, a_r$ are such that (1), (2), and (3) hold. Then

$$(c_i q_i)x = (c_i q_i)(\Sigma\, a_j) = \Sigma\, \{(c_i q_i)a_j \mid 1 \leq j \leq r\}$$
$$= (c_i q_i)a_i = (1 - \Sigma\, \{c_j q_j \mid j \neq i\})a_i = a_i.$$

Existence. Let $a_i = (c_i q_i)x$. By Theorem 5.1.4,

$$o(a_i) = \frac{\pi n_j}{(\pi n_j, c_i q_i)} = \frac{n_i}{(n_i, c_i)} = n_i,$$

for if a prime $p \mid n_i$ and $p \mid c_i$, then $p \mid \Sigma c_j q_j$ so that $p \mid 1$, which is impossible. Therefore (1) is true. Again

$$\Sigma a_i = \Sigma (c_i q_i)x = (\Sigma c_i q_i)x = x,$$

and (2) is true. (3) is obvious.

5.1.6. *If G is an Abelian group and T is its torsion subgroup, then $T = \Sigma \{G_p \mid p \in \mathscr{P}\}$.*

Proof. It is obvious that $G_p \subset T$ for each p, hence $\langle G_p \rangle \subset T$. If $x \in T$, then there are distinct primes $p_1, \ldots, p_n$ and integers $f_i \geqq 0$ such that $o(x) = \pi p_i^{f_i}$. By the preceding theorem, $\exists\, a_i \in G$ such that $o(a_i) = p_i^{f_i}$ and $x = \Sigma a_i$. Hence $x \in \langle G_p \rangle$. Therefore $T = \langle G_p \rangle$. If $y \in G_p \cap \langle G_q \mid q \neq p \rangle$, then $\exists$ primes $p_1, \ldots, p_n$, different from p, and $y_i \in G_{p_i}$ such that $y = \Sigma y_i$. If $r = \pi o(y_i)$, then $ry = 0$. Since $p \nmid r$, $y = 0$. Hence $G_p \cap \langle G_q \mid q \neq p \rangle = E$. Therefore $T = \Sigma G_p$. ∥

The group G_p is called the *p-component* of G. The p-components of G for $p \in \mathscr{P}$ are called *primary components*.

5.1.7. (*Decomposition of Abelian periodic group into primary components.*) *If G is an Abelian periodic group, then $G = \Sigma G_p$.* ∥

A *p-group* ($p \in \mathscr{P}$) is a group all of whose elements have orders a power of p. A group is *primary* iff it is a p-group for some $p \in \mathscr{P}$.
The decomposition given in Theorem 5.1.7 is unique in a strong sense.

5.1.8. *If G is an Abelian group, $G = \Sigma \{H(p) \mid p \in \mathscr{P}\}$, and each $H(p)$ is a p-group, then each $H(p) = G_p$.*

Proof. By its definition, G_p is the maximum p-subgroup of G. Hence each $H(p) \subset G_p$. If $\exists\, p \in \mathscr{P}$ and $x \in G_p \backslash H(p)$, then it is clear (see Exercise 5.1.19) that $G = \Sigma H(p) < \Sigma G_p = G$, a contradiction. Hence $H(p) = G_p$ for all primes p.

5.1.9. *If G is an Abelian p-group ($p \in \mathscr{P}$) such that $px = 0$ for all $x \in G$, then $G = \Sigma H_i$ where each $o(H_i) = p$.*

Proof. By Theorem 4.4.1 (with $H = E$), there is a maximal set S of subgroups of G such that $M = \Sigma \{K \mid K \in S\}$ exists and such that if $K \in S$ then $o(K) = p$. If $x \in G \backslash M$, then $\langle x \rangle + M$ exists, contradicting the maximality of S. Hence $G = M = \Sigma \{K \mid K \in S\}$. (We are using the convention that E is the direct sum of the empty set of subgroups.) ∥

An *elementary* Abelian group is a group which is the direct sum of subgroups of (possibly different) prime orders. Thus the group in Theorem 5.1.9 is an elementary Abelian p-group.

If G is an Abelian group and $n \in J$, then nG will denote the set of elements nx with $x \in G$.

5.1.10. *If G is an Abelian group and $n \in J$, then nG is a fully characteristic subgroup of G.* ‖

The *exponent* of a group G is the smallest positive integer n such that $nx = 0$ for all $x \in G$ if such an n exists; otherwise it is ∞. It will be denoted by $\mathrm{Exp}(G)$.

5.1.11. *If G is an Abelian group of exponent p^n, $p \in \mathscr{P}$, and $n \in \mathscr{N}$, then $G = \Sigma\, H_i$, where each H_i is cyclic of order p^{n_i}.*

Proof. If $n = 1$, this is Theorem 5.1.9. Induct on n. Now pG is a subgroup by Theorem 5.1.10, and, since $p^{n-1}(pG) = E$, $pG = \Sigma\,\{\langle y_i\rangle \mid i \in S\}$ with $y_i \in G^{\#}$, by the inductive hypothesis. For each $i \in S$, $\exists\, x_i \in G$ such that $px_i = y_i$.

We assert that $K = \Sigma\,\{\langle x_i\rangle \mid i \in S\}$ exists. If this is false, then there are distinct $i_1, \dots, i_r$ in S and $n_j \in J$ such that $\Sigma\, n_j x_{i_j} = 0$, but each $n_j x_{i_j} \neq 0$. Suppose first that $p \mid n_j$ for each j. Then the last equation implies that $\Sigma\,(n_j/p)y_{i_j} = 0$. But since $\Sigma\,\langle y_{i_j}\rangle$ exists, each $(n_j/p)y_{i_j} = 0$. Hence each $n_j x_{i_j} = 0$, a contradiction. Therefore *WLOG* $p \nmid n_1$. Then, multiplying $\Sigma\, n_j x_{i_j}$ by p, we have $\Sigma\, n_j y_{i_j} = 0$. This implies that $n_1 y_{i_1} = 0$, and, since $p \nmid n_1$, $y_{i_1} = 0$, a contradiction. Hence the assertion is true.

By Theorem 4.4.1, $\exists$ a maximal set T of subgroups L such that $o(L) = p$ and such that $K + \Sigma\,\{L \mid L \in T\} = M$ exists. If $z \in G\backslash M$, then $pz \in pG$, so that

$$pz = m_1 y_{k_1} + \dots + m_s y_{k_s} = p(m_1 x_{k_1} + \dots + m_s x_{k_s}) = pu,$$

where $u \in M$. Now $z - u \in G\backslash M$ and $p(z - u) = 0$. Therefore $M + \langle z - u\rangle$ exists and the set T is not maximal. This contradiction proves that $G = K + \Sigma\,\{L \mid L \in T\}$, so that G is the direct sum of cyclic groups whose orders are powers of p.

5.1.12. *If G is an Abelian group with finite exponent, then G is the direct sum of cyclic groups of prime power orders.*

Proof. This follows from Theorems 5.1.7 and 5.1.11.

5.1.13. (*Fundamental theorem of finite Abelian groups, part 1.*) *If G is a finite Abelian group, then G is the direct sum of cyclic groups of prime power orders.*

5.1.14. *If $p \in \mathscr{P}$, G is an elementary Abelian p-group, and*

$$G = \sum \{H \mid H \in S\} = \sum \{K \mid K \in T\},$$

where all summands are of order p, then $o(S) = o(T)$. In fact, $o(G) = p^{o(S)}$ if S is finite, and $o(G) = o(S)$ if S is infinite.

Proof. If $o(S) = n$ is finite, then $o(G) = p^n$. This settles the finite case. Now suppose that $o(S) = A$ is infinite. Then there are at most $(pA)^m = A$ elements of G with exactly m nonzero components relative to the first decomposition. Hence there are at most $A\aleph_0 = A$ elements of G. But there are at least A elements of G (one in each $H \in S$). Hence $o(G) = A$. Thus $o(S) = A = o(T)$, and the theorem is valid in the infinite case also. ‖

It is easy to verify that there is no strong uniqueness theorem for the direct decomposition in Theorem 5.1.11 (or 5.1.12 or 5.1.13) as there was in the case of the decomposition of a periodic Abelian group into primary components. In fact, if G is the direct sum of two cyclic groups of order 2, $G = H + K$, then there is a third subgroup L of order 2, and also $G = H + L = K + L$. However, the *number* of direct summands of a given prime power order is independent of the decomposition. This can actually be proved for a more general setting than that of Theorem 5.1.11.

5.1.15. *If $p \in \mathscr{P}$ and*

$$G = \sum \{\sum \{H \mid H \in S_i\} \mid i \in \mathscr{N}\}$$
$$= \sum \{\sum \{K \mid K \in T_i\} \mid i \in \mathscr{N}\},$$

where each $H \in S_i$ and each $K \in T_i$ is cyclic of order p^i, then $o(S_i) = o(T_i)$ for all i.

Proof. Let $P = \{x \in G \mid px = 0\}$ and $L_i = P \cap p^i G$. Then P and L_i are subgroups of G. Now

$$p^n G = \sum \{\sum \{p^n H \mid H \in S_i\} \mid i > n\},$$
$$L_n = \sum \{\sum \{p^{i-1} H \mid H \in S_i\} \mid i > n\},$$
$$\frac{L_{n-1}}{L_n} \cong \sum \{p^{n-1} H \mid H \in S_n\}.$$

Similarly,

$$\frac{L_{n-1}}{L_n} \cong \sum \{p^{n-1} K \mid K \in T_n\}.$$

Now $o(p^{n-1}H) = p$ if $H \in S_n$. Hence by Theorem 5.1.14, $o(S_n) = o(T_n)$.

5.1.16. *If a group G is expressed as the direct sum of cyclic groups of prime power orders in two ways, then there are the same number of summands of order p^n in the one decomposition as in the other for each prime p and each n.*

Proof. This follows from Theorems 5.1.8, and 5.1.15.

5.1.17. (*Fundamental theorem of finite Abelian groups, part* 2). *If G is a finite Abelian group and $G = \Sigma\, H_i = \Sigma\, K_j$, where H_i and K_j are cyclic of prime power order (not 1), then there are the same number of summands in each decomposition, and after rearrangement, $o(H_i) = o(K_i)$.*

Proof. This is a corollary of Theorem 5.1.16 (or of the Remak theorem, 4.6.2). ‖

The fundamental theorem of finite Abelian groups enables one to determine the number of isomorphism classes of Abelian groups of order n for each natural number n. A *partition* of $k \in \mathcal{N}$ is an r-tuple $(k_1, \ldots, k_r)$ of natural numbers such that $k = \Sigma\, k_i$ and each $k_i \geq k_{i+1}$.

5.1.18. *If $P(i)$ is the number of partitions of i, $n = \pi p_j^{k_j}$, $p_j \in \mathscr{P}$, and $p_i \neq p_j$ if $i \neq j$, then the number of isomorphism classes of Abelian groups of order n is $\pi P(k_j)$.*

Proof. This follows from the fundamental theorem of finite Abelian groups (Theorems 5.1.13 and 5.1.17). ‖

Example. Let $n = 1440 = 2^5 3^2 5^1$. Now $P(1) = 1$, $P(2) = 2$, and $P(5) = 7$. Hence there are fourteen isomorphism classes of Abelian groups of order 1440. Representatives of these fourteen classes are:

$J_{2^5} \times J_{3^2} \times J_5,$

$J_{2^5} \times J_{3^1} \times J_{3^1} \times J_5,$

$J_{2^4} \times J_{2^1} \times J_{3^2} \times J_5,$

$J_{2^4} \times J_{2^1} \times J_{3^1} \times J_{3^1} \times J_5,$

$J_{2^3} \times J_{2^2} \times J_{3^2} \times J_5,$

$J_{2^3} \times J_{2^2} \times J_{3^1} \times J_{3^1} \times J_5,$

$J_{2^3} \times J_{2^1} \times J_{2^1} \times J_{3^2} \times J_5,$

$J_{2^3} \times J_{2^1} \times J_{2^1} \times J_{3^1} \times J_{3^1} \times J_5,$

$J_{2^2} \times J_{2^2} \times J_{2^1} \times J_{3^2} \times J_5,$

$J_{2^2} \times J_{2^2} \times J_{2^1} \times J_{3^1} \times J_{3^1} \times J_5,$

$J_{2^2} \times J_{2^1} \times J_{2^1} \times J_{2^1} \times J_{3^2} \times J_5,$

$J_{2^2} \times J_{2^1} \times J_{2^1} \times J_{2^1} \times J_{3^1} \times J_{3^1} \times J_5,$

$J_{2^1} \times J_{2^1} \times J_{2^1} \times J_{2^1} \times J_{2^1} \times J_{3^2} \times J_5,$

$J_{2^1} \times J_{2^1} \times J_{2^1} \times J_{2^1} \times J_{2^1} \times J_{3^1} \times J_{3^1} \times J_5.$

EXERCISES

5.1.19. Prove that if $G = \Sigma \{H \mid H \in S\}$ and $K_H \subseteq H$ for each $H \in S$ with strict inclusion for at least one H, then $\Sigma \{K_H \mid H \in S\} < G$.

5.1.20. How many isomorphism classes of Abelian groups of order 1024 are there?

5.1.21. A direct sum of two cyclic groups of orders m and n is cyclic iff $(m, n) = 1$.

5.1.22. Let $p \in \mathscr{P}$ and let G be the direct sum of n groups of order p. How many subgroups of order p does G have?

5.1.23. A direct sum of p-groups is a p-group.

5.1.24. (a) Show that a direct product of p-groups need not be a p-group.

   (b) Under what conditions is a direct product of p-groups again a p-group?

   (c) Under what conditions is a direct product of finite cyclic groups a direct sum of finite cyclic groups?

5.1.25. It is false that if G is finite Abelian and $H \subset G$, then $\exists$ cyclic subgroups A_i and B_i such that

$$G = A_1 + \ldots + A_n, \; H = B_1 + \ldots + B_n, \; B_i \subseteq A_i.$$

(Let $G = J_2 \times J_8$.)

5.1.26. If G is an Abelian p-group, $p \in \mathscr{P}$, $n \in J$, $p \nmid n$, and $xT = nx$ for $x \in G$ then $T \in \text{Aut}(G)$.

5.1.27. Let G be an Abelian group.

   (a) G is completely reducible iff every subgroup is a direct summand.

   (b) G is completely reducible iff G is elementary.

5.1.28. If G is an elementary Abelian group, $H \subset G$, and $K \subset G$, then there is a subset B of K such that $\langle H, K \rangle = H + \Sigma \{\langle b \rangle \mid b \in B\}$.

5.1.29. Show that there are nonisomorphic Abelian groups G and H such that each is a homomorphic image of the other. [Let G be the external direct sum of $\aleph_0$ cyclic groups of order 4 and $H = G + K$, $o(K) = 2$.]

5.2 Divisible groups

A group G is *divisible* iff $x \in G$ and $n \in \mathscr{N}$ imply that $\exists \; y \in G$ such that $ny = x$. The group $\mathscr{R}$ of rationals under addition is divisible. A finite group with more than one element is not divisible.

Let $p \in \mathscr{P}$ and let G be the group of rationals r such that $0 \leq r < 1$ and such that $r = i/p^n$ for some nonnegative integers i and n, under addition *mod* 1. (Another way of describing G is to write $G = (\mathscr{R}/J)_p$.) The group G is a divisible Abelian p-group. Any group isomorphic to G will be called a *p^∞-group*.

5.2.1. *If G is an Abelian group, then there is a maximum divisible subgroup D of G.*

Proof. E is a divisible subgroup of G. Let D be the subgroup generated by all divisible subgroups of G. If $x \in D$, then $\exists$ divisible subgroups $H_1, \ldots, H_r$ of G and $x_i \in H_i$ such that $x = \Sigma\, x_i$. If $n \in \mathscr{N}$, then $\exists\, y_i \in H_i$ such that $ny_i = x_i$. But then $\Sigma\, y_i \in D$ and $n(\Sigma\, y_i) = x$. Thus D is divisible. It is obvious that D is the maximum divisible subgroup of G.

5.2.2. *If G is an Abelian group and D is a divisible subgroup of G, then D is a direct summand of G.*

Proof. An easy application of Zorn's lemma shows (Exercise 5.2.17) that there is a maximal subgroup M of G such that $M \cap D = E$. Then $M + D$ exists. Suppose $M + D < G$.

Case 1. $M = E$. Now $\exists\, x \in G\backslash D$. Since $D \cap \langle x \rangle \neq E$ by assumption, some multiple of x is in D. WLOG, $px \in D$ for some $p \in \mathscr{P}$. Since D is divisible, $\exists\, y \in D$ such that $py = px$. Then $p(x - y) = 0$ and $x - y \notin D$. Thus $D \cap \langle x - y \rangle = E$ although $\langle x - y \rangle \neq E$, a contradiction.

Case 2. $M > E$. Then $(D + M)/M$ is a proper divisible subgroup of G/M since $(D + M)/M \cong D$. Since Case 1 was impossible, $\exists\, K > M$ such that $((D + M)/M) \cap (K/M) = E$. Hence $(D + M) \cap K \subseteq M$. Therefore $D \cap K \subseteq M$, and this implies that $D \cap K \subseteq D \cap M = E$. This contradicts the maximality of M.

Hence $G = M + D$. ‖

A *reduced* Abelian group is one which contains no divisible subgroup except E.

5.2.3. *If G is an Abelian group, then $G = D + R$, where D is divisible and R is reduced.*

Proof. By Theorem 5.2.1, there is a maximum divisible subgroup D of G. By Theorem 5.2.2, $G = D + R$ for some subgroup R of G. Since D contains all divisible subgroups of G, R is reduced. ‖

The divisible summand in the divisible-reduced decomposition Theorem 5.2.3 is unique, but the reduced summand is not (see Exercise 5.2.18).

5.2.4. *If G is an Abelian group and $G = D + R$ where D is divisible and R is reduced, then D is the maximum divisible subgroup of G.*

Proof. Let D^* be the maximum divisible subgroup of G. By Exercise 4.1.13, $D^* = D + (D^* \cap R)$. By Exercise 5.2.20, $D^* \cap R$ is divisible. Since R is reduced, this implies that $D^* \cap R = E$. Hence $D^* = D$.

5.2.5. *If G is an Abelian torsion-free group, n a nonzero integer, $x \in G$, $y \in G$, and $nx = ny$, then $x = y$.*

Proof. For $n(x - y) = 0$, and, since G is torsion-free, $x - y = 0$, $x = y$.

5.2.6. *If G is a torsion-free, divisible Abelian group and $x \in G^{\#}$, then there is a unique isomorphism T of $\mathscr{R}$ into G such that $1T = x$.*

Proof. Let $m \in J$ and $n \in J^{\#}$. By Theorem 5.2.5, $\exists \mid y \in G$ such that $ny = mx$. If, now, the required isomorphism T exists and $(m/n)T = u$, then

$$ny = mx = m(1T) = mT = (n(m/n))T = n((m/n)T) = nu,$$

so $(m/n)T = u = y$. Hence there is at most one isomorphism of the required type.

Suppose that $m'/n' = m/n$. Then $mn' = m'n$. Therefore $mn'y = m'ny = m'mx$, so that, if $m \neq 0$, $n'y = m'x$. If, on the other hand, $m = 0$, then $ny = 0x = 0$, $y = 0$, and $m' = 0$, so that $n'y = m'x$ again. Thus a function T may be defined from $\mathscr{R}$ into G as follows:

if $ny = mx$, then $(m/n)T = y$.

Now suppose that $ny = mx$ and $sz = rx$. Then

$$ns(y + z) = smx + nrx = (ms + nr)x.$$

Since $(m/n) + (r/s) = (ms + nr)/ns$, this shows that

$$((m/n) + (r/s))T = (m/n)T + (r/s)T.$$

Hence T is a homomorphism. If $(m/n)T = 0$, then $n0 = mx$, hence $m = 0$ (G is torsion-free). Therefore T is an isomorphism of $\mathscr{R}$ into G. Setting $m = n = 1$, we see that $1T = x$.

5.2.7. *A torsion-free divisible Abelian group is the direct sum of groups each isomorphic to the rationals.*

Proof. Let G be a torsion-free divisible Abelian group. By Theorem 4.4.1, $\exists$ a maximal subset S of subgroups of G each isomorphic to $\mathscr{R}$ such that $M = \Sigma \{H \mid H \in S\}$ exists. If $\exists x \in G \backslash M$, then, by Theorem 5.2.6, there is an isomorphism T of $\mathscr{R}$ into G such that $1T = x$. If $m \in J$, $n \in J^{\#}$, and $(m/n)T \in M$, then $mT = n((m/n)T) \in M$. Since M is divisible (Exercise 5.2.20) it follows that $mx = m(1T) = mT = my$, with $y \in M$. By Theorem 5.2.5, $m = 0$. Thus $(\mathscr{R}T) \cap M = E$, so $M + \mathscr{R}T$ exists, contradicting the maximality of S. Therefore $M = G$, proving the theorem.

5.2.8. *If G is an Abelian p-group and H a cyclic subgroup of maximum order, then H is a direct summand of G.*

Proof. (Note the similarity of this proof to that of Theorem 5.2.2.) By Zorn, $\exists$ a maximal subgroup K such that $H \cap K = E$. Suppose that $G \neq H + K$.

CASE 1. $K = E$. Now $\exists x \in G\backslash H$ such that $px \in H$. Let $o(H) = p^k$. Since $o(x) \leq p^k$, $o(px) < p^k$, so $\exists y \in H$ such that $px = py$. Then $p(x - y) = 0$ and $x - y \notin H$. Hence $\langle x - y \rangle \cap H = E$ and $\langle x - y \rangle \neq E$, a contradiction.

CASE 2. $K > E$. Then $(H + K)/K \cong H$, so $(H + K)/K$ is a cyclic subgroup of maximum order in G/K, and $(H + K)/K < G/K$. By the impossibility of Case 1, $\exists L > K$ such that $(H + K)/K \cap L/K = E$. Hence

$$L \cap H \subset (L \cap (H + K)) \cap H \subset K \cap H = E.$$

This contradicts the maximality of K.
Hence $G = H + K$.

5.2.9. *If G is a nonzero Abelian p-group such that $pG = G$, then there is a p^∞-subgroup of G.*

Proof. $\exists x_1 \in G^\#$ such that $px_1 = 0$. Inductively, $\exists x_i$ such that $px_i = x_{i-1}$. Then (inductively), $o(x_i) = p^i$. There is a function T such that $(a/p^i)T = ax_i$; for if $a/p^i = b/p^{i+r}$, then $b = ap^r$ and $bx_{i+r} = ap^r x_{i+r} = ax_i$. Let S be the group of all rationals which, in reduced form, have denominator a power of p. Then T is a function from S into G. Any two elements of S may be represented by fractions with equal denominators, and

$$\left(\frac{a}{p^i} + \frac{b}{p^i}\right)T = \left(\frac{a+b}{p^i}\right)T = (a + b)x_i = ax_i + bx_i$$

$$= \left(\frac{a}{p^i}\right)T + \left(\frac{b}{p^i}\right)T.$$

Hence T is a homomorphism. Now $(a/p^i)T = 0$ iff $ax_i = 0$, hence iff $p^i \mid a$, i.e., iff $a/p^i \in J$. Thus T induces an isomorphism of S/J into G. But S/J is a p^∞-group.

5.2.10. (*Kulikov* [1].) *If G is an Abelian group which is not torsion-free, then G has a direct summand which is either cyclic of prime power order or a p^∞-group for some $p \in \mathscr{P}$. Hence G is indecomposable iff it is cyclic of prime power order or a p^∞-group.*

Proof. If G is cyclic of prime power order, then G is indecomposable by the uniqueness half of the fundamental theorem of finite Abelian groups

(Theorem 5.1.17). If G is a p^∞-group, then it has only one subgroup of order p, whereas if $G = H + K$ with $H \neq E$ and $K \neq E$, then both H and K would have a subgroup of order p, and G would have at least two. Hence G is indecomposable in this case also. (For hints of other possible arguments in the p^∞ case, see Exercises 5.2.22 and 5.2.23.)

Now suppose that G is neither a cyclic group of prime power order nor a p^∞-group for some p. Since G is not torsion-free, the torsion subgroup $T \neq E$. Hence $\exists\, p \in \mathscr{P}$ such that $T_p \neq E$. If $pT_p = T_p$, then by Theorem 5.2.9 there is a p^∞-subgroup H of T_p, hence of G. Since H is divisible, it is a proper direct summand of G (Theorem 5.2.2). Therefore $WLOG$, $pT_p < T_p$. Let x be an element of smallest order in $T_p \backslash pT_p$. Let $o(x) = p^k$. If $\exists\, y \in G$ such that $p^{k-1}x = p^k y$, then $p^{k-1}(x - py) = 0$, and $x - py \in T_p \backslash pT_p$. This contradicts the minimality of $o(x)$. Hence $p^{k-1}x \notin p^k G$, so $\langle x \rangle \cap p^k G = E$. Therefore $G/p^k G$ is an Abelian p-group with $\langle x + p^k G \rangle$ as cyclic subgroup of maximum order p^k. By Theorem 5.2.8, $\exists\, H/p^k G$ such that

$$G/p^k G = \langle x + p^k G \rangle + (H/p^k G).$$

Hence $\langle x, H \rangle = G$, and

$$\langle x \rangle \cap H \subset \langle x \rangle \cap ((\langle x \rangle + p^k G) \cap H) \subset \langle x \rangle \cap p^k G = E.$$

Therefore $G = \langle x \rangle + H$.

5.2.11. *If G is a divisible Abelian p-group, then G is the direct sum of p^∞-groups.*

Proof. Let S be the set of p^∞-subgroups of G. By Theorem 4.4.1, there is a maximal subset T of S such that $M = \Sigma \{H \mid H \in T\}$ exists. Since M is divisible (5.2.20), $\exists\, K$ such that $G = M + K$. If $K \neq E$, then $K = L + Q$, where L is a p^∞-group or a finite cyclic group, by Theorem 5.2.10. But L cannot be a p^∞-group by the maximality of T. If L is a finite cyclic group, then the group $G = M + L + Q$ is divisible while L is not divisible, contradicting Exercise 5.2.20. Hence $K = E$, and G is the direct sum of p^∞-groups. $\|$

It is now possible to characterize divisible Abelian groups.

5.2.12. (*Decomposition of divisible Abelian groups.*) *If G is a divisible Abelian group, then $G = \Sigma \{H \mid H \in S\}$, where each summand is a p^∞-group for some prime p or is isomorphic to $\mathscr{R}$.*

Proof. The torsion subgroup T of G is divisible. Hence, by Theorems 5.2.2 and 5.1.2, and Exercise 5.2.20, $G = T + U$, where U is torsion-free and divisible. By Theorem 5.2.7, U is the direct sum of groups each isomorphic to $\mathscr{R}$. By Theorem 5.1.6, $T = \Sigma\, G_p$, and each G_p is divisible by Exercise 5.2.20. By Theorem 5.2.11, each G_p is the direct sum of p^∞-groups. $\|$

The question of uniqueness of this decomposition arises. The direct summands themselves are not unique, but the number of summands of various types is determined by G. The proof of this fact occupies the remainder of this section.

5.2.13. *Let $p \in \mathscr{P}$ and let*

$$G = \Sigma \{H \mid H \in S\} = \Sigma \{K \mid K \in T\},$$

where each $H \in S$ and each $K \in T$ is a p^∞-group. Then $o(S) = o(T)$.

Proof. Let $P = \{x \in G \mid px = 0\}$. Then $P \cap H$ is of order p for each $H \in S$, and

$$P = \Sigma \{P \cap H \mid H \in S\} = \Sigma \{P \cap K \mid K \in T\}.$$

By Theorem 5.1.14, $o(S) = o(T)$.

5.2.14. *Let*

$$G = \Sigma \{H \mid H \in S\} = \Sigma \{K \mid K \in T\},$$

where each $H \in S$ and each $K \in T$ is isomorphic to $\mathscr{R}$. Then $o(S) = o(T)$.

Proof. If $o(S) = 1$, then G contains no nontrivial divisible subgroup, hence $o(T) = 1$ also. Now suppose that $o(S)$ is finite and use induction. Let $H_1 \in S$ and let $S^* = S \backslash \{H_1\}$. By Theorem 4.4.1, there is a maximal subset T^* of T such that $M = H_1 + \Sigma \{K \mid K \in T^*\}$ exists. Since (Exercise 5.2.24) the intersection $M \cap K$ is divisible for each $K \in T$, $M \cap K = K$ if $K \in T \backslash T^*$ (for otherwise (see Exercise 5.2.16) T^* is not maximal). Hence

$$M \supset \Sigma \{K \mid K \in T\} = G,$$

so $M = G$. Thus

$$G/H_1 \cong \Sigma \{K \mid H \in S^*\} \cong \Sigma \{K \mid K \in T^*\}.$$

Therefore, by induction, $o(S^*) = o(T^*)$. Now

$$H_1 \cong G/\Sigma \{K \mid K \in T^*\} \cong \Sigma \{K \mid K \in T \backslash T^*\}.$$

By the first case discussed, $o(T \backslash T^*) = 1$. Hence

$$o(S) = o(S^*) + 1 = o(T^*) + 1 = o(T).$$

Finally, suppose that $o(S)$ is infinite. Then by symmetry and the last paragraph, $o(T)$ is infinite also. A computation similar to that in the proof of Theorem 5.1.14 shows that $o(S) = o(G) = o(T)$.

5.2.15. *If*

$$G = \sum \{H \mid H \in S\} = \sum \{K \mid K \in T\},$$

where each $H \in S$ and each $K \in T$ is either a p^∞-group for some prime p or is isomorphic to $\mathcal{R}$, then

(i) *for each $p \in \mathcal{P}$, the number of p^∞-groups in S equals the number in T,*

(ii) *the number of $H \in S$ is isomorphic to $\mathcal{R}$ equals the number of $K \in T$ isomorphic to $\mathcal{R}$.*

   Proof. We have

$$G_p = \sum \{H \mid H \in S \text{ and } H \text{ is a } p^\infty\text{-group}\}$$
$$= \sum \{K \mid K \in T \text{ and } K \text{ is a } p^\infty\text{-group}\}.$$

By Theorem 5.2.13, statement (i) follows. If M is the torsion subgroup of G, then $M = \sum G_p$, hence

$$\frac{G}{M} \cong \sum \{H \mid H \in S \text{ and } H \cong \mathcal{R}\} \cong \sum \{K \mid K \in T \text{ and } K \cong \mathcal{R}\}.$$

By Theorem 5.2.14, statement (ii) follows.

EXERCISES

5.2.16. Prove that $\mathcal{R}$ has no nontrivial divisible subgroups.

5.2.17. Show that if G is any group and H a subgroup, then $\exists$ a maximal subgroup K of G such that $H \cap K = E$.

5.2.18. Let $G = A + B$, where A is a p^∞-group and B cyclic of order p ($p \in \mathcal{P}$). Prove that there is a subgroup $H \neq B$ such that $G = A + H$. (This proves the nonuniqueness of the reduced summand in a divisible-reduced decomposition of G.)

5.2.19. A homomorphic image of a divisible group is divisible.

5.2.20. A direct sum of groups is divisible iff each summand is divisible.

5.2.21. Prove that a p^∞-group is divisible.

5.2.22. The lattice of subgroups of a p^∞-group is a chain.

5.2.23. Every proper subgroup of a p^∞-group is finite cyclic.

5.2.24. If G is a torsion-free Abelian group and H is a divisible subgroup for each $H \in S$, then $\cap \{H \mid H \in S\}$ is divisible.

5.2.25. $\pi\{H \mid H \in S\}$ is divisible iff each $H \in S$ is divisible.

5.2.26. (a) Improve Theorem 5.2.2 to read: if G is Abelian, D a divisible subgroup, $H \subset G$, and $D \cap H = E$, then $G = D + M$ where $H \subset M$.

(b) The following theorem is false. If G is Abelian and $H \subset G$, then $G = D + R$ and $H = D' + R'$, where D and D' are divisible, R and R' reduced, $D' \subset D$, and $R' \subset R$.

5.2.27. (a) If G is an Abelian group which is not divisible, then G has a subgroup of prime index.

(b) If G is an infinite divisible Abelian group, then there is no subgroup of finite index, but there is a subgroup of index $\aleph_0$.

5.2.28. If G is an infinite Abelian group such that all proper subgroups have smaller order, then G is a p^∞-group, and conversely. (See also Exercise 3.3.16.)

5.3   Free Abelian groups

A *free* Abelian group is a group which is the direct sum of infinite cyclic groups. A *basis* of a free Abelian group F is a subset B of $F^\#$ such that $F = \Sigma \{\langle x \rangle \mid x \in B\}$.

Let S be a subset of an Abelian group G. A formal sum $\Sigma \{n_x x \mid x \in S\}$, where $n_x \in J$ and all but a finite number of n_x are equal to 0, has an obvious meaning as an element of G (if S is empty, the sum is interpreted to be 0). Formal sums add and are acted on by homomorphisms in the same way that finite sums are.

5.3.1.   *If B is a basis of a free Abelian group F, G an Abelian group, and T a function from B into G, then $\exists \mid U \in \mathrm{Hom}(F, G)$ such that $(U \mid B) = T$.*

Proof. Any $y \in F$ can be written uniquely as a formal sum $\Sigma \{n_x x \mid x \in B\}$, where $n_x \in J$ and all but a finite number are 0. If U exists, it must be given by the rule:

$$(\Sigma \, n_x x)U = \Sigma \, n_x(xT).$$

Conversely, let U be defined by this rule. Then U is a function from F into G such that $(U \mid B) = T$. If $y \in F$ and $z \in F$, then with $y = \Sigma \, n_x x$ and $z = \Sigma \, m_x x$,

$$(y + z)U = (\Sigma \, (n_x + m_x)x)U = \Sigma \, (n_x + m_x)(xT)$$
$$= \Sigma \, n_x(xT) + \Sigma \, m_x(xT) = yU + zU.$$

Hence U is a homomorphism from F into G. ‖

The converse of Theorem 5.3.1 is also true.

5.3.2. Let F be an Abelian group and B a subset of F such that if G is an Abelian group and T is a function from B into G, then $\exists \mid U \in \text{Hom}(F, G)$ such that $U \mid B = T$. Then F is free Abelian with basis B.

Proof. First suppose that B does not generate F, i.e., that $\langle B \rangle = H < F$. By assumption, $\exists U \in \text{Hom}(F, H)$ such that $U \mid B = I_B$. Hence I_B has two distinct extensions to homomorphisms from F into F, namely U and I_F. This contradicts the hypotheses of the theorem. Therefore $\langle B \rangle = F$.

Let G be the group of finitely nonzero functions from B into J. If $b \in B$ and $b' \in B$, let

$$b'(bT) = \begin{cases} 1 & \text{if} \quad b = b', \\ 0 & \text{if} \quad b \neq b'. \end{cases}$$

Then it is readily verified that G is free Abelian with basis BT. By assumption, $\exists U \in \text{Hom}(F, G)$ such that $U \mid B = T$. If F is not free, then, since $\langle B \rangle = F$, $\exists$ distinct elements $b_1, \ldots, b_n$ in B and nonzero integers $i_1, \ldots, i_n$ such that $\Sigma i_j b_j = 0$. But

$$b_1((\Sigma i_j b_j)U) = b_1(\Sigma i_j(b_j U)) = \Sigma i_j(b_1(b_j U))$$
$$= \Sigma i_j(b_1(b_j T)) = i_1 \neq 0.$$

Hence $0U = (\Sigma i_j b_j)U \neq 0$, a contradiction. Hence F is free with basis B.

5.3.3. Any Abelian group is a homomorphic image of a free Abelian group.

Proof. Let G be an Abelian group. Let F be a free Abelian group with basis B such that $o(B) = o(G)$. (It is clear from the proof of Theorem 5.3.2 that there is such a free Abelian group F.) Then $\exists$ a 1–1 function T from B onto G. By Theorem 5.3.1, $\exists$ a homomorphism U of F onto G. $\parallel$

A set S is *well ordered* iff it is ordered and every nonempty subset T of S has a smallest element. Although it will not be proved here, Zorn's lemma is equivalent to the statement "any set S can be well-ordered." This fact will be used in the proof of the following theorem.

5.3.4. A subgroup of a free Abelian group is a free Abelian group.

Proof. Let B be a basis of the free Abelian group G, and let $H \subset G$. Since E is a free Abelian group with empty basis, it may be assumed that $H \neq E$. Well-order B. If $h \in H^{\#}$, let $f(h)$ be the largest element of B which has a nonzero coefficient in the canonical expression for h. For each $b \in B$ such that $\exists h \in H$ with $f(h) = b$, let h_b be an element of H with $f(h_b) = b$ and having the smallest possible positive coefficient of b in its canonical decomposition. Note that if $f(h) = b$, but h has negative coefficient of b, then $f(-h) = b$, and $-h$ has positive coefficient of b. Let S be the set of all h_b thus defined. We assert that S is a basis for H.

(a) If $h_{b_1}, \ldots, h_{b_n}$ are distinct elements of S, $c_1, \ldots, c_n$ are nonzero integers, and $b_1 > b_2 > \ldots > b_n$, then when $h = c_1 h_{b_1} + \ldots + c_n h_{b_n}$ is expressed in terms of B, its b_1 coefficient is not 0, so $h \neq 0$. Therefore, the direct sum $K = \Sigma \langle h_b \rangle$ exists.

(b) Suppose that $K \neq H$. Then the set T of $b \in B$ such that $\exists\, h \in H \backslash K$ with $f(h) = b$ is not empty. Since B is well-ordered, T has a first element b', But then, $h_{b'} \in K$ is defined, and $\exists\, h \in H \backslash K$ with $f(h) = b'$. We then have

$$h = c_1 b' + \ldots, \qquad h_{b'} = c_2 b' + \ldots,$$

where the omitted terms involve elements of B preceding b'. Now $\exists$ integers q and r such that $c_1 = q c_2 + r, 0 \leq r < c_2$. Then $h - q h_{b'}$ is in H, involves no terms beyond b', and has r as coefficient of b'. By the definition of $h_{b'}$, $r = 0$. Therefore, $f(h - q h_{b'}) < b'$, so, by the definition of S, $h - q h_{b'} \in K$. But this implies that $h \in K$, a contradiction. Hence $K = H$. Therefore H is a free Abelian group.

5.3.5. *If G is a free Abelian group, $H \subset G$, and B and D are bases of G and H respectively, then $o(B) \geq o(D)$.*

Proof. First suppose that B is finite, say $B = \{x_1, \ldots, x_n\}$. If

$$o(D) > o(B),$$

then $\exists$ distinct elements $y_1, \ldots, y_{n+1}$ in D. Since $H \subset G$, $\exists\, a_{ij} \in J$ such that $y_i = \Sigma\, a_{ij} x_j$. In the group $\Sigma_E \{ \mathcal{R} \mid 1 \leq j \leq n \}$, let $z_i = (a_{i1}, \ldots, a_{in})$, $i = 1, \ldots, n + 1$. If $r \in \mathcal{R}$, then $r z_i$ has an obvious meaning. If the z_i are $\mathcal{R}$-independent, i.e., if $\Sigma\, r_i z_i = 0, r_i \in \mathcal{R}$, implies all $r_i = 0$, then the theorem 5.2.14, asserting the invariance of the number of rational summands of a divisible group would be violated. Hence there are rationals r_i not all 0 such that $\Sigma\, r_i z_i = 0$. Clearing fractions, $\exists\, c_i \in J$ such that $\Sigma\, c_i z_i = 0$, and not all c_i are 0. But then

$$\Sigma\, c_i y_i = \Sigma_i \left(c_i \Sigma_j a_{ij} x_j \right) = \Sigma_j \left(\Sigma_i c_i a_{ij} \right) x_j = 0,$$

contradicting the fact that D is a basis of H. Hence $o(D) \leq o(B)$ in case $o(B)$ is finite.

Next suppose that B is infinite and $o(D) > o(B)$. Then

$$o(G) = \Sigma\, \aleph_0^n o(B)^n = o(B) < o(D) = o(H),$$

a contradiction. Hence $o(D) \leq o(B)$ in this case also.

5.3.6. (*Invariance of rank.*) *If G is a free Abelian group with bases B and B', then $o(B) = o(B')$.*

Proof. Use 5.3.5 with $H = G$, and symmetry. ‖

This theorem permits the following definition. The *rank* of a free Abelian group G is $o(B)$ where B is any basis of G.

Theorem 5.3.3 has the following dual.

5.3.7. *Any Abelian group is a subgroup of a divisible Abelian group.*

Proof. Let G be an Abelian group. By Theorem 5.3.3, there is a free Abelian group F such that $G \cong F/H$ for some H. If B is a basis of F, then $F \cong \Sigma_E \{J \mid b \in B\}$. Hence $G \cong \Sigma_E \{J \mid b \in B\}/K$ for some K. Since $J \subset \mathscr{R}$,

$$\Sigma_E \{J \mid b \in B\} \subset \Sigma_E \{\mathscr{R} \mid b \in B\},$$

which is divisible and equals D, say. Hence $G \cong D/K$, which is divisible by Exercise 5.2.19. By Exercise 2.1.36, G is a subgroup of some divisible group.

EXERCISES

5.3.8. Let $F = J \times J \times J$, let x_i, y_i, and z_i be integers, $i = 1, 2, 3$, and let

$B = \{(x_1, y_1, z_1), (x_2, y_2, z_2), (x_3, y_3, z_3)\}$.

State a necessary and sufficient condition that B be a basis of F in terms of $x_1, \ldots, z_3$. Generalize.

5.3.9. (See Fuchs [1, p. 45].) A more conceptual proof of Theorem 5.3.4 will be sketched.

(a)   If S is a well-ordered set, $x \in S$, and x is not the last element, then $\exists\, y > x$ such that if $z > x$, then $y \leqq z$. Call y the successor of x.

(b)   A *well-ordered ascending normal series* of a group G is a set S of subgroups, well-ordered by inclusion, starting with E, ending with G, such that $A \lhd B$ if B is the successor of A, and such that

$A = \cup \{H \mid H \in S, H < A\}$

if $A \neq E$ and A is not the successor of any element of S (A is a *limit* element of S). A *factor* of S is a group B/A where B is the successor of A.

(c)   An Abelian group G is free iff it has a well-ordered, ascending (normal) series with all factors infinite cyclic.

(d)   A subgroup H of a free Abelian group G is free Abelian. [Use (c) and Theorem 2.5.9.]

5.4   Finitely generated Abelian groups

In this section it will be shown that any finitely generated Abelian group is the direct sum of a finite number of cyclic groups, and the appropriate uniqueness theorem will be proved.

5.4.1. *If G is a torsion-free Abelian group, $[G:H]$ is finite, and H is free Abelian, then G is free Abelian, and* $\text{rank}(G) = \text{rank}(H)$.

Proof. Let $[G:H] = n$ and $xT = nx$ for $x \in G$. Then T is an isomorphism of G into H, and GT is free Abelian, since it is a subgroup of a free Abelian group (Theorem 5.3.4). Hence G is free Abelian also. Moreover, by Theorem 5.3.5,

$$\text{rank}(G) = \text{rank}(GT) \leq \text{rank}(H) \leq \text{rank}(G),$$

hence $\text{rank}(G) = \text{rank}(H)$. ‖

A group G is *finitely generated* iff ∃ a finite generating subset S: $G = \langle S \rangle$. The structure of finitely generated (even 2-generated) non-Abelian groups can be very complicated. For Abelian groups, the structure is completely determined by the following theorem.

5.4.2. *An Abelian group is finitely generated iff it is the direct sum of a finite number of cyclic groups.*

Proof. It is trivial that a direct sum of a finite number of cyclic groups is finitely generated.

Conversely, let G be Abelian and $G = \langle S \rangle$ where S is finite. If $o(S) = 1$, then G itself is cyclic and the theorem is true. Induct on $o(S)$. Let $S = S^* \cup \{y\}$. By the inductive hypothesis $H = \langle S^* \rangle$ is the direct sum of a finite number of cyclic groups. If $y \in H$, then $H = G$, and we are done. If $\langle y \rangle \cap H = E$, then $G = H + \langle y \rangle$ is of the desired type. If $\langle y \rangle \cap H \neq E$, then it is sufficient to consider the case where $py \in H$ with $p \in \mathscr{P}$ (because one can induct). Hence $[G:H] = p$. Now the decomposition of H gives $H = T + F$, where F is torsion-free, and T is the torsion subgroup of H and is the direct sum of the finite cyclic subgroups in the given decomposition of H, hence is finite. Let U be the torsion subgroup of G. If $U > T$, then $U \cap F = E$, $U + F > T + F$, so $G = U + F$. Moreover $[U:T] = p$, so U is finite; hence, by the fundamental theorem of finite Abelian groups, U is the direct sum of cyclic groups, so G has the desired form. Finally suppose that $U = T$. Then $[G/T:H/T] = p$ and $H/T \cong F$ is free Abelian of finite rank. By Theorem 5.4.1, G/T is free Abelian of finite rank. Let $(x_1 + T, \ldots, x_n + T)$ be a basis of G/T, $x_1 \in G$. Then $K = \langle x_1, \ldots, x_n \rangle$ is free Abelian, and $G = T + K$ is a direct sum of a finite number of cyclic groups. ‖

Such uniqueness as there is about the decomposition of a finitely generated Abelian group will follow from the next theorem.

5.4.3. *Let*

$$G = \sum \{\sum \{H \mid H \in S_n\} \mid n \in \mathscr{P}^\infty \text{ or } n = \infty\}$$
$$= \sum \{\sum \{K \mid K \in T_n\} \mid n \in \mathscr{P}^\infty \text{ or } n = \infty\}$$

where each $H \in S_n$ and each $K \in T_n$ is cyclic of order n, then $o(S_n) = o(T_n)$ for all n.

Proof. Let T be the torsion subgroup of G, and, for each $p \in \mathscr{P}$, T_p the p-component of G. Then

$$T_p = \sum \{\sum \{H \mid H \in S_{p^i}\} \mid i \in \mathscr{N}\}$$
$$= \sum \{\sum \{K \mid K \in T_{p^i}\} \mid i \in \mathscr{N}\}.$$

By Theorem 5.1.15, $o(S_{p^i}) = o(T_{p^i})$ for all $i \in \mathscr{N}$ and all $p \in \mathscr{P}$. Factoring out T yields

$$G/T \cong \sum \{H \mid H \in S_\infty\} \cong \sum \{K \mid K \in T_\infty\},$$

i.e., G/T is free Abelian of rank $o(S_\infty)$ and of rank $o(T_\infty)$. By the invariance of rank (Theorem 5.3.6), $o(S_\infty) = o(T_\infty)$.

5.4.4. *If G is a finitely generated Abelian group, then $G = H_1 + \ldots + H_n$, where each H_i is cyclic of prime power or infinite order. If also $G = K_1 + \ldots + K_m$, where each K_j is cyclic of prime power or infinite order, then $m = n$ and, after rearrangement, $o(H_i) = o(K_i)$.*

EXERCISES

5.4.5. There are only a countable number of isomorphism classes of finitely generated Abelian groups.

5.4.6. Let $G = J_2 \times J$. Show that there is more than one direct decomposition of G into cyclic subgroups. (This proves the nonuniqueness of the torsion-free summand in the decomposition theorems of this section.) Find an element x of infinite order such that $x = 2y = 2z$ with $y \neq z$.

5.4.7. If G is a finitely generated Abelian group, then

(a) $\text{Aut}(G)$ is finite iff there is at most one summand isomorphic to J in a cyclic decomposition of G;

(b) $\text{Aut}(G)$ is countable in any case.

5.4.8. If G is a finitely generated Abelian group, then its torsion subgroup is a direct summand.

5.5   Direct sums of cyclic groups

The principal theorem in this section states that a subgroup of a direct sum of cyclic groups is again a direct sum of cyclic groups. This theorem is due to Kulikov [2], and requires a somewhat more complicated proof than the other decomposition theorems of this chapter.

Let G be an Abelian p-group, $p \in \mathscr{P}$. Let $x \in G$ and define the *height* of x to be

$$Ht(x) = \begin{cases} n & \text{if } x \in p^n G \backslash p^{n+1} G, \\ \infty & \text{if } x \in \cap p^n G. \end{cases}$$

If $H \subset G$, let $Ht_G(H) = \max\{Ht_G(x) \mid x \in H^{\#}\}$, or ∞ if there is no finite maximum.

5.5.1. *An Abelian p-group G is a direct sum of cyclic groups iff $\exists$ a sequence $\{H_n \mid n \in \mathscr{N}\}$ of subgroups such that* (i) $H_n \subset H_{n+1}$ *for all n,* (ii) $G = \cup H_n$, *and* (iii) $Ht_G(H_n) < \infty$ *for all n.*

Proof. If $G = \Sigma \{A \mid A \in S\}$ where each A is cyclic, let

$$H_n = \Sigma \{A \in S \mid o(A) \leq p^n\}.$$

Then (i) and (ii) are obvious, while $Ht_G(H_n) \leq n - 1$ is nearly so (see Exercise 5.5.13).

Now suppose that G satisfies (i), (ii), and (iii). Let $i_n = Ht_G(H_n)$, $P = \{x \in G \mid px = 0\}$, and $R_{nj} = P \cap H_n \cap p^j G$, $0 \leq j \leq i_n$. Order the set of subscripts (n, j) by the rule

$$(m, j) < (n, k) \quad \text{iff} \quad \begin{cases} m < n, & \text{or} \\ m = n & \text{and} \quad j > k. \end{cases}$$

Let $S_{nj} = \langle R_{mk} \mid (m, k) \leq (n, j) \rangle$. Since $S_{n0} \supset R_{n0} = P \cap H_n$, $\langle S_{nj} \rangle = P$ by (ii). Since P is elementary Abelian, by Exercise 5.1.28 for each (n, j) there is a subset B_{nj} of R_{nj} such that

$$S_{nj} = \Sigma \{\Sigma \{\langle b \rangle \mid b \in B_{mk}\} \mid (m, k) \leq (n, j)\}.$$

Consequently, if $B = \cup B_{nj}$, then $P = \Sigma \{\langle b \rangle \mid b \in B\}$. If $x \in B$, then $\exists y \in G$ such that $p^r y = x$ where $r = Ht(x)$. Let B^* be the set of y so obtained (one for each $x \in B$). If $y_1, \ldots, y_n$ are distinct elements of B^*, $a_h \in J$, $\Sigma a_h y_h = 0$, and all $a_h y_h \neq 0$, then, multiplying by a suitable power of p, one has

$$a_{h_1} x_1 + \ldots + a_{h_s} x_s = 0, \qquad x_j \in B,$$

with some $a_{h_i} x_i \neq 0$ and with distinct x_j, a contradiction. Hence $K = \Sigma \{\langle y \rangle \mid y \in B^*\}$ exists.

If $K = G$, then the theorem is true. Suppose $K \neq G$. For any $h \in G \backslash K$, $p^t h \in P$ for some t, hence $p^t h \in S_{nj}$ for some (n, j). Let $g \in G \backslash K$ be such that $p^t g \in S_{nj}$ for some t and minimum (n, j). Now

$$p^t g = u_1 b_1 + \ldots + u_r b_r, \qquad u_h \in J, \qquad b_h \in B.$$

Further assume that g is such that r is a minimum (with (n, j) unchanged). By the decomposition of S_{nj} and the minimality of (n, j), all $b_h \in B_{mk}$ with $(m, k) \leq (n, j)$, and some b_h, say b_1, is in B_{nj}. Since $p^t g \in P \cap p^t G$ and $p^t g \in S_{n0} = H_n \cap P$, we have

$$p^t g \in P \cap p^t G \cap H_n = R_{nt} \subset S_{nt}.$$

Therefore $j \geq t$. Since $b_1 \in B_{nj}$, b_1 is in R_{nj} but not in any R_{nk} with $k > j$. Hence $b_1 = p^j y$ with $y \in B^*$. Therefore $b_1 = p^t z$, where $z \in K$. But then,

$$p^t(g - u_1 z) = u_2 b_2 + \ldots + u_r b_r \in S_{nj},$$

and $g - u_1 z \in G \backslash K$, contradicting the minimality of either (n, j) or r.

5.5.2. *A subgroup of a direct sum of cyclic groups is a direct sum of cyclic groups.*

Proof. Let G be a direct sum of cyclic groups. Each finite summand is a direct sum of cyclic groups of prime power order by the fundamental theorem of finite Abelian groups. Hence (associativity of direct sums), G is the direct sum of cyclic groups of prime power or infinite order.

Let $H \subset G$, and let G and H have torsion subgroups T and U, respectively. Then, for each $p \in \mathscr{P}$, $U_p \subset T_p$, and T_p is the direct sum of cyclic groups, namely the direct sum of the summands of G whose orders are powers of p. Fix p for the moment. By Theorem 5.5.1, $\exists \{G_n \mid n \in \mathscr{N}\}$ such that

(i) $G_n \subset G_{n+1}$,

(ii) $T_p = \cup G_n$,

and

(iii) $Ht_{T_p}(G_n) < \infty$.

Let $H_n = G_n \cap H$. Then

(i) $H_n \subset H_{n+1}$,

(ii) $\cup H_n = (\cup G_n) \cap H = T_p \cap H = U_p$,

and

(iii) $Ht_{U_p}(H_n) \leq Ht_{T_p}(H_n) \leq Ht_{T_p}(G_n) < \infty$.

By Theorem 5.5.1 again, U_p is a direct sum of cyclic groups for all p. Hence (Theorem 5.1.6) U is a direct sum of cyclic groups.

From the direct decomposition of G we see that $G = T + F$, where F is a direct sum of infinite cyclic groups; i.e., F is free Abelian. Also, by the isomorphism theorem,

$$\frac{H}{U} = \frac{H}{T \cap H} \cong \frac{HT}{T} \subset \frac{G}{T} \cong F,$$

so that, by Theorem 5.3.4, H/U is free Abelian. Hence there is a subset B of H such that

$$\frac{H}{U} = \Sigma \{\langle b + U \rangle \mid b \in B\}.$$

Then $W = \Sigma \{\langle b \rangle \mid b \in B\}$ exists, $U \cap W = E$, and $H = \langle U, W \rangle$. Therefore, $H = U + W$, and H is a direct sum of cyclic groups. ‖

Let G be a direct sum of cyclic groups and $H \subset G$. We wish to study the relation between the cyclic decomposition of G and that of H. Instead of deriving one complicated over-all theorem, let us note that the last paragraph of the proof of the preceding theorem shows that the torsion-free part of H is isomorphic to a subgroup of the torsion-free part of G. Therefore, Theorem 5.3.5 yields the following theorem.

5.5.3. *If G is a direct sum of cyclic groups and $H \subset G$, then the number of infinite summands in a cyclic decomposition of H is less than or equal to the corresponding number for G.* ‖

This theorem plus the standard technique reduces the problem to one for p-groups. Here the result is as follows.

5.5.4. *If $p \in \mathscr{P}$,*

$$G = \Sigma \{\Sigma \{A \mid A \in S_n\} \mid n \in \mathscr{N}\},$$

$H \subset G$, and

$$H = \Sigma \{\Sigma \{B \mid B \in T_n\} \mid n \in \mathscr{N}\},$$

where each $A \in S_n$ and $B \in T_n$ is cyclic of order p^n, then for each $i \geq 0$,

$$\Sigma \{o(T_n) \mid n > i\} \leq \Sigma \{o(S_n) \mid n > i\}.$$

Proof. Let $P = \{x \in G \mid px = 0\}$. Then for each i,

$$P \cap p^i G = \Sigma \{\Sigma \{p^{n-1}A \mid A \in S_n\} \mid n > i\},$$
$$P \cap p^i H = \Sigma \{\Sigma \{p^{n-1}B \mid B \in T_n\} \mid n > i\},$$

and $P \cap p^i G \supset P \cap p^i H$. Since $P \cap p^i G$ is an elementary Abelian p-group, it follows from Theorem 5.1.14 that

$$\Sigma \{o(T_n) \mid n > i\} \leq \Sigma \{o(S_n) \mid n > i\}. ‖$$

The converse of this theorem is also true, as we shall prove. The main part of the proof is contained in the following set-theoretical lemma.

5.5.5. *If* $U = \cup \{U_n \mid n \in \mathcal{N}\}$, $S = \cup \{S_n \mid n \in \mathcal{N}\}$, *and*

$$\sum \{o(U_n) \mid n > i\} \leq \sum \{o(S_n) \mid n > i\}$$

for all $i \geq 0$, *then* $\exists$ *a* 1–1 *function* f *from* U *into* S *such that if* $u \in U_n$ *and* $f(u) \in S_m$, *then* $m \geq n$.

Proof. CASE 1. S is finite. Then U is finite, and

$$S = \{x_i \mid 1 \leq i \leq r\}, \quad x_i \in S_{n_i}, \quad n_1 \geq n_2 \geq \ldots \geq n_r,$$
$$U = \{y_i \mid 1 \leq i \leq s\}, \quad y_i \in U_{m_i}, \quad m_1 \geq m_2 \geq \ldots \geq m_s.$$

Since

$$o(U) = \sum o(U_i) \leq \sum o(S_i) = o(S),$$

$s \leq r$. Hence the function f such that $f(y_i) = x_i$ is 1–1 from U into S, and we need only verify that $m_i \leq n_i$ for all i. Suppose, on the contrary, that some $m_i > n_i$. Then

$$\sum \{o(U_j) \mid j > n_i\} \geq i, \quad \sum \{o(S_j) \mid j > n_i\} < i,$$

a contradiction.

CASE 2. S is infinite, and each $o(S_n) < o(S)$. Then $o(U_n) \leq o(U) \leq o(S)$ for all n. There is a subsequence $\{S_{n_i}\}$ such that if $\{S_{n_{i_j}} \mid j \in \mathcal{N}\}$ is a subsequence of $\{S_{n_i}\}$, then $o(\cup \{S_{n_{i_j}} \mid j \in \mathcal{N}\}) = o(S)$. In fact, if $o(S) = \aleph_0$, one may let $\{S_{n_i}\}$ be the set of those S_n which are nonempty, and if $o(S) > \aleph_0$, one may take the set of S_n such that $o(S_n) > o(S_m)$ if $n > m$. Since $n_i \geq i$, there is no loss of generality in assuming that $\{S_n\}$ itself has the property required of $\{S_{n_i}\}$.

Let $\mathcal{N} = \cup \{R_i \mid i \in \mathcal{N}\}$ be any partition of $\mathcal{N}$ into $\aleph_0$ disjoint infinite subsets. Let

$$V_i = \cup \{S_n \mid n \in R_i \text{ and } n \geq i\}.$$

Then $o(V_i) = o(S)$. Hence $\exists$ a 1–1 function f_i from U_i into V_i and clearly $f_i(u) \in S_n$ with $n \geq i$. The function $f = \cup \{f_i \mid i \in \mathcal{N}\}$ therefore has the desired properties.

CASE 3. S is infinite and $o(S_n) = o(S)$ for some n. Then $o(U_i) \leq o(S_n)$ for all i. There is a partition $S_n = \cup \{V_i \mid 1 \leq i \leq n\}$ with $o(V_i) = o(S_n)$. There is a 1–1 function f_i from U_i into V_i, hence $g = \cup \{f_i \mid 1 \leq i \leq n\}$ is a 1–1 function from $\cup \{U_i \mid 1 \leq i \leq n\}$ into S_n such that if $u \in U_i$, then $g(u) \in S_n$, $i \leq n$.

Now consider the sets $U' = \dot{\cup} \{U_j \mid j > n\}$ and $S' = \dot{\cup} \{S_j \mid j > n\}$. Except for a minor matter of notation, U' and S' satisfy the hypotheses of the theorem, and we can proceed as before. One of two things occurs. It may happen that after k steps in each of which we are faced with Case 3, we arrive at a Case 1 or Case 2 situation. If $g_1, \ldots, g_k$, and h are the partial functions arising from the separate steps, then $f = g_1 \dot{\cup} \ldots \dot{\cup} g_k \dot{\cup} h$ has the desired properties. If we always have a Case 3 situation, then $f = \dot{\cup} g_n$ works (with a little care, this last possibility can be avoided, but it doesn't matter).

5.5.6. *If $p \in \mathscr{P}$,*

$$G = \Sigma \{\Sigma \{A \mid A \in S_n\} \mid n \in \mathcal{N}\}$$

where each $A \in S_n$ is cyclic of order p^n, $\{c_n \mid n \in \mathcal{N}\}$ is a sequence of cardinal numbers, and

$$\Sigma \{c_n \mid n > i\} \leq \Sigma \{o(S_n) \mid n > i\}$$

for all $i \geq 0$, then $\exists\ H \subset G$ such that

$$H = \Sigma \{\Sigma \{B \mid B \in T_n\} \mid n \in \mathcal{N}\},$$

where each $B \in T_n$ is cyclic of order p^n, and $o(T_n) = c_n$ for all n.

Proof. Let $\{U_n\}$ be a sequence of disjoint sets such that $o(U_n) = c_n$, and let $U = \dot{\cup} U_n$. By the lemma, $\exists$ a 1–1 function f from U into $S = \dot{\cup} S_n$ such that if $u \in U_n$ and $f(u) \in S_m$, then $n \leq m$. For all n and all $u \in U_n$, let B_u be the unique subgroup of $f(u)$ of order p^n. Let $T_n = \{B_u \mid u \in U_n\}$. Then

$$H = \Sigma \{\Sigma \{B \mid B \in T_n\} \mid n \in \mathcal{N}\}$$

exists and has the desired properties. ‖

As an example of what the last theorem says, let $G = \Sigma A_n$ where A_n is cyclic of order p^n. Then G has a subgroup $H = \Sigma B_n$ where each B_n is the direct sum of $\aleph_0$ cyclic groups of order p^n.

5.5.7. *If G is a finite Abelian group, $H \subset G$,*

$$G = \Sigma \{A_i \mid 1 \leq i \leq m\}, \quad H = \Sigma \{B_j \mid 1 \leq j \leq n\},$$

A_i and B_j cyclic of prime power order, then $m \geq n$, and after rearrangement, $o(B_i) \mid o(A_i)$ for $1 \leq i \leq n$.

Proof. Since $G = \Sigma G_p$, $H = \Sigma H_p$, and $H_p \subset G_p$, it follows immediately from Theorem 5.5.4 that the number of summands B_j of order a power of p is at most the number of summands A_i of order a power of p. Hence $m \geq n$. The remainder of the theorem follows from Theorems 5.5.4 and 5.5.5.

5.5.8. *If G is a finitely generated Abelian group and $H \subset G$, then H is finitely generated.*

Proof. Since G is finitely generated Abelian, it is a direct sum of a finite number of cyclic groups by Theorem 5.4.2. By Theorem 5.5.2, H is a direct sum of cyclic groups. By Theorems 5.5.3 and 5.5.4 the number of summands (in a standard prime-power or infinite decomposition) of H is at most that for G. Hence H is finitely generated.

EXERCISES

5.5.9. If G is a finitely generated Abelian group, then G has only a finite number of isomorphism classes of subgroups.

5.5.10. If G is a finite Abelian group, then $G = \Sigma A_i$ where A_i is cyclic and $1 \neq o(A_i) \mid o(A_{i+1})$ for all i. The A_i are not (necessarily) unique, but their orders are.

5.5.11. A factor group of a direct sum of cyclic groups is not necessarily a direct sum of cyclic groups.

5.5.12. If G is an Abelian p-group, then for $x \in G$, $y \in G$,

   (a) $Ht(0) = \infty$,

   (b) $Ht(x + y) \geqq \min(Ht(x), Ht(y))$,

   (c) $Ht(x + y) = \min(Ht(x), Ht(y))$ if $Ht(x) \neq Ht(y)$.

5.5.13. Let G be an Abelian p-group and $G = \Sigma \{A \mid A \in S\}$.

   (a)   If $x \in A \in S$, then $Ht_A(x) = Ht_G(x)$.

   (b)   If $x_i \in A_i \in S$ for distinct $A_1, \ldots, A_n$, then $Ht(x_1 + \ldots + x_n) = \min(Ht(x_i))$.

5.5.14. Let $p \in \mathscr{P}$ and $G = \Sigma \{A_n \mid n \in \mathscr{N}\}$, where each A_n is cyclic of order p^n. Find the structure of all subgroups of G, and show that there are $2^{\aleph_0}$ different isomorphism classes of subgroups.

5.5.15. (a)   A group G satisfies the maximal condition for subgroups iff every subgroup of G is finitely generated.

   (b)   An Abelian group G satisfies the maximal condition for subgroups iff G is finitely generated.

5.5.16. (a)   If G satisfies the maximal condition for normal subgroups, then any endomorphism of G onto G is an automorphism (in different language: G is not isomorphic to any proper homomorphic image).

   (b)   If G is a finitely generated Abelian group, then any endomorphism of G onto G is an automorphism.

   (c)   Give an example of an Abelian group G and an endomorphism of G onto G which is not an automorphism.

5.6   Vector spaces

In this section, vector spaces will be discussed up to and including the idea of dimension. It will not, however, be possible to make the book self-contained in this respect, and some facts will have to be assumed without proof later.

Let R be a ring with identity 1. A *left R-module* is an ordered pair $(G, *)$ such that G is an Abelian group, and $*$ is a function from $R \times G$ into G such that if $a \in R$, $b \in R$, $x \in G$, and $y \in G$, then

$$\begin{cases} a * (x + y) = a * x + a * y, \\ (a + b) * x = a * x + b * x, \end{cases} \qquad \text{(distributive laws)}$$

$$(ab) * x = a * (b * x), \qquad \text{(associative law)}$$

$$1 * x = x.$$

Both the ring addition and the group operation have been denoted by $+$. We shall drop the $*$ immediately, and write ax instead of $a * x$. A *right R-module* is defined similarly. A module is a group with a ring of operators satisfying certain additional natural conditions. In particular, the theory of operator groups (Sections 2.7 to 2.11) applies to modules.

Example. Any Abelian group G is a (left) *J*-module in a natural way. In fact, if $x \in G$ and $n \in J$, then nx has a meaning, and one can readily verify the four requirements. Thus the theory of Abelian groups is a special case of the theory of modules. (See Kaplansky [1].)

A *division ring* (also called a skew field, sfield, and field) is a ring D such that $D^{\#}$ is a group under multiplication. A *field* is a commutative division ring.

A (left) *vector space* over a division ring D is a left D-module. Right vector space is defined similarly, but we shall seldom need the concept, and a vector space will mean a left vector space. A *subspace* is a D-subgroup (in the operator sense), i.e., a subgroup H of the vector space V such that if $h \in H$ and $d \in D$, then $dh \in H$. The terminology of operator groups can now be adopted, but we shall normally drop the prefix D. Thus, a vector space is simple iff it has no nontrivial subspace. Let us also agree that the zero space E is not simple.

5.6.1. *If V is a vector space over a division ring D, then the simple subspaces of V are precisely the $Dx = \{dx \mid d \in D\}$ with $x \neq 0$, $x \in V$.*

Proof. Let $x \in V^{\#}$. Then Dx is a subgroup by one of the distributive laws, a subspace by the associative law, and is not E since $x = 1x \in Dx$. If

$W \neq E$ is a subspace of V and $W \subset Dx$, then $\exists\ y \in W^{\#}$. Hence $y = dx$, $d \in D$, and $d \neq 0$ (Exercise 5.6.6). Therefore $x = 1x = d^{-1}(dx) = d^{-1}y \in W$, so $Dx \subset W$ and $W = Dx$. Hence Dx is a simple subspace.

Next, let U be a simple subspace of V. By agreement, $\exists\ x \in U^{\#}$. Hence Dx is a subspace of U, and by the simplicity of U, $U = Dx$.

5.6.2. *If V is a vector space over a division ring D and $W \subset V$ (this now means: "W is a subspace of V"), then W is a direct summand of V.*

Proof. By Zorn, $\exists$ a maximal subspace X such that $W \cap X = E$. If $y \in V \backslash (W + X)$, then, by Theorem 5.6.1, $Dy \cap (W + X) = E$. Hence $(W + X) + Dy$ exists. Hence $W + (X + Dy)$ exists, contradicting the maximality of X. Therefore $V = W + X$.

5.6.3. *A vector space V over a division ring D is characteristically simple.*

Proof. Deny, and let W be a nontrivial characteristic subspace of V. Then $\exists\ x \in W^{\#}$, $y \in V \backslash W$. Now $Dx \cap Dy = E$, so that $Dx + Dy$ exists and is a subspace. By Theorem 5.6.2, $\exists\ Z \subset V$ such that $V = Dx + Dy + Z$. Any $v \in V$ has a unique expression of the form: $v = ax + by + z$, $a \in D$, $b \in D$, $z \in Z$. Let

$$(ax + by + z)T = ay + bx + z.$$

Then, with obvious notation,

$$\begin{aligned}
[(ax + by + z) &+ (a'x + b'y + z')]T \\
&= [(a + a')x + (b + b')y + (z + z')]T \\
&= (a + a')y + (b + b')x + (z + z') \\
&= (ay + bx + z) + (a'y + b'x + z') \\
&= (ax + by + z)T + (a'x + b'y + z')T;
\end{aligned}$$

$$\begin{aligned}
[c(ax + by + z)]T &= (cax + cby + cz)T = cay + cbx + cz \\
&= c(ay + bx + z) = c(ax + by + z)T.
\end{aligned}$$

Hence T is an endomorphism of V. If $(ax + by + z)T = 0$, then $ay + bx + z = 0$. Since we are dealing with a direct sum $Dx + Dy + Z$, $ay = 0$, $bx = 0$ and $z = 0$. Since $y \neq 0$, $a = 0$; since $x \neq 0$, $b = 0$. Hence $ax + by + z = 0$, and T is an isomorphism. It is clear that T maps V onto V.

Now T is an automorphism of V and $xT = y$, $x \in W$, $y \notin W$. Therefore W is not characteristic, a contradiction.

5.6.4. *If V is a vector space over a division ring D, then there is a subset B of $V^{\#}$ such that $V = \Sigma \{Dx \mid x \in B\}$.*

Proof. V has a minimal subspace, namely any Dx, $x \neq 0$. Also V is characteristically simple by Theorem 5.6.3. By Theorem 4.4.2, $V = \Sigma \, W_i$, where the W_i are simple subspaces. By Theorem 5.6.1, each $W_i = Dx_i$ for some $x_i \neq 0$. ‖

A subset B of $V^\#$ such that $V = \Sigma \, \{Dx \mid x \in B\}$ is called a *basis* of V. If some basis B of V is finite, then $V = Dx_1 + \ldots + Dx_n$, $x_i \in B$. Let $V_0 = E$, $V_i = Dx_1 + \ldots + Dx_i$, $1 \leq i \leq n$. Then the sequence $(V_0, V_1, \ldots, V_n)$ is a composition series of V. By Theorem 4.7.2, V satisfies the ascending and descending chain conditions. If B' is a second basis of V, then B' is finite (or there would be an infinite ascending chain formed as above). The Jordan-Hölder theorem, 2.10.2, or the Remak-Schmidt theorem, 4.7.8, says that $o(B) = o(B')$.

5.6.5. *If a vector space V has a finite basis, then any two bases have the same order.* ‖

The same theorem is true in general (Exercise 5.6.8). The order of a basis of a vector space V is called the *dimension* of V and written $\mathrm{Dim}(V)$.

EXERCISES

5.6.6. If V is a vector space over a division ring D, $a \in D$, and $x \in V$, then

    (a) $a0 = 0$,

    (b) $0x = 0$,

    (c) $ax = 0$ iff $a = 0$ or $x = 0$,

    (d) $(-a)x = -(ax) = a(-x)$.

5.6.7. If R is a ring with identity and G a left R-module, then (a), (b), and (d) of Exercise 5.6.6 are true. Give an example where (c) is false.

5.6.8. Let V be a vector space over a division ring D without finite basis, and let B and B' be two bases of V. Prove that $o(B) = o(B')$. By symmetry, it is sufficient to prove that $o(B) \leq o(B')$. If $x \in B'$, let $f(x)$ be the set of all elements of B appearing in the expansion of x via B. Every element of V is expressible by B', hence by $\cup \{f(x) \mid x \in B'\}$, so $\cup \{f(x) \mid x \in B'\} = B$. Now count B.

5.6.9. If K is a subdivision ring of a division ring D, then D is both a right and a left vector space over K in a natural way.

5.6.10. (a)   If K is a subdivision ring of a divison ring D and V is a left vector space over D, then V is a left vector space over K.

      (b)   In the above situation, treating all spaces as left spaces and using obvious notation, $\mathrm{Dim}_K(V) = \mathrm{Dim}_K(D) \cdot \mathrm{Dim}_D(V)$.

5.7 Automorphism groups of cyclic groups

The automorphism group of a cyclic group will be determined in this section. In addition, some information will be obtained about the automorphism group of an elementary Abelian group.

5.7.1. *If* $G = \Sigma \{H \mid H \in S\}$ *and each* $H \in S$ *is characteristic in* G, *then* $\text{Aut}(G) \cong \pi\{\text{Aut}(H) \mid H \in S\}$.

Proof. Let f be the function from $\text{Aut}(G)$ into $\pi\{\text{Aut}(H) \mid H \in S\}$ defined by $H(Tf) = (T \mid H)$, $T \in \text{Aut}(G)$, $H \in S$. Since each $H \in S$ is characteristic in G, $(T \mid H) \in \text{Aut}(H)$. If $T \in \text{Aut}(G)$, $U \in \text{Aut}(G)$, and $H \in S$, then

$$H((TU)f) = (TU) \mid H = (T \mid H)(U \mid H) = (H(Tf))(H(Uf)).$$

Because of the definition of addition in the direct product, $(TU)f = (Tf) + (Uf)$, and f is a homomorphism.

If $Tf = I$, then each $T \mid H = I_H$. But if $g \in G$, then $g = \Sigma h_i$, $h_i \in H_i \in S$, and

$$gT = \Sigma (h_i T) = \Sigma h_i = g,$$

so $T = I_G$. Therefore, f is an isomorphism.

Finally, it must be shown that $\text{Rng}(f) = \pi \text{Aut}(H)$. Let $s \in \pi \text{Aut}(H)$. Then, for all $H \in S$, $Hs \in \text{Aut}(H)$. Each $g \in G$ is a unique formal sum $g = \Sigma \{g_H \mid H \in S\}$, where $g_H \in H$ and only finitely many $g_H \neq 0$. Define U by the rule

$$gU = \Sigma \{g_H(Hs) \mid H \in S\}.$$

If $g_H = 0$, then $g_H(Hs) = 0$, so again only a finite number of terms are not 0. Hence $gU \in G$. One verifies readily that $U \in \text{Aut}(G)$. Moreover, $H(Uf) = (U \mid H) = Hs$ for all H, so that $Uf = s$. ‖

This theorem is false if the $H \in S$ are not characteristic, as one sees by looking at the case $G = J_2 \times J_2$, for example. For Abelian groups, the theorem yields two immediate corollaries.

5.7.2. *If* G *is a periodic Abelian group, then*

$$\text{Aut}(G) \cong \pi \{\text{Aut}(G_p) \mid p \in \mathscr{P}\}.$$

5.7.3. *If* G *is a finite Abelian group of order* $p_1^{i_1} \cdots p_n^{i_n}$, *where* $p_1, \ldots, p_n$ *are distinct primes, then*

$$\text{Aut}(G) \cong \text{Aut}(G_{p_1}) \times \ldots \times \text{Aut}(G_{p_n}).$$

5.7.4. *J has exactly two automorphisms, the identity I and* $-I$: $n(-I) =$
$-n$. *Hence* Aut$(J) \cong J_2$.

Proof. If $T \in$ Aut(J), then $\langle 1T \rangle = J$. Hence $1T = 1$ or -1. The first
case gives $T = I$, the second $T = -I$. Since $o(\text{Aut}(J)) = 2$, Aut$(J) \cong J_2$. ‖

Before tackling the finite case, we need some preliminary remarks and
lemmas. If R is a ring with identity 1, a *unit* is an element x of R which has
both a right inverse y and a left inverse z. These are then equal, since

$$z = z1 = zxy = 1y = y.$$

Any two right inverses of a unit x are equal since both are equal to a left
inverse z. Similarly, any two left inverses of a unit are equal. One may
therefore speak of *the* inverse of a unit x, and denote it by x^{-1} as usual.

5.7.5. *If R is a ring with identity* 1, *then the set U of units of R is a group
under multiplication.*

Proof. Let $x \in U$ and $y \in U$. Then

$$(xy)(y^{-1}x^{-1}) = x1x^{-1} = xx^{-1} = 1,$$

$$(y^{-1}x^{-1})(xy) = y^{-1}1y = 1.$$

Hence $xy \in U$. Therefore multiplication is an associative operation on U.
Since $1 \cdot 1 = 1$, $1 \in U$. Moreover, 1 is an identity for U. If $x \in U$, then
$xx^{-1} = x^{-1}x = 1$, so $x^{-1} \in U$. Also, x^{-1} is an inverse for x. Hence U is a
group. ‖

Remark. No use has been made of addition in the above proof. Therefore, the theorem is actually valid for associative multiplicative systems.

5.7.6. *If G is a finite group, such that if* $n \in \mathcal{N}$ *then there are at most n
elements* $x \in G$ *such that* $x^n = e$, *then G is cyclic.*

Proof. Let A be a cyclic group with $o(A) = o(G)$. Let $n \in \mathcal{N}$. If $\exists\, g \in G$
with $o(g) = n$, then $\langle g \rangle$ furnishes n solutions of $x^n = e$, hence, by hypothesis,
all solutions. Since $n \mid o(G) = o(A)$ and A is cyclic, $\exists\, a \in A$ with $o(a) = n$.
Therefore $\langle g \rangle \cong \langle a \rangle$, so that A has at least as many elements of order n as G
has. Since G and A have the same total number of elements, G must have
the same number of elements of order m as A, for all m. But A has an element
of order $o(A) = o(G)$, and therefore G does also. Hence G is cyclic.

5.7.7. *If F is a field, and* $c_0, \ldots, c_n$ *are in F with* $c_n \neq 0$, *then there are
at most n elements x of F such that* $\Sigma\, c_i x^i = 0$.

Proof. This is true for $n = 0$. Induct on n. Suppose that $x, y_1, \ldots, y_n$ are distinct elements of F such that $\Sigma\, c_i x^i = 0$, $\Sigma\, c_i y_j^i = 0$. Subtraction yields

$$(x - y_j)(c_n(x^{n-1} + x^{n-2}y_j + x^{n-3}y_j^2 + \ldots + y_j^{n-1}) + \ldots + c_1) = 0,$$

or $\Sigma\, \{b_i y_j^i \mid 0 \leq i \leq n - 1\} = 0$ with $b_{n-1} = c_n \neq 0$. This contradicts the inductive hypothesis. Hence the theorem holds.

5.7.8. *The multiplicative group of a finite field is cyclic.*

Proof. If F is a field and $n \in \mathcal{N}$, then, by Theorem 5.7.7, there are at most n elements of F such that $x^n = 1$. Therefore, by Theorem 5.7.6, the multiplicative group $F^{\#}$ of F is cyclic. $\|$

In Section 2.4 on cyclic groups,

$$J_n = \{[0], [1], \ldots, [n - 1]\}$$

was given the structure of an additive group. It is, in fact, a commutative ring with identity $[1]$, in a natural way. If $n = 12$, for example, then $[4][5] = [8]$. By Theorem 5.7.5, the set J_n^* of units of this ring forms a multiplicative group.

Example. If $n = 12$, then $J_n^* = \{[1], [5], [7], [11]\}$, and is isomorphic to the 4-group.

5.7.9. *If $n \geq 2$, then*

$$J_n^* = \{[m] \mid 0 < m < n, (m, n) = 1\}, \tag{1}$$
$$J_n^* = \{[m] \mid o([m]) = n \quad in \quad (J_n, +)\}. \tag{2}$$

Proof. Clearly, $[0] \notin J_n^*$. Let $0 < m < n$.

If $[m] \in J_n^*$, then $\exists\, r$ such that $[m][r] = [1]$. Therefore, $\exists\, s \in J$ such that $mr + sn = 1$. Hence $(m, n) = 1$.

If $(m, n) = 1$, then $\exists\, r \in J$ and $s \in J$ such that $mr + ns = 1$. Hence $mr \equiv 1 \pmod{n}$. If $o([m]) = t$ in J_n, then $t[m] = [0]$, so $tm \equiv 0 \pmod{n}$. Therefore,

$$t \equiv t \cdot 1 \equiv t(mr) \equiv (tm)r \equiv 0 \pmod{n}.$$

Hence $t \geq n$, so $t = n$.

Finally, if $o([m]) = n$ in J_n, then the n elements $[m], 2[m], \ldots, n[m]$ are all distinct. Hence $\exists\, t$ such that $t[m] = [1]$, so that $[t][m] = [1]$. Therefore $[m] \in J_n^*$.

5.7.10. *J_p is a field, $p \in \mathcal{P}$.*

Proof. If $0 < m < p$, then $(m, p) = 1$. Therefore J_p^* consists of all non-0 elements of J_p, by Theorem 5.7.9. Hence J_p is a field.

5.7.11. $\mathrm{Aut}(J_n) \cong J_n^*$, if $n \geq 2$.

Proof. For all $T \in \mathrm{Aut}(J_n)$, let $Tf = [1]T$. Now $o([1]T) = o([1]) = n$. Therefore, by Theorem 5.7.9, f is a function from $\mathrm{Aut}(J_n)$ into J_n^*. If also $U \in \mathrm{Aut}(J_n)$ and $[1]T = [1]U$, then $[i]T = [i]U$ for all i, so that $T = U$. This means that f is 1–1. If $[i] \in J_n^*$, then the function T given by $[j]T = [ij]$ is an automorphism of J_n (see Exercise 5.7.23), and $Tf = [1]T = [i]$. Therefore f is onto J_n^*. Finally, if $[1]T = [i]$ and $[1]U = [k]$ with T and U in $\mathrm{Aut}(J_n)$, then

$$[1]TU = [i]U = (i[1])U = i([1]U) = i[k] = [i][k].$$

Hence $(TU)f = (Tf)(Uf)$. Therefore f is an isomorphism of $\mathrm{Aut}(J_n)$ onto J_n^*.

5.7.12. If $n = 2^i p_1^{j_1} \ldots p_r^{j_r}$, where $p_1, \ldots, p_r$ are distinct odd primes, then

$$\mathrm{Aut}(J_n) \cong H_2 \times H_{p_1} \times \ldots \times H_{p_r},$$

$$H_2 = \begin{cases} J_1 & \text{if } i = 0, \\ J_{2^{i-1}} & \text{if } i = 1 \text{ or } 2, \\ J_2 \times J_{2^{i-2}} & \text{if } i \geq 3, \end{cases} \tag{1}$$

$$H_p = J_{p^{j-1}(p-1)} \qquad \text{for } p = p_i > 2, j = j_i. \tag{2}$$

Proof. By Theorem 5.1.7, J_n is the direct sum of its p-components, themselves cyclic by Theorem 2.4.4. By Theorem 5.7.3, the problem is reduced to that of determining $\mathrm{Aut}(J_{p^t})$, $p \in \mathscr{P}$.

By Theorem 5.7.10, J_p is a field, hence by Theorem 5.7.8, J_p^* is cyclic. Therefore $\exists \, a \in \mathscr{N}$ whose multiplicative order (mod p) is $p - 1$. Now

$$o(\mathrm{Aut}(J_{p^t})) = p^{t-1}(p - 1)$$

by Theorems 5.7.9 and 5.7.11, and an easy counting argument. If $a^i \equiv 1$ (mod p^t), then $a^i \equiv 1$ (mod p) and $p - 1 \mid i$. Therefore $o(a)$ in $J_{p^t}^*$ is a multiple of $p - 1$. Hence a suitable power of a, say b, has order $p - 1$ in $J_{p^t}^*$.

If $c = 1 + kp^m$ and $m \geq 1$, then

$$c^p \equiv 1 + kp^{m+1} + \frac{p(p - 1)}{2} k^2 p^{2m} \pmod{p^{m+2}}, \tag{3}$$

since $3m \geq m + 2$, and the binomial coefficients are integers. By (3), if $c \equiv 1$ (mod p^m), then $c^p \equiv 1$ (mod p^{m+1}). Also, except in the case $p = 2$ and $m = 1$, if $c \equiv 1$ (mod p^m) but $c \not\equiv 1$ (mod p^{m+1}), then $c^p \not\equiv 1$ (mod p^{m+2}).

If, now, $p > 2$, then $c = 1 + p$ has multiplicative order p^{t-1} (mod p^t) by the above remarks and induction. Since $J_{p^t}^*$ is Abelian and $(o(b), o(c)) = 1$,

$o(bc) = o(b)o(c) = p^{t-1}(p-1)$. Therefore $J_{p^t}^*$ is cyclic, hence $\mathrm{Aut}(J_{p^t})$ is cyclic.

Now let $p = 2$. By the remarks after (3), 5 is of order 2^{t-2} in $J_{2^t}^*$ if $t \geq 2$. But $-1 \notin \langle 5 \rangle$, since $5 \equiv 1 \not\equiv -1 \pmod 4$. Since $o(-1) = 2$ in $J_{2^t}^*$ and $o(J_{2^t}^*) = 2^{t-1}$, this proves (1) for $i \geq 3$. The rest of (1) is trivial. ‖

Actually, slightly more was proved than was stated in the theorem, since in the case $p = 2$, generators of $\mathrm{Aut}(J_{2^t})$ were found. Let us record this fact.

5.7.13. *If $t \geq 3$, then $\mathrm{Aut}(J_{2^t}) \cong H + K$ where H is cyclic of order 2^{t-2} and is generated by $5I$, K is cyclic of order 2 and is generated by $-I$, $x(5I) = 5x$, and $x(-I) = -x$.* ‖

The remainder of the section is concerned primarily with various groups associated with a vector space.

If D is a division ring with identity element e, the *characteristic* of D is 0 if $o(e) = \infty$, and is $o(e)$ otherwise. It is clear that if K is a subdivision ring of D, then their characteristics are equal.

5.7.14. *If D is a division ring, then*

(i) *if the characteristic of D is 0, then $(D, +)$ is the direct sum of groups isomorphic to $\mathscr{R}$,*

(ii) *if the characteristic of D is not 0, then it is a prime p and $(D, +)$ is the direct sum of groups of order p.*

Proof. (i) Let $x \in D$, $n \in \mathscr{N}$. Then $ne \in D^\#$, so $(ne)^{-1}x \in D$. Thus $n((ne)^{-1}x) = (ne)((ne)^{-1}x) = x$. Hence $(D, +)$ is a divisible Abelian group. If $x \neq 0$ and $n \in \mathscr{N}$, then $nx = (ne)x \neq 0$. Therefore D is torsion-free. By Theorem 5.2.7, D is a direct sum of groups isomorphic to $\mathscr{R}$.

(ii) Let $o(e) = p$. If $p = ij$, then $(ie)(je) = (ij)e = 0$, so $ie = 0$ or $je = 0$ since D is a division ring. Therefore p is prime. If $x \in D^\#$, then $px = p(ex) = (pe)x = 0$, so that $o(x) = p$ also. Thus $(D, +)$ is an Abelian group with $px = 0$ for all $x \in D$. By Theorem 5.1.9, $(D, +)$ is the direct sum of groups of order p.

5.7.15. *Let D be a division ring with identity element e. Then*

(i) *If the characteristic of D is $p \in \mathscr{P}$ and*

$$P = \{0, e, 2e, \ldots, (p-1)e\},$$

then P is a subfield of every subdivision ring of D and $P \cong J_p$ (as a field);

(ii)   *if the characteristic of D is 0 and P = {(m/n)e}, then again P is a subfield
of every subdivision ring of D and P ≅ ℛ (as a field).*

Proof. Let K be any subdivision ring of D. Then $K^\#$ is a subgroup of
$D^\#$, hence $e \in K^\#$. Therefore $ne \in K$ for all $n \in J$. If D has characteristic 0,
so does K, and $(K, +)$ is divisible by Theorem 5.7.14. In this case, by Theorem
5.2.6, there is a unique isomorphism of $(\mathscr{R}, +)$ into D sending 1 into e. This
justifies the notation $(m/n)e$ in the statement of the theorem. Moreover P is
contained in K for all subdivision rings K in the case of either characteristic.

It is clear that $(me)(ne) = (mn)e$ for all integers m and n. In J_p a similar
rule holds: $[m][n] = [mn]$, and $[n] = 0$ iff $ne = 0$. It follows that if the
characteristic of D is p, then $P \cong J_p$ as a field. The corresponding verification
for the characteristic 0 case is routine and will be omitted. ‖

The subfield P in Theorem 5.7.15 is called the *prime* subfield of D.

5.7.16. *If V is a vector space over a division ring D, then*

(i)   *if the characteristic of D is 0, then (as a group) V is the direct sum of
groups isomorphic to ℛ,*

(ii)   *if the characteristic of D is p ≠ 0, then V is the direct sum of groups of
order p.*

Proof. By Theorem 5.6.10, V is a vector space over the prime subfield
P of D. By Theorem 5.6.4, there is a basis B of V such that

$$V = \Sigma \{Px \mid x \in B\}.$$

The function $f : rf = rx$ is an isomorphism of $(P, +)$ onto the group $(Px, +)$.
The theorem now follows from Theorem 5.7.15. ‖

Let F be a field, V a vector space over F, End(V) the ring of F-endo-
morphisms of V, and Aut(V) the group of F-automorphisms of V. We first
remark that Aut(V) is just the group of units of End(V). Next, we note that
End(V) is even a vector space over F in a natural way. Let $c \in F$, $T \in$ End(V),
$x \in V$, and define $x(cT) = c(xT)$. Then

$$(x + y)(cT) = c((x + y)T) = c(xT + yT) = c(xT) + c(yT)$$
$$= x(cT) + y(cT),$$

and if $d \in F$ also, then

$$(dx)(cT) = c((dx)T) = c(d(xT)) = (cd)(xT) = d(c(xT)) = d(x(cT)).$$

Hence $cT \in$ End(V). The verification that End(V) thus becomes a vector
space is left for Exercise 5.7.25.

Actually, End(V) has a somewhat stronger structure. An *algebra A* over a field F is a set with operations of addition, multiplication, and left multiplication by elements of F such that

(a) A is a ring,

(b) A is a vector space over F,

(c) if $c \in F$, $x \in A$, and $y \in A$, then $c(xy) = (cx)y = x(cy)$.

One can then verify that End(V) is an algebra with identity over F.

There is another way of looking at End(V), at least up to isomorphism. For the sake of simplicity, let Dim(V) be finite. Then there is a finite basis B of V. A *matrix* (over F with respect to B) is a function f from $B \times B$ into F. The set of matrices with respect to B will be denoted by $M(B)$ (F being fixed). Addition, multiplication, and scalar multiplication in $M(B)$ are defined as follows. Let $f \in M(B)$, $g \in M(B)$, $x \in B$, $y \in B$, and $c \in F$, and define

$$(f + g)(x, y) = f(x, y) + g(x, y),$$
$$(fg)(x, y) = \Sigma \{f(x, z)g(z, y) \mid z \in B\},$$
$$(cf)(x, y) = c(f(x, y)).$$

One can then verify that $M(B)$ is also an algebra over F (Exercise 5.7.26).

If $T \in$ End(V) and B is a finite basis of V, then there is a natural way of associating a matrix with T. For, if $x \in B$, then

$$xT = \Sigma \{f(x, y)y \mid y \in B\}, \quad f \in M(B).$$

Let u be the function, $Tu = f$, defined in this way.

5.7.17. *If V is a finite-dimensional vector space over a field F with basis B, then u (defined above) is an isomorphism of the algebra* End(V) *onto* $M(B)$.

Proof. If $T \in$ End(V), $U \in$ End(V), $c \in F$, $x \in B$, $Tu = f$, and $Uu = g$, then

$$x(T + U) = xT + xU = \Sigma f(x, y)y + \Sigma g(x, y)y$$
$$= \Sigma (f(x, y) + g(x, y))y = \Sigma (f + g)(x, y)y,$$

so $(T + U)u = f + g = Tu + Uu$;

$$x(TU) = (\Sigma f(x, y)y)U = \Sigma f(x, y)(yU)$$
$$= \Sigma \{f(x, y) \Sigma \{g(y, z)z \mid z \in B\} \mid y \in B\}$$
$$= \Sigma \{\Sigma \{f(x, y)g(y, z) \mid y \in B\} \mid z \in B\}$$
$$= \Sigma \{(fg)(x, z) \mid z \in B\},$$

so $(TU)u = fg = (Tu)(Uu)$;

$$x(cT) = c(xT) = c \Sigma f(x, y)y = \Sigma cf(x, y)y = \Sigma (cf)(x, y)y,$$

so $(cT)u = cf = c(Tu)$. Hence u is a homomorphism of $\mathrm{End}(V)$ into $M(B)$. If $Tu = Uu = f$, then for $v \in V$, $v = \sum \{a_x x \mid x \in B\}$, and

$$vT = \sum \{a_x(xT) \mid x \in B\} = \sum \{a_x \sum \{f(x, y)y \mid y \in B\} \mid x \in B\}$$
$$= \sum \{a_x(xU) \mid x \in B\} = vU.$$

Therefore $T = U$, and u is 1–1.

Finally, if $f \in M(B)$, define T by the rule:

$$(\sum a_x x)T = \sum \{a_x f(x, y)y \mid x \in B, y \in B\}.$$

Then

$$\begin{aligned}(\sum a_x x + \sum b_x x)T &= (\sum (a_x + b_x)x)T \\ &= \sum \{(a_x + b_x)f(x, y)y \mid x \in B, y \in B\} \\ &= \sum \{a_x f(x, y)y \mid x \in B, y \in B\} \\ &\quad + \sum \{b_x f(x, y)y \mid x \in B, y \in B\} \\ &= (\sum a_x x)T + (\sum b_x x)T.\end{aligned}$$

Again,

$$\begin{aligned}(c(\sum a_x x))T &= (\sum ca_x x)T = \sum ca_x f(x, y)y \\ &= c \sum a_x f(x, y)y = c(\sum a_x x)T.\end{aligned}$$

Hence $T \in \mathrm{End}(V)$. Moreover, if $x \in B$, then

$$xT = (\sum \{\delta_{xy}y \mid y \in B\})T = \sum \{\delta_{xy}f(y, z)z \mid y \in B, z \in B\}$$
$$= \sum \{f(x, z)z \mid z \in B\},$$

so that $Tu = f$. Therefore u is onto $M(B)$. ‖

The isomorphism u clearly depends on the choice of B.

If $\mathrm{Dim}(V) = n$, then the group $\mathrm{Aut}(V)$ is called the *general linear* group, and is denoted by $GL(V)$ or $GL(n, F)$. If F is a finite field of order q, it is also denoted by $GL(n, q)$. By the above discussion, $GL(n, F)$ is isomorphic to the group of nonsingular n by n matrices over F. We shall next determine the order of $GL(n, q)$.

5.7.18. *If V is a vector space over a finite field F, $o(F) = q$, and $\mathrm{Dim}(V) = n$, then $o(V) = q^n$.*

Proof. There is a basis B of V over F with $o(B) = n$. Each element $x \in V$ has a unique expression $x = \sum \{c_y y \mid y \in B\}$ (hence is determined by a function from B into F). There are clearly just $o(F)^{o(B)} = q^n$ such expressions (the number of functions from B into F).

5.7.19. *If F is a finite field of order q and characteristic p, then $q = p^n$ for some n.*

Proof. The prime subfield P of F has order p. The theorem now follows from Exercise 5.6.9 and Theorem 5.7.18. ‖

We assume without proof the converse: there is a field F of order p^n, unique up to isomorphism.

5.7.20. $o(GL(n, q)) = \pi\{q^n - q^i \mid 0 \leqq i \leqq n - 1\}.$

Proof. Let V be a vector space of dimension n over a finite field with q elements. Let $(x_1, \ldots, x_n)$ be an ordered basis of V. If $T \in GL(n, q)$, then $(x_1 T, \ldots, x_n T)$ is again an ordered basis of V (Exercise 5.7.27). Conversely (Exercise 5.7.30) if $(y_1, \ldots, y_n)$ is any ordered basis of V then $\exists \mid T \in GL(n, q)$ such that $x_i T = y_i$ for all i. Thus $o(GL(n, q))$ equals the number of ordered bases of V. In constructing an ordered basis, if elements $y_1, \ldots, y_{r-1}$ have been chosen, then y_r may be chosen as any element in V outside the $r - 1$ dimensional subspace W generated by $y_1, \ldots, y_{r-1}$. By Theorem 5.7.18, $o(W) = q^{r-1}$, and the theorem follows. ‖

The *special linear group* $SL(n, F)$ is the group of all n by n matrices of determinant 1 with elements in F.

5.7.21. (i) $GL(n, F)/SL(n, F) \cong F^{\#}.$

(ii) $o(SL(n, q)) = (\pi\{q^n - q^i \mid 0 \leqq i \leqq n - 1\})/(q - 1).$

Proof. Consider $GL(n, F)$ as the group of all nonsingular n by n matrices A. Since $\mathrm{Det}(AB) = \mathrm{Det}(A) \cdot \mathrm{Det}(B)$, Det is a homomorphism of $GL(n, F)$ into $F^{\#}$. If $x \in F^{\#}$ and A is a diagonal matrix with all diagonal entries but one equal to 1 and that entry equal to x, then $\mathrm{Det}(A) = x$. Therefore Det is onto $F^{\#}$, and clearly has kernel $SL(n, F)$. Part (i) follows, and (ii) follows from (i) and Theorem 5.7.20. ‖

Let G be the direct sum of groups, each of order the same prime p. Then G is an Abelian group, and consequently, is a J-module. If $[i] \in J_p$ and $x \in G$, define $[i]x = ix$. This definition is a valid one, since if $[i] = [j]$, then $i \equiv j \pmod{p}$, hence $ix = jx$ [since $o(x) = 1$ or p]. It follows rather easily that G is thereby a J_p-module, i.e., a vector space over J_p. An automorphism of G as group is also an automorphism of G as vector space. The converse is obvious. For the finite case, we therefore have

5.7.22. *If a group G is the direct sum of n cyclic groups of order $p \in \mathscr{P}$, then* $\mathrm{Aut}(G) \cong GL(n, p)$.

EXERCISES

5.7.23. End(J_n, $+$) is isomorphic to the ring J_n.

5.7.24. If G is an Abelian group, then Aut(G) is the group of units of the ring End(G).

5.7.25. Verify that if V is a vector space over a field F, then End(V) is an algebra with identity over F.

5.7.26. (a) Verify that if B is a finite set and F is a field, then the set $M(B)$ of matrices is an algebra over F.

(b) How much of this remains true if the field F is replaced by a division ring D?

5.7.27. If S is a generating subset of a vector space V over a field F, then S contains a basis B of V.

5.7.28. If $G = \Sigma \{H \mid H \in S\}$, then $\pi \{\text{Aut}(H) \mid H \in S\} \cong \text{Aut}(G)$.

5.7.29. (a) If F is a free Abelian group of finite rank r and $F = \langle S \rangle$, then $o(S) \geqq r$. (Imbed F in a vector space of dimension r over $\mathscr{R}$ and use Exercise 5.7.27 and Theorem 5.6.5.)

(b) Part (a) is also true in case r is infinite.

5.7.30. If V is a vector space over a division ring D and $(x_1, \ldots, x_n)$ and $(y_1, \ldots y_n)$ are ordered bases of V, then $\exists \mid T \in GL(V)$ such that $x_i T = y_i$ for all i.

5.7.31. Let A be an infinite cardinal. Show that there is a group of order A with an Abelian group of automorphisms (not the full automorphism group) of order 2^A as follows. Let S be a set of order A and G_i a group of order 3 for $i \in S$.

(a) $o(\Sigma_E \{G_i \mid i \in S\}) = A$.

(b) $\text{Aut}(\Sigma_E G_i) \cong \pi \text{Aut}(G_i)$.

(c) $\pi \text{Aut}(G_i)$ is an Abelian group of order 2^A.

5.8   Hom(A, B)

In this section, Hom(A, B) will be determined in case both A and B are finitely generated Abelian groups, and in some other cases.

5.8.1. *If B and $\Sigma \{H \mid H \in S\}$ are Abelian groups, then*

$$\text{Hom}(\Sigma \{H \mid H \in S\}, B) \cong \pi \{\text{Hom}(H, B) \mid H \in S\}.$$

More fully, the function f defined by the rule: $H(Tf) = (T \mid H)$, is an isomorphism of Hom(ΣH, B) *onto* π Hom(H, B).

Proof. Let $H \in S$ and $T \in \text{Hom}(\Sigma \{K \mid K \in S\}, B)$. Then

$$(T \mid H) \in \text{Hom}(H, B).$$

Hence, $Tf \in \pi \{\text{Hom}(H, B) \mid H \in S\}$. If also $T' \in \text{Hom}(\Sigma K, B)$, then

$$H((T + T')f) = (T + T') \mid H = (T \mid H) + (T' \mid H)$$
$$= H(Tf) + H(T'f) = H(Tf + T'f)$$

because of the rules of operation in the direct product. Hence $(T + T')f = Tf + T'f$, and f is a homomorphism. If $Tf = O$, then each $(T \mid H) = O$. Therefore $\text{Ker}(T)$ contains all $H \in S$, hence also ΣH. Thus $T = O$, so that f is an isomorphism. Finally, if $U \in \pi \{\text{Hom}(H, B) \mid H \in S\}$, define T by the rule:

$$(\Sigma x_H)T = \Sigma \{x_H(HU) \mid H \in S\}, x_H \in H.$$

The formal sum on the left has all but a finite number of terms equal to 0, and, since $HU \in \text{Hom}(H, B)$, the same is true on the right. We have

$$(\Sigma x_H + \Sigma y_H)T = (\Sigma (x_H + y_H))T = \Sigma (x_H + y_H)(HU)$$
$$= \Sigma (x_H(HU) + y_H(HU))$$
$$= \Sigma x_H(HU) + \Sigma y_H(HU) = (\Sigma x_H)T + (\Sigma y_H)T.$$

Hence $T \in \text{Hom}(\Sigma H, B)$. Moreover, if $h \in H$, then

$$h(H(Tf)) = h(T \mid H) = hT = h(HU)$$

by the definition of T. Thus $H(Tf) = HU$ for all $H \in S$. Therefore $Tf = U$, so that f is an isomorphism of $\text{Hom}(\Sigma H, B)$ onto $\pi (\text{Hom}(H, B))$.

5.8.2. *If A and $\pi \{H \mid H \in S\}$ are Abelian groups, then*

$$\text{Hom}(A, \pi \{H \mid H \in S\}) \cong \pi \{\text{Hom}(A, H) \mid H \in S\}.$$

More fully, the function f defined by the rule

$$a(H(Tf)) = H(aT), a \in A, H \in S, T \in \text{Hom}(A, \pi H),$$

is an isomorphism of $\text{Hom}(A, \pi H)$ onto $\pi (\text{Hom}(A, H))$.

Proof. First note that $a(H(Tf)) = H(aT) \in H$. If also $a' \in A$, then

$$(a + a')(H(Tf)) = H((a + a')T) = H(aT + a'T) = H(aT) + H(a'T)$$
$$= a(H(Tf)) + a'(H(Tf)).$$

Hence, $H(Tf) \in \text{Hom}(A, H)$, so that $Tf \in \pi \{\text{Hom}(A, H) \mid H \in S\}$.
If also $T' \in \text{Hom}(A, \pi H)$, then

$$a(H((T + T')f)) = H(a(T + T')) = H(aT + aT') = HaT + HaT'$$
$$= a(H(Tf)) + a(H(T'f)) = a(H(Tf) + H(T'f)).$$

Since this is true for all $a \in A$, $H((T + T')f) = H(Tf) + H(T'f)$ for all $H \in S$. Hence $(T + T')f = Tf + T'f$, and f is a homomorphism of $\mathrm{Hom}(A, \pi H)$ into $\pi \mathrm{Hom}(A, H)$.

If $Tf = O$, then $H(Tf) = O$ for all $H \in S$, and $H(aT) = a(H(Tf)) = 0$ for all $a \in A$. Hence $aT = O$ for all $a \in A$, so $T = O$. Therefore f is an isomorphism.

If $U \in \pi \{\mathrm{Hom}(A, H)\}$, define T by the rule $H(aT) = a(HU)$. Then $HU \in \mathrm{Hom}(A, H)$, so $H(aT) \in H$ for all $a \in A$. Hence $aT \in \pi \{H \mid H \in S\}$. If $a' \in A$ also, then

$$H((a + a')T) = (a + a')(HU) = a(HU) + a'(HU)$$
$$= H(aT) + H(a'T) = H(aT + a'T).$$

Therefore, $(a + a')T = aT + a'T$, and $T \in \mathrm{Hom}(A, \pi H)$. Also $a(H(Tf)) = H(aT) = a(HU)$ for all $a \in A$, so that $H(Tf) = HU$ for all $H \in S$, hence $Tf = U$. Thus f is onto $\pi \{\mathrm{Hom}(A, H)\}$. ∥

Theorems 5.8.1 and 5.8.2 evidently reduce the question of finding $\mathrm{Hom}(A, B)$ for finitely generated Abelian A and B to the corresponding question for cyclic groups. We have (somewhat more generally than needed):

5.8.3. *If A is an Abelian group, then* $\mathrm{Hom}(J, A) \cong A$.

Proof. Since J is free Abelian on $\{1\}$, if $a \in A$ then $\exists \mid T \in \mathrm{Hom}(J, A)$ such that $1T = a$ (Theorem 5.3.1). If T and U are homomorphisms, then $1(T + U) = 1T + 1U$. Hence $\mathrm{Hom}(J, A) \cong A$. ∥

This takes care of the case where J is the first component of the pair (A, B) in $\mathrm{Hom}(A, B)$.

5.8.4. *If A is periodic Abelian and B is torsion-free Abelian, then*

$$\mathrm{Hom}(A, B) = E.$$

Proof. This is obvious.

5.8.5. *Let $p \in \mathscr{P}$, $q \in \mathscr{P}$. Then*

(i) $\mathrm{Hom}(J_{p^m}, J_{q^n}) = E$ if $p \neq q$,

(ii) $\mathrm{Hom}(J_{p^n}, J_{p^m}) \cong J_{p^{\min(m,n)}}$.

Proof. There is a homomorphism T of a finite cyclic group $\langle a \rangle$ onto a cyclic group $\langle b \rangle$ such that $aT = b$ iff $o(b) \mid o(a)$, and if this condition is met, then the homomorphism is unique and is given by $(ia)T = ib$. Statement (i) follows immediately. If $n \geq m$, then statement (ii) follows. If $n < m$, then any homomorphism of J_{p^n} into J_{p^m} has image in the unique cyclic subgroup of order p^n, and the statement (ii) follows from the case $n = m$. ∥

It is clear that if A and B are finitely generated Abelian groups, then Hom(A, B) can be calculated from the preceding theorems. A precise statement is left as an exercise.

The preceding theory can be used to prove the following theorem: "the lattice of subgroups Lat(G) of a finite Abelian group G looks the same upside down as right side up." If A and B are lattices, a *duality* of A onto B is a 1–1 function f from A onto B such that for all x and y in A, $x \leq y$ iff $yf \leq xf$.

5.8.6. *If G is a finite Abelian group, then* Lat(G) *is self-dual.*

Proof. Let H be a cyclic group of order $n = \mathrm{Exp}(G)$, and $K = \mathrm{Hom}(G, H)$. By Theorems 5.8.1, 5.8.2, and 5.8.5, $K \cong G$. Therefore, Lat(K) $\cong$ Lat(G). Define a function f by the rule:

$$Af = \{ T \in K \mid AT = E \}, \qquad A \subset G.$$

One verifies that $Af \subset K$. It is clear that if $A \subset B \subset G$, then $Bf \subset Af$. It remains only to prove that f is 1–1.

Suppose that $Af = Bf$ with $A \neq B$. If $T \in Af$, then $\langle A, B \rangle T = E$. Hence $\langle A, B \rangle f \supset Af$, and therefore $\langle A, B \rangle f = Af$. A brief examination of cases then shows that *WLOG* it may be assumed that $A < B$ and $Af = Bf$. The group $G/A = \Sigma\, D_i$ with D_i cyclic. Now $\exists\, b \in B \backslash A$, so that $b + A = d_1 + \ldots + d_s$, $d_i \in D_i$, where, say, $d_1 \neq 0$. There is then a homomorphism U of G/A into H which maps $D_2, \ldots, D_s$ onto E and D_1 isomorphically. If T is the natural homomorphism of G onto G/A, then $TU \in K$, and TU has kernel containing A but not B. Hence $Bf < Af$, a contradiction. Therefore f is 1–1. Hence f is a duality of Lat(G) onto Lat(K). If f' is an isomorphism of Lat(K) onto Lat(G), then ff' is a self-duality of Lat(G). $\parallel$

The next result is not directly connected with this line of thought.

5.8.7. *If D is a divisible Abelian group, G is Abelian, $H \subset G$, and $T \in \mathrm{Hom}(H, D)$, then $\exists\, U \in \mathrm{Hom}(G, D)$ such that $U \mid H = T$.*

Proof. Let S be the set of pairs (K, V) such that $H \subset K \subset G$, $V \in \mathrm{Hom}(K, D)$, and $V \mid H = T$. Since $(H, T) \in S$, $S \neq \varnothing$. Partially order S by the rule $(K, V) < (K', V')$ iff $K \subset K'$ and $V' \mid K = V$. It is readily verified (Exercise 5.8.9) that the hypotheses of Zorn's lemma are satisfied. Hence there is a maximal element (K, V) of S. If $K = G$, we are done. If $\exists\, x \in G \backslash K$ such that $K + \langle x \rangle = L$ exists, it is clear how to extend the homomorphism V to L (map x onto 0), hence a contradiction is reached. If there is no such x, then $\exists\, y \in G \backslash K$ such that $py \in K$ for some $p \in \mathscr{P}$. Then $(py)V \in D$, so by the divisibility of D, $\exists\, z \in D$ such that $pz = (py)V$. Define W on the group $L = \langle K, y \rangle$ to be the relation

$$W = \{ (k + ry, kV + rz) \mid k \in K, r \in J \}.$$

If $k_1 + r_1 y = k_2 + r_2 y$, then $k_1 - k_2 = (r_2 - r_1)y \in K$, so that $r_2 - r_1 = jp$ for some $j \in J$. Hence

$$k_1 - k_2 = j(py),$$
$$(k_1 - k_2)V = j((py)V) = j(pz) = (jp)z = r_2 z - r_1 z,$$
$$k_1 V + r_1 z = k_2 V + r_2 z.$$

Hence W is a function. It is now easy to check that W is a homomorphism. Moreover

$$kW = (k + 0y)W = kV + 0z = kV,$$

so $W \mid K = V$. Hence $W \mid H = V \mid H = T$. Therefore $(K, V) < (L, W)$ and $K \neq L$. This contradicts the maximality of (K, V). Therefore the theorem is true.

EXERCISES

5.8.8. State a theorem which describes $\mathrm{Hom}(A, B)$ completely for finitely generated Abelian groups A and B.

5.8.9. Verify that the hypotheses of Zorn's lemma are satisfied by the partially ordered set S constructed in the proof of Theorem 5.8.7.

5.8.10. $\mathrm{Hom}\,(J_n, B) \cong B_n = \{x \in B \mid nx = 0\}$.

5.8.11. (a) Describe $\mathrm{Hom}\,(A, B)$ for A a direct sum of cyclic groups and B Abelian.
 (b) Do the same more explicitly for B a direct sum of cyclic groups.

REFERENCES FOR CHAPTER 5

For the entire chapter, Fuchs [1] and Kaplansky [1]; for Theorem 5.1.17, Frobenius and Stickelberger [1]; Section 5.7, Zassenhaus [4].

p-GROUPS AND p-SUBGROUPS

6.1 Sylow theorems

6.1.1. *A subgroup or factor group of a p-group is a p-group.*

Proof. For a subgroup this is obvious. For a factor group, it follows from Exercise 2.4.12.

6.1.2. *If $H \lhd G$, then G is a p-group iff H and G/H are p-groups.*

Proof. If G is a p-group, then H and G/H are p-groups by Theorem 6.1.1. Conversely, if H and G/H are p-groups and $g \in G$, then $g^{p^n} \in H$ for some n. Hence, for some r, $g^{p^{n+r}} = e$, so $o(g) = p^s$, and G is a p-group.

6.1.3. *If G is a group, then $\exists$ a subset S of G such that*

$$G = Z \overset{.}{\cup} (\overset{.}{\cup} \{\mathrm{Cl}(x) \mid x \in S\}),$$

$$o(G) = o(Z) + \Sigma \{o(\mathrm{Cl}(x)) \mid x \in S\}.$$

Proof. Since conjugacy is an equivalence relation, G is the disjoint union of conjugate classes of elements. If $x \in G$, then $\{x\}$ is a conjugate class iff $x \in Z$. The theorem follows. ‖

The last equation in Theorem 6.1.3 is the *class equation* of G.

6.1.4. *If G is a finite group, $o(G) = p^r m$, and $(p, m) = 1$, then there is an $H \subset G$ such that $o(H) = p^r$.*

Proof. Induct on $o(G)$. If $\exists K < G$ such that $p^r \mid o(K)$, then the conclusion is true by the inductive hypothesis. In the contrary case, if $o(\mathrm{Cl}(x)) > 1$, then $p^r \nmid o(C(x))$, hence $p \mid [G:C(x)]$, so that $p \mid o(\mathrm{Cl}(x))$ by Theorem 3.3.3. From the class equation $p \mid o(Z)$. Since Z is a finite Abelian group, $\exists A \subset Z$ with $o(A) = p$ (Theorem 5.1.13). Clearly $A \lhd G$. By the induction assumption, $\exists L \subset G/A$ with $o(L) = p^{r-1}$. The inverse image H of L under the natural homomorphism of G onto G/A is then of order p^r.

6.1.5. *If G is a finite group and $p \in \mathscr{P}$, then G is a p-group iff $o(G) = p^r$ for some r.*

Proof. If $o(G) = p^r$, then G is a p-group by Theorem 2.4.3. Conversely, if G is a p-group but $q \mid o(G)$ for some prime $q \neq p$, then by Theorem 6.1.4, $\exists H \subset G$ with $o(H) = q^s$, $s > 0$. By the first part of the proof, H is not a p-group, contradicting Theorem 6.1.1.

6.1.6. *If G is a p-group and H a subgroup of finite index, then $\exists r$ such that $[G:H] = p^r$.*

Proof. By Theorem 3.3.5, $\exists K \subset H$ such that $K \lhd G$ and G/K is finite. By Theorems 6.1.1 and 6.1.5, $[G:K] = p^s$ for some s. By Theorem 1.7.11, $[G:H] = p^r$ for some r.

6.1.7. *If $p \in \mathscr{P}$ and H is a p-subgroup of a group G, then there is a maximal p-subgroup of G containing H.*

Proof. By Zorn. ‖

A *Sylow p-subgroup* of G is a maximal p-subgroup of G, $p \in \mathscr{P}$. For each $p \in \mathscr{P}$, the set of Sylow p-subgroups of G will be denoted by $\mathrm{Syl}_p(G)$. The set of all Sylow p-subgroups of G for all p, i.e., $\cup \{\mathrm{Syl}_p(G) \mid p \in \mathscr{P}\}$, will be denoted by $\mathrm{Syl}(G)$. It is clear that any conjugate of a Sylow p-subgroup is again a Sylow p-subgroup. G_p will denote the Sylow p-subgroup in case there is just one.

6.1.8. *If G is a group, $o(G) = p^r m$, $p \in \mathscr{P}$, $(p, m) = 1$, $H \subset G$, and $o(H) = p^r$, then $H \in \mathrm{Syl}_p(G)$.*

Proof. This follows from Theorem 6.1.5.

6.1.9. *If G is a group, $p \in \mathscr{P}$, $H \in \mathrm{Syl}_p(G)$, $x \in G \backslash H$, and $o(x) = p^n$, then $x \notin N(H)$.*

Proof. Deny the theorem. By Theorem 6.1.2, $\langle H, x \rangle / H$ is not a p-group. It is, however, cyclic, being generated by xH. Since (Exercise 2.4.12.) $o(xH) \mid o(x)$, this is a contradiction.

6.1.10. *If K is a Sylow p-subgroup of G with a finite number n_p of conjugates, then all Sylow p-subgroups of G are conjugate, and their number $n_p \equiv 1 \pmod{p}$.*

Proof. Let H be a Sylow p-subgroup (perhaps K). Partition $\mathrm{Cl}(K)$ into equivalence classes $\mathrm{Cl}'(L)$ as follows: $M \in \mathrm{Cl}'(L)$ iff $M = L^h$ for some $h \in H$. Then if $L \neq H$, $o(\mathrm{Cl}'(L)) = [H : H \cap N(L)]$ (see proof of Theorem 3.3.2) is finite and not 1 (Theorem 6.1.9). By Theorem 6.1.6, $p \mid o(\mathrm{Cl}'(L))$ if $L \neq H$, and clearly $o(\mathrm{Cl}'(H)) = 1$ if $H \in \mathrm{Cl}(K)$. Thus

$$o(\mathrm{Cl}(K)) \equiv \begin{cases} 0 \pmod{p} & \text{if } H \notin \mathrm{Cl}(K), \\ 1 \pmod{p} & \text{if } H \in \mathrm{Cl}(K). \end{cases}$$

The case $H = K$ shows that $o(\mathrm{Cl}(K)) \equiv 1 \pmod{p}$. Hence the possibility that $H \notin \mathrm{Cl}(K)$ cannot arise, and therefore $H \in \mathrm{Cl}(K)$ for all Sylow p-subgroups H. $\parallel$

Let us gather the foregoing results together for the case where G is finite.

6.1.11. *(Sylow's theorem.) If G is a group of order $p^r m$, $p \in \mathscr{P}$, and $(p, m) = 1$, then*

(a)   *the number n_p of Sylow p-subgroups is such that $n_p \equiv 1 \pmod{p}$,*
(b) $n_p \mid m$,
(c)   *all Sylow p-subgroups are conjugate,*
(d)   *if $H \subset G$, then H is a Sylow p-subgroup iff $o(H) = p^r$.*

Proof. Conclusions (a) and (c) follow from Theorem 6.1.10. By (c) and Theorem 3.3.2, $n_p \mid o(G)$, hence by (a), $n_p \mid m$. If $o(H) = p^r$ then $H \in \mathrm{Syl}_p(G)$ by Theorem 6.1.8. By Theorem 6.1.4, there is a Sylow p-subgroup of order p^r, hence by (c) all of them are of order p^r.

6.1.12. *If $o(G) = p^r m$, $p \in \mathscr{P}$, $(p, m) = 1$, $H \subset G$, and $o(H) = p^s$, then H is contained in some (Sylow) subgroup of order p^r.*

Proof. H is contained in a maximal p-subgroup K. By definition K is a Sylow p-subgroup of G. By Theorem 6.1.11, $o(K) = p^r$.

6.1.13. *If H is a subgroup of G, $p \in \mathscr{P}$, P and Q are distinct Sylow p-subgroups of H, $P^* \supset P$, $Q^* \supset Q$, and P^* and Q^* are Sylow p-subgroups of G, then $P^* \neq Q^*$.*

Proof. If $P^* = Q^*$, then $\langle P, Q \rangle$ is a p-subgroup of H such that $\langle P, Q \rangle > P$, a contradiction.

6.1.14. *If $H \subset G$ and $p \in \mathscr{P}$, then $n_p(H) \leqq n_p(G)$.*

Proof. If $P \in \mathrm{Syl}_p(H)$, then, by Theorem 6.1.7, $\exists\, P^* \in \mathrm{Syl}_p(G)$ such that $P^* \supset P$. Now use Theorem 6.1.13.

6.1.15. *If G is a group, $p \in \mathscr{P}$, $n_p(G)$ is finite, $A \lhd G$, and $H \in \mathrm{Syl}_p(G)$, then $A \cap H \in \mathrm{Syl}_p(A)$.*

Proof. Deny the theorem. Then $\exists\, P \in \mathrm{Syl}_p(A)$ and $Q \in \mathrm{Syl}_p(G)$ such that $A \cap H < P \subset Q$. By Theorem 6.1.10, $\exists\, x \in G$ such that $Q^x = H$. Then

$$P^x \subset (A \cap Q)^x = A^x \cap Q^x = A \cap H < P.$$

Since $P^x \in \mathrm{Syl}_p(A)$, this is a contradiction.

6.1.16. *If G is a finite group, $p \in \mathscr{P}$, $H \lhd G$, and $L \subset G/H$, then $L \in \mathrm{Syl}_p(G/H)$ iff $\exists\, P \in \mathrm{Syl}_p(G)$ such that $L = PH/H$.*

Proof. If $\exists\, P \in \mathrm{Syl}_p(G)$ such that $L = PH/H$, then by the isomorphism theorem and Theorem 6.1.1, L is a p-subgroup of G/H. By Theorem 6.1.11, $p \nmid [G{:}P]$. By Theorem 1.7.11, $p \nmid [G{:}PH] = [G/H{:}PH/H]$. Hence

$$PH/H \in \mathrm{Syl}_p(G/H)$$

(Theorem 6.1.11 again).

Conversely, let $L \in \mathrm{Syl}_p(G/H)$, let T be the natural homomorphism of G onto G/H, and let $K = LT^{-1}$. Then $[G{:}K] = [G/H{:}L]$ is not divisible by p. Therefore, if $Q \in \mathrm{Syl}_p(K)$, then $Q \in \mathrm{Syl}_p(G)$. Now $QH/H = QT \subset KT = L$. Since $QH/H \in \mathrm{Syl}_p(G/H)$ by the first half of the proof, we must have $L = QH/H$.

6.1.17. *If G is a finite group, $p \in \mathscr{P}$, and $H \lhd G$, then $n_p(G/H) \leqq n_p(G)$.*

Proof. By Theorem 6.1.16.

6.1.18. *If G is a group, $p \in \mathscr{P}$, $n_p(G)$ is finite, $P \in \mathrm{Syl}_p(G)$, and $H \lhd \lhd G$, then $H \cap P \in \mathrm{Syl}_p(H)$.*

Proof. Since H is subnormal, there is a normal series $H \lhd A_1 \lhd \ldots \lhd A_r = G$. If $r = 1$, then $H \lhd G$, and the theorem follows from Theorem 6.1.15. Induct on r. By the inductive hypothesis, $A_1 \cap P \in \mathrm{Syl}_p(A_1)$. By Theorem 6.1.14, $n_p(A_1)$ is finite. Since $H \lhd A_1$, it follows from Theorem 6.1.15 that $H \cap P = H \cap (A_1 \cap P) \in \mathrm{Syl}_p(H)$.

6.1.19. (*Brodkey* [1].) *If $n_p(G)$ is finite, $p \in \mathscr{P}$, a Sylow p-subgroup of G is Abelian, and the intersection of all Sylow p-subgroups of G is E, then there are two Sylow p-subgroups whose intersection is E.*

Proof. Deny, and let G be a counterexample with smallest n_p. There are distinct Sylow p-subgroups $P_1, \ldots, P_n$ such that $\cap P_i = E$ and $E \neq A = \cap \{P_i \mid i > 1\}$. If $P_1 \subset C(A)$, then $P_1 + A$ exists and is a p-subgroup of G properly containing P_1, a contradiction. Hence $P_1 \not\subset C(A)$. However, $P_i \subset C(A)$ for $i > 1$ since P_i is Abelian. Since $n_p(C(A))$ is finite (Theorem 6.1.14), all Sylow p-subgroups of $C(A)$ are conjugate (Theorem 6.1.10), hence all Sylow p-subgroups of $C(A)$ are also Sylow p-subgroups of G (since P_2 is, for example). Since $P_1 \not\subset C(A)$, $n_p(C(A)) < n_p(G)$. Now the Sylow subgroups of $C(A)/A$ are precisely those S/A such that $S \in \mathrm{Syl}_p(C(A))$. Hence

$$n_p(C(A)/A) < n_p(G).$$

Since the intersection of the Sylow p-subgroups of $C(A)/A$ is E, it follows from the minimality of $n_p(G)$ that the intersection of some two Sylow p-subgroups of $C(A)/A$ is E. Hence the intersection of some two Sylow p-subgroups of $C(A)$ is A. Now $P_1 \cap C(A) \subset Q \in \mathrm{Syl}_p(C(A))$ for some Q. By conjugation (Theorem 6.1.10), $Q \cap R = A$ for some $R \in \mathrm{Syl}_p(C(A))$. Since also $R \in \mathrm{Syl}_p(G)$.

$$E < P_1 \cap R \subset P_1 \cap R \cap P_1 \cap C(A) \subset P_1 \cap R \cap Q = P_1 \cap A = E,$$

a contradiction. ‖

From time to time, application will be made of various theorems to the determination of properties of groups of small orders. In particular, the non-existence of simple groups of various orders will be used for examples or exercises. Sylow's theorem is frequently useful in this respect (see Exercises 6.1.22 and 6.1.23).

Counterexamples to various possible generalizations of the theorems of this section are given in Section 8.3, Exercises 9.2.20, 9.2.21, and 10.2.16, and the exercises below.

EXERCISES

6.1.20. If G has a finite number of Sylow p-subgroups including H and K, and if $Z(H) \lhd K$, then $Z(H) = Z(K)$.

6.1.21. If H is a normal Sylow subgroup of G, then H is fully characteristic in G.

6.1.22. There are no simple groups of order 28 or 312. (Use Sylow.)

6.1.23. There are no simple groups of order 12 or 56. (Use Sylow and count elements.)

6.1.24. Any group of order 15 is cyclic. (Use Sylow.)

6.1.25. Give an example of a finite group G, $A \subset G$, $p \in \mathscr{P}$, $H \in \mathrm{Syl}_p(G)$, but $A \cap H \notin \mathrm{Syl}_p(A)$. (Hence normality of A is needed in Theorem 6.1.15.)

6.1.26. Theorem 6.1.16 is false for infinite groups G even if $n_p(G) = 1$. (Let $G = J$ and $G/H = J_p$.)

6.1.27. Sym(3) has 3 Sylow 2-subgroups, but has a factor group with only one Sylow 2-subgroup (hence Theorem 6.1.17 cannot be improved to equality).

6.1.28. In the group Sym(3) × Sym(3), there are distinct Abelian Sylow 2-subgroups H, K, and L such that $H \cap K = E$ but $H \cap L \neq E$. (Compare with Theorem 6.1.19.)

6.2   Normalizers of Sylow subgroups

Let $p \in \mathscr{P}$ throughout this section.

6.2.1. *If H is a p-subgroup of G which is contained in exactly one Sylow p-subgroup P of G, then $N(H) \subset N(P)$.*

Proof. Let $x \in N(H)$. Then $P^x \in \mathrm{Syl}_p(G)$, and $H = H^x \subset P^x$. Hence $P^x = P$ and $x \in N(P)$.

6.2.2. *If $n_p(G)$ is finite, $P \in \mathrm{Syl}_p(G)$, and $H \supset N(P)$, then $N(H) = H$.*

Proof. Let $x \in N(H)$. By Theorem 6.1.14, H has a finite number of Sylow p-subgroups, among them P and P^x. By Theorem 6.1.10, $\exists\, y \in H$ such that $P^{xy} = (P^x)^y = P$. Hence $xy \in N(P) \subset H$, so that $x \in H$. ∥

6.2.3. *If $n_p(G)$ is finite, $P \in \mathrm{Syl}_p(G)$, and $H \supset N(P)$, then $[G:H] \equiv 1$ (mod p).*

Proof. By Theorem 6.1.14, $n_p(H)$ is finite, and by Theorem 6.1.10,

$$n_p(G) \equiv n_p(H) \equiv 1 (\mathrm{mod}\ p).$$

By 3.3.2,

$$[G:N(P)] \equiv [H:N(P)] \equiv 1 (\mathrm{mod}\ p).$$

Since $[G:N(P)] = [G:H][H:N(P)]$, it follows that $[G:H] \equiv 1 (\mathrm{mod}\ p)$.

6.2.4. *If G is a group, $H \lhd G$, $n_p(H)$ is finite, and $P \in \mathrm{Syl}_p(H)$, then $G = HN(P)$.*

Proof. Let $g \in G$. Then $P^g \subset H^g = H$. Since $n_p(H)$ is finite, $\exists \ x \in H$ such that $P^{gx} = P$. Therefore $gx \in N(P)$, so $g \in N(P)H = HN(P)$. Hence $G = HN(P)$. ‖

A subset S of a group G will be called *normal* iff S^x is a subset of S for all $x \in G$. (It follows that $S^x = S$ for all $x \in G$.)

6.2.5. *If $n_p(G)$ is finite, $P \in \mathrm{Syl}_p(G)$, and L and M are normal subsets of P which are conjugate in G, then they are conjugate in $N(P)$.*

Proof. $\exists \ x \in G$ such that $M = L^x$. Then $M = L^x$ is a normal subset of P^x. Hence P and P^x are Sylow p-subgroups of $N(M)$. Therefore $\exists \ z \in N(M)$ such that $P^{xz} = P$. Since $L^{xz} = M^z = M$, we are done. ‖

The following two corollaries are immediate.

6.2.6. *If $n_p(G)$ is finite, $P \in \mathrm{Syl}_p(G)$, and L and M are normal subgroups of P which are conjugate in G, then they are conjugate in $N(P)$.*

6.2.7. *If $n_p(G)$ is finite, $P \in \mathrm{Syl}_p(G)$, and x and y are elements of $Z(P)$ which are conjugate in G, then they are conjugate in $N(P)$.* ‖

The following generalization of Theorem 6.2.7 has a similar proof.

6.2.8. *If $n_p(G)$ is finite, $P \in \mathrm{Syl}_p(G)$, and x and y are elements of $C(P)$ which are conjugate in G, then they are conjugate in $N(P)$.*

Proof. $\exists \ u \in G$ such that $y = x^u$. Since $x \in C(P)$, $y \in C(P^u)$. Then P and P^u are Sylow p-subgroups of $C(y)$. By Theorem 6.1.14, $n_p(C(y))$ is finite, hence by Sylow (Theorem 6.1.10), $\exists \ z \in C(y)$ such that $P^{uz} = P$. Thus $uz \in N(P)$, and $x^{uz} = y^z = y$. ‖

A subgroup K of G is a *complement* of a subgroup H of G iff $G = HK$ and $H \cap K = E$. If K is a complement of H, then H is a complement of K.

6.2.9. *(Burnside's theorem.) If G is a finite group, $P \in \mathrm{Syl}_p(G)$, and $N(P) = C(P)$, then P has a normal complement in G.*

Proof. Since $C(P) = N(P) \supset P$, P is Abelian. Let T be the transfer of G into P. Let $G = \dot{\cup} \{xP \mid x \in S\}$ and $y \in P^{\#}$. By Theorem 3.5.6, there is a subset S' of S such that

$$yT = \pi\{x^{-1}y^{n_x}x \mid x \in S'\}$$

where $\Sigma \ n_x = [G:P]$ and $x^{-1}y^{n_x}x \in P$. By Theorem 6.2.7, $\exists \ z \in N(P)$ such that $z^{-1}y^{n_x}z = x^{-1}y^{n_x}x$. Since $N(P) = C(P)$, $x^{-1}y^{n_x}x = y^{n_x}$. Therefore, since $([G:P], p) = 1$,

$$yT = \pi y^{n_x} = y^{\Sigma n_x} = y^{[G:P]} \neq e.$$

Hence $\text{Ker}(T) \cap P = E$. Therefore $PT = P$, $G/\text{Ker}(T) \cong P$, and $o(G) = o(P)o(\text{Ker}(T))$. It follows that $G = \text{Ker}(T)P$, and $\text{Ker}(T)$ is a normal complement of P.

6.2.10. *If G is a finite group, $P \in \text{Syl}_p(G)$, and $N(P) = C(P)$, then $p \nmid o(G^1)$.*

Proof. By Burnside's theorem, P has a normal complement H: $P \cap H = E$ and $G = PH$. Hence $G/H \cong P$ is Abelian, so $H \supset G^1$. Since $p \nmid o(H)$, $p \nmid o(G^1)$. ∥

Some applications of Burnside's theorem will now be given.

6.2.11. *If G is a finite group, p is the smallest prime dividing $o(G)$, and a Sylow p-subgroup P of G is cyclic, then P has a normal p-complement.*

Proof. By the N/C theorem, 3.2.3, $N(P)/C(P) \cong \text{Aut}(P)$. By Theorem 5.7.12, $o(\text{Aut}(P)) = p^{n-1}(p-1)$ where $o(P) = p^n$. Since $C(P) \supset P$, $p \nmid o(N(P)/C(P))$. Therefore, $o(N(P)/C(P)) \mid p-1$. Since p is the smallest prime divisor of $o(G)$, $N(P) = C(P)$. By Burnside, P has a normal complement.

6.2.12. *If G is a simple group of order $pm > p$, $p \nmid m$, and $P \in \text{Syl}_p(G)$, then $C(P) < N(P) < G$, and $[N(P):C(P)] \mid p-1$.*

Proof. Since G is simple, $N(P) < G$. By Burnside's theorem, $C(P) < N(P)$. By the N/C theorem, 3.2.3, $N(P)/C(P) \cong \text{Aut}(P)$. By Theorem 5.7.12, $\text{Aut}(P)$ is (cyclic) of order $p-1$. The theorem follows. ∥

Example. There is no simple group of order 396. In fact, suppose that G is a simple group of order $396 = 2^2 3^2 11$. By Sylow, $n_{11}(G) = 12$, so that if $P \in \text{Syl}_{11}(G)$, then $o(N(P)) = 3 \cdot 11$. By Theorem 6.2.12, $o(C(P)) = 11$ and $3 \mid 10$, a contradiction.

EXERCISES

6.2.13. If G is a finite group, $H \lhd G$, and $P \in \text{Syl}_p(H)$, then $\exists\, Q \in \text{Syl}_p(G)$ such that $Q \subset N(P)$.

6.2.14. There are no simple groups of order 616.

6.2.15. There are no simple groups of order $2^3 \cdot 3 \cdot 7 \cdot 23$.

6.2.16. There are no simple groups of order $3^3 \cdot 5 \cdot 7$.

6.2.17. If $n_p(G)$ is finite, $P \in \mathrm{Syl}_p(G)$, and $H \supset N(P)$, then $H^G = G$. (Use Theorem 6.2.2.)

6.2.18. If $G \neq E$ is a finite solvable group, then there is at most one prime p such that if $P \in \mathrm{Syl}_p(G)$, then $N(P) = P$. (Use Theorem 6.2.4.)

6.2.19. If a Sylow p-subgroup of a finite group G has a normal p-complement H, then H is fully characteristic in G.

6.3 p-Groups

The principal theorems for finite p-groups will be proved in this section. Several of the theorems will be generalized in later sections. Let $p \in \mathscr{P}$ throughout this section.

6.3.1. *If G is a p-group, $E < H \triangleleft G$, and H is finite, then $Z(G) \cap H \neq E$.*

Proof. There is a subset S of H such that

$$o(H) = o(Z \cap H) + \Sigma \{o(\mathrm{Cl}(x)) \mid x \in S\}.$$

If $x \in S$, then $\mathrm{Cl}(x)$ is finite, so that, by Theorem 6.1.6,

$$o(\mathrm{Cl}(x)) = [G:C(x)] = p^r, \qquad r > 0.$$

Since H is a p-group, $p \mid o(H)$. Hence $p \mid o(Z \cap H)$.

6.3.2. *If $G \neq E$ is a finite p-group, then $Z \neq E$.*

Proof. Set $H = G$ in Theorem 6.3.1. ‖

This theorem is false for infinite p-groups (Section 9.2, Example 5).

6.3.3. *If G is a p-group, $H \triangleleft G$, and $o(H) = p$, then $H \subset Z$.*

Proof. By Theorem 6.3.1.

6.3.4. *If $o(G) = p^n$, then $o(Z) \neq p^{n-1}$.*

Proof. Otherwise G/Z is cyclic and G is non-Abelian, contradicting Theorem 3.2.8.

6.3.5. *If $o(G) = p^2$, then (a) G is Abelian, and (b) $G \cong J_{p^2}$ or $J_p \times J_p$.*

Proof. By Lagrange, Theorems 6.3.2 and 6.3.4, $o(Z) = p^2$, so that G is Abelian. Statement (b) now follows from the fundamental theorem of finite Abelian groups. ‖

The *upper central series* of a group G is the sequence $(E = Z_0, Z_1, \ldots)$, where

$$Z_{n+1} = \{x \in G \mid [x, y] \in Z_n \text{ for all } y \in G\}.$$

6.3.6. *If $(Z_0, Z_1, \ldots)$ is the upper central series of a group G, then $Z_{n+1}/Z_n = Z(G/Z_n)$ for all n.*

Proof. Inductively assume that $Z_n \lhd G$. Then $x \in Z_{n+1}$ iff $[xZ_n, yZ_n] = Z_n$ for all $yZ_n \in G/Z_n$. But this means that $Z_{n+1}/Z_n = Z(G/Z_n)$. Therefore, $Z_{n+1}/Z_n \lhd G/Z_n$, hence $Z_{n+1} \lhd G$. ‖

A group is *nilpotent* iff $G = Z_n$ for some n. Nilpotent groups will be studied more fully in the next section. Returning to p-groups, we have the result:

6.3.7. *If G is a finite p-group, then G is nilpotent.*

Proof. If $Z_n < G$, then G/Z_n is a p-group with nontrivial center (Theorem 6.3.2), hence (Theorem 6.3.6), $Z_{n+1} > Z_n$. Therefore some $Z_i = G$, and G is nilpotent.

6.3.8. *If G is a finite p-group, then G is solvable.*

Proof. The upper central series of G has Abelian factors.

6.3.9. *If G is a finite p-group and $H < G$, then $N(H) > H$.*

Proof. Let n be maximal such that $Z_n \subseteq H$. Then $\exists\; x \in Z_{n+1}\backslash H$. If $h \in H$, then $h^x = h[h, x] \in HZ_n = H$. Hence $H^x = H$, $x \in N(H)$, and $H < N(H)$.

6.3.10. *If G is a finite p-group and $H \subseteq G$, then H is a member of some composition series of G.*

Proof. By Theorem 6.3.9 and induction, there is a normal series $S = (E, H, N(H), N(N(H)), \ldots, G)$ of G. By Theorem 2.5.6, there is a composition series of G which is a refinement of S.

6.3.11. *If G is a finite p-group, then a principal series of G has factors of order p.*

Proof. The upper central series of G has Abelian factors Z_{i+1}/Z_i. Now Z_{i+1}/Z_i has a composition series whose factors are of order p by Theorem 2.6.4. Hence, by the lattice theorem, there is a normal series $(E = H_0, \ldots,$

$H_r = G$) of G whose factors are of order p, and which refines the upper central series. For each H_i, $\exists\, j$ such that $Z_j \subset H_i \subset Z_{j+1}$. Then

$$\frac{H_i}{Z_j} \subset \frac{Z_{j+1}}{Z_j} = Z\!\left(\frac{G}{Z_j}\right).$$

Hence $H_i/Z_j \lhd G/Z_j$, so that $H_i \lhd G$. Therefore the normal series which has been constructed is actually a principal series whose factors are of order p.

EXERCISES

6.3.12. If $o(G) = p^n$, $n \geq 2$, then $o(G^1) \leq p^{n-2}$.

6.3.13. If $o(G) = 36$, then G has a normal Sylow subgroup. (*Hint*: a Sylow 3-subgroup P is Abelian, and if it is not normal, then Burnside applies).

6.3.14. If $o(G) = p^2$, $p \in \mathscr{P}$, then

$$o(\mathrm{Aut}(G)) = \begin{cases} p(p-1) & \text{if } G \text{ is cyclic,} \\ (p+1)p(p-1)^2 & \text{if } G \text{ is not cyclic.} \end{cases}$$

6.3.15. If G is finite, p is the smallest prime dividing $o(G)$, $P \in \mathrm{Syl}_p(G)$, and $o(P) = p^2$, then either

  (a) P has a normal p-complement, or

  (b) $p = 2$, P is not cyclic, and $o(N(P)/C(P)) = 3$ (hence $12 \mid o(G)$).

6.3.16. A simple non-Abelian group of odd order has order divisible by the cube of its smallest prime divisor.

6.3.17. If $o(G) = p^n$, $T \in \mathrm{Aut}(G)$, and $o(T) = p^m$, then $m < n$. (T is a permutation of G. Examine its cycle structure.)

6.4 Nilpotent groups

An *ascending central series* of a group G is an invariant series ($A_0 = E$, $A_1, \ldots$) (not necessarily reaching G) in which each $A_{i+1}/A_i \subset Z(G/A_i)$. Equivalently it could be required that $[A_{i+1}, G] \subset A_i$. *Descending central series* is defined similarly. It is clear that the upper central series is actually an ascending central series. Next, it will be shown that it is really "upper."

6.4.1. *If $(A_0, A_1, \ldots)$ is an ascending central series for G, then $A_n \subset Z_n$ for all n.*

Proof. Induct on n. By the induction hypothesis, $[A_{n+1}, G] \subset A_n \subset Z_n$, so that, by definition of Z_{n+1}, $A_{n+1} \subset Z_{n+1}$.

6.4.2. *Each term Z_n of the upper central series is characteristic in G.*

Proof. Induct on n. Then $Z_{n+1}/Z_n \in \text{Char}(G/Z_n)$ (Theorem 3.2.5), and by the inductive hypothesis, $Z_n \in \text{Char}(G)$. Therefore (Exercise 2.11.22) $Z_{n+1} \in \text{Char}(G)$. ‖

The *lower central series* of a group G is the sequence

$$(Z^0 = G, Z^1, \ldots), Z^{n+1} = [G, Z^n] \quad \text{for} \quad n \geqq 0.$$

The lower central series is a descending central series. Moreover, it is actually "lower" because of the following theorem.

6.4.3. *If $(A_0 = G, A_1, \ldots)$ is a descending central series of G, then each $Z^n \subset A_n$.*

Proof. If, inductively, $A_n \supset Z^n$, then

$$A_{n+1} \supset [G, A_n] \supset [G, Z^n] = Z^{n+1}. ‖$$

If G is nilpotent, its *class* is the smallest integer n such that $Z_n = G$. Thus the class is the length (one less than the number of terms) of the upper central series. The next theorem shows that the class is also the length of the lower central series.

6.4.4. *If G is a nilpotent group of class n, then Z^n is the first term of the lower central series which equals E.*

Proof. $(Z_n, Z_{n-1}, \ldots, Z_0)$ is a descending central series of G. By Theorem 6.4.3, $Z^n \subset Z_0 = E$, so $Z^n = E$. If $Z^{n-1} = E$, then $(Z^{n-1}, \ldots, Z^0)$ is an ascending central series of G, so by Theorem 6.4.1, $G = Z^0 \subset Z_{n-1}$, a contradiction. Therefore $Z^{n-1} \neq E$. ‖

The inheritance properties of the class of nilpotent groups will be considered next.

6.4.5. *A subgroup of a nilpotent group (of class n) is nilpotent (of class $\leqq n$).*

Proof. Let G be nilpotent of class n, and $H \subset G$. Then $Z^0(H) = H \subset G = Z^0(G)$. If, inductively, $Z^i(H) \subset Z^i(G)$, then

$$Z^{i+1}(H) = [H, Z^i(H)] \subset [G, Z^i(G)] = Z^{i+1}(G).$$

Hence, $Z^n(H) \subset Z^n(G)$, so $Z^n(H) = E$.

6.4.6. *A homomorphic image of a nilpotent group (of class n) is nilpotent (of class $\leqq n$). In fact, if $T \in \text{Hom}(G)$, then GT has lower central series $\{Z^n(G)T\}$.*

Proof. Inductively,

$$Z^{n+1}T = [G, Z^n]T = [GT, Z^nT] = [GT, Z^n(GT)] = Z^{n+1}(GT).$$

Both statements now follow. ‖

The next two theorems can be proved by copying the proofs of the corresponding theorems for p-groups (Theorems 6.3.8 and 6.3.9).

6.4.7. *If G is a nilpotent group, then G is solvable.*

6.4.8. *If G is a nilpotent group and $H < G$, then $N(H) > H$.*

6.4.9. *If G is a nilpotent group and H is a maximal proper subgroup, then $H \lhd G$.*

Proof. This follows from Theorem 6.4.8.

6.4.10. *If G is a nilpotent group and $H \subset G$, then H is subnormal in G.*

Proof. We assert that, in fact,

$$H = HZ_0 \lhd HZ_1 \lhd \ldots \lhd HZ_n = G,$$

where n is the nilpotence class of G. To prove this, note that if $h \in H$ and $z \in Z_{i+1}$, then $h^z = h[h, z] \in HZ_i$, so that $Z_{i+1} \subset N(HZ_i)$. Clearly, $H \subset N(HZ_i)$, and therefore $HZ_i \lhd HZ_{i+1}$ as asserted.

6.4.11. *If G is a periodic group such that, for each $p \in \mathscr{P}$, a Sylow p-subgroup G_p is normal in G, then $G = \Sigma \, G_p$.*

Proof. By Theorem 6.1.10, there is only one Sylow p-subgroup G_p for each prime p. By Theorem 6.1.7, if $g \in G$ and $o(g) = p^i$, then $g \in G_p$. If $x \in G$, then by Theorem 5.1.5, $x \in \langle G_p \rangle$, so that $G = \langle G_p \rangle$. Since $G_p \cap G_q = E$ if $p \neq q$, Theorem 4.1.1 implies that $G_p \subset C(G_q)$. Hence if $x \in \langle G_q \mid q \neq p \rangle$, then $p \nmid o(x)$. Therefore $G_p \cap \langle G_q \mid q \neq p \rangle = E$. But this means that $G = \Sigma \, G_p$.

6.4.12. *If G is a group such that $H < G$ implies $H < N(H)$, then*
(i)   *every Sylow subgroup of G is normal,*
(ii) *the set T of elements of finite order in G is a (fully characteristic) subgroup,*
(iii) $T = \Sigma \, \{G_p \mid p \in \mathscr{P}\}.$

Proof. Let $p \in \mathscr{P}$ and $G_p \in \mathrm{Syl}_p(G)$. If G_p is not normal, then $G_p < N(G_p) < N(N(G_p))$. Since G_p is the only Sylow p-subgroup of $N(G_p)$ (Theorem 6.1.10), G_p is characteristic in $N(G_p)$, hence normal in $N(N(G_p))$, a contradiction. Therefore $G_p \lhd G$. Hence $n_p(G) = 1$ and G_p contains all elements of order a power of p.

If $x \in T$, then $x = x_1 \cdots x_n$, where $o(x_i) = p_i^{f_i}$ by Theorem 5.1.5. Thus each $x_i \in G_{p_i}$, and therefore $T \subset \langle G_p \rangle$. Since $G_p \subset T$ for all p, $T = \langle G_p \rangle$. By Theorem 6.4.11, $T = \Sigma \, G_p$. T is fully characteristic by 5.1.1. ‖

The hypothesis of Theorem 6.4.12 is satisfied if G is nilpotent (Theorem 6.4.8). Therefore, we have the following theorem.

6.4.13. *If G is a nilpotent group, then every Sylow subgroup G_p of G is normal, and $\Sigma \, G_p$ exists and equals the subgroup of all elements of finite order in G.* ‖

There are several ways of characterizing finite nilpotent groups, three of which will now be given.

6.4.14. *If G is a finite group, then the following conditions are equivalent:*

(i)   *G is nilpotent,*

(ii) *$H < G$ implies $H < N(H)$,*

(iii) *if M is a maximal proper subgroup of G, then $M \lhd G$,*

(iv) *$G = \Sigma \, \{ G_p \mid G_p \in \mathrm{Syl}_p(G), p \in \mathscr{P} \}$.*

Proof. (i) implies (ii) by Theorem 6.4.8. Obviously, (ii) implies (iii). Assume (iii) holds, and let $G_p \in \mathrm{Syl}_p(G)$. If $N(G_p) < G$, then $\exists$ a maximal proper subgroup M of G containing $N(G_p)$. By assumption, $N(M) = G$, but by Theorem 6.2.2, $N(M) = M$, a contradiction. Hence all Sylow subgroups are normal, and (iv) follows. Finally, assume (iv). An easy induction gives

$$Z_n(G) = \Sigma \, \{ Z_n(G_p) \mid p \in \mathscr{P} \text{ and } p \mid o(G) \}.$$

Since each G_p is nilpotent (Theorem 6.3.7) and there are only a finite number of relevant primes, some $Z_n(G) = G$ and G is nilpotent.

EXERCISES

6.4.15. Each term of the upper central series of G is strictly characteristic in G.

6.4.16. Each term of the lower central series is fully characteristic.

6.4.17. It is false that if H and G/H are nilpotent, then G is nilpotent [try Sym(3)].

6.4.18. The fact that $Z(G) = E$ does not imply that the first term G^1 of the lower central series is G. (It is even true that there is a group G whose center is E, but $\cap \, Z^n = E$. See Theorem 8.4.16.)

6.4.19. If $A_1, \ldots, A_n$ are nilpotent, so is $\Sigma_E \, A_i$.

6.4.20. (Transfinite nilpotence.) One can continue the upper central series of a group G so as to obtain a well-ordered sequence $\{Z_i \mid i \in S\}$, where S is a set of ordinal numbers, such that $Z_{i+1}/Z_i = Z(G/Z_i)$ for all $i \in S$ and $Z_i = \cup \{Z_j \mid j < i\}$ for limit ordinals i. Making additional definitions where necessary, prove the analogues of Theorems 6.4.1, 6.4.2, 6.4.5, 6.4.6 (first part), 6.4.7. 6.4.8, 6.4.9, 6.4.10, and 6.4.13.

6.4.21. Define a transfinite version of the lower central series and prove the analogues of Theorems 6.4.3, 6.4.5, and 6.4.7.

6.4.22. (a) If $Z^2(G) < Z^1(G)(=G^1)$, then G/G^1 is not cyclic. (*Hint*: G/Z^2 is nilpotent of class 2. Now use Theorem 3.2.8.)

(b) If G is a nilpotent non-Abelian group, then G/G^1 is not cyclic.

6.4.23. If G is a finite group, then there is a nilpotent subgroup H such that $H^G = G$. [Use Theorem 6.4.14 (iii) and induct.]

6.4.24. If G is nilpotent and $E < H \lhd G$, then $H \cap Z(G) \neq E$. Consider first Z_i intersecting H, and compute $[G, H \cap Z_i]$.

6.4.25. If G is nilpotent and A is a maximal normal Abelian subgroup, then A is a maximal Abelian subgroup. [$C(A)/A \lhd G/A$. Now use Exercise 6.4.24.]

6.4.26. Let G be a group.

(a) Let $x \in Z_2 \backslash Z_1$, and define T by the rule $yT = [y, x]$ for $y \in G$. Then T is a homomorphism of G into Z_1.

(b) $[G^1, Z_2] = E$. [Use (a).]

(c) If G/Z_1 is periodic, then Z_2/Z_1 has no elements of infinite height in itself (except e).

(d) If Z_2/Z_1 is not torsion-free, then G has a subgroup of finite index. ($\exists x \in Z_2 \backslash Z_1$ with $x^n \in Z_1$. Now examine GT.)

(e) If Z_1 is torsion-free, then all Z_{n+1}/Z_n are torsion-free (even the transfinite terms).

(f) If $Z_1 < Z_2$, then $G^1 < G$.

6.4.27. Let F be a field, $G = GL(k, F)$ with $k > 1$, D the set of diagonal matrices in G, S the set of scalar matrices in G, T the set of upper triangular matrices in G (all entries below main diagonal are 0),

$H = \{I + A \mid A$ is strictly upper triangular$\}$, and

$H_n = \{I + A \mid A$ is strictly triangular, and the first $n - 1$ diagonals of A above the main diagonal are zero$\}$.

Then

(a) $S \subset D \subset T \subset G$ and $H_n \subset H \subset T$.

(b) If $I + A \in H$, then $(I + A)^{-1} = I - A + A^2 - A^3 + \ldots$.

(c) $[H_r, H_s] = H_{r+s}$.

(d) $Z_n(H) = H_{k-n}$.

(e) H is nilpotent of class $k - 1$.

(f) $C(H) = Z(H) + S = H_{k-1} + S$.

(g) $S = Z(G)$.

(h) $N(H) = T = HD$, $H \cap D = E$.

(i) $[T, H_n] = H_n$ if $o(F) \neq 2$. (What happens if $o(F) = 2$?)

(j) $D \cong \Sigma \{F^{\#} \mid 1 \leq i \leq k\}$.

(k) $H_n/H_{n+1} \cong \Sigma\{F^+ \mid 1 \leq i \leq k - n\}$.

(l) $T^1 = H$ if $o(F) \neq 2$.

(m) T is solvable of length the least integer which is as large as $1 + \log_2 k$.

(n)   If F is infinite, then $o(\text{Cl}(H)) = o(F) = o(G)$.

(o)   If $o(F) = p^n$, $p \in \mathcal{P}$, then $H \in \text{Syl}_p(G)$.

(p)   If $o(F) = p^n = q$, $p \in \mathcal{P}$, then

$$o(\text{Cl}(H)) = (\pi\{q^i - 1 \mid 1 \leq i \leq k\})/(q - 1)^k.$$

(q)   If the characteristic of F is 0, then H is torsion-free.

6.5 Applications

In this section, some applications of the preceding sections (mainly Burnside's theorem) will be given.

6.5.1. *If $o(G) = p^n$, $p \in \mathcal{P}$, $n \geq 4$, then there is an Abelian normal subgroup H of G with $o(H) = p^3$.*

Proof. By 6.3.11, $\exists K \lhd G$ with $o(K) = p^2$. By Exercise 6.3.14, $p^2 \nmid o(\text{Aut}(K))$. By the N/C theorem,

$$[G:C(K)] = [N(K):C(K)] \mid o(\text{Aut}(K)).$$

Hence $p^3 \mid o(C(K))$. The invariant series $(E, K, C(K), G)$ has a refinement $(E, A, K, H, \ldots, C(K), \ldots, G)$ which is a principal series. Thus $o(H) = p^3$ and $H \lhd G$. Now $\exists x \in H \backslash K$ such that $H = \langle K, x \rangle$. Since $x \in C(K)$ and K is Abelian, H is Abelian. ‖

For a generalization of Theorem 6.5.1 see Exercise 7.3.27.

6.5.2. *If G is a group of order p^2q^2 with p and q distinct primes, then G has a normal Sylow subgroup.*

Proof. WLOG $p < q$. If $n_q = 1$, then $G_q \lhd G$. If $n_q > 1$, then by Sylow, $n_q = p^2$. Therefore, if $P \in \text{Syl}_q(G)$, then $N(P) = C(P) = P$. By Burnside, $G_p \lhd G$.

6.5.3. *If G is a finite group, $p \in \mathscr{P}$, $i \in \mathscr{N}$, and $n_p \not\equiv 1 (\bmod\, p^i)$, then there are distinct Sylow p-subgroups H and K such that $[H:H \cap K] < p^i$.*

Proof. Let $H \in \mathrm{Syl}_p(G)$. As in the proof of Theorem 6.1.10, $\mathrm{Syl}_p(G)$ breaks up into equivalence classes with $L \sim M$ iff $M = L^h$ for some $h \in H$. The size of each equivalence class $Cl'(L)$ is $[H:H \cap N(L)]$, which equals $[H:H \cap L]$ by Theorem 6.1.9. If the theorem were false, then all $Cl'(L)$ except $Cl'(H)$ would have p^j, $j \geqq i$ members. Hence we would have

$$n_p = 1 + p^{j_1} + \ldots + p^{j_r} \equiv 1 (\bmod\, p^i),$$

a contradiction. The theorem follows. ‖

In particular, for $i = 2$, the theorem reads:

6.5.4. *If G is a finite group and $n_p \not\equiv 1 (\bmod\, p^2)$, then there are Sylow subgroups H and K such that $o(H/H \cap K) = p$.*

6.5.5. *If $o(G) = p^2 q^3$, $p \in \mathscr{P}$, $q \in \mathscr{P}$, and $p < q$, then G has a normal Sylow subgroup.*

Proof. Deny. As in Theorem 6.5.2, $n_q = p^2$, $q \,|\, p^2 - 1$, $p = 2$, $q = 3$, and $n_3 = 4$. Let $Q \in \mathrm{Syl}_3(G)$. If Q were Abelian, Burnside would give a contradiction. Hence $o(Z(Q)) = 3$, by Theorems 6.3.2 and 6.3.4. By Theorem 6.5.4, $\exists\, Q^* \in \mathrm{Syl}_3(G)$ such that if $H = Q \cap Q^*$, then $o(H) = 9$. By Theorem 6.3.9, $H \lhd Q$ and $H \lhd Q^*$, so that $N(H)$ has more than one Sylow 3-subgroup. Therefore $n_3(N(H)) = 4$, $2^2 3^3 \,|\, o(N(H))$, and $H \lhd G$. By Theorem 6.3.1, $Z(Q) \subset H$.

If $\exists\, K \subset G$ with $o(K) = 54$, then K would contain a Sylow 3-subgroup Q' of G. By Sylow $Q' \lhd K$, and $n_3(G) \neq 4$, a contradiction. Hence there is no subgroup of order 54.

Suppose that $N(Z(Q)) = G$. By the N/C theorem, either $C(Z(Q)) = G$ or $o(C(Z(Q))) = 54$, which we have just shown to be impossible. Hence $Z(Q) \subset Z(G)$. By Theorem 6.5.2, $G/Z(Q)$ has a normal subgroup of order 4 or 9. The second possibility would yield a normal subgroup of G of order 3^3, a contradiction. The first gives a normal subgroup L of G of order 12 such that $L \supset Z(Q)$. Let $M \in \mathrm{Syl}_2(L)$. Then $M \subset G = C(Z(Q))$, hence $M \in \mathrm{Char}(L)$ (since M is the only Sylow 2-subgroup). Since $L \lhd G$, it follows that $M \lhd G$, and we are done.

Suppose that $N(Z(Q)) < G$. Since there is no subgroup of order 54, it follows that $N(Z(Q)) = Q$. Therefore $Z(Q)$ has four conjugates, all in H since $Z(Q) \subset H \lhd G$. This means that every element of H different from e is in exactly one of the conjugates of $Z(Q)$. Therefore, if $o(x) = 2$ and $y \in H\#$, then $y^x \notin \langle y \rangle$. But $y^x = b$, say, and $b^x = y$, hence $(yb)^x = by = yb$, a contradiction. This proves the theorem.

6.5.6. *Let G be a finite group such that if A is an Abelian subgroup of G, then $N(A) = C(A)$. Then G is Abelian.*

Proof. Let $P \in \mathrm{Syl}(G)$, and let A be a maximal Abelian subgroup of P. If $A < P$, then (Theorem 6.3.9) $N_P(A) > A$. Hence by assumption, $\exists\ x \in P \cap C(A)$ with $x \notin A$. Then $\langle A, x \rangle$ is an Abelian subgroup of P which is larger than A, a contradiction. Therefore $A = P$; i.e., every Sylow subgroup is Abelian. Moreover, $N(P) = C(P)$ by assumption. It follows from corollary 6.2.10 to Burnside's theorem that $p \nmid o(G^1)$ for all primes p. Hence $G^1 = E$, so that G is Abelian.

6.5.7. *If G is a finite nonnilpotent group all of whose proper subgroups are nilpotent, then $\exists\ P \in \mathrm{Syl}_p(G)$ and $Q \in \mathrm{Syl}_q(G)$ for some distinct primes p and q such that* (i) $G = PQ$, (ii) $Q \lhd G$, (iii) *P is cyclic, and* (iv) *G is solvable.*

Proof. Induct on $o(G)$. First suppose that G is simple. If there is just one maximal proper subgroup H, then H is normal, hence $H = E$ and G is cyclic. Therefore there are at least two maximal proper subgroups. Let H and K be such with a maximal intersection. If $H \cap K \neq E$, then $N(H \cap K) < G$ by the simplicity of G. Therefore there is a maximal proper subgroup $L \supset N(H \cap K)$. Since H is nilpotent $N(H \cap K) \cap H > H \cap K$ (Theorem 6.4.8), so that $L \cap H > H \cap K$. Therefore $L = H$, and similarly, $L = K$, a contradiction. Hence any two distinct maximal subgroups intersect in E.

Let H be a maximal proper subgroup of G. Then $N(H) = H$ and $H \cap H^x = E$ for $x \notin H$. Therefore the number of elements in conjugates of $H^\#$ is

$$(o(H) - 1)[G:H] = o(G) - [G:H] \geq o(G)/2.$$

Since $[G:H] \geq 2$, there is at least one element $x \neq e$ outside all conjugates of H, and therefore a maximal subgroup K containing x not in $\mathrm{Cl}(H)$. By the intersection property of maximal subgroups, the elements in conjugates of $K^\#$ are disjoint from those in conjugates of $H^\#$. Thus, counting e, at least $2(o(G)/2) + 1$ elements of G have been located, an impossibility.

Therefore G is not simple. Thus $\exists\ K \lhd G$ with $E < K < G$. Now all subgroups of G/K are nilpotent, hence by the induction hypothesis, G/K is solvable. Since K is nilpotent, G is solvable. If all Sylow subgroups are normal, then G is nilpotent. Hence $\exists\ P \in \mathrm{Syl}_p(G)$, say, such that P is not normal in G. Since G is solvable, $\exists\ H \lhd G$ such that $[G:H]$ is prime. Since H is nilpotent, any Sylow subgroup of H is characteristic in H, hence normal in G. Therefore $P \nsubseteq H$, so that $[G:H] = p$. Thus all $Q \in \mathrm{Syl}_q(G)$ with $q \neq p$ are normal in G. Now $\exists\ Q \in \mathrm{Syl}_q(G)$, $q \neq p$, such that $Q \nsubseteq C(P)$ (otherwise $P \lhd G$). Then $PQ \subset G$ and PQ is not nilpotent. Therefore $G = PQ$, so that (i) and (ii) hold. If $\langle x \rangle < P$, then $Q\langle x \rangle$ is nilpotent, so $Q \subset C(x)$. If P is not cyclic, then $C(P) \supset Q$, a contradiction. Hence P is cyclic and (iii) holds. ‖

Further information on such groups is contained in Exercise 7.3.28.

6.5.8. *If G is a finite non-Abelian group all of whose proper subgroups are Abelian, then either* (i) *G is a p-group for some prime p, or* (ii) *the conclusions of Theorem* 6.5.7 *hold and, in addition, Q is elementary Abelian.*

Proof. If G is nilpotent but not a p-group, then G is the direct sum of its Sylow subgroups. Since the latter are Abelian, so is G, a contradiction. Hence if G is nilpotent it is a p-group for some prime p.

If G is not nilpotent, then Theorem 6.5.7 applies. Also Q is Abelian by assumption. Suppose that Q is not elementary, let $P = \langle x \rangle$, and let T be the automorphism of Q induced by x. Any proper characteristic subgroup L of Q is normal in G, so PL is Abelian, hence T fixes all elements of L. Two such characteristic subgroups are Q_1, the subgroup of qth powers of elements of Q, and Q_2, the subgroup of all elements of order q or 1. Since $Q \nsubseteq C(P)$, $Q \nsubseteq C(x)$. Let $y \in Q \backslash C(x)$. Now $y^q \in Q_1$, so $y^q T = y^q$. Hence $(yT)^q = y^q$. But $y^q = z^q$ iff $y = zu$ where $u \in Q_2$. Hence T permutes the elements of yQ_2. But $o(yQ_2) = o(Q_2) = q^i$ for some $i > 0$. Since $Q\langle x^p \rangle$ is Abelian, $o(T) = p$, and T must fix some element z of yQ_2 (for T is a product of p-cycles and 1-cycles). Since T fixes all elements of Q_2, if $y = zu$ with $u \in Q_2$, then $yT = y$, a contradiction. Hence, Q is elementary.

EXERCISES

6.5.9. If $p \in \mathscr{P}$ and $p^2 \mid o(G)$, then $p \mid o(\mathrm{Aut}(G))$. (Use Theorem 3.2.4, Burnside, the structure of Abelian groups, Theorems 5.7.12 and 5.7.20.)

6.5.10. The hypothesis $p < q$ cannot be omitted in Theorem 6.5.5. [Try Sym(4) $\times$ J_3.]

6.5.11. If G is a group whose order is odd and less than 1000, then G is solvable.

REFERENCES FOR CHAPTER 6

For the entire chapter, Zassenhaus [4] and M. Hall [1]; for Section 6.1, Sylow [1] and Baer [5]; Theorem 6.1.19, Brodkey [1] and Itô [5]; Exercise 6.3.17, Baer [8]; Exercise 6.4.23, Szep and Itô [1]; Theorem 6.5.6, Zassenhaus [3]; Theorem 6.5.7, Iwasawa [2]; Theorem 6.5.8, Miller and Moreno [1]; Exercise 6.5.9, Herstein and Adney [1] (for generalizations, see Ledermann and Neumann [1], Green [1], and Howarth [1]).

SEVEN

SUPERSOLVABLE GROUPS

7.1 M-groups

A *normal M-series* of a group G is a finite normal series $(A_0 = E, A_1, \ldots,$ $A_n = G)$ whose factors are infinite cyclic or finite. An *M-group* is a group G which has a normal M-series. Thus all finite groups are M-groups.

7.1.1. *Subgroups and factor groups of M-groups are M-groups.*

Proof. Let $(A_0, \ldots, A_n)$ be a normal M-series of an M-group G. If $H \subseteq G$, then $(A_0 \cap H, \ldots, A_n \cap H)$ is a normal M-series of H since $(A_{i+1} \cap H)/(A_i \cap H) \cong A_{i+1}/A_i$ by Theorem 2.5.9. Therefore, H is an M-group.

If $H \triangleleft G$, then the normal series (E, H, G) and $(A_0, \ldots, A_n)$ have equivalent refinements

$$R = (E, \ldots, H = B_0, B_1, \ldots, B_m = G)$$

and S. A refinement of an M-series is clearly an M-series, so that S, and therefore R, are M-series. Therefore, G/H has the normal M-series $(B_0/H,$ $B_1/H, \ldots, G/H)$.

7.1.2. *If H and G/H are M-groups, so is G.*

Proof. Let S_1 be a normal M-series of H, S_2 a normal M-series of G/H, and T the natural homomorphism of G onto G/H. Then the series formed by following S_1 by $S_2 T^{-1}$ is a normal M-series of G.

7.1.3. *If H and G/H are groups satisfying the maximal condition for subgroups, so is G.*

Proof. Let S be a nonempty set of subgroups of G. The set $S_H = \{K \cap H \mid K \in S\}$ has a maximal element A. The set

$$S^* = \{KH/H \mid K \in S \quad \text{and} \quad K \cap H = A\}$$

has a maximal element B. Then $\exists\, K \in S$ such that $K \cap H = A$ and $KH/H = B$. If K is not maximal in S, then $\exists\, K' \in S$ such that $K' > K$, $K' \cap H = A$, and $K'H/H = B$. Let $x \in K' \backslash K$. Then $xH \in B$, so $\exists\, y \in K$ such that $xH = yH$. Thus $x = yh$ with $h \in H$, and $y^{-1}x \in H \cap K' = A = H \cap K$. Therefore $x \in K$, a contradiction,

7.1.4. *An M-group G satisfies the maximal condition for subgroups.*

Proof. G has a normal M-series $(A_0, \ldots, A_n)$. Since J satisfies the maximal condition for subgroups, each A_{i+1}/A_i does also. By Theorem 7.1.3 and induction, G satisfies the maximal condition.

7.1.5. *If G is an M-group, then any two normal M-series of G have the same number of infinite factors.*

Proof. The two series have isomorphic refinements. Now if $A_{i+1}/A_i \cong J$ and

$$(\ldots, A_i = B_0, \ldots, B_m = A_{i+1}, \ldots)$$

is a normal series of G with all B_{j+1}/B_j nontrivial, then B_1/B_0 is infinite cyclic while B_m/B_1 is finite, since any nontrivial subgroup of J is of finite index. It follows that a normal M-series and its refinement have the same number of infinite factors. Hence the original series have the same number of infinite factors.

7.1.6. *If G is a finitely generated group and $n \in \mathcal{N}$, then there are only a finite number of subgroups of G of index n.*

Proof. If $[G:H] = n$, then H has at most n conjugates, so

$$[G:\mathrm{Core}(H)] \leq n^n.$$

By Cayley's theorem, 3.1.1, applied to $G/\mathrm{Core}(H)$, there is a homomorphism T_H of G into $\mathrm{Sym}(n^n)$ with $\mathrm{Ker}(T_H) = \mathrm{Core}(H)$. If $G = \langle x_1, \ldots, x_m \rangle$, then T_H is determined by $x_1 T_H, \ldots, x_m T_H$. Hence there are only a finite number of homomorphisms from G into $\text{Sym}(n^n)$, therefore only a finite number of subgroups which can act as $\text{Core}(H^*)$ for some subgroup H^* of index n.

If $K = \text{Core}(H)$, then H/K is a subgroup of the finite group G/K. Hence K serves as the core of only a finite number of subgroups. Therefore there are only a finite number of subgroups of index n.

7.1.7. *If G is a finitely generated group and $[G:H]$ is finite, then $\exists\, K \in$ $\text{Char}(G)$ such that $K \subset H$ and $[G:K]$ is finite.*

Proof. Let $K = \cap \{L \subset G \mid [G:L] = [G:H]\}$. Clearly, $K \subset H$ and $K \in \text{Char}(G)$. By Theorem 7.1.6 (and Poincaré's theorem), G/K is finite.

7.1.8. *If G is an M-group, then G has a characteristic series whose factors are finite or free Abelian of finite rank.*

Proof. Let $(A_0, \ldots, A_n)$ be a shortest normal M-series of G, and induct on n. By Theorem 7.1.4 and Exercise 5.5.15, G is finitely generated. If A_n/A_{n-1} is finite, then by Theorem 7.1.7, there is a characteristic subgroup H of finite index in G such that $H \subset A_{n-1}$.

If A_n/A_{n-1} is infinite cyclic, let H be the characteristic core of A_{n-1}: $H = \cap \{A_{n-1}T \mid T \in \text{Aut}(G)\}$. Since G/A_{n-1} is Abelian, $A_{n-1}T \supset G^1$. Therefore $H \supset G^1$ and G/H is Abelian. Since G/H is finitely generated, it is the direct sum of a finite number of cyclic groups. Let $x \in G \backslash H$. Then $x \notin A_{n-1}T$ for some $T \in \text{Aut}(G)$, and since $G/A_{n-1}T \cong G/A_{n-1} \cong J$, $x^r \notin A_{n-1}T$ for all $r \in \mathcal{N}$. Hence $x^r \notin H$ and G/H is torsion-free. Therefore G/H is a free Abelian group of finite rank.

Thus, in any case, there is a characteristic subgroup H of G such that $H \subset A_{n-1}$ and G/H is finite or free Abelian of finite rank. Now $(A_0 \cap H, \ldots, A_{n-1} \cap H)$ is a normal M-series of H. Hence, by induction, H has a characteristic series S of the required type. By Exercise 2.11.12, the series S^* obtained by adding G at the end of S is of the desired kind. ∥

We wish to show that if G is an M-group, then it has a normal M-series in which all of the infinite factors occur before any of the finite factors. The proof uses a "rearrangement" lemma of some interest in itself.

7.1.9. *If H is a finite normal subgroup of a group G and G/H is free Abelian of finite rank r, then there is a free Abelian rank r subgroup K of G such that K is characteristic and G/K is finite.*

Proof. By Exercise 5.5.15, G/H satisfies the maximal condition. Therefore, by Theorem 7.1.3, G satisfies the maximal condition. It is sufficient to prove that there is a free Abelian subgroup of finite index in G. For if L is such, then by Theorem 7.1.7, there is a characteristic subgroup K such that

G/K is finite. As a subgroup of a free Abelian group, K is free Abelian (Theorem 5.3.4). Hence $K \cong HK/H$, which is of finite index in G/H. By Theorem 5.4.1, K is of rank r.

By the N/C theorem, $G/C(H)$ is finite. Now $C(H)$ has a finite central subgroup $C(H) \cap H$, and

$$\frac{C(H)}{C(H) \cap H} \cong \frac{H(C(H))}{H},$$

which is a subgroup of finite index in G/H. Hence, as above, $C(H)/(C(H) \cap H)$ is free Abelian of rank r. Therefore by the first paragraph, $WLOG\ G = C(H)$.

Let $o(H) = n$, $x \in G$, and $y \in G$. Since $G^1 \subset H$, $x^{-1}y^{-1}xy = h \in H$, $y^{-1}xy = xh$, $(y^{-1}xy)^n = x^n$, so $y \in C(x^n)$. Therefore the subgroup $L = \langle z^n \mid z \in G \rangle$ is Abelian. Since G satisfies the maximal condition for subgroups, L is finitely generated. Therefore L is the direct sum of a finite group and a free Abelian group K. Hence $[L:K]$ is finite. Again, $G/LH \cong (G/H)/(LH/H)$ is an Abelian, periodic, finitely generated group, hence is finite. Since $LH/L \cong H/(H \cap L)$, LH/L also is finite. Therefore K is a free Abelian subgroup of finite index in G.

7.1.10. *If G is an M-group, then G has a characteristic series $(A_0, \ldots, A_n)$ such that A_n/A_{n-1} is finite and all other factors are free Abelian of finite rank.*

Proof. By Theorem 7.1.8, G has a characteristic series whose factors are finite or free Abelian of finite rank. Let $(B_0, \ldots, B_m)$ be such a series such that the number k of infinite factors occurring after the first finite factor B_{j+1}/B_j is a minimum. If $k = 0$, we are done. Suppose $k > 0$. Then $WLOG$, B_{j+2}/B_{j+1} is free Abelian of finite rank. By Theorem 7.1.9, B_{j+2}/B_j has a characteristic free Abelian subgroup B'_{j+1}/B_j of finite rank such that B_{j+2}/B'_{j+1} is finite. Then $B'_{j+1} \in \mathrm{Char}(G)$. Thus the characteristic series

$$(B_0, \ldots, B_j, B'_{j+1}, B_{j+2}, \ldots, B_m)$$

of G, whose factors are finite or free Abelian of finite rank, has $k - 1$ infinite factors after the first finite factor, a contradiction. $\|$

The next theorem is an immediate corollary.

7.1.11. *If G is an M-group, then G has a characteristic torsion-free subgroup H of finite index.*

7.1.12. *If an M-group G is not nilpotent, then it has a finite, nonnilpotent, homomorphic image.*

Proof. Deny the theorem. By Theorem 7.1.10, there is an invariant series $(A_0, \ldots, A_n)$ of G with A_n/A_{n-1} finite and all other factors free Abelian of finite rank. Passing to a factor group if necessary, it may be assumed that G/A_1 is nilpotent. Then $Z^j(G) \subset A_1$ for some j. Let $r = \mathrm{rank}(A_1)$.

If $p \in \mathscr{P}$, then the subgroup $A_1^p = \langle x^p \mid x \in A_1 \rangle$ is characteristic in A_1, hence normal in G. If G/A_1^p is nilpotent and $Z^k(G) \nsubseteq A_1^p$, then $Z^k(G)A_1^p > Z^{k+1}(G)A_1^p$. Since $o(A_1/A_1^p) = p^r$, a chain of subgroups from A_1 to A_1^p has length at most r. Therefore $Z^{j+r}(G) \subset A_1^p$. If G/A_1^p is nilpotent for all $p \in \mathscr{P}$, then

$$Z^{j+r}(G) \subset \cap \{A_1^p \mid p \in \mathscr{P}\} = E,$$

and G is nilpotent, contrary to assumption. Hence $\exists\, p$ such that G/A_1^p is not nilpotent. Thus *WLOG* there is an invariant series $E < B < G$ with B finite and G/B nilpotent, hence infinite. By Theorem 7.1.11, there is a normal torsion-free subgroup D of finite index in G. By assumption, G/D is nilpotent. Therefore $\exists\, k$ such that $Z^k(G) \subset B \cap D$. But B is finite and D is torsion-free, so that $B \cap D = E$. Therefore G is nilpotent, contrary to assumption.

EXERCISES

7.1.13. (Hirsch [1], [2], and [3].) A *cyclic* normal series of a group G is a normal series whose factors are cyclic. A group is *polycyclic* iff it has a cyclic normal series.

    (a)   A polycyclic group is an M-group.

    (b)   Subgroups and factor groups of polycyclic groups are polycyclic.

    (c)   If H and G/H are polycyclic, so is G.

    (d)   A group is polycyclic iff it is solvable and satisfies the maximal condition for subgroups.

    (e)   If G is polycyclic, then any two cyclic normal series of G have the same number of infinite factors.

    (f)   If G is polycyclic, then G has a characteristic series whose factors are free Abelian of finite rank or finite, elementary, primary Abelian groups, and such that the infinite factors occur first.

7.1.14. If an infinite M-group G has a normal M-series without repetition of length n, then it has a normal M-series without repetition of length m for all $m > n$.

7.1.15. Improve Theorem 7.1.8 by requiring that all finite factors be the direct sum of isomorphic simple groups (see Exercise 4.4.11). Make a similar improvement in Theorem 7.1.10.

7.2 Supersolvable groups

A *cyclic invariant series* of a group G is an invariant series of G whose factors are cyclic. A group G is *supersolvable* iff it has a cyclic invariant series.

7.2.1. *A group G is supersolvable iff it has an invariant series whose factors are cyclic of prime or infinite order.*

Proof. Suppose that G is supersolvable and that $(A_0, \ldots, A_n)$ is a cyclic invariant series of G. If A_{i+1}/A_i is finite, then there is a chain $(A_i = B_0, B_1, \ldots, B_m = A_{i+1})$ with factors of prime order (Theorem 2.6.4). Each $B_j/A_i \in \text{Char}(A_{i+1}/A_i)$ (Exercise 2.11.10), hence $B_j/A_i \lhd G/A_i$, so $B_j \lhd G$. Making simultaneous refinements of this type, one obtains an invariant series of G whose factors are cyclic of prime or infinite order. The converse is clear.

7.2.2. *A supersolvable group is solvable.*

7.2.3. *A supersolvable group is an M-group.*

7.2.4. *Subgroups and factor groups of supersolvable groups are supersolvable.*

Proof. The proof is the same as for M-groups (Theorem 7.1.1) with minor changes. ∥

It is not true that if H and G/H are supersolvable, then G is supersolvable (see Exercise 7.2.18).

7.2.5. *The direct product of a finite number of supersolvable groups is supersolvable.*

Proof. By induction, it is sufficient to consider the case of two factors. Let $(A_0, \ldots, A_m)$ and $(B_0, \ldots, B_n)$ be cyclic invariant series of G and H, respectively. Then

$$(A_0 \times B_0, A_1 \times B_0, \ldots, A_m \times B_0, A_m \times B_1, \ldots, A_m \times B_n)$$

is a cyclic invariant series of $G \times H$.

7.2.6. *A finite nilpotent group is supersolvable.*

Proof. If G is a finite nilpotent group, then G is the direct sum of its Sylow subgroups (Theorem 6.4.14). The Sylow subgroups are supersolvable (Theorem 6.3.11), hence G is supersolvable (Theorem 7.2.5). ∥

With slightly more work, one can prove a generalization.

7.2.7. *A nilpotent group G is supersolvable iff it satisfies the maximal condition for subgroups.*

Proof. If G is supersolvable, it satisfies the maximal condition by Theorems 7.1.4 and 7.2.3. Conversely, suppose that G satisfies the maximal condition. Since G is nilpotent, $Z_n = G$ for some n. Each Z_{i+1}/Z_i satisfies the maximal condition, hence is a finitely generated Abelian group. Hence, Z_{i+1}/Z_i has a cyclic (invariant) series

$$\left(\frac{B_0}{Z_i}, \frac{B_1}{Z_i}, \ldots, \frac{B_m}{Z_i} = \frac{Z_{i+1}}{Z_i} \right).$$

Since $B_j/Z_i \subset Z(G/Z_i)$, $B_j/Z_i \vartriangleleft G/Z_i$, and $B_j \vartriangleleft G$. Therefore G has a cyclic invariant series, hence is supersolvable. ‖

Any supersolvable group is countable, hence there are certainly Abelian groups which are not supersolvable. However, for finite groups, we have the following hierarchy of classes of groups:

Cyclic $\subset$ Abelian $\subset$ Nilpotent $\subset$ Supersolvable $\subset$ Solvable $\subset$ Group.

It will be shown eventually that all inclusions are proper.

7.2.8. *If G is a supersolvable group and H is a maximal proper subgroup, then $[G:H]$ is a prime.*

Proof. If $H \vartriangleleft G$, then this is obvious. Suppose that H is not normal, and let $K = \mathrm{Core}(H)$. Since H/K is a maximal proper subgroup of the supersolvable group G/K, *WLOG* $K = E$. Since G is supersolvable, $\exists \, A \vartriangleleft G$ with A cyclic of prime or infinite order. Now $A \cap H = E$ because every subgroup of A is normal in G and $\mathrm{Core}(H) = E$. If A is infinite, it has a proper subgroup B, and $H < BH < AH$, contradicting the maximality of H. Hence $o(A) \in \mathscr{P}$. Since $H < AH$, $G = AH$. Therefore by Exercise 2.3.7,

$$[G:H] = [AH:H] = [A:A \cap H] = o(A) \in \mathscr{P}.$$

7.2.9. *If G is a supersolvable group, then any two cyclic invariant series of G have the same number of infinite factors.*

Proof. This follows from Theorems 7.1.5 and 7.2.3.

7.2.10. *If $E < H < K < G$ is an invariant series of a group G, H is finite, and K/H is infinite cyclic, then there is an invariant series $E < R < K < G$ with R infinite cyclic and K/R finite.*

Proof. This follows directly from the rearrangement lemma (Theorem 7.1.9).

7.2.11. *If G is a supersolvable group, then it has a cyclic invariant series with all the infinite factors appearing first.*

Proof. Deny, and let $(B_0, \ldots, B_n)$ be a cyclic invariant series of G such that B_{i+1}/B_i is infinite for $i < r$, where r is a maximum. One then has B_{i+1}/B_i finite for $i = r, \ldots, s - 1$, while B_{s+1}/B_s is infinite. Going over to G/B_r, one has the conditions of the lemma (Theorem 7.2.10) with $E = B_r/B_r$, $H = B_s/B_r$ and $K = B_{s+1}/B_r$. Therefore the subgroup R of the lemma exists. Since $(G/B_r)/R$ is supersolvable, one readily constructs a cyclic invariant series

$$(B_0, \ldots, B_r, B_{r+1}^*, \ldots, B_m)$$

of G with $r + 1$ infinite factors at the beginning, contradicting the maximality of r.

7.2.12. *If G is a finite supersolvable group, then it has a cyclic invariant series $(A_0, \ldots, A_n)$ such that each factor has prime order, and*

$$o\left(\frac{A_i}{A_{i-1}}\right) \geq o\left(\frac{A_{i+1}}{A_i}\right) \qquad \text{for all } i.$$

Proof. By Theorem 7.2.1, a finite supersolvable group has a cyclic invariant series with factors of prime orders. As in the preceding proof, it is sufficient to show that if $E < H < K < L$ is an invariant series of L, $o(H) = p \in \mathscr{P}$, $o(K/H) = q \in \mathscr{P}$, and $p < q$, then there is an invariant series $E < R < K < L$ with $o(R) = q$. But the assumed conditions imply that $o(K) = pq$, so that, if $R \in \mathrm{Syl}_q(K)$, then $R \lhd K$, by Sylow. Hence $R \in \mathrm{Char}(K)$, so that $R \lhd L$ and we are done. $\|$

Theorems 7.2.11 and 7.2.12 can, of course, be combined into a single theorem, but this will be left for the exercises.

7.2.13. *If G is a supersolvable group, then G^1 is nilpotent.*

Proof. Let $(A_0, \ldots, A_n)$ be a cyclic invariant series of G, and $B_i = A_i \cap G^1$. Then $(B_0, \ldots, B_n)$ is a cyclic invariant series of G^1. Each $g \in G$ induces an automorphism of B_{i+1}/B_i, so

$$\frac{G}{M} \tilde{\subseteq} \mathrm{Aut}\left(\frac{B_{i+1}}{B_i}\right),$$

where

$$\frac{M}{B_i} = C_{G/B_i}\left(\frac{B_{i+1}}{B_i}\right).$$

By Theorems 5.7.4 and 5.7.12, the automorphism group of a cyclic group is Abelian. Hence G/M is Abelian. Therefore $M \supset G^1$, so that $B_{i+1}/B_i \subset Z(G^1/B_i)$. But this means that $(B_0, \ldots, B_n)$ is an ascending central series for G^1, and G^1 is nilpotent.

7.2.14. *If G/H is supersolvable and H is cyclic, then G is supersolvable.*

Proof. If T is the natural homomorphism of G onto G/H and S is a cyclic invariant series of G/H, then E followed by ST^{-1} is a cyclic invariant series of G. ‖

The section will be closed with a semi-numerical theorem (not best possible).

7.2.15. *If $o(G) = 2p^n$, $p \in \mathscr{P}$, then G is supersolvable.*

Proof. Induct on n. The statement is obvious if $n = 0$ or $p = 2$ (Theorem 7.2.6). Let p be odd, $P \in \mathrm{Syl}_p(G)$, $Q \in \mathrm{Syl}_2(G)$, and H a minimal non-E normal subgroup of G. If $o(H) = 2$ or p, the theorem follows from Theorem 7.2.14. Now $P \lhd G$ by Sylow, and P is solvable, hence G is solvable. Therefore, by Theorem 4.4.5, H is an elementary Abelian p-group of order p^r, $r > 1$. Since $Z(P)$ is characteristic in P, $Z(P) \lhd G$. By Theorem 6.3.1, $H \cap Z(P) \neq E$. By the minimality of H, $H \subset Z(P)$. Also $HQ \subset G$.

If $HQ < G$, then HQ is supersolvable by the inductive hypothesis. Any two principle series of HQ are equivalent by the (generalized) Jordan-Hölder theorem, one principal series has factors of prime order by supersolvability, and one principal series contains H. Thus there is a normal subgroup R of HQ of order p. Since $R < H \subset Z(P)$, $N(R) \supset PQ = G$, in contradiction to the minimality of H.

Hence $HQ = G$, i.e., $H = P$. Let $e \neq x \in Q$, $y \in P$. If $y^x \in \langle y \rangle$, then $\langle y \rangle \lhd G$, contradicting the minimality of H. If $y^x \notin \langle y \rangle$, then $y^x \in P$, $y^x y \neq e$, and

$$(y^x y)^x = y^{x^2} y^x = yy^x = y^x y.$$

Hence $\langle y^x y \rangle \lhd G$, and a contradiction is again reached.

EXERCISES

7.2.16. State and prove a single theorem containing both Theorems 7.2.11 and 7.2.12.

7.2.17. There is a finite supersolvable group which is not nilpotent.

7.2.18. (a) Sym(4) is solvable but not supersolvable.

(b) Show that Sym(4)^1 is not nilpotent (see Theorem 7.2.13).

(c) Show that Sym(4) has a normal supersolvable subgroup H such that Sym(4)$/H$ is supersolvable.

7.2.19. (Sylow tower theorem.) Let G be a supersolvable group of order $\pi\{p_i^{n_i} \mid 1 \leq i \leq r\}$, with $p_i \in \mathscr{P}$ and $p_i > p_{i+1}$ for all i, and let $P_i \in \mathrm{Syl}_{p_i}(G)$. Prove that, for each k, $P_1P_2 \cdots P_k$ is a normal subgroup of G. [Use Theorem 7.2.12. The converse is false (see Exercise 9.2.13).]

7.2.20. If G is a supersolvable group and A is a maximal normal Abelian subgroup, then A is a maximal Abelian subgroup. (If not, refine $(E, A, C(A), G)$ to a cyclic invariant series, and reach a contradiction.)

7.2.21. Improve Theorem 7.2.13 to read: If G is supersolvable, then $\exists \, H \subset G$ such that $H \supset G^1$, G/H is finite, and H is nilpotent. (Use the proof of Theorem 7.2.13 and $\mathrm{Aut}(J) \cong J_2$.)

7.2.22. If G/H and G/K are supersolvable, then $G/(H \cap K)$ is supersolvable.

7.2.23. If H is a normal, nilpotent, supersolvable subgroup of G and K is a normal, supersolvable subgroup of G, then HK is a normal, supersolvable subgroup of G. (One cannot omit "nilpotent" in this theorem. See Exercise 9.2.19.)

7.2.24. Give an example of a finite, nonsupersolvable group, all of whose proper subgroups are nilpotent. (Compare with Theorem 6.5.7.)

7.3   Frattini subgroup

The *Frattini subgroup*, $\mathrm{Fr}(G)$, of a group G is the intersection of all maximal proper subgroups of G (and $\mathrm{Fr}(G) = G$ if there are no maximal proper subgroups of G).

7.3.1. $\mathrm{Fr}(G) \in \mathrm{Char}(G)$. ‖

An element x of G is a *nongenerator* iff

(*)   if S is a subset of G such that $\langle S, x \rangle = G$, then $\langle S \rangle = G$.

7.3.2. *If G is a group, then $\mathrm{Fr}(G)$ is the set of nongenerators of G.*

Proof. First let x be a nongenerator and M a maximal proper subgroup. If $x \notin M$, then $\langle M, x \rangle = G$, hence $M = \langle M \rangle = G$, a contradiction. Therefore $x \in M$, so that $x \in \mathrm{Fr}(G)$.

Next, let $x \in \mathrm{Fr}(G)$, and let S be a subset of G such that $\langle S, x \rangle = G$. Suppose that $\langle S \rangle \neq G$. By Zorn, $\exists$ a subgroup M maximal with respect to (i) $M \supset \langle S \rangle$ and (ii) $x \notin M$. If $M < H \subset G$, then $x \in H$, so that $H \supset \langle S, x \rangle = G$. Hence M is a maximal proper subgroup of G. Since $x \notin M$, $x \notin \mathrm{Fr}(G)$, a contradiction. Therefore $\langle S \rangle = G$, (*) is satisfied, and x is a nongenerator.

7.3.3. *If G is a group, then $\mathrm{Fr}(G) \supset G^1$ iff all maximal proper subgroups of G are normal.*

Proof. If all maximal proper subgroups of G are normal and M is a maximal proper subgroup, then G/M is of prime order, hence Abelian,

$M \supset G^1$, and $\mathrm{Fr}(G) \supset G^1$. If, conversely, $\mathrm{Fr}(G) \supset G^1$ and M is a maximal proper subgroup of G, then $M \supset G^1$, hence M is a normal subgroup of G (Theorem 3.4.11).

7.3.4. *If G is a nilpotent group, then $\mathrm{Fr}(G) \supset G^1$.*

Proof. By Theorem 6.4.9, all maximal proper subgroups of G are normal. Therefore by Theorem 7.3.3, $\mathrm{Fr}(G) \supset G^1$. ‖

The next theorem should be compared with Theorem 6.4.14.

7.3.5. *If G is an M-group, then the following conditions are equivalent:*
(1)   *G is nilpotent,*
(2)   *if $H < G$, then $H < N(H)$,*
(3)   *if M is a maximal proper subgroup of G, then $M \lhd G$,*
(4) $\mathrm{Fr}(G) \supset G^1$.

Proof. By Theorem 6.4.8, (1) implies (2). It is obvious that (2) implies (3). Suppose that (3) holds, but (1) is false. Then by Theorem 7.1.12, some finite factor group G/H is not nilpotent. By Theorem 6.4.14, ∃ a nonnormal maximal proper subgroup K/H of G/H. But then, K is a nonnormal, maximal, proper subgroup of G, contrary to hypothesis. Therefore (3) implies (1). Finally, (3) and (4) are equivalent by Theorem 7.3.3.

7.3.6. *If G is an elementary Abelian p-group ($p \in \mathscr{P}$), then $\mathrm{Fr}(G) = E$.*

Proof. $G = \Sigma \{H \mid H \in S\}$, $o(H) = p$ for all $H \in S$. The subgroup

$$K_H = \Sigma \{H^* \mid H^* \in S, H^* \neq H\}$$

is a maximal proper subgroup of G for all $H \in S$ since $G = K_H + H$ and $o(G/K_H) = o(H) = p$. Therefore,

$$\mathrm{Fr}(G) \subset \cap \{K_H \mid H \in S\} = E.$$

7.3.7. *If G is a finite p-group and $H \subset G$, then $\mathrm{Fr}(G) \subset H$ iff $H \lhd G$ and G/H is elementary Abelian.*

Proof. If G/H is elementary Abelian, then $\mathrm{Fr}(G) \subset H$ by Theorem 7.3.6 and the lattice theorem. Now suppose that $\mathrm{Fr}(G) \subset H$. Since G is nilpotent, $\mathrm{Fr}(G) \supset G^1$ by Theorem 7.3.4, hence G/H is Abelian. Let $x \in G$ and let M be a maximal proper subgroup of G. Then $o(G/M) = p$, so that $x^p \in M$. Therefore, $x^p \in \mathrm{Fr}(G) \subset H$. Hence G/H is elementary Abelian.

7.3.8. *If G is a group, $H \subset G$, $\mathrm{Fr}(G)$ is finitely generated, and $G = \mathrm{Fr}(G)H$, then $H = G$.*

Proof. Let $\mathrm{Fr}(G) = \langle x_1, \ldots, x_n \rangle$. Then $G = \langle H, x_1, \ldots, x_n \rangle$, so that, by Theorem 7.3.2 applied n times, $G = \langle H \rangle = H$.

7.3.9. *If G is a group, $F = \mathrm{Fr}(G)$ is finitely generated, and $G/F = \langle Fx \mid x \in S \rangle$, then $G = \langle S \rangle$.*

Proof. Let $H = \langle S \rangle$. Since HF contains F and an element from each coset of F, $HF = G$. By Theorem 7.3.8, $\langle S \rangle = H = G$. ‖

If G is a finite p-group, then $G/\mathrm{Fr}(G)$ is an elementary Abelian group (Theorem 7.3.7) of order p^n, say. Hence, $G/\mathrm{Fr}(G)$ is generated by some set of n elements but by no set with fewer than n elements. Moreover, a standard argument (see Exercise 5.7.27) shows that any generating set contains a generating set with exactly n elements.

7.3.10. *(Burnside basis theorem.)* *If G is a finite p-group, $p \in \mathscr{P}$, $F = \mathrm{Fr}(G)$, $o(G/F) = p^n$, $G/F = \langle Fx_1, \ldots, Fx_n \rangle$, and $G = \langle y_1, \ldots, y_r \rangle$, then*

(i) $G = \langle x_1, \ldots, x_n \rangle$,

(ii) $G/F = \langle Fy_{i_1}, \ldots, Fy_{i_n} \rangle$ *for some* $i_1, \ldots, i_n$.

Proof. Statement (i) follows from Theorem 7.3.9. It is clear that $G/F = \langle Fy_1, \ldots, Fy_r \rangle$. Statement (ii) then follows from the discussion preceding the statement of the theorem.

7.3.11. *If $o(G) = p^n$, $p \in \mathscr{P}$, and $o(\mathrm{Fr}(G)) = p^{n-r}$, then*

$$o(\mathrm{Aut}(G)) \mid p^{r(n-r)} \, \pi\{p^r - p^i \mid 0 \leq i \leq r - 1\}.$$

Proof. Any automorphism of G induces an automorphism of G/F, $F = \mathrm{Fr}(G)$. There results a natural homomorphism T from $\mathrm{Aut}(G)$ into $\mathrm{Aut}(G/F)$. By Theorem 7.3.7, G/F is elementary Abelian of order p^r, and by Theorems 5.7.20 and 5.7.22,

$$o(\mathrm{Aut}(G/F)) = \pi \{p^r - p^i \mid 0 \leq i \leq r - 1\} = m.$$

Therefore, $o(\mathrm{Aut}(G)) \mid o(\mathrm{Ker}(T))m$, and it now remains to show that $o(\mathrm{Ker}(T)) \mid p^{r(n-r)}$.

By the Burnside basis theorem, there is an ordered generating set $(x_1, \ldots, x_r)$ of G. Let

$$S = \{(y_1, \ldots, y_r) \mid y_i \in x_i F\}.$$

By the Burnside basis theorem, any element of S is an ordered generating set of G. If $U \in \mathrm{Ker}(T)$, then U induces a permutation U^* of S, and if $U \neq e$, then $\mathrm{Ch}(U^*) = 0$, so that, in particular, $*$ is an isomorphism. Now U^* is the product of disjoint cycles of the same length (or $\mathrm{Ch}(U^{*i}) > 0$ for some

$U^{*i} \neq e$). Hence $o(U^*) \mid o(S)$. Since $o(S) = o(F)^r = p^{r(n-r)}$, $o(U^*) = p^i$ for some i. Therefore $\text{Ker}(T)$ is a p-group. If $a \in S$, $U_i \in \text{Ker}(T)$, and $aU_1^* = aU_2^*$, then $aU_1^*U_2^{*-1} = a$, $U_1^*U_2^{*-1} = e$, $U_1^* = U_2^*$, and $U_1 = U_2$. Hence $o(\text{Ker}(T)) \leq o(S)$. Therefore $o(\text{Ker}(T)) \mid p^{r(n-r)}$. ∥

A glance at the last half of the proof shows that the following theorem has been proved.

7.3.12. *If G is a finite p-group, $p \in \mathscr{P}$, and U an automorphism of G inducing the identity automorphism on $G/\text{Fr}(G)$, then $o(U) = p^i$ for some i.*

7.3.13. *If $F = \text{Fr}(G)$ is an M-group, then $\text{Fr}(G)$ is nilpotent.*

Proof. Deny the theorem. By Theorem 7.1.12, ∃ H such that F/H is finite and not nilpotent. By Theorem 7.1.7, ∃ $K \in \text{Char}(F)$ such that F/K is finite and not nilpotent. Now $K \lhd G$ and $\text{Fr}(G/K) = F/K$ is finite and not nilpotent. Since G/K is again an M-group, $WLOG$ F is finite and not nilpotent.

Let $P \in \text{Syl}(F)$. By Theorem 6.2.4, $G = N(P)F$. By Theorem 7.3.8, $N(P) = G$. Since all Sylow subgroups of F are normal, F is nilpotent, a contradiction. ∥

In particular, we have

7.3.14. *If G, or even $\text{Fr}(G)$, is a finite group, then $\text{Fr}(G)$ is nilpotent.*

7.3.15. *If G, or even $\text{Fr}(G)$, is supersolvable, then $\text{Fr}(G)$ is nilpotent.* ∥

It is not true that if $H \subset G$, then $\text{Fr}(H) \subset \text{Fr}(G)$ (see Exercise 7.3.32). This statement is true if G is finite and $H \lhd G$, as we shall prove.

7.3.16. *If G is a group, $H \lhd G$, $K \subset G$, $H \subset \text{Fr}(K)$, and $\text{Fr}(K)$ is finitely generated, then $H \subset \text{Fr}(G)$.*

Proof. Deny the theorem. There is a maximal proper subgroup M of G such that $H \nsubseteq M$. Hence $K \nsubseteq M$ and $K \cap M < K$. Now $G = HM$, since M is maximal proper. By Exercise 1.6.15, $K = H(K \cap M)$. Since $H \subset \text{Fr}(K) \subset K$, also $K = \text{Fr}(K)(K \cap M)$. By Theorem 7.3.8, $K \cap M = K$, a contradiction.

7.3.17. *If H is a normal subgroup of G and $\text{Fr}(H)$ is finitely generated, then $\text{Fr}(H) \subset \text{Fr}(G)$.*

Proof. Since $\text{Fr}(H) \in \text{Char}(H)$, $\text{Fr}(H) \lhd G$. The theorem now follows from Theorem 7.3.16 if, in that theorem, H is replaced by $\text{Fr}(H)$ and K by H.

7.3.18. *If H is a normal subgroup of a finite group G contained in* Fr(*G*), *then* Inn(*H*) ⊂ Fr(Aut(*H*)).

Proof. Each element g of G induces an automorphism gT of H. Thus T is a homomorphism from G into Aut(*H*). Now

$$HT = \text{Inn}(H) \subset (\text{Fr}(G))T \subset \text{Fr}(GT),$$

the last inclusion being Exercise 7.3.30. Since Inn(*H*) ◁ Aut(*H*) (Theorem 2.11.4), it follows from Theorem 7.3.16 that Inn(*H*) ⊂ Fr(Aut(*H*)).

7.3.19. *If H* = Fr(*G*) *and G is finite, then* Inn(*H*) ⊂ Fr(Aut(*H*)).

Proof. In Theorem 7.3.18, let *H* = Fr(*G*). ‖

This theorem gives a criterion which enables one to show that certain nilpotent groups are not the Frattini subgroup of any finite group (see Exercise 8.2.17).

The remainder of the section is concerned with the question whether

$$\text{Fr}(\Sigma \{H \mid H \in S\}) = \Sigma \{\text{Fr}(H) \mid H \in S\}.$$

7.3.20. Fr(Σ {*H* | *H* ∈ *S*}) ⊂ Σ {Fr(*H*) | *H* ∈ *S*}.

Proof. If $G = A + B$ and M is a maximal proper subgroup of A, then $M + B$ is a maximal proper subgroup of G (Exercise 4.1.13). Hence $M + B \supset \text{Fr}(G)$. Therefore, $\text{Fr}(A) + B \supset \text{Fr}(G)$. It follows that $\text{Fr}(\Sigma H) \subset \Sigma \text{Fr}(H)$.

7.3.21. *If* $G = \Sigma \{H \mid H \in S\}$ *and* Fr(*G*) ∩ *K* < Fr(*K*) *for some K* ∈ *S*, *then* ∃ *M* ◁ *K* *such that K/M is an infinite simple group and* Fr(*K*/*M*) = *K*/*M*.

Proof. Let $L = \Sigma \{H \mid H \in S, H \neq K\}$. Then $G = K + L$. By assumption, one of the maximal proper subgroups R of G is such that $R \cap K \nsubseteq \text{Fr}(K)$. Hence $R \cap K$ is neither K nor a maximal proper subgroup of K.

First suppose that R is not a subdirect sum of K and L. Then there are subgroups K_1 of K and L_1 of L such that $R \subset K_1 + L_1 < K + L$. By the maximality of R, $R = K_1 + L_1$. From this fact and earlier remarks, $R \cap K = K_1 < U < K$ for some subgroup U of K. But then

$$R = K_1 + L_1 < U + L_1 < K + L,$$

contradicting the maximality of R.

Hence R is a subdirect sum of K and L. By Theorem 4.3.1, there are subgroups $M \lhd K$ and $P \lhd L$ and an isomorphism U of K/M onto L/P such that $R = \{ab \mid (aM)U = bP\}$. Clearly $M < K$. Suppose that K/M has a nontrivial normal subgroup V/M. Let $W/P = (V/M)U$. Then the function $U^*: (aV)U^* = ((aM)U)W$ is an isomorphism of K/V onto L/W such that if $(aM)U = bP$, then $(aV)U^* = bW$. It follows that the associated subdirect sum $R^* = \{ab \mid (aV)U^* = bW\}$ is such that $R < R^* < K + L$. Hence K/M is a simple group. If $\mathrm{Fr}(K/M) \neq K/M$, then by the simplicity of K/M, $\mathrm{Fr}(K/M) = E$, and $\mathrm{Fr}(K) \subset M = R \cap K$, a contradiction. Hence $\mathrm{Fr}(K/M) = K/M$. Therefore K/M is infinite.

7.3.22. *The statement*

$$\mathrm{Fr}(\Sigma \{H \mid H \in S\}) = \Sigma \{\mathrm{Fr}(H) \mid H \in S\}$$

is universally true iff there is no simple infinite group B *such that* $\mathrm{Fr}(B) = B$.

Proof. If, for some S,

$$\mathrm{Fr}(\Sigma \{H \mid H \in S\}) \neq \Sigma \{\mathrm{Fr}(H) \mid H \in S\},$$

then by Theorem 7.3.20, for some $K \in S$, $\mathrm{Fr}(\Sigma\ H) \cap K < \mathrm{Fr}(K)$. Hence by Theorem 7.3.21, there is an infinite simple group which coincides with its Frattini subgroup.

Conversely, suppose that B is an infinite simple group such that $\mathrm{Fr}(B) = B$. Let $G = B \times B$, and let $D = \{(b, b) \mid b \in B\}$ be the diagonal subgroup. If $D < H \subset G$, then H contains an element (a, b) with $a \neq b$, hence an element (c, e) with $c \neq e$. Since H is a subdirect product of $B \times B$, it follows from Theorem 4.3.1 that $H = B \times B$. Hence, D is a maximal proper subgroup of G. Since $\mathrm{Fr}(G)$ is a normal subgroup of G contained in D, and since the only normal subgroups of G are G, $B \times E$, $E \times B$, and E (Theorem 4.4.6), $\mathrm{Fr}(G) = E$. Thus

$$\mathrm{Fr}(B \times B) \neq \mathrm{Fr}(B) \times \mathrm{Fr}(B).$$

A change to the direct sum notation now gives the other half of the theorem.

7.3.23. *If* $G = \Sigma \{H \mid H \in S\}$ *and for all* $H \in S$ *either* (i) H *is solvable,* (ii) H *is finitely generated, or* (iii) $\mathrm{Fr}(H)$ *is finitely generated, then* $\mathrm{Fr}(G) = \Sigma \{\mathrm{Fr}(H) \mid H \in S\}$.

Proof. By Theorem 7.3.20, it suffices to show that if $G = A + B$ and A satisfies (i), (ii), or (iii), then $\mathrm{Fr}(G) \cap A \supset \mathrm{Fr}(A)$. Suppose this is not the case. By Theorem 7.3.21, some factor group A/M of A is an infinite simple group such that $\mathrm{Fr}(A/M) = A/M$. It follows immediately that A is not solvable. Suppose that $A/M = \langle x_1, \ldots, x_n \rangle$. Then there is a subgroup L of A/M maximal with respect to the property of omitting at least one x_i. Any larger subgroup would contain all the x_i, hence would equal A/M. Therefore L is a maximal proper subgroup of A/M and $\mathrm{Fr}(A/M) < A/M$, a contradiction. Hence the third possibility must hold: $\mathrm{Fr}(A)$ is finitely generated. Since $\mathrm{Fr}(A) \not\subset \mathrm{Fr}(G)$, there is a subgroup P of G not containing $\mathrm{Fr}(A)$ such that P is a maximal proper subgroup of G. Since $\mathrm{Fr}(A) \lhd G$, $G = \mathrm{Fr}(A)P$.

Hence $A = (A \cap P)\mathrm{Fr}(A)$. By Theorem 7.3.8, $A \cap P = A$, so that $P \supset A \supset \mathrm{Fr}(A)$, a contradiction. ‖

It is an unsolved problem whether there exists an infinite simple group without maximal proper subgroups. If there is an uncountable group G containing no proper subgroups of order $o(G)$, then Exercise 3.3.16 shows that there is a simple group H of the same order with the same property. Such an H could have no maximal proper subgroup. For if K were such and $x \in H \backslash K$, then $o(\langle K, x \rangle) = \max(o(K), \aleph_0) < o(G)$.

EXERCISES

7.3.24. (a) If G is an additive Abelian group, then
$\mathrm{Fr}(G) = \cap \{pG \mid p \in \mathscr{P}\}$.

     (b) Determine the Frattini subgroup for all finite Abelian groups.

     (c) In $\mathscr{R}$, let $G = \langle 1/p \mid p \in \mathscr{P} \rangle$. Prove that $\mathrm{Fr}(G) = J$.

     (d) Show that if G is an Abelian group, then there is an Abelian group H such that $\mathrm{Fr}(H) = G$. [Use Theorem 5.3.3, (c), and Theorem 7.3.23.]

7.3.25. Give an example of a finite group G, normal subgroup H, and maximal, proper subgroup M of G such that $H \cap M$ is neither H nor a maximal, proper subgroup of H.

7.3.26. If G is a group, then $\mathrm{Fr}(G) \supset Z(G) \cap G^1$.

7.3.27. If $o(G) = p^m$, $p \in \mathscr{P}$, and A is a maximal normal Abelian subgroup of order p^n, say, then $n(n + 1) \geq 2m$. (This generalizes Theorem 6.5.1. Use Exercise 6.4.25 and Theorem 7.3.11.)

7.3.28. Let G be a finite nonnilpotent group, all of whose proper subgroups are nilpotent. Then in the notation of 6.5.7, G/Q^1 is non-Abelian, but has all proper subgroups Abelian. (Use Theorems 6.5.7, 7.3.4, and 7.3.12 for the first part.)

7.3.29. (a) If G is finite, then G is cyclic iff $G/\mathrm{Fr}(G)$ is cyclic. (Use Theorem 7.3.9.)

     (b) If G is finite, then G is solvable iff $G/\mathrm{Fr}(G)$ is solvable. (Theorem 7.3.14.)

     (The corresponding questions for Abelian, nilpotent, and supersolvable groups are answered in Theorem 7.4.10 and Exercises 8.2.14 and 9.3.18.)

7.3.30. If $T \in \mathrm{Hom}(G)$, then $\mathrm{Fr}(GT) \supset (\mathrm{Fr}(G))T$.

7.3.31. $\mathrm{Fr}(J) = E$ and $\mathrm{Fr}(J_4) \neq E$, hence equality need not hold in the previous exercise. (See also Exercise 9.2.23.)

7.3.32. (a) $\mathrm{Fr}(\mathrm{Sym}(4)) = E$.

        (b) Sym(4) has a cyclic subgroup of order 4.

        (c) It is not true that if $H \subset G$, then $\mathrm{Fr}(H) \subset \mathrm{Fr}(G)$.

7.3.33. (See Theorem 7.3.4.) Give an example of a finite supersolvable group G such that $\mathrm{Fr}(G) \not\supset G^1$.

7.4   Fitting subgroup

The subgroup generated by two nilpotent subgroups of a group need not be nilpotent, even if one of the subgroups is normal. For example, if $G = \mathrm{Sym}(3)$, $A \in \mathrm{Syl}_2(G)$, and $B \in \mathrm{Syl}_3(G)$, then A and B are nilpotent, B is normal, but $AB = G$ is not nilpotent. However, if one sticks to normal nilpotent subgroups, the product is nilpotent.

    7.4.1. If A and B are normal, nilpotent subgroups of a group G, then AB is also a normal, nilpotent subgroup of G.

    Proof. Certainly $AB \lhd G$. If $H \lhd G$, let $f_i(H) = [H, \ldots, H]$, where there are i terms H. Assume inductively that

$$f_i(AB) = \langle [K_1, \ldots, K_i] \mid \text{each } K_j = A \text{ or } B \rangle.$$

By Theorem 3.4.2 and the normality of all subgroups involved (Theorem 3.4.5), this equation holds if i is replaced by $i + 1$, and therefore for all i. Let A and B be of class m and n, respectively. Then $f_{m+n+1}(AB)$ is the product of subgroups $[K_1, \ldots, K_{m+n+1}]$, where either at least $m + 1$ of the K_i equal A or at least $n+1$ of the K_i equal B. If the expression $[K_1, \ldots, K_{m+n+1}]$ contains $m+1$ A's, for example, then by the normality of A and B, the other terms may be dropped (Theorem 3.4.5). Therefore,

$$[K_1, \ldots, K_{m+n+1}] \subset f_{m+1}(A) = E.$$

Hence $f_{m+n+1}(AB) = E$, and AB is nilpotent. ‖

    The union of a chain of normal nilpotent subgroups is normal, but need not be nilpotent (see Exercise 9.2.32). Therefore a group need not have a maximal normal nilpotent subgroup. If infinite ascending chains of this type cannot occur, then the preceding theorem shows that there is a maximum normal nilpotent subgroup.

    7.4.2. If a group satisfies the maximal condition for normal subgroups, then it has a maximum normal nilpotent subgroup. ‖

This maximum normal nilpotent subgroup is called the *Fitting subgroup*, and will be denoted by Fit(G). It is clear that Fit(G) is characteristic in G. Note that a finite group always has a Fitting subgroup. In this case, Fit(G) has a nice characterization.

7.4.3. *If G is a finite group and $K_p = \cap \{G_p \mid G_p \in \mathrm{Syl}_p(G)\}$ for each $p \in \mathscr{P}$ such that $p \mid o(G)$, then* Fit(G) $= \Sigma\, K_p$.

Proof. Since each $\mathrm{Syl}_p(G)$ is a conjugate class of subgroups, K_p is just the core of each $G_p \in \mathrm{Syl}_p(G)$, and is normal in G. Hence $K_p \subset$ Fit(G), so that $\Sigma\, K_p \subset$ Fit(G). Conversely, a Sylow p-subgroup P of Fit(G) is contained in some $Q \in \mathrm{Syl}_p(G)$, hence, as a normal subgroup of G, $P \subset \mathrm{Core}(Q) = K_p$. Therefore Fit($G$) $= \Sigma\, K_p$.

7.4.4. *If G is a finite group, then* Fit(G) $\supset$ Fr(G).

Proof. By Theorem 7.3.14, Fr(G) is a (normal) nilpotent subgroup of G.

7.4.5. *If a group G has a Fitting subgroup (in particular, if G is finite) and M is a minimal normal non-E subgroup of G, then* Fit(G) $\subset C(M)$.

Proof. Let $F = $ Fit(G). If $M \cap F = E$, then $M \times F$ exists, and $F \subset C(M)$. If $M \cap F \neq E$, then by the minimality of M, $M \subset F$. Now $M \cap Z(F) \lhd G$. By Exercise 6.4.24, $M \cap Z(F) \neq E$. Therefore, $M \subset Z(F)$, so that $F \subset C(M)$.

7.4.6. *If G is a solvable group having a maximum characteristic nilpotent subgroup H, then $H \supset C(H)$.*

Proof. The theorem is obvious if G is Abelian. Induct on the solvable length of G. For some i, $G^i = E$, while $A = G^{i-1} \neq E$. A is an Abelian characteristic subgroup of G, and $C(A)$ is a characteristic subgroup. If K/A is a characteristic nilpotent subgroup of $C(A)/A$, then K is characteristic in G, and, since $A \subset Z(K)$, K is nilpotent. Therefore $K \subset H$. This implies that the union of a chain of characteristic nilpotent subgroups of $C(A)/A$ is again a characteristic nilpotent subgroup of $C(A)/A$. By Zorn and Theorem 7.4.1, $C(A)/A$ has a maximum characteristic nilpotent subgroup L/A. Moreover $L \subset H$. Now $C(L) \subset C(A)$ since $A \subset L$. Since $C(A)/A$ has solvable length less than i, it follows from the induction assumption that $C_{C(A)/A}(L/A) \subset L/A$. Therefore $C(L) \subset L$. Hence, finally, $H \supset L \supset C(L) \supset C(H)$. $\|$

The hypotheses of this theorem are satisfied if G is a solvable group having a Fitting subgroup F, for F is then the maximum characteristic nilpotent subgroup of G. In particular, we have

7.4.7. *If G is a finite solvable group, then* Fit(G) $\supset C($Fit(G)). $\|$

There is a relation which we shall now develop between the Fitting and Frattini subgroups of a finite group. This relation, and in fact the results in the rest of this section, are due to Gaschütz [2].

7.4.8. *If G is a finite group, $A \lhd G$, $B \lhd G$, $B \subset \mathrm{Fr}(G)$, and A/B is nilpotent, then A is nilpotent.*

Proof. Let $P \in \mathrm{Syl}_p(A)$. Then $BP/B \in \mathrm{Syl}_p(A/B)$ (Theorem 6.1.16). Since A/B is nilpotent, $BP/B \in \mathrm{Char}(A/B)$, hence $BP \lhd G$. Also $P \in \mathrm{Syl}_p(BP)$. By Theorem 6.2.4, $G = N(P)B \subset N(P)\mathrm{Fr}(G)$. By Theorem 7.3.8, $G = N(P)$. Since all Sylow subgroups of A are normal, A is nilpotent.

7.4.9. *If G is a finite group, then*

$$\mathrm{Fit}(G/\mathrm{Fr}(G)) = \mathrm{Fit}(G)/\mathrm{Fr}(G).$$

Proof. By 7.4.8, with $B = \mathrm{Fr}(G)$ and $A/B = \mathrm{Fit}(G/\mathrm{Fr}(G))$, A is nilpotent. Hence $\mathrm{Fit}(G/\mathrm{Fr}(G)) \subset \mathrm{Fit}(G)/\mathrm{Fr}(G)$. But since $\mathrm{Fit}(G)$ is nilpotent, $\mathrm{Fit}(G)/\mathrm{Fr}(G)$ is a nilpotent normal subgroup of $G/\mathrm{Fr}(G)$. Hence

$$\mathrm{Fit}(G)/\mathrm{Fr}(G) \subset \mathrm{Fit}(G/\mathrm{Fr}(G)),$$

and the conclusion follows.

7.4.10. *If G is a finite group and $G/\mathrm{Fr}(G)$ is nilpotent, then G is nilpotent.*

Proof. This follows from Theorem 7.4.9 (or 7.4.8). ‖

It is clear that $\mathrm{Fr}(G/\mathrm{Fr}(G)) = E$. We will obtain (Theorem 7.4.15) more precise information about the structure of Fitting subgroups of finite groups H such that $\mathrm{Fr}(H) = E$. In turn, Theorem 7.4.9 will then give more information about $\mathrm{Fit}(G)$. In order to do all this, some new concepts will be introduced and studied briefly.

The *sockel*,* $\mathrm{Soc}(G)$, of a group G is the union of its minimal normal non-E subgroups. It is a characteristic subgroup of G.

7.4.11. *$\mathrm{Soc}(G)$ is a direct sum of some minimal normal non-E subgroups of G.*

Proof. We proceed as in Theorem 4.4.2. There is a maximal set S of minimal normal non-E subgroups such that $\Sigma \{H \mid H \in S\}$ exists. If $\mathrm{Soc}(G) > U = \Sigma \{H \mid H \in S\}$, then there is a minimal normal non-E subgroup K of G such that $K \not\subseteq U$. Since $U \cap K \lhd G$, $U \cap K = E$ and $K + U$ exists, contradicting the maximality of S. Hence $\mathrm{Soc}(G) = U$.

---

* This German word is customarily not translated.

7.4.12. *If G is a finite group, then* Soc(G) *is a direct sum of simple groups and is completely reducible.*

Proof. This follows from Theorems 7.4.11, 4.4.4, and 4.4.7. ‖

There are various ways of decomposing the sockel. We will describe the most important one of these. The *Abelian sockel* of G, call it A, is the subgroup generated by the Abelian, minimal normal non-E subgroups of G. One proves, just as in Theorem 7.4.11, that A is the direct sum of some Abelian, minimal normal non-E subgroups. It is an Abelian characteristic subgroup of G. Similarly one defines the *non-Abelian sockel B* of G. It is the direct sum of some (see Exercise 7.4.17) non-Abelian, minimal normal non-E subgroups of G, and is also characteristic in G. By the definitions just made, $AB =$ Soc(G).

7.4.13. *If G is a group, then its sockel is the direct sum of its Abelian sockel and its non-Abelian sockel.*

Proof. There is a maximal set S^* of non-Abelian, minimal normal non-E subgroups of G such that $A + \Sigma \{H \mid H \in S^*\} = U$ exists (Theorem 4.4.1). If $U < $ Soc(G), one reaches a contradiction as usual, since all Abelian, minimal normal non-E subgroups are contained in A. Hence

$$\text{Soc}(G) = A + \Sigma \{H \mid H \in S^*\}, \tag{1}$$

and $\Sigma \{H \mid H \in S^*\} \subset B$. If $A \cap B \neq E$, then there is a non-Abelian, minimal normal non-E subgroup H of G such that its projection T on A, with respect to the decomposition (1), is not 0. Hence $H/\text{Ker}(T)$ is an Abelian non-E group. Therefore $E < H^1 < H$. But H^1 is characteristic in H, hence normal in G, contradicting the minimality of H. Therefore $A \cap B = E$, so that Soc $G = A + B$.

7.4.14. *If A is an Abelian, normal subgroup of a finite group G and* Fr(G) $= E$, *then A has a complement in G.*

Proof. Induct on $o(A)$. If $A = E$, then G is a complement of A. Suppose $A \neq E$. Since Fr(G) $= E$, $\exists M$ such that $G = AM$ and M is a maximal proper subgroup of G. Now $A \cap M \vartriangleleft A$ (since A is Abelian), and $A \cap M \vartriangleleft M$. Hence $A \cap M \vartriangleleft AM = G$. Since $A \cap M < A$, the induction hypothesis implies that $\exists K \subset G$ such that

$$G = (A \cap M)K \quad \text{and} \quad (A \cap M) \cap K = E.$$

If $H = M \cap K$, it follows that

$$M = (A \cap M)H \quad \text{and} \quad (A \cap M) \cap H = E.$$

Therefore

$$G = AM = A(A \cap M)H = AH,$$
$$A \cap H = A \cap (M \cap K) = E.$$

This proves the result.

7.4.15. *If G is a finite group such that $\mathrm{Fr}(G) = E$, then $\mathrm{Fit}(G)$ is the Abelian sockel of G.*

Proof. The Abelian sockel A of G is certainly a nilpotent normal subgroup, hence $A \subset F = \mathrm{Fit}(G)$. Since F is nilpotent, $F = \Sigma \{F_p \mid F_p \in \mathrm{Syl}_p(F)\}$. Hence by Theorems 7.3.17 and 7.3.23,

$$E = \mathrm{Fr}(F) = \Sigma \mathrm{Fr}(F_p), \qquad \mathrm{Fr}(F_p) = E.$$

By Theorem 7.3.7, F_p is elementary Abelian, hence F is Abelian. By Theorem 7.4.14, $\exists\, H \subset G$ such that $G = AH$ and $A \cap H = E$. By Exercise 1.6.15, $F = A(H \cap F)$ and $A \cap (H \cap F) = E$. Now $H \cap F \lhd H$ since $F \lhd G$, and $N(H \cap F) \supset A$ since F is Abelian. Hence $N(H \cap F) \supset AH = G$, $H \cap F \lhd G$. Therefore if $H \cap F \neq E$, then $H \cap F$ contains an Abelian, minimal normal non-E subgroup K of G, contrary to $A \cap (H \cap F) = E$. Hence, $H \cap F = E$ and $F = A$.

7.4.16. *If G is a finite group, then $\mathrm{Fit}(G)/\mathrm{Fr}(G)$ is the Abelian sockel of $G/\mathrm{Fr}(G)$.*

Proof. This follows from Theorem 7.4.15, since $\mathrm{Fr}(G/\mathrm{Fr}(G)) = E$ and $\mathrm{Fit}(G/\mathrm{Fr}(G)) = \mathrm{Fit}(G)/\mathrm{Fr}(G)$ by Theorem 7.4.9.

EXERCISES

7.4.17. The non-Abelian sockel of a group G is the direct sum of *all* non-Abelian, minimal normal non-E subgroups of G.

7.4.18. If $\mathrm{Fit}(G)$ exists, then $\mathrm{Fit}(G) \subset C(\mathrm{Soc}(G))$.

7.4.19. If G is a polycyclic group (see Exercise 7.1.13), then G has a Fitting subgroup F and $F \supset C(F)$.

7.4.20. If H is a maximal nilpotent subgroup of a group G, then $N(N(H)) = N(H)$.

7.4.21. (a) If G/A and G/B are nilpotent, so is $G/(A \cap B)$.

  (b) If G satisfies the minimal condition for normal subgroups, then there is a minimum normal subgroup M of G such that G/M is nilpotent.

7.4.22. Let G be a group satisfying the maximal and minimal conditions for normal subgroups.

    (a)   Define ascending, descending, upper, and lower (invariant) nilpotent series.

    (b)   Prove the analogues of Theorems 6.4.1 through 6.4.6.

7.5 Regular p-groups

A p-group G is *regular* iff $a \in G$, $b \in G$, and $n \in \mathcal{N}$ imply that

$$(1) \quad (ab)^{p^n} = a^{p^n} b^{p^n} x_1^{p^n} \cdots x_r^{p^n}$$

for some $x_i \in \langle a, b \rangle^1$.

7.5.1. *If G is a regular p-group and $H \subset G$, then H is regular.*

7.5.2. *If G is a regular p-group and $H \lhd G$, then G/H is regular.*

Proof. A homomorphism T of G applied to Equation (1) and the condition $x_i \in \langle a, b \rangle^1$ yields a similar equation and the condition

$$x_i T \in (\langle a, b \rangle^1) T = (\langle a, b \rangle T)^1 = \langle aT, bT \rangle^1.$$

7.5.3. *An Abelian p-group is regular.*

7.5.4. *If G is a finite 2-group, then G is regular iff G is Abelian.*

Proof. If G is Abelian, then it is regular. Now suppose that G is a regular non-Abelian 2-group of least order. If $o(G^1) > 2$, then G^1 contains a subgroup $H \subset Z(G)$ of order 2 (Theorem 6.3.1), so that G/H is non-Abelian and regular (Theorem 7.5.2), contradicting the minimality of G. Hence $o(G^1) = 2$. Therefore by Equation (1), for all $a \in G$ and $b \in G$,

$$(ab)^2 = a^2 b^2 x_1^2 \cdots x_r^2 = a^2 b^2, \qquad x_i \in G^1,$$

whence $ba = ab$ and G is Abelian, a contradiction. Hence the theorem is true. ‖

If H is a group and $n \in J$, let $nH = \{x^n \mid x \in H\}$.

7.5.5. *Let G be a finite p-group. Of the conditions:*

(2) *G is regular,*

(3) *For all $a \in G$ and $b \in G$, $\exists \, c \in \langle a, b \rangle^1$ such that $(ab)^p = a^p b^p c^p$,*

(4) *If $H \subset G$ and $n \in \mathcal{N}$, then $p^n H \subset G$ and $(p^n H)^1 \subset p^n H^1$,*

either (2) or (3) implies the other two.

Proof. Deny, and let G be a smallest counterexample. Then (2) or (3) holds for G, and (2), (3), and (4) for all proper subgroups of G.

(3) implies (4). Let us first show that pH is a subgroup. Only closure need be checked. Let $a \in H$ and $b \in H$. WLOG $H = \langle a, b \rangle$. If H is cyclic, then $pH \subset G$. If H is not cyclic, then

$$a^p b^p = (ab)^p c^p, \qquad c \in H^1 \subset \mathrm{Fr}(H),$$

hence c is a nongenerator of H (Theorem 7.3.2). Therefore $\langle ab, c \rangle < H$ (for H is not cyclic), and by induction,

$$a^p b^p = (ab)^p c^p = d^p, \qquad d \in \langle ab, c \rangle \subset H.$$

Therefore $pH \subset G$.

Again let $a \in H$ and $b \in H$. By (3) (with some terms transposed) and the fact that $pH^1 \subset G$,

$$
\begin{aligned}
a^{-p} b^{-p} a^p b^p &= d^p (a^{-1} b^{-1})^p (ab)^p c^p, \qquad c \in H^1, d \in H^1, \\
&= d^p (a^{-1} b^{-1} ab)^p g^p c^p, \qquad g \in \langle a^{-1} b^{-1}, ab \rangle^1 \subset H^1, \\
&= f^p, \qquad f \in H^1.
\end{aligned}
$$

Now $(pH)^1$ is generated by commutators of the form $[a^p, b^p]$ and $[a^p, b^p] \in pH^1$ by the preceding argument. It follows that $(pH)^1 \subset pH^1$.

Now induct on n. Since $p^n H$ is a subgroup of G,

$$p^{n+1} H = p(p^n H) \subset p^n H \subset G,$$
$$(p^{n+1} H)^1 = (p(p^n H))^1 \subset p(p^n H)^1 \subset p p^n H^1 = p^{n+1} H^1.$$

(3) implies (2). WLOG, $G = \langle a, b \rangle$. Equation (1) holds for $n = 1$ by assumption. Induct on n. Then, by (2) inductively, (3), and both statements in (4),

$$
\begin{aligned}
(ab)^{p^{n+1}} &= (a^{p^n} b^{p^n} x_1^{p^n} \cdots x_r^{p^n})^p, \qquad x_i \in G^1 \\
&= (a^{p^n} b^{p^n} c^{p^n})^p, \qquad c \in G^1 \\
&= a^{p^{n+1}} (b^{p^n} c^{p^n})^p d^p, \qquad d \in (p^n G)^1 \\
&= a^{p^{n+1}} b^{p^{n+1}} c^{p^{n+1}} g^p d^p, \qquad g \in (p^n G)^1 \\
&= a^{p^{n+1}} b^{p^{n+1}} c^{p^{n+1}} x^{p^{n+1}} y^{p^{n+1}}, \qquad x, y \in G^1 = \langle a, b \rangle^1.
\end{aligned}
$$

Hence (1) holds for all n, and G is regular.

(2) implies (3). Again, WLOG $G = \langle a, b \rangle$. By (1), and (4) inductively,

$$
\begin{aligned}
(ab)^p &= a^p b^p x_1^p \cdots x_r^p, \qquad x_j \in G^1 \\
&= a^p b^p c^p \qquad c \in G^1. \; \|
\end{aligned}
$$

In particular:

7.5.6. (*Kemhadze* [1].) *A finite p-group G is regular iff for all $a \in G$ and* $b \in G$, $\exists\ c \in \langle a, b \rangle^1$ *such that* $(ab)^p = a^p b^p c^p$. ‖

An example of an irregular p-group (p arbitrary) is given in Exercise 9.2.30.

EXERCISE

7.5.7. If p is an odd prime, then a group of order p^3 is regular.

REFERENCES FOR CHAPTER 7

For Section 7.1, Hirsch [1], [2], [3]; Theorems 7.1.6 and 7.1.7, Baer [8]; Section 7.2, M. Hall [1]; Exercise 7.2.21, P. Hall [5]; Exercise 7.2.22, Huppert [2]; Section 7.3, Gaschütz [2] (see also P. Hall [7] and Hobby [1]); Theorem 7.3.13, Hirsch [4]; Theorems 7.3.20 to 7.3.23, Dlab and Kořínek [1]; Exercise 7.3.24, Dlab [1]; Exercise 7.3.27, Zassenhaus [4]; Exercise 7.3.28, Iwasawa [2]; Section 7.4, Fitting [1] and Gaschütz [2]; Theorem 7.4.1, Hirsch [3]; Section 7.5, Kemhadze [1] and [2] and M. Hall [1].

EIGHT

FREE GROUPS AND FREE PRODUCTS

8.1 Definitions and existence

If G is a group, then a subset S of $G^\#$ *freely generates* G iff every $g \in G^\#$ can be written in exactly one way in the form

$$g = x_1^{n_1} \cdots x_k^{n_k}, \ x_i \in S, \ x_i \neq x_{i+1}, \ 0 \neq n_i \in J. \tag{1}$$

A group G is *free* iff G is freely generated by some subset S. One also says that G is free on S.

It should be noted that in (1), x_1 may equal x_3, for example. The group J is free, being freely generated by $\{1\}$ and also by $\{-1\}$. It will be shown later (Theorem 8.1.10) that, given a set S, there is a group G which is free on S. The group E is free on the empty set $\varnothing$.

8.1.1. *If G is freely generated by S, H is a group, and T is a function from S into H, then $\exists \, | \, U \in \operatorname{Hom}(G, H)$ such that $U \, | \, S = T$.*

Proof. Suppose that such a U exists, and let $g \in G$ be given by (1). Then $eU = e_H$, and

$$gU = \pi(x_i \, U)^{n_i} = \pi(x_i T)^{n_i}. \tag{2}$$

Hence there is at most one U with the required properties.

Now let U be defined by (2) and $eU = e_H$, where g is given uniquely by (1). Then certainly $U \mid S = T$. It is clear that

$$(eg)U = gU = (eU)(gU), (ge)U = (gU)(eU), (ee)U = (eU)(eU).$$

Let $h = y^m$, $y \in S$, $m \neq 0$. If $y \neq x_k$, then

$$gh = x_1^{n_1} \cdots x_k^{n_k} y^m,$$
$$(gh)U = (x_1 T)^{n_1} \cdots (x_k T)^{n_k} (yT)^m = (gU)(hU).$$

If $y = x_k$ and $m + n_k \neq 0$, then

$$gh = x_1^{n_1} \cdots x_k^{m+n_k},$$
$$(gh)U = (x_1 T)^{n_1} \cdots (x_{k-1} T)^{n_{k-1}} (x_k T)^{m+n_k}$$
$$= (x_1 T)^{n_1} \cdots (x_k T)^{n_k} (yT)^m = (gU)(hU).$$

If $y = x_k$ and $m + n_k = 0$, then

$$gh = x_1^{n_1} \cdots x_{k-1}^{n_{k-1}} \quad \text{(or } e \text{ if } k = 1),$$
$$(gh)U = (x_1 T)^{n_1} \cdots (x_{k-1} T)^{n_{k-1}} = (x_1 T)^{n_1} \cdots (x_k T)^{n_k} (yT)^m$$
$$= (gU)(hU).$$

Therefore $(gh)U = (gU)(hU)$ if $h = y^m$, $y \in S$, and $m \neq 0$. Now suppose that

$$h = y_1^{m_1} \cdots y_r^{m_r}, \ y_i \in S, \ y_i \neq y_{i+1}, \ 0 \neq m_i \in J,$$

and induct on r. Then $h = h_1 h_2$ where $h_2 = y_r^{m_r}$. By the inductive hypothesis,

$$(gh)U = (gh_1 h_2)U = ((gh_1)U)(h_2 U) = (gU)(h_1 U)(h_2 U)$$
$$= (gU)((h_1 h_2)U) = (gU)(hU).$$

Hence, by induction, $U \in \operatorname{Hom}(G, H)$. $\|$

A converse of Theorem 8.1.1 is true (see Exercise 8.1.18). If S is a set, then there is essentially just one group freely generated by S. The uniqueness part of this statement will now be formulated and proved.

8.1.2. *If G and H are groups both freely generated by S, then $G \cong H$. In fact, there is an isomorphism of G onto H fixing all elements of S.*

Proof. By Theorem 8.1.1, $\exists \ U \in \operatorname{Hom}(G, H)$ such that $U \mid S = I_S$. If $*$ and $\circ$ are the operations in G and H, respectively, then

$$(x_1^{n_1} * \ldots * x_k^{n_k})U = x_1^{n_1} \circ \cdots \circ x_k^{n_k},$$

so that (by definition of "freely generates") U is 1–1 from G onto H, i.e., $G \cong H$. $\|$

If S is a nonempty set of subgroups of a group G, then G is the *free product* $\overset{*}{\pi}\{H \mid H \in S\}$ of the set S of subgroups iff each $g \in G^{\#}$ can be written in exactly one way in the form

$$g = x_1 \cdots x_k, e \neq x_i \in H_i \in S, H_i \neq H_{i+1}. \tag{3}$$

One assigns a similar meaning to $G = A_1 * \ldots * A_n$, where $A_i \subset G$. It is clear that a free group is a free product of a set of infinite cyclic subgroups. Hence the concept of free product is a generalization of the concept of free group. Theorem 8.1.1 has an analogue:

8.1.3. *If* $G = \overset{*}{\pi}\{H \mid H \in S\}$, K *is a group, and* $T_H \in \mathrm{Hom}(H, K)$ *for all* $H \in S$, *then* $\exists \mid T \in \mathrm{Hom}(G, K)$ *such that* $T \mid H = T_H$.

Proof. Suppose that T exists, and let $g \in G$ have the form (3). Then $eT = e_K$, and

$$gT = \pi(x_i T) = \pi(x_i T_{H_i}). \tag{4}$$

Hence there is at most one T with the required properties.

Now let T be defined by (4) (and $eT = e$), where g is given by (3). Then $T \mid H = T_H$ for all $H \in S$. Also $(gh)T = (gT)(hT)$ if at least one of the two factors is e. Let g be given by (3) and let $e \neq h \in H \in S$. Then

$$gh = \begin{cases} x_1 \cdots x_k h & \text{if } H_k \neq H, \\ x_1 \cdots x_{k-1}(x_k h) & \text{if } H_k = H \text{ and } x_k h \neq e, \\ x_1 \cdots x_{k-1} & \text{if } H_k = H \text{ and } x_k h = e \end{cases}$$

is in standard form. One readily checks that $(gh)T = (gT)(hT)$ in all three cases, using the fact that T_{H_k} is a homomorphism. Finally, if $h = y_1 \ldots y_r$ and $e \neq y_i \in H_i' \in S$, then an induction on r shows, just as in Theorem 8.1.1, that $(gh)T = (gT)(hT)$.

8.1.4. *If* G *and* K *are both free products of the same set* S *of subgroups, then* $G \cong K$.

Proof. By Theorem 8.1.3, $\exists \, T \in \mathrm{Hom}(G, K)$ such that $T \mid H = I_H$ for all $H \in S$. If $*$ and $\circ$ are the operations in G and K, respectively, then

$$(x_1 * \ldots * x_k)T = x_1 \circ \ldots \circ x_k, \qquad x_i \in H_i^{\#}, H_i \neq H_{i+1}.$$

By the definition of free product, T is 1–1 from G onto K, so that $G \cong K$. ∥

A generalization of the concept of free product will now be introduced. A preliminary lemma is needed to clear up various uniqueness questions.

8.1.5. *Let* G *be a group,* $A \subset G$, S *a nonempty set of subgroups of* G, *and, for each* $H \in S$,

$$H = A \, \dot{\cup} \, (\dot{\cup} \, \{Ax_{H,i} \mid i \in S_H\}),$$

$$Ax_{H,i} = Ay_{H,i} \qquad \text{for } i \in S_H.$$

Then

(i) *if $g \in G$, then g can be written in the form*

$$ax_{H_1,i_1} \cdots x_{H_n,i_n}, a \in A, H_k \neq H_{k+1}, n = 0, 1, 2, \ldots, \qquad (5)$$

iff $g \in \langle H \mid H \in S \rangle$.

(ii) *if $g = h_1 \cdots h_r$ with $h_k \in H'_k \backslash A$ and $H'_k \neq H'_{k+1}$, then g has an expression (5) with $n = r$, $H_k = H'_k$ for all k, and so $Ah_r = Ax_{H_n,i_n}$.*

(iii) *Every element of G can be written in exactly one way in the form (5) iff every element of G can be written in exactly one way in the form*

$$by_{K_1,j_1} \cdots y_{K_m,j_m}, b \in A, K_k \neq K_{k+1}, K_k \in S, m \geq 0. \qquad (6)$$

In this case, if g has the representations (5) and (6), then $m = n$, $H_k = K_k$ for all k, and $i_n = j_n$.

Proof. (i) and (ii). If g can be written in the form (5), then $x_{H_k,i_k} \in H_k \in S$ and $a \in H_1$ (or $a \in H \in S$ if $n = 0$), so that $g \in \langle H \mid H \in S \rangle$.

Conversely, assume that $g = h_1 \cdots h_r$ with $h_k \in H_k \in S$ and r minimal. If some $h_k \in A$, then it can be combined with a neighboring factor to contradict the minimality of r if $r \neq 1$, while if $r = 1$, then g has the form (5) already. Hence we need only prove (ii). For $r = 1$, this follows from the given coset decomposition of H_1 with respect to A. Induct on r. Now $h_r = cx_{H_r,i_r}$, $c \in A$. Hence

$$g = h_1 \cdots h_{r-2}(h_{r-1}c)x_{H_r,i_r}, h_{r-1}c \in H_{r-1}\backslash A, Ah_r = Ax_{H_r,i_r}.$$

The conclusion (ii) now follows from the inductive hypothesis.

(iii) It follows from the hypotheses that, for each $H \in S$,

$$H = A \,\dot\cup\, (\dot\cup \{Ay_{H,i} \mid i \in S_H\}).$$

Conclusion (ii) implies that if g has a decomposition (5) (not necessarily unique), then it has a decomposition (6) with $m = n$, $H_k = K_k$ for all k, and $i_n = j_n$ (this is also trivially true if $n = 0$). This proves the assertion in the final sentence of the theorem. By symmetry, if g has a decomposition (6), then it has a decomposition (5) with $m = n$.

Now assume that every element of G has a unique representation (5). It follows from the preceding paragraph that all representations (5) or (6) of $g \in G$ have the same length n, and that all representations (6) have the same final term. If the theorem is false, then there is a shortest g which has two distinct representations (6). These representations have the same final term y_n, so that gy_n^{-1} is shorter than g and has two distinct representations (6). ∥

Let G be a group, $A \subseteq G$, S a nonempty set of subgroups of G, and, for each $H \in S$,

$$H = A \ \dot\cup \ (\dot\cup \ \{Ax_{H,i} \mid i \in S_H\}).$$

Then G is the *free product* of S *with amalgamated subgroup* A, denoted $G = (\overset{*}{\pi}\{H \mid H \in S\})_A$, iff each $g \in G$ can be represented in exactly one way in the form (5). Theorem 8.1.5 shows that this definition is independent of the choice of $x_{H,i}$ within the coset $Ax_{H,i}$. It, moreover, justifies defining the *length* Len(g) to be n if g has the representation (5) or, equivalently, if $g = h_1 \ldots h_n$ with $h_k \in H_k \backslash A$, $H_k \in S$.

Finally, it should be remarked that if $G = (\overset{*}{\pi}\{H \mid H \in S\})_A$, $H \in S$, and $H \neq K \in S$, then $H \cap K = A$.

8.1.6. $G = \overset{*}{\pi}\{H \mid H \in S\}$ *iff* $G = (\overset{*}{\pi}\{H \mid H \in S\})_E$. ‖

Thus the free product with amalgamated subgroup is a generalization of the free product.

8.1.7. *If* $G = (\overset{*}{\pi}\{H \mid H \in S\})_A$, K *is a group,* $T_H \in \text{Hom}(H, K)$ *for all* $H \in S$, $T_A \in \text{Hom}(A, K)$, *and* $T_H \mid A = T_A$ *for all* $H \in S$, *then*

$$\exists \mid U \in \text{Hom}(G, K)$$

such that $U \mid H = T_H$ *for all* $H \in S$.

Proof. Suppose that $U \in \text{Hom}(G, K)$ and $U \mid H = T_H$ for all H. If g has the form (5), then

$$gU = (a(\pi x_{H_k, i_k}))U = (aT_A)\pi(x_{H_k, i_k}T_{H_k}). \tag{7}$$

Hence there is at most one U with the required properties.

Conversely, let U be defined by (7). Then U is a function from G into K, and $U \mid H = T_H$ for all $H \in S$. Also [Theorem 8.1.5 (i)], $G = \langle H \mid H \in S \rangle$. We assert that if $g = h_1 \cdots h_n$ with $h_i \in H_i \in S$ (H_i may equal H_{i+1}), then $gU = \pi(h_iU)$. This is obvious if $n = 1$. Induct on n. Now $h_n = bx_n$ where $b \in A$ and x_n is e or one of the coset representatives x_{H_n, i_n}. Then [Theorem 8.1.5 (ii)],

$$h_1 \cdots h_{n-2}(h_{n-1}b) = ax_1 \cdots x_r, \qquad r \leqq n - 1,$$

in the standard form (5). If x_r and x_n are not in the same $H \in S$, then, with obvious notation,

$$\begin{aligned}
(h_1 \ldots h_n)U &= (aT_A)(x_1T_1) \cdots (x_rT_r)(x_nT_n) \\
&= ((ax_1 \cdots x_r)U)(x_nU) = ((h_1 \cdots h_{n-2}(h_{n-1}b))U)(x_nU) \\
&= (h_1U) \cdots (h_{n-2}U)((h_{n-1}b)U)(x_nU) \\
&= (h_1U) \cdots (h_{n-1}U)(bU)(x_nU) \\
&= (h_1U) \cdots (h_{n-1}U)((bx_n)U) = (h_1U) \cdots (h_nU).
\end{aligned}$$

If x_r and x_n are in the same $H \in S$, then

$$h_1 \cdots h_n = (ax_1) \cdots x_{r-1}(x_r x_n),$$
$$(h_1 \cdots h_n)U = ((ax_1)U) \cdots (x_{r-1}U)((x_r x_n)U)$$
$$= ((ax_1)U) \cdots (x_r U)(x_n U)$$
$$= ((ax_1)U) \cdots ((x_r b^{-1})U)((bx_n)U)$$
$$= ((ax_1 \cdots x_r b^{-1})U)((bx_n)U) = ((h_1 \cdots h_{n-1})U)(h_n U)$$
$$= (h_1 U) \cdots (h_{n-1}U)(h_n U).$$

Hence, $(h_1 \cdots h_n)U = (h_1 U) \cdots (h_n U)$ for all n. It follows that

$$((h_1 \cdots h_n)(h_{n+1} \cdots h_m))U = \pi(h_i U)$$
$$= ((h_1 \cdots h_n)U)((h_{n+1} \cdots h_m)U),$$

so that U is a homomorphism. ‖

The generalization of Theorem 8.1.4 to free products with amalgamated subgroup is true and is left as an Exercise, 8.1.16.

Let us consider the question of existence of free groups, free products, and free products with amalgamated subgroup. Since the last concept is the most general, considering it will suffice.

Let S be a set of groups. There does not always exist a group G containing all $H \in S$ as subgroups; for example two of the groups might have the same set of elements but different operations. For this reason, a somewhat more complex setup is required.

8.1.8. *Let A be a group, S a set of groups, and U_H an isomorphism of A into H for all $H \in S$. Then there is a group G containing A as a subgroup, and isomorphisms T_H of H into G for $H \in S$ such that*

(i) $U_H T_H = I_A,$

(ii) $G = (\#\pi\{HT_H \mid H \in S\})_A.$

Proof. By Exercise 2.1.36 and a little extra argument, it may be assumed that $A \subseteq H$, $U_H = I_A$ for all $H \in S$, and $H \cap K = A$ for distinct H and K in S. By Theorem 2.1.16, it is sufficient to construct a group G and isomorphisms T_H of H into G and T_A of A into G such that

(iii) $T_H \mid A = T_A$ for all $H \in S,$

(iv) $G = (\overset{*}{\pi}\{HT_H \mid H \in S\})_{AT_A}.$

Since $A \subseteq H$, $\exists\, S_H$ and $x_{H,i}$ such that

$$H = A \,\dot{\cup}\, (\dot{\cup} \,\{Ax_{H,i} \mid i \in S_H\})$$

for each $H \in S$. Let

$$R = \{x_{H,i} \mid H \in S, i \in S_H\},$$

and let M be the set of objects

$$(a, x_1, \ldots, x_n), a \in A, n \geq 0, x_i \in R,$$

x_i and x_{i+1} not in the same $H \in S$.

Let $m = (a, x_1, \ldots, x_n) \in M$, and let $h \in H \in S$. If $x_n \in H$, then

$$x_n h = b_n \in A, \tag{8}$$

or

$$x_n h = b_n x_n', \qquad (b_n \in A, x_n' \in R \cap H). \tag{9}$$

If $x_n \notin H$, then

$$h = bx, \qquad (b \in A, x \in R \cap H, x_n \notin H), \tag{10}$$

and

$$x_n b = b_n x_n', \qquad (b_n \in A, x_n' \in R). \tag{11}$$

In any case,

$$x_{n-1} b_n = b_{n-1} x_{n-1}', \ldots, x_1 b_2 = b_1 x_1', \qquad b_i \in A, x_i' \in R. \tag{12}$$

Also x_i' is in the same H as x_i. Now define

$$m(hf) = \begin{cases} (ab_1, x_1', \ldots, x_{n-1}') & \text{if (8) and (12)}, \\ (ab_1, x_1', \ldots, x_n') & \text{if (9) and (12)}, \\ (ab_1, x_1', \ldots, x_n', x) & \text{if (10), (11), and (12)}, \\ (ab) & \text{if } m = (a) \text{ and } h = b \in A, \\ (ab, x) & \text{if } m = (a) \text{ and } h = bx, b \in A, x \in R. \end{cases} \tag{13}$$

Thus, hf is a function from M into M.

Let $k \in H$ also. We wish to show that

$$m(hf)(kf) = m((hk)f). \tag{14}$$

The case $n = 0$, and the case $n = 1$ and $x_1 \in H$ are left as Exercise 8.1.17. Assume, therefore, that $n > 1$ or $n = 1$ and $x_1 \notin H$.

CASE 1. $x_n \in H$. Then

$$x_n hk = b_n' \in A \tag{8'}$$

or

$$x_n hk = b_n' x_n'', \qquad (b_n' \in A, x_n'' \in R), \tag{9'}$$

and

$$x_{n-1} b_n' = b_{n-1}' x_{n-1}'', \ldots, x_1 b_2' = b_1' x_1'', \qquad (b_i' \in A, x_i'' \in R). \tag{12'}$$

Therefore

$$m((hk)f) = \begin{cases} (ab_1', x_1'', \ldots, x_{n-1}'') & \text{if (8)' and (12)'}, \\ (ab_1', x_1'', \ldots, x_n'') & \text{if (9)' and (12)'}. \end{cases}$$

Suppose that (9) holds. Then

$$x'_n k = b_n^{-1} x_n h k = \begin{cases} b_n^{-1} b'_n & \text{if (8)}', \\ b_n^{-1} b'_n x''_n & \text{if (9)}'; \end{cases}$$

and

$$x'_{i-1}(b_i^{-1} b'_i) = (b_{i-1}^{-1} x_{i-1} b_i)(b_i^{-1} b'_i) = b_{i-1}^{-1} x_{i-1} b'_i$$
$$= b_{i-1}^{-1} b'_{i-1} x''_{i-1}, \qquad (2 \le i \le n). \tag{15}$$

Hence

$$m(hf)(kf) = \begin{cases} (ab_1 b_1^{-1} b'_1, x''_1, \ldots, x''_{n-1}) & \text{if (8)}', \\ (ab_1 b_1^{-1} b'_1, x''_1, \ldots, x''_n) & \text{if (9)}', \end{cases}$$

so that (14) is satisfied if (9) holds.

Next, suppose that (8) is true. If (8)' holds, then $k \in A$, and

$$x'_{n-1} k = b_{n-1}^{-1} x_{n-1} b_n k = b_{n-1}^{-1} x_{n-1} x_n h k = b_{n-1}^{-1} x_{n-1} b'_n$$
$$= b_{n-1}^{-1} b'_{n-1} x''_{n-1},$$

and (15) holds for $2 \le i \le n - 1$. Hence

$$m(hf)(kf) = (ab'_1, x''_1, \ldots, x''_{n-1}) = m((hk)f).$$

If (9)' is satisfied, then $k \notin A$, and

$$k = (x_n h)^{-1} b'_n x''_n, \qquad (x_n h)^{-1} b'_n \in A;$$
$$x'_{n-1}(x_n h)^{-1} b'_n = x'_{n-1} b_n^{-1} b'_n = b_{n-1}^{-1} x_{n-1} b'_n = b_{n-1}^{-1} b'_{n-1} x''_{n-1}.$$

Moreover (15) holds for $2 \le i \le n - 1$. Hence

$$m(hf)(kf) = (ab'_1, x''_1, \ldots, x''_n) = m((hk)f).$$

CASE 2. $x_n \notin H$. Then (10), (11), and (12) hold, and
$$m(hf) = (ab_1, x'_1, \ldots, x'_n, x).$$

First, suppose that $hk \in A$. Then $bxk \in A$, so $xk \in A$. Hence

$$x'_n(xk) = b'_n x''_n, \qquad\qquad (b'_n \in A, x''_n \in R),$$
$$x'_{n-1} b'_n = b'_{n-1} x''_{n-1}, \ldots, x'_1 b'_2 = b'_1 x''_1 \qquad (b'_i \in A, x''_i \in R).$$

Then

$$m(hf)(kf) = (ab_1 b'_1, x''_1, \ldots, x''_n).$$

Also

$$x_n h k = x_n b x k = b_n x'_n x k = b_n b'_n x''_n,$$
$$x_{n-1} b_n b'_n = b_{n-1} x'_{n-1} b'_n = b_{n-1} b'_{n-1} x''_{n-1}, \ldots,$$
$$x_1 b_2 b'_2 = b_1 x'_1 b'_2 = b_1 b'_1 x''_1.$$

Therefore,

$$m((hk)f) = (ab_1 b'_1, x''_1, \ldots, x''_n) = m(hf)(kf).$$

Finally, suppose that $hk \notin A$. Then $xk \notin A$. Hence,

$$xk = b'_{n+1}x''_{n+1}, \qquad b'_{n+1} \in A, \ x''_{n+1} \in R,$$
$$x'_n b'_{n+1} = b'_n x''_n, \dots, x'_1 b'_2 = b'_1 x''_1 \qquad (b'_i \in A, \ x''_i \in R).$$

Therefore,

$$m(hf)(kf) = (ab_1 b'_1, x''_1, \dots, x''_n, x''_{n+1}).$$

Again,

$$hk = bxk = bb'_{n+1}x''_{n+1},$$
$$x_n bb'_{n+1} = b_n x'_n b'_{n+1} = b_n b'_n x''_n, \dots,$$
$$x_1 b_2 b'_2 = b_1 x'_1 b'_2 = b_1 b'_1 x''_1,$$
$$m((hk)f) = (ab_1 b'_1, x''_1, \dots, x''_{n+1}) = m(hf)(kf).$$

It follows that $(hk)f = (hf)(kf)$ for all $h \in H$, $k \in H$. Now set $h = e$. Equations (9) and (12) imply that

$$b_n = e = b_{n-1} = \dots = b_1, \ x'_i = x_i.$$

By (13), ef is the identity function on M. Since

$$ef = (hh^{-1})f = (hf)(h^{-1}f) = (h^{-1}f)(hf)$$

for all $h \in H$, hf is both 1–1 and onto M, hence $hf \in \mathrm{Sym}(M)$ for all $h \in H$. Therefore f is a homomorphism from H into $\mathrm{Sym}(M)$. If $h \neq e$, then $(e)(hf) = (h)$ or (b, x) according as $h \in A$ or $h = bx \notin A$. In any case, hf is not the identity, so that f is an isomorphism. Let $G = \langle Hf \mid H \in S \rangle$. Then (iii) is satisfied if $T_H = (f \mid H)$. Any element g of G has the form

$$(af)(x_1 f) \cdots (x_n f), \ a \in A, \ x_i \in R,$$

where consecutive x_i's are in different H's. Moreover

$$(e)((af)(x_1 f) \cdots (x_n f)) = (a, x_1, \dots, x_n),$$

so that g cannot have two different representations of this form. It follows that

$$G = (\overset{*}{\pi}\{Hf \mid H \in S\})_{Af} = (\overset{*}{\pi}\{HT_H \mid H \in S\})_{AT_A}.$$

8.1.9. *If S is a set of groups, then $\exists$ a group G and isomorphisms T_H from H into G such that $G = \overset{*}{\pi}\{HT_H \mid H \in S\}$.*

Proof. Let A be a group of order 1 and U_H the unique isomorphism of A into H for all $H \in S$. By Theorem 8.1.8, there are a group $G \supset A$ and isomorphisms T_H of H into G such that $G = (\overset{*}{\pi}\{HT_H \mid H \in S\})_A$. But A is the identity subgroup E of G, so that by Theorem 8.1.6, $G = \overset{*}{\pi}\{HT_H \mid H \in S\}$.

8.1.10. *If S is a set, then there is a group G freely generated by S.*

Proof. If $x \in S$, then there is an infinite cyclic group H_x generated by x (Theorem 2.1.16). With a little care, $H_x \neq H_y$ if $x = y$. By Theorem 8.1.9, there are a group G and isomorphisms T_x from H_x into G such that $G = \frac{*}{\pi}\{H_x T_x \mid x \in S\}$. By Theorem 2.1.16 again, there are a group K containing S and an isomorphism U of K onto G such that $xU = xT_x$. It follows readily that $K = \frac{*}{\pi}\{\langle x \rangle \mid x \in S\}$. An examination of the definitions involved then shows that K is a free group, freely generated by S. ‖

Converses of Theorems 8.1.1, 8.1.3, and 8.1.7 are true. Only the converse of Theorem 8.1.7 will be proved, the others being left as exercises.

8.1.11. *If G is a group, $A \subset G$, S is a set of subgroups of G each containing A, and if*

(∗) *if K is a group, $T_A \in \mathrm{Hom}(A, K)$, $T_H \in \mathrm{Hom}(H, K)$ and $T_H \mid A = T_A$ for all $H \in S$, then $\exists \mid U \in \mathrm{Hom}(G, K)$ such that $U \mid H = T_H$ for all $H \in S$,*

then $G = (\frac{*}{\pi}\{H \mid H \in S\})_A$.

Proof. Let $L = \langle H \mid H \in S \rangle$. By assumption, $\exists \, U \in \mathrm{Hom}(G, L)$ such that $U \mid H = I_H$ for all $H \in S$. Also $I_G \mid H = I_H$. Since both U and I_G are homomorphisms from G into G, by (∗), $U = I_G$. Hence $L = G$, i.e., $\langle H \mid H \in S \rangle = G$.

By Theorem 8.1.8, $\exists \, K \supset A$ and isomorphisms T_H from H into K such that each $T_H \mid A = I_A$ and $K = (\frac{*}{\pi}\{HT_H \mid H \in S\})_A$. By (∗), $\exists \, U \in \mathrm{Hom}(G, K)$ such that $U \mid H = T_H$. If $e \neq g \in \mathrm{Ker}(U)$, then, writing g in the form (5) (Theorem 8.1.5), we have

$$e = gU = a(\pi x_{H_k, i_k} T_{H_k}), \qquad H_k \neq H_{k+1}.$$

Since $K = (\frac{*}{\pi}\{HT_H \mid H \in S\})_A$, this is a contradiction. Hence U is an isomorphism. Since $K = \langle HT_H \rangle = \langle HU \mid H \in S \rangle$, U is an isomorphism of G onto K. It follows readily from the definition that $G = (\frac{*}{\pi}\{H \mid H \in S\})_A$.

8.1.12. *If $G = (\frac{*}{\pi}\{H \mid H \in S\})_A$, $K_H \subset H$, and $K_H \cap A = E$ for all $H \in S$, then*

$$\langle K_H \mid H \in S \rangle = \frac{*}{\pi}\{K_H \mid H \in S\}.$$

Proof. Since $K_H \cap A = E$ and $K_H \subset H$, $Ax \neq Ay$ if x and y are distinct elements of K_H. Hence there is a subset S_H of $H^\#$ such that

(i) $K_H^\#$ is a subset of S_H,

(ii) $H = A \, \dot\cup \, (\dot\cup \{Ax \mid x \in S_H\})$.

Any $z \in \langle K_H \rangle^\#$ has a representation in the form

$$z = x_1 \cdots x_n, \qquad x_k \in K_{H_k}^\#, \qquad H_k \neq H_{k+1}.$$

Since $x_k \in S_{H_k}$ by (i), and since $G = (\overset{*}{\pi}\{H \mid H \in S\})_A$, it follows that the representation of z is unique. Hence $\langle K_H \rangle = \overset{*}{\pi}K_H$. ‖

It will be useful to define still another product, analogous to, but simpler than, the free product with amalgamated subgroup. A group G is the *direct product of H and K with amalgamated subgroup A*, denoted by $(H \times K)_A$, iff

$$H \subset G, \ K \subset G, \ G = HK, \ H \cap K = A \quad \text{and} \quad H \subset C(K).$$

8.1.13. *If $G = (H \times K)_A$, then $A \subset Z(G)$ and $G/A = (H/A) + (K/A)$.*

Proof. If $a \in A$, then $a \in K$, hence $H \subset C(a)$. Thus, $C(A) \supset H$ and, similarly, $C(A) \supset K$. Since $G = HK$, $A \subset Z(G)$. Since elements of H and K commute, so do those of H/A and K/A. Hence $H/A \lhd G/A$ and $K/A \lhd G/A$. Since $G = HK$, $G/A = (H/A)(K/A)$. Since $H \cap K = A$, $(H/A) \cap (K/A) = E$. Therefore, $G/A = (H/A) + (K/A)$.

8.1.14. *If $G = (H \times K)_A$ and $L = (H \times K)_A$, then there is an isomorphism T of G onto L such that $T \mid H = I_H$ and $T \mid K = I_K$.*

Proof. One readily (Exercise 8.1.19) verifies that if G and L have operations $*$ and $\circ$, respectively, then the relation

$$T = \{(h * k, h \circ k) \mid h \in H, k \in K\}$$

does the job.

8.1.15. *Let A, H, and K be groups, U_H an isomorphism of A into $Z(H)$, and U_K an isomorphism of A into $Z(K)$. Then there are a group G containing A as a subgroup, and isomorphisms T_H of H into G and T_K of K into G such that*

(i) $U_H T_H = I_A = U_K T_K$,

(ii) $G = (HT_H \times KT_K)_A$.

Proof. By Exercise 2.1.36 and a little extra argument, it may be assumed that $A \subset Z(H)$, $A \subset Z(K)$, $H \cap K = A$, and $U_H = I_A = U_K$. By Theorem 2.1.16, it is sufficient to construct a group G and isomorphisms T_H and T_K of H and K, respectively, into G, such that

(iii) $T_H \mid A = T_K \mid A$,

(iv) $G = (HT_H \times KT_K)_{AT_H}$.

To this end, let $D = \{(a, a^{-1}) \mid a \in A\}$. Since $A \subset Z(H)$ and $A \subset Z(K)$, $D \subset Z(H \times K)$. Let $G = (H \times K)/D$. Let T_H and T_K be defined by the equations:

$$hT_H = (h, e)D, \ kT_K = (e, k)D, \qquad h \in H, k \in K.$$

It is then clear that T_H and T_K are isomorphisms into G. If $a \in A$, then $(a, a^{-1}) \in D$, hence $(a, e)D = (e, a)D$. Therefore $T_H \mid A = T_K \mid A$. It is readily seen that

$$HT_H = \frac{H \times A}{D}, \qquad KT_K = \frac{A \times K}{D}, \qquad AT_H = \frac{A \times A}{D},$$

$$G = (HT_H)(KT_K), \; (HT_H) \cap (KT_K) = AT_H,$$

$$HT_H \subseteq C(KT_K).$$

Hence (iv) is satisfied, and the theorem is proved.

EXERCISES

8.1.16. Prove the analogue of Theorem 8.1.4 for free products with amalgamated subgroup.

8.1.17. Prove that (14) is valid in the case $n = 0$, and in the case $n = 1$ and $x_1 \in H$.

8.1.18. State and prove the analogues of Theorem 8.1.11 for free groups and free products.

8.1.19. Complete the proof of Theorem 8.1.14.

8.1.20. If $G = (\overset{*}{\pi}\{H \mid H \in S\})_A$ and T is a subset of S, then

$$\langle H \mid H \in T \rangle = (\overset{*}{\pi}\{H \mid H \in T\})_A.$$

8.1.21. (Internal associativity of free products with amalgamated subgroup.) If $G = (\overset{*}{\pi}\{H \mid H \in S\})_A$,

$$S = \overset{.}{\cup}\{S_i \mid i \in T\}, \quad \text{and} \quad K_i = \langle H \mid H \in S_i \rangle,$$

then

$$G = (\overset{*}{\pi}\{K_i \mid i \in T\})_A.$$

8.1.22. (External associativity of free products with amalgamated subgroup.) Let S be a set of groups, A a group, U_H an isomorphism of A into H for all $H \in S$, and $S = \overset{.}{\cup}\{S_i \mid i \in M\}$. Call a group G an external free product of S with amalgamated subgroup A iff there are isomorphisms T_H of H into G such that $(U_H T_H) \mid A = (U_K T_K) \mid A$ for all H and K in S, and such that

$$G = (\overset{*}{\pi}\{HT_H \mid H \in S\})_{AU_H T_H}.$$

Prove that if G is an external free product of S with amalgamated subgroup A, and if G_i is an external free product of S_i with amalgamated subgroup A for all $i \in M$, then G is an external free product of $\{G_i \mid i \in M\}$ with amalgamated subgroup A.

8.1.23. (a)   If S is a set of groups each containing A as a subgroup, and if $H \cap K = A$ for $H \in S$, $K \in S$, $H \neq K$, then there is a group G such that

$$G = (\overset{*}{\pi}\{H \mid H \in S\})_A.$$

(b)   If S is a set of groups such that $H \cap K = E_H$ for all $H \in S$, $K \in S$, $H \neq K$, then there is a group G such that

$$G = \overset{*}{\pi}\{H \mid H \in S\}.$$

(c)   If A, H, and K are groups such that $H \cap K = A$, $A \subset Z(H)$, and $A \subset Z(K)$, then there is a group G such that $G = (H \times K)_A$.

8.1.24. (a)   If S is a set of groups, G and K are groups, T_H and U_H are isomorphisms of H into G and K, respectively, for all $H \in S$,

$$G = \overset{*}{\pi}\{HT_H \mid H \in S\},$$

and $K = \overset{*}{\pi}\{HU_H \mid H \in S\}$, then $G \cong K$.

(b)   State and prove an analogous theorem for free products with amalgamated subgroup.

8.1.25. (a)   If $G = (\overset{*}{\pi}\{H \mid H \in S\})_A$ and $A \lhd H$ for all $H \in S$, then $A \lhd G$ and $G/A \cong \overset{*}{\pi}\{H/A \mid H \in S\}$.

(b)   If $G = (\overset{*}{\pi}\{H \mid H \in S\})_A$ and each $H \in S$ is Abelian, then $A \lhd G$ and $G/A \cong \overset{*}{\pi}\{H/A \mid H \in S\}$.

8.1.26. (a)   Let $G = (H \times K)_A$, $T \in \mathrm{Aut}(H)$, $U \in \mathrm{Aut}(K)$, and $T \mid A = U \mid A \in \mathrm{Aut}(A)$. Then $\exists \mid V \in \mathrm{Aut}(G)$ such that $V \mid H = T$ and $V \mid K = U$. In fact, if $h \in H$ and $k \in K$, then $(hk)V = (hT)(kU)$.

(b)   If $G = (\overset{*}{\pi}\{H \mid H \in S\})_A$, $T_H \in \mathrm{Aut}(H)$ for $H \in S$, and

$$T_H \mid A = T_K \mid A \in \mathrm{Aut}(A)$$

for $H \in S$ and $K \in S$, then $\exists \mid V \in \mathrm{Aut}(G)$ such that $V \mid H = T_H$ for all $H \in S$.

(c)   If $G = \overset{*}{\pi}\{H \mid H \in S\}$ and $T_H \in \mathrm{Aut}(H)$ for $H \in S$, then $\exists \mid V \in \mathrm{Aut}(G)$ such that $V \mid H = T_H$ for all $H \in S$.

8.1.27.   If $G = \overset{*}{\pi}\{H \mid H \in S\}$, then there is a homomorphism T of G onto H such that $T \mid H = I_H$ and $T \mid K = O$ for $H \neq K \in S$. Construct T explicitly.

8.1.28.   If $G = H * K$, $H \neq E$, and $K \neq E$, then $H^G < G$. (Use Exercise 8.1.27.)

8.1.29.   If $G = (H \times K)_A$ and $A = Z(H) = Z(K)$, then $A = Z(G)$.

8.1.30.   If F is a free group and $G/H \cong F$, then H has a complement in G (use Theorem 8.1.1).

8.1.31.   Generalize Theorem 8.1.12 to read: if $G = (\overset{*}{\pi}\{H \mid H \in S\})_A$, $B \subset A$, and $K_H \subset H$ and $K_H \cap A = B$ for all $H \in S$, then

$$\langle K_H \mid H \in S \rangle = (\overset{*}{\pi}\{K_H \mid H \in S\})_B.$$

8.2 Presentations

8.2.1. *Any group is isomorphic to a factor group of a free group.*

Proof. Let G be a group and S a generating subset. There is a free group H freely generated by S (Theorem 8.1.10). By 8.1.1, $\exists\ U \in \mathrm{Hom}(H, G)$ such that $U \mid S = I_S$. Since $G = \langle S \rangle = \langle SU \rangle$, U is a homomorphism from H onto G. By the homomorphism theorem, $G \cong H/R$ where $R = \mathrm{Ker}(U)$.

8.2.2. *A finitely generated group is isomorphic to a factor group of a finitely generated free group.* ‖

Let S be a set and T a set of words in the elements of S. If $(F, *)$ is free on S and

$$t = s_1^{n_1} \cdots s_k^{n_k} \in T, \qquad s_i \in S, \tag{1}$$

let $t_F = s_1^{n_1} * \ldots * s_k^{n_k}$. If $(H, \circ)$ is also free on S, then there is a unique (and obvious) isomorphism U of F onto H such that $U \mid S = I_S$. Moreover, for each $t \in T$, $t_F U = t_H$. If $T_F = \{t_F \mid t \in T\}$ and $T_H = \{t_H \mid t \in T\}$, then it follows that $(T_F^F)U = T_H^H$, so that $F/T_F^F \cong H/T_H^H$. We shall usually identify t with t_F and T with T_F, and write T^F for T_F^F.

A group G has *presentation* $(S; T)$ (or, equivalently, has *generators* S and *relations* T) iff S is a set, T a set of words in the elements of S, and $G \cong F/T^F$, where F is free on S. It follows from the above remarks that this definition is independent of the choice of F. It follows from Theorem 8.2.1 that any group has a presentation, actually many presentations. It follows from the definition that any pair $(S; T)$, where S is a set and T is a set of relations in S, is a presentation of some group. If two groups have presentation $(S; T)$, then they are isomorphic.

Let us use the abbreviation $G = (S; T; F; u)$ to mean that F is freely generated by S and u is a homomorphism of F onto G with kernel T^F. If $G = (S; T; F; u)$, then G has presentation $(S; T)$, and if G has presentation $(S; T)$, then $\exists\ F$ and u such that $G = (S; T; F; u)$. If $G = (S; T; F; u)$, then $G = \langle Su \rangle$, and if t has the form (1), then

$$tu = (s_1 u)^{n_1} \cdots (s_k u)^{n_k} = e.$$

It is in this sense that G has generators S and relations T. However, G is generated by Su, not S, and u need not be 1–1.

Let S be a set and T a set of words in S. If K is a group such that there is a function f from S into K for which (i) $K = \langle Sf \rangle$, and (ii) if $\pi s_i^{n_i} \in T$, $s_i \in S$, then $\pi (s_i f)^{n_i} = e$, then K will be said to *satisfy* the relations T (with respect to f).

8.2.3. *If $G = (S; T; F; u)$ and K is a group satisfying the relations T with respect to f: $S \rightarrow K$, then there is a homomorphism v of G onto K such that $(uv) \mid S = f$.*

Proof. By 8.1.1, $\exists f' \in \text{Hom}(F, K)$ such that $f' \mid S = f$. By (i), $Ff' = K$. By (ii), if $t \in T$, then $tf' = e$. Hence T is contained in $\text{Ker}(f')$. Therefore $T^F \subset \text{Ker}(f')$. Let $v = \{(xu, xf') \mid x \in F\}$. Since $\text{Ker}(u) = T^F \subset \text{Ker}(f')$, v is a function. Since u is a homomorphism onto G and f' a homomorphism onto K, v is a homomorphism from G onto K. If $s \in S$, then $suv = sf' = sf$. Hence $(uv) \mid S = f$.

8.2.4. *If a finite group G has presentation $(S; T)$, K is a group satisfying the relations T, and $o(K) \geq o(G)$, then $K \cong G$.*

Proof. By Theorem 8.2.3, there is a homomorphism v from G onto K. Since $o(K) \geq o(G)$ is finite, v is an isomorphism. ∥

One now has a device for defining as many groups as desired—simply write down presentations at random. The trouble is that the properties of a group are not always evident from its presentation. Even such a simple matter as whether the group is finite may be difficult to decide. Some examples where the structure can be determined will now be given.

Example 1. Dihedral group. Let G be a group with presentation $(\{x, y\}; \{x^n, y^2, (yx)^2\})$, where $x \neq y, n \in \mathcal{N}, n > 1$. Then $\exists a$ and b in G (not necessarily distinct) such that $G = \langle a, b \rangle$ and

$$a^n = e, b^2 = e, (ba)^2 = e.^*$$

It follows that $b = b^{-1}$, $b^{-1}aba = e$, $b^{-1}ab = a^{-1}$. Thus $\langle a \rangle = M \triangleleft G$, since $N(M)$ contains both a and b. Now $G/M = \langle bM \rangle$ and $b^2 = e$. Moreover $o(M) = o(a) \mid n$. Therefore $o(G) \leq 2n$. One now suspects that $o(G) = 2n$, but it should be emphasized that this has not yet been proved. To prove that $o(G) = 2n$, it suffices, by Theorem 8.2.4, to exhibit a group K of order at least $2n$ generated by two elements satisfying the above relations. To this end, let $K = \langle c, d \rangle$, where

$$c = (1, 2, \ldots, n)(n + 1, \ldots, 2n),$$
$$d = (1, 2n)(2, 2n - 1) \ldots (n, n + 1)$$
$$= \pi\{(i, 2n - i + 1) \mid 1 \leq i \leq n\}.$$

---

* The more usual procedure is to presume that $G = \langle x, y \rangle$, etc. However, it appears to the author that in this instance one cannot have his cake and eat it too. Let $S = \{x, y\}$, $x \neq y$. Then, if $T = \{x^n, y^2, (yx)^2\}$, a group G with presentation $(S; T)$ could indeed be generated by x and y if desired, but if $T' = \{x, y\}$, then a group G' with presentation $(S; T')$ has only one element, so could not be generated by x and y.

Then $c^n = e$, $d^2 = e$, and

$$dcd = d^{-1}cd = (2n, 2n - 1, \ldots, n + 1)(n, \ldots, 1) = c^{-1},$$

so that $(dc)^2 = e$. Thus K satisfies the given relations. Since $o(c) = n$, $n \mid o(K)$. Since $1d = 2n \neq 1c^i$ for all i, $d \notin \langle c \rangle$, so $o(K) \neq n$. Hence $o(K) \geqq 2n$. Therefore $o(G)$ (and $o(K)$) is $2n$.

Example 2. Elementary Abelian p-group. Let $p \in \mathscr{P}$, let S be a set, and let G be a group with presentation

$$(S; x^p, [x, y], (x \in S, y \in S)).$$

Let $'$ denote the resulting homomorphism of a free group F on S onto G. Then $G = \langle S' \rangle$ and $x'^p = e$, $[x', y'] = e$ for all x and y in S. It follows first that G is Abelian; then that G is an elementary Abelian p-group.

There is an elementary Abelian p-group L which is the direct sum of $o(S)$ cyclic groups of order p. By Theorem 2.1.16, there is even a group $K = \Sigma \{ \langle x \rangle \mid x \in S \}$ where each $o(\langle x \rangle) = p$. The group K satisfies the given relations. By Theorem 8.2.3, there is a homomorphism U of G onto K such that if $x \in S$ then $x'U = x$. Let $e \neq y \in \text{Ker}(U)$. Since G is Abelian and generated by S',

$$y = x_1'^{m_1} \ldots x_r'^{m_r}, \qquad x_i \in S, \ x_i' \neq x_j' \ \text{if} \ i \neq j, 0 < m_i < p.$$

Hence $x_i \neq x_j$ if $i \neq j$, and $yU = \pi x_i^{m_i}$. The structure of K now shows that $yU \neq e_K$, a contradiction. Hence $G \cong K$, i.e., G is the direct sum of $o(S)$ cyclic groups of order p.

Example 3. Dicyclic group. Let G be a group with presentation $(x, y; x^{2n}, x^n y^{-2}, y^{-1} xyx)$, where $x \neq y$, $n \in \mathscr{N}$, $n > 1$. Then there are $a \in G$, $b \in G$ such that $G = \langle a, b \rangle$, $a^{2n} = a^n b^{-2} = b^{-1} aba = e$. Thus $b^{-1} ab = a^{-1}$ and $b^2 = a^n$. Hence $M = \langle a \rangle \lhd G$, $G/M = \langle bM \rangle$, and $(bM)^2 = M$. Therefore $o(G) \leqq 4n$.

Now let

$$c = (1, \ldots, 2n)(2n + 1, \ldots, 4n)$$
$$d = \pi\{(i, 4n - (i - 1), n + i, 3n - (i - 1)) \mid 1 \leqq i \leqq n\}.$$

One can then check that $K = \langle c, d \rangle$ satisfies the given relations, and has order at least $4n$. Therefore $o(G) = o(K) = 4n$.

Example 4. Quaternion group. In Example 3, let $n = 2$. Thus $G = \langle a, b \rangle$, $a^4 = e$, $a^2 = b^2$, and $b^{-1} ab = a^{-1}$. In different notation,

$$G = \{1, i, j, k, -1, -i, -j, -k\},$$

where $i^2 = j^2 = k^2 = -1$, $ij = k$, $jk = i$, $ki = j$, $ji = -k$, etc. (The second interpretation accounts for the name of this group.)

Example 5. Let G be a group with presentation

$$(x, y; x^3, y^4, (xy)^2), \quad x \neq y.$$

Then there are elements a and b in G such that $G = \langle a, b \rangle$ and

$$a^3 = b^4 = (ab)^2 = e.$$

This time the discussion is more complicated since there is no normal subgroup in plain sight. Let $H = \langle b \rangle$. H has the right cosets

$$H, Ha, Ha^2, Ha^2b, Ha^2b^2, Ha^2b^2a.$$

Multiplication by a or b yields no new right cosets, since

$$abab = e, \quad ba = a^2b^3, \quad ab = b^3a^2,$$
$$(Ha)b = Hb^3a^2 = Ha^2,$$
$$(Ha^2b^2)b = Ha^2b^3 = Hba = Ha,$$
$$(Ha^2b)a = Ha^2a^2b^3 = Hab^3 = Habb^2 = Ha^2b^2,$$
$$(Ha^2b^2a)b = Ha^2b^2b^3a^2 = Ha^2ba^2 = (Ha^2ba)a = Ha^2b^2a,$$
$$(Ha^2b^2a)a = (Ha^2b^2)b(b^3a^2) = (Ha)ab = Ha^2b.$$

Hence there are at most six right cosets of H, and $o(G) \leq 24$.

Let $K = \langle c, d \rangle$, $c = (1, 2, 3)$, $d = (1, 3, 2, 4)$. Then

$$c^3 = d^4 = (cd)^2 = e.$$

Also

$$dc^2d^2c = (1, 3, 2, 4)(1, 3, 2)(1, 2)(3, 4)(1, 2, 3) = (1, 2) = z,$$
$$z^{-1} dz = (2, 3, 1, 4) = d^{-1}.$$

Hence $\langle d, z \rangle$ is a Sylow 2-subgroup of K of order 8. Since $o(c) = 3, o(K) = 24$, and $K = \text{Sym}(4)$. Hence $G \cong \text{Sym}(4)$. ‖

A systematic method (which we omit) for handling computations of the above type is given in Coxeter and Moser [1, pp. 12–18].

Example 6. If G is a free group freely generated by S, then G has presentation $(S; \varnothing)$. For $\text{Ker}(I_G) = E = \varnothing^G$.

The following theorem is a special case of Exercise 8.1.21.

8.2.5. *If H is freely generated by S, $S = \cup \{S_i \mid i \in M\}$, and $H_i = \langle S_i \rangle$, then each H_i is freely generated by S_i and $H = \#\{H_i \mid i \in M\}$.*

Proof. If $g \in H_i^\#$, then $g = \pi x_j^{n_j}$ with $x_j \in S_i$, $n_j \neq 0$, and $x_j \neq x_{j+1}$. But g cannot have two such representations since H is free on S. Hence H_i

is freely generated by S_i. Also, $H = \langle H_i \rangle$. If $h \in H^\#$ has two different standard representations:

$$h = h_1 \cdots h_n, \qquad h_i \in H_i^\#, H_i \neq H_{i+1},$$
$$h = h'_1 \cdots h'_m, \qquad h'_i \in H_i'^\#, H'_i \neq H'_{i+1},$$

then each h_i and h'_i may be replaced by its representation in terms of the elements of $S_i(S'_i)$ to obtain two distinct standard representations of h in terms of S. This contradicts the freeness of H. Hence h has only one representation, and $H = \overset{*}{\pi} H_i$.

8.2.6. *If* $G = \overset{*}{\pi}\{G_i \mid i \in M\}$, G_i *has presentation* $(S_i; T_i)$ *for each* $i \in M$, *and* $S = \overset{.}{\cup} \{S_i \mid i \in M\}$, *then* G *has presentation* $(S; \overset{.}{\cup} T_i)$.

Remark. Given $\{G_i \mid i \in M\}$, there are presentations $(S_i; T_i)$ of G_i such that the union $S = \cup S_i$ is disjoint. Moreover, the free product of some isomorphic copies of the G_i always exists. Hence all of the assumptions are essentially notational.

Proof. It is clear that the T_i are disjoint. There is a free group H freely generated by S. By Theorem 8.2.5, its subgroups $H_i = \langle S_i \rangle$ are freely generated by S_i, and $H = \overset{*}{\pi} H_i$. By assumption, there is a homomorphism U_i of H_i onto G_i with $\mathrm{Ker}(U_i) = T_i^{H_i}$. By Theorem 8.1.3, $\exists\ U \in \mathrm{Hom}(H, G)$ such that $U \mid H_i = U_i$. Since $G = \langle G_i \rangle$, U is a homomorphism of H onto G. The kernel R of U contains all $T_i^{H_i}$, hence all T_i, hence $T = \overset{.}{\cup} T_i$. Therefore $R \supset T^H$. If $R = T^H$, then G has presentation $(S; T)$ and we are done.

Suppose $T^H < R$, and let y be an element of $R \backslash T^H$ with shortest standard representation:

$$y = x_1 \cdots x_n, \qquad x_i \in H_i^\#, H_i \neq H_{i+1}.$$

If some $x_i \in T_i^{H_i}$, then

$$y = (x_1 \cdots x_{i-1} x_i x_{i-1}^{-1} \cdots x_1^{-1})(x_1 \cdots x_{i-1} x_{i+1} \cdots x_n)$$

and the first factor is in T^H, so the second is in $R \backslash T^H$ and is shorter than y, a contradiction. Hence $x_i \notin T_i^{H_i}$ for all i. But then

$$yU = (x_1 U) \cdots (x_n U), x_i U = x_i U_i \in G_i^\#, G_i \neq G_{i+1}.$$

Since $G = \overset{*}{\pi} G_i$, this implies that $yU \neq e$, a contradiction.

8.2.7. *If* $G = \langle S \rangle = \langle T \rangle$ *and* S *is finite, then there is a finite subset* U *of* T *such that* $G = \langle U \rangle$.

Proof. If $s \in S$, then there is a finite subset U_s of T such that $s \in \langle U_s \rangle$. If $U = \cup \{U_s \mid s \in S\}$, then U is finite and

$$G = \langle S \rangle \subset \langle U_s \mid s \in S \rangle = \langle U \rangle.$$

8.2.8. *If $G = (S; T; F; u)$, $S' \cap F = \varnothing$, $w_{s'} \in F$ for all $s' \in S'$, and $T' = \{s'w_{s'}^{-1} \mid s' \in S'\}$, then*

$$G = (S \cup S'; T \cup T'; \underline{\qquad}; u'')$$

where $u'' \mid S = u$ *and* $s'u'' = w_{s'}u$, $s' \in S'$.

Proof. There is a group F'' which is free on $S \cup S'$ and which contains F as a subgroup (Theorems 2.1.16 and 8.1.10). Then $F' = \langle S' \rangle$ is free on S' and $F'' = F * F'$ (Theorem 8.2.5). By Theorem 8.1.1, $\exists\, u' \in \text{Hom}(F', G)$ such that $s'u' = w_{s'}u$ for $s' \in S'$. By Theorem 8.1.3, $\exists\, u'' \in \text{Hom}(F'', G)$ such that $u'' \mid F = u$ and $u'' \mid F' = u'$. Since u is onto G, so is u''. If $t \in T$, then $tu'' = tu = e$. If $s' \in S'$, then

$$(s'w_{s'}^{-1})u'' = (s'u')(w_{s'}u)^{-1} = e.$$

Hence

$$\text{Ker}(u'') \supset (T \cup T')^{F''} = R,$$

say.

Suppose $\text{Ker}(u'') \neq R$. Let $y = x_1 \cdots x_n$, $x_i \in F$ or F', x_i and x_{i+1} not both in F or both in F', be a shortest element of $\text{Ker}(u'') \backslash R$. If $n = 1$ and $x_1 \in F$, then $x_1 u = e$, so that $x_1 \in T^F$, hence $y \in R$, a contradiction. Therefore some $x_i \in F'$, $i = 1$ or 2. Thus

$$x_i = s_1'^{n_1} \cdots s_r'^{n_r}, \qquad s_j' \in S',$$
$$x_i = \pi(s_j'w_{s'_j}^{-1}w_{s'_j})^{n_j} = za, \qquad a \in R, z \in F,$$

by normality of R. If $n = 1$, then $z \in F \backslash R$ is a shortest element of $\text{Ker}(u'') \backslash R$; but this has been shown to be impossible. If $n > 1$ and $i = 2$ (the case $i = 1$ is similar), then

$$y = b(x_1 z x_3) \cdots x_n, \qquad b \in R,$$

and $y' = (x_1 z x_3) \cdots x_n$ is a shorter element of $\text{Ker}(u'') \backslash R$ than y. This contradiction shows that $\text{Ker}(u'') = R$. The theorem follows. $\parallel$

The converse of Theorem 8.2.8 is true and will be needed.

8.2.9. *If F is free on S, T is a subset of F, $S' \cap F = \varnothing$, $w_{s'} \in F$ for $s' \in S'$, $T' = \{s'w_{s'}^{-1} \mid s' \in S'\}$, and G has presentation $(S \cup S'; T \cup T')$, then G has presentation $(S; T)$.*

Proof. In any case, some group H has presentation $(S; T)$. By Theorem 8.2.8, H also has presentation $(S \cup S'; T \cup T')$. Hence $H \cong G$. Therefore G has presentation $(S; T)$.

8.2.10. *If G has presentations $(S; T)$ and $(S'; T')$ where S, T, and S' are finite, then there is a finite subset T'' of T' such that G has presentation (S', T'').*

Proof. $G = (S; T; F; u)$ for some F and u. *WLOG*, $S' \cap F = \varnothing$. Then $G = (S'; T'; F'; u')$ with $F \cap F' = E$. There are $w_s \in F'$ and $w_{s'} \in F$ such that

$$su = w_s u', \quad s'u' = w_{s'} u, \qquad s \in S, s' \in S'.$$

Let $U = \{sw_s^{-1} \mid s \in S\}$ and $U' = \{s'w_{s'}^{-1} \mid s' \in S'\}$. By Theorem 8.2.8,

$$G = (S \cup S'; T \cup U'; F''; v)$$
$$= (S \cup S'; T' \cup U; F''; v'),$$

where, for $s \in S$ and $s' \in S'$,

$$sv = su, \ s'v = w_{s'} u, \ sv' = w_s u' = su, \ s'v' = s'u' = w_{s'} u.$$

Hence $v = v'$. Therefore $(T \cup U')^{F''} = (T' \cup U)^{F''}$. Now $T \cup U'$ is finite. By Exercise 8.2.24, there is a finite subset T'' of T' such that G has presentation $(S \cup S; T'' \cup U)$. By Theorem 8.2.9, G has presentation $(S'; T'')$.

EXERCISES

8.2.11. What is the group G with presentation (for $p \in \mathscr{P}$) $(\{x_n \mid n \in \mathscr{N}\}; x_1^p, \{x_{n+1}^p x_n^{-1} \mid n \in \mathscr{N}\})$?

8.2.12. If $G = (S; T), x \in S$, and x does not appear in the S-expansion of any $t \in T$, then G is infinite. (Find an infinite group K satisfying the relations T and apply Theorem 8.2.3.)

8.2.13. The dihedral group of order 16 contains a nonnormal maximal Abelian subgroup of order 4. (Compare with Exercise 6.4.25.)

8.2.14. The quaternion group is a finite non-Abelian group G such that $G/\mathrm{Fr}(G)$ is Abelian. (See Exercise 7.3.29.)

8.2.15. (a) If $G = (S; T; F; u)$, $u \mid S = I_S$, f is a function from S into G, and G satisfies the relations T with respect to f, then f can be extended uniquely to a homomorphism of G onto G.

(b) If, in addition, G is finite, then f can be extended uniquely to an automorphism of G.

8.2.16. (a) Show that the quaternion group Q has an automorphism of order 3. [Use Exercise 8.2.15 (b).]

(b) If $T \in \mathrm{Aut}(Q)$ and $o(T) = 3$, then T fixes exactly two elements, e and the element of order 2.

8.2.17. Let G be dihedral of order 8.

(a) $\mathrm{Aut}(G) \cong G$ (obtain generators and relations for $\mathrm{Aut}(G)$).

(b) $o(\mathrm{Fr}(G)) = 2$, $o(\mathrm{Inn}(G)) = 4$.

(c) G is not the Frattini subgroup of a finite group H (use Theorem 7.3.19).

8.2.18. In the dihedral group of Example 1, all elements outside $\langle x \rangle$ have order 2.

8.2.19. In the dicyclic group of Example 3, all elements outside $\langle x \rangle$ have order 4, and have square x^n.

8.2.20. Discuss the infinite dihedral group with presentation $(x, y; y^2, (xy)^2)$.

8.2.21. The quaternion group is not the Frattini subgroup of a finite group H (use Theorem 7.3.19).

8.2.22. There is no group H such that $H/Z(H)$ is isomorphic to a dicyclic group (use Theorem 3.2.10 and Exercise 8.2.19).

8.2.23. (a)   (Group table presentation) Let $(G, *)$ be a group, and let $T = \{xyz^{-1} \mid x * y = z\}$. Then G has presentation $(G; T)$.

     (b)   A finite group G has a finite presentation $(S; T)$ (i.e., one with S and T finite).

8.2.24. If S and T are subsets of a group G with S finite and $T^G = S^G$, then there is a finite subset U of T such that $U^G = S^G$.

8.3   Some examples

In this section, the free product will be used to construct some examples showing that the Sylow theorems are not true in general.

8.3.1. *If $G = (\overset{*}{\pi}\{H \mid H \in S\})_A$, $g = ax_1 \cdots x_n$ and $h = by_1 \cdots y_m$ are elements of G in standard form, and x_n and y_1 are not in the same $H \in S$, then* $\text{Len}(gh) = m + n$.

Proof. This follows from Theorem 8.1.5(ii).

8.3.2. *If $G = (\overset{*}{\pi}\{H \mid H \in S\})_A$ and $g \in G$, then $o(g)$ is finite iff g is conjugate to an element of finite order in some $H \in S$.*

Proof. If $g = y^{-1}hy$, $h \in H \in S$, $y \in G$, and $o(h)$ is finite, then $o(g) = o(h)$, so $o(g)$ is finite.

Conversely, assume that $g^r = e$, $r \in \mathcal{N}$, and let $g = ax_1 \cdots x_n$ in standard form. It may be assumed that g has minimum length among the elements of $\text{Cl}(g)$. If $n = 1$, we are done. Suppose $n > 1$. If x_1 and x_n belong to the same $H \in S$, then

$$x_n g x_n^{-1} = x_n a x_1 \cdots x_{n-1} = b x_1' x_2 \cdots x_{n-1}$$

has length less than n, a contradiction. Hence x_1 and x_n belong to distinct H's. It follows from Theorem 8.3.1 that the length of g^r is rn. Hence $g^r \neq e$, a contradiction. ∥

In particular, this theorem applies if G is a free product of a set S of subgroups.

8.3.3. *If $G = H * K$ and H ($\neq E$) is periodic, then H is a maximal periodic subgroup of G.*

Proof. Suppose that there is a periodic subgroup L of G such that $L > H$. Let $y \in L\backslash H$ and $x \in H^{\#}$. By Theorem 8.3.2, y is conjugate to an element of H or K. Hence, in standard form,

$$y = x_1 \ldots x_j z x_j^{-1} \ldots x_1^{-1}, \qquad z \in H^{\#} \text{ or } K^{\#}. \tag{1}$$

But $xy \in L\backslash H$ also, and, if $j \neq 0$, then the standard form of xy ends with x_1^{-1} but does not begin with x_1, a contradiction. If $j = 0$, then $y = z \in K$, and $xy = xz$ is again not in the form (1). This contradiction proves the theorem. ‖

Example 1. Let $p \in \mathscr{P}$ and $G = H * K$, where H and K are cyclic groups of order p and p^2, respectively (see Theorem 8.1.9 for the existence of such a group G). By Theorem 8.3.3, H and K are Sylow p-subgroups of G. But $H \ncong K$, hence certainly H and K are not conjugate.

Example 2. Let $p \in \mathscr{P}$ and $G = H * K$ where H and K are each of order p. Again H and K are Sylow p-subgroups of G. If $z \in H$ and y is a conjugate of z, then the standard form of y is given by (1) (using the fact that H is Abelian, so that conjugation by an element of H does nothing to z). Hence $y \notin K$, and H and K are not conjugate. By Theorem 8.3.2, there are just two conjugate classes of Sylow p-subgroups, $\text{Cl}(H)$ and $\text{Cl}(K)$. Hence all Sylow p-subgroups are isomorphic, but not all are conjugate.

Example 3. Let $p \in \mathscr{P}$ and $G = H * K$ where H is of order p and K is infinite cyclic. This time, all Sylow p-subgroups are conjugate to H. There are an infinite number of Sylow p-subgroups, since, if $x \in H^{\#}$, $y \in K^{\#}$, and $z_{2i} = y$ and $z_{2i+1} = x$ for all i, then

$$z_1 \ldots z_{2i} x z_{2i}^{-1} \ldots z_1^{-1}$$

is in standard form and is of order p (more generally, see Exercises 8.3.6 and 8.3.7). Hence it is possible for all Sylow p-subgroups to be conjugate even though their number is infinite. ‖

The concept of free product with amalgamated subgroup has been quite useful in the construction of examples of groups with various properties (see, for example, B. Neumann [3]). A simple illustration is contained in the next two theorems, due to Neumann [1].

8.3.4. *If G is a group, $x \in G$, and $n \in \mathscr{N}$, then $\exists H \supset G$ and $y \in H$ such that $y^n = x$.*

Proof. If $o(x) = \infty$, then there is an isomorphism of $\langle x \rangle = A$ onto J carrying x onto 1. In $\mathcal{R}$, $n(1/n) = 1$, so by Exercise 2.1.36, there is a cyclic group $K = \langle y \rangle$ containing A such that $y^n = x$ and such that $K \cap G = A$. If $o(x) = r$, one works in J_{rn} instead of $\mathcal{R}$, and reaches the same conclusion. By Exercise 8.1.23, there is a group H such that $H = (G * K)_A$. Then $G \subset H$, $y \in K \subset H$, and $y^n = x$. ‖

This theorem can be generalized.

8.3.5. *If G is a group, then $\exists\, H \supset G$ such that if $h \in H$ and $n \in \mathcal{N}$, then $\exists\, y \in H$ such that $y^n = h$.*

Proof. Well-order the pairs (x, n) with $x \in G$ and $n \in \mathcal{N}$. By Theorem 8.3.4 and the usual transfinite induction argument, one can adjoin an nth root of x at stage (x, n). After all of this is done, one has a group G_1 containing an nth root of x for all $x \in G = G_0$ and all $n \in \mathcal{N}$. Construct G_2 from G_1 in the same manner, etc. Let $H = \cup\, G_n$. If $h \in H$ and $n \in \mathcal{N}$, then $h \in G_i$ for some i, hence $\exists\, y \in G_{i+1}$ such that $y^n = h$. Thus H has the desired property.

EXERCISES

8.3.6. (a) If $G = \overset{*}{\pi}\{H \mid H \in S\}$ and $K \in S$, then $N(K) = K$. In fact, if $x \in G \backslash K$, then $K^x \cap K = E$.

  (b) If $G = \overset{*}{\pi}\{H \mid H \in S\}$, $K \in S$, and $K \neq L \in S$, then K and L are not conjugate.

8.3.7. (a) If $G = H * K$, and $o(H) = o(K) = 2$, then there are non-e elements which have just two conjugates.

  (b) If $G = H * K$, $o(H) > 2$, and $o(K) > 1$, then any $g \in G^{\#}$ has an infinite number of conjugates.

8.3.8. (a) If $G = A * B$ where $A \neq E$ and $B \neq E$, then $Z(G) = E$.

  (b) What is the center of a free product with amalgamated subgroup A?

8.3.9. (Compare with Theorem 6.1.15.) There is a group G, $H \lhd G$, $P \in \mathrm{Syl}(G)$, but $H \cap P \notin \mathrm{Syl}(H)$. (Let G be the free product of two groups of order $p \in \mathcal{P}$, and apply Exercise 8.1.28.)

8.3.10. Let $G = P * J$ where $o(P) = p \in \mathcal{P}$. Then all Sylow p-subgroups are conjugate to P (Example 3). Let $H = P^G$.

  (a) (Compare with Theorem 6.2.2.) $H \supset N(P)$ but $N(H) > H$.

  (b) (Compare with Theorem 6.2.4.) $H \lhd G$ but $G \neq N(P)H$.

8.4 Subgroups of free groups

Of the several theorems in this section, the most striking are (i) any two freely generating subsets of a free group F have the same order (called the rank of F), and (ii) any subgroup H of a free group F is free. In addition, in (ii), a formula for rank (H) is given in the case where both rank(F) and $[F:H]$ are finite. The theorems (i) and (ii) and formula are due to Schreier [2]. We shall follow an exposition by Weir [1] (Theorems 8.4.6 through 8.4.14).

8.4.1. *If $G = (S; T; F; u)$ and U is a subset of F, then $G/(Uu)^G$ has presentation $(S; T \cup U)$.*

Proof. For some K and f, $K = (S; T \cup U; F; f)$. By Theorem 8.2.3, there is a homomorphism v of G onto K such that $(uv) \mid S = f$. Since u, v, and f are homomorphisms and $F = \langle S \rangle$, $uv = f$. Now $\mathrm{Ker}(f) = (T \cup U)^F$. Hence (Exercise 3.3.19),

$$(\mathrm{Ker}(f))u = ((T \cup U)u)^G = (\{e\} \cup (Uu))^G = (Uu)^G.$$

Since $uv = f$, $(Uu)^G = \mathrm{Ker}(v)$. Therefore $G/(Uu)^G \cong K$.

8.4.2. *If a group G has presentation $(S; T)$, then G/G^1 has presentation $(S; T \cup \{[a, b] \mid a \in S, b \in S\})$.*

Proof. There are F and u such that $G = (S; T; F; u)$. Then $G = \langle Su \rangle$, so

$$G^1 = \{[a, b]u \mid a \in S, b \in S\}^G$$

(see Exercise 3.4.18). The result now follows from Theorem 8.4.1.

8.4.3. *If G has presentation $(S; \{[a, b] \mid a \in S, b \in S\})$, then G is free Abelian of rank $o(S)$.*

Proof. There is a free Abelian group F with basis S. Let $G = (S; \{[a,b]\}; H; u)$. Since F satisfies the given relations, by Theorem 8.2.3, there is a homomorphism v of G onto F such that $(uv) \mid S = I_S$. Now G is Abelian (Theorem 8.4.2 with $T = \varnothing$), so by Theorem 5.3.1, $\exists\, w \in \mathrm{Hom}(F, G)$ such that $w \mid S = u \mid S$. Then $(wv) \mid S = I_S$, hence $wv = I_F$. Since

$$(wv) \mid S = (uv) \mid S = I_S,$$

v is 1–1 from Su onto S and w is 1–1 from S onto Su. Hence $(vw) \mid Su = I_{Su}$, and since $G = \langle Su \rangle$, $vw = I_G$. Therefore, w is an isomorphism of the free Abelian group F of rank $o(S)$ onto G.

8.4.4. *If a free group G is freely generated by both S and S', then o(S) = o(S').*

Proof. G has presentation $(S; \varnothing)$ (see Example 6 of Section 8.2). By Theorem 8.4.2, G/G^1 has presentation $(S; \{[a, b] \mid a \in S, b \in S\})$. By 8.4.3, G/G^1 is free Abelian of rank $o(S)$. Similarly, G/G^1 is free Abelian of rank $o(S')$. Hence (Theorem 5.3.6), $o(S) = o(S')$. ‖

This theorem justifies the following definition. The *rank* of a free group F is $o(S)$, where S freely generates F. Note that we have proved

8.4.5. *If G is a free group of rank A, then G/G^1 is free Abelian of rank A.*

The next several lemmas are all of a technical nature.

8.4.6. *If F is free on S, $H \subset F$, B is the set of right cosets of H in F, G a group, and $y_{b,s} \in G$ for $b \in B$ and $s \in S$, then there is a unique set $\{b^* \mid b \in B\}$ of functions from F into G such that, for all $b \in B$, $s \in S$, $u \in F$, and $v \in F$,*

$$sb^* = y_{b,s}, \tag{1}$$
$$(uv)b^* = (ub^*)(v(bu)^*), \tag{2}$$
$$eb^* = e, \tag{3}$$
$$u^{-1}b^* = (u(bu^{-1})^*)^{-1}. \tag{4}$$

Proof. Suppose that $\{b^* \mid b \in B\}$ is a set of functions such that (1) and (2) hold. Then by (2),

$$eb^* = (ee)b^* = (eb^*)(e(be)^*) = (eb^*)(eb^*),$$

so (3) holds. Hence

$$e = (u^{-1}u)b^* = (u^{-1}b^*)(u(bu^{-1})^*), \ u^{-1}b^* = (u(bu^{-1})^*)^{-1},$$

and (4) holds. Each element u of F has a unique expression

$$u = s_1^{i_1} \ldots s_n^{i_n}, \qquad s_j \in S, \ i_j = \pm 1, \qquad \text{and if } x_j = x_{j+1} \text{ then } i_j = i_{j+1}. \tag{5}$$

Call n the *length* Len(u) of u. Equations (3), (1), and (4) determine ub^* in case Len(u) = 0 or 1. Suppose that uniqueness of vb^* has been shown in case Len(v) < n. Then, if Len(u) = n, $u = vs^i$ where Len(v) = $n - 1$, $s \in S$, $i = \pm 1$, and by (2),

$$ub^* = (vs^i)b^* = (vb^*)(s^i(bu)^*)$$

which is determined, by the inductive hypothesis. Hence there is at most one such family of functions.

Existence of such a family must now be shown. Accordingly, define $\{b^* \mid b \in B\}$ by (3), (1),

$$s^{-1}b^* = (s(bs^{-1})^*)^{-1}, \qquad (s \in S), \tag{4'}$$

and, inductively, if $\text{Len}(u) = n - 1$, $\text{Len}(us^i) = n$, $s \in S$, and $i = \pm 1$, then

$$(us^i)b^* = (ub^*)(s^i(bu)^*). \tag{2'}$$

Since (1) is part of the definition, and since it has been shown that (3) and (4) follow from (1) and (2), it remains only to prove (2). By (3), (2) holds for $u = e$ or $v = e$.

Next, suppose that $v = s^i$, and let u have the form (5). If $s_n^{i_n} \neq s^{-i}$, then (2) is satisfied by (2)'. If $s_n^{i_n} = s^{-i}$, then, since $u = (uv)v^{-1}$ and $\text{Len}(uv) = n - 1$, it follows by induction that

$$ub^* = ((uv)b^*)(s^{-i}(buv)^*),\ (uv)b^* = (ub^*)(s^{-i}(buv)^*)^{-1}.$$

If $i = 1$, then by (4)',

$$(s^{-i}(buv)^*)^{-1} = s(buvs^{-1})^* = s^i(bu)^*$$

and (2) holds. If $i = -1$, then, by (4)',

$$(s^{-i}(buv)^*)^{-1} = (s(bus^{-1})^*)^{-1} = s^{-1}(bu)^* = v(bu)^*,$$

and (2) again holds.

Now induct on $\text{Len}(v)$, and let $v = v_1 s^i$, $\text{Len}(v_1) < \text{Len}(v)$. Then

$$\begin{aligned}
(uv)b^* &= (uv_1 s^i)b^* = ((uv_1)b^*)(s^i(buv_1)^*) \\
&= (ub^*)(v_1(bu)^*)(s^i(buv_1)^*) = (ub^*)((v_1 s^i)(bu)^*) \\
&= (ub^*)(v(bu)^*).
\end{aligned}$$

Hence (2) holds in general. This proves the lemma.

8.4.7. *If S, F, H, and B are as in Theorem 8.4.6, G is free on $B \times S$, $\{b^* \mid b \in B\}$ is the family of functions guaranteed by Theorem 8.4.6 with $(b, s) = y_{b,s}$, $b' \in b$ for all $b \in B$, and $H' = e$, then there is a homomorphism T of G onto H such that for all $b \in B$ and $u \in F$,*

$$(ub^*)T = (b'u)(bu)'^{-1}, \tag{6}$$

$$(H^*T) \mid H = I_H. \tag{7}$$

Proof. If $b \in B$ and $s \in S$, then $b's \in bs$, $b's = h(bs)'$ where $h \in H$, and $(b's)(bs)'^{-1} \in H$. Hence (Theorem 8.1.1) $\exists \mid T \in \text{Hom}(G, H)$ such that

$$(b, s)T = (b's)(bs)'^{-1}. \tag{8}$$

If $u = e$, then by (3),

$$(ub^*)T = eT = e_H = (b'u)(bu)'^{-1}.$$

If $u = s \in S$, then by (1) and (8),

$$(ub^*)T = (b, s)T = (b's)(bs)'^{-1} = (b'u)(bu)'^{-1},$$

and (6) again holds. If $u = s^{-1}$, then by (4), (1), and (8),

$$(ub^*)T = (s(bs^{-1})^*)^{-1}T = ((bs^{-1}, s)T)^{-1}$$
$$= (bs^{-1}s)'((bs^{-1})'s)^{-1} = b'((bs^{-1})'s)^{-1} = (b's^{-1})(bs^{-1})'^{-1},$$

and (6) is true. Finally, assume that (6) holds for u and v. Then by (2),

$$((uv)b^*)T = ((ub^*)(v(bu)^*))T = ((ub^*)T)((v(bu)^*)T)$$
$$= (b'u)(bu)'^{-1}(bu)'v(buv)'^{-1} = (b'uv)(buv)'^{-1}.$$

Hence, by induction on Len(u), (6) holds for all $u \in F$.

Let $U = H^* \,|\, H$. By (2), $U \in \mathrm{Hom}(H, G)$. If $h \in H$, then by (6) and hypothesis,

$$(hU)T = (H'h)(Hh)'^{-1} = (eh)e^{-1} = h.$$

Hence $UT = I_H$, so (7) is true. Therefore T is a homomorphism of G onto H, and the theorem is proved.

8.4.8. *If T is an idempotent endomorphism of G, $G = \langle S \rangle$, and $L = \{s(s^{-1}T) \,|\, s \in S\}^G$, then $\mathrm{Ker}(T) = L$.*

Proof. Since

$$(s(s^{-1}T))T = (sT)(s^{-1}T^2) = (sT)(s^{-1}T) = e,$$

$\mathrm{Ker}(T) \supset L$. Let $k = \pi s_i^{n_i} \in \mathrm{Ker}(T)$, $s_i \in S$. If $U \in \mathrm{Hom}(G)$ is such that $(s(s^{-1}T))U = e$ for all $s \in S$, then $sU = sTU$ for all $s \in S$, and

$$kU = \pi(s_i U)^{n_i} = \pi(s_i TU)^{n_i} = (\pi s_i^{n_i})TU = kTU = e.$$

Hence $k \in L$ (Exercise 3.3.20), so $\mathrm{Ker}(T) \subset L$. Therefore $\mathrm{Ker}(T) = L$.

8.4.9. *If S, F, H, B, G, T, $'$, and $*$ are as in Theorem 8.4.7, then*

$$H = (B \times S; \{(b, s)((b, s)^{-1}TH^*) \,|\, b \in B, s \in S\}; G; T),$$
$$H = (B \times S; \{b'H^* \,|\, b \in B\}; G; T).$$

Proof. Again let $U = H^* \,|\, H$. By (7), $UT = I_H$. Therefore U is an isomorphism of H into G. Hence $\mathrm{Ker}(TU) = \mathrm{Ker}(T) = K$, say. Also, $(TU)^2 = T(UT)U = TU$. By Theorem 8.4.8,

$$K = \{(b, s)((b, s)^{-1}TU) \,|\, b \in B, s \in S\}^G.$$

By Theorem 8.4.7, T has range H. Therefore H has the first form listed in the theorem.

Let $b \in B$. Then by (6),

$$b'H^*T = (H'b')(Hb')'^{-1} = b'b'^{-1} = e,$$

and $b'H^* \in K$. Hence $K \supset \{b'H^* \mid b \in B\}^G$. By (1), (6), (2), and (4),

$$
\begin{aligned}
(b, s)TU &= ((b's)(bs)'^{-1})U = (b'H^*)(s(bs)'^{-1})(Hb')^* \\
&= (b'H^*)(s(Hb')^*)((bs)'^{-1}(Hb's)^*) \\
&= (b'H^*)(Hb', s)((bs)'(Hb's(bs)'^{-1})^*)^{-1} \\
&= (b'H^*)(b, s)((bs)'H^*)^{-1}.
\end{aligned}
$$

Therefore any homomorphism of G sending $b'H^*$ into e for all $b \in B$ sends all $(b, s)((b, s)^{-1}TU)$ into e, and therefore sends K into E. Hence (Exercise 3.3.20), $\{b'H^* \mid b \in B\}^G \supset K$. Thus $K = \{b'H^* \mid b \in B\}^G$, and H has the second form also.

8.4.10. *If* $L = (S; X; F; d)$, $M \subset L$, $H = Md^{-1}$, *and* B *is the set of right cosets of* H *in* F, *then* M *has a presentation*

$$(B \times S; \{b'H^* \mid b \in B\} \cup \{xb^* \mid x \in X, b \in B\})$$

where $b' \in b$ *for all* $b \in B$ *and* $H' = e$, *and* $*$ *satisfies* (1) *(with* $y_{b,s} = (b, s)$*),* (2), (3), *and* (4).

Proof. Since $X^F = \text{Ker}(d) \subset H$,

$$X^F = \{b'xb'^{-1} \mid b \in B, x \in X\}^H$$

(Exercise 8.4.21). If $x \in X$ and $b \in B$, then, since $x \in H$ and $X^F \lhd F$, we have

$$(bx)' = ((bxb^{-1})b)' = (hb)' = b' \qquad (h \in H).$$

Now introduce G and T as in Theorem 8.4.7. Hence by (6),

$$b'xb'^{-1} = b'x(bx)'^{-1} = (xb^*)T.$$

By Theorem 8.4.9,

$$H = (B \times S; \{b'H^* \mid b \in B\}; G; T).$$

Therefore by Theorem 8.4.1, H/X^F has presentation

$$(B \times S; \{b'H^* \mid b \in B\} \cup \{xb^* \mid x \in X, b \in B\}).$$

Since $M \cong H/X^F$, we are done.

8.4.11. *If* L *is a group which has a finite presentation and* $[L:M]$ *is finite, then* M *has a finite presentation.*

Proof. In Theorem 8.4.10, S, B, and X are all finite in the present situation. Hence the presentation of M given there is finite. $\parallel$

Example. Let $L = \mathrm{Sym}(4)$, $S = \{x, y\}$ with $x \neq y$, and $X = \{x^3, y^4,$ $(xy)^2\}$. Then L has presentation $(S; X)$ (see Example 5 of Section 8.2). Let F be a free group on S, and let the corresponding homomorphism take x onto $(1, 2, 3)$ and y onto $(1, 3, 2, 4)$ (see Example 5 of Section 8.2 again). Let $M = \mathrm{Alt}(4)$ (see Exercise 2.3.11). Then using earlier notation, $F = H \cup Hy$, and we may let $B' = \{e, y\}$. The relations in the presentation of M given in Theorem 8.4.10 may now be computed from Equations (1) through (4), and the facts that $x \in H$ and $y^2 \in H$. These relations are:

$$eH^* = e, \; yH^* = (H, y), \; x^3 H^* = (H, x)^3,$$
$$y^4 H^* = (y^2 H^*)^2 = ((yH^*)(y(Hy)^*))^2 = ((H, y)(Hy, y))^2,$$
$$(xy)^2 H^* = (H, x)(H, y)(Hy, x)(Hy, y),$$
$$x^3 (Hy)^* = (Hy, x)^3$$
$$y^4 (Hy)^* = (Hy, y)(H, y)(Hy, y)(H, y),$$
$$(xy)^2 (Hy)^* = (Hy, x)(Hy, y)(H, x)(H, y).$$

To simplify the notation, let $a = (H, x)$, $b = (H, y)$, $c = (Hy, x)$, and $d = (Hy, y)$. Then by Theorem 8.4.10, M has presentation

$$(a, b, c, d; b, a^3, (bd)^2, abcd, c^3, (db)^2, cdab). \tag{9}$$

There is a simpler presentation which can be obtained directly from this one. Intuitively, one would say that the relation $b = e$ simplifies the relation $(bd)^2 = e$ to $d^2 = e$, etc. More formally, the homomorphism i of the free group G onto M associated with (9) can be factored $i = jk$, where j is the canonical homomorphism of G onto its free subgroup generated by $\{a, c, d\}$ with kernel b^G. In any case, one obtains the presentation

$$(a, c, d; a^3, d^2, acd, c^3, cda).$$

However, $cda = a^{-1}(acd)a$, so the normal closure of the set of relations is unaltered if cda is omitted. Thus, finally, $\mathrm{Alt}(4)$ has presentation

$$(a, c, d; a^3, d^2, acd, c^3).$$

(Exercise 8.4.23 shows that none of the relations may be omitted.) ‖

Let F be freely generated by S. A *Schreier system* is a nonempty subset Q of F such that if $a = bs^i \in Q$, $s \in S$, $i = \pm 1$, and $\mathrm{Len}(b) < \mathrm{Len}(a)$, then $b \in Q$.

8.4.12. *If F is freely generated by S and $H \subset F$, then there is a Schreier system Q such that $F = \cup \{Ha \mid a \in Q\}$.*

Proof. Let B be the set of right cosets of H in F. For $b \in B$, let Len(b) be the length of a shortest element of b. Define $H' = e$. Inductively suppose that $b' \in b$ is defined for all $b \in B$ for which Len(b) $< n$ in such a way that

(i)   Len(b') = Len(b), and

(ii) if $b' = as^i$, $s \in S$, $i = \pm 1$, and Len(a) $<$ Len(b') $< n$, then $a = a''$, where $a'' = (Ha)'$.

Now let Len(b) $= n$. Then $\exists\ c \in b$ such that $c = ds^i$, $i = \pm 1$ and Len(d) $= n - 1$. If Len(d'') $<$ Len(d), then $d''s^i \in Hds^i = b$ and Len($d''s^i$) $\leqq n - 1$, a contradiction. Hence Len(d'') = Len(d). Define $b' = d''s^i$. Then $b' \in Hds^i = b$, and Len(b') $= n =$ Len(b), so that (i) is valid. By its very definition, $b' = d''s^i$ where Len(d'') $<$ Len(b') and, of course, $(d'')'' = d''$, so that (ii) holds when n is replaced by $n + 1$. Therefore there is a function ' such that (i) and (ii) hold universally. The theorem follows if we let $Q = B'$.

8.4.13. *If H is a subgroup of a free group F, then H is free. If $[F:H]$ is finite, then*

$$\text{rank}(H) = 1 + (\text{rank}(F) - 1)[F:H].$$

Proof. Using the same notation as before, H has the two presentations given in Theorem 8.4.9. By Theorem 8.4.12, ' may be chosen in such a way that B' is a Schreier system. Let $K = \text{Ker}(T)$. Now $(b, s) \in K$ iff

$$e = (b, s)T = b's(bs)'^{-1},$$

i.e., iff $(bs)' = b's$. This occurs iff one of the two elements b' and $(bs)'$ is obtainable from the other by deletion of the last letter [if b' ends in s^{-1}, then b' is the longer of the two elements; otherwise, $(bs)' = b's$ is the longer]. It is now clear that there is a 1–1 function from $K \cap (B \times S)$ into B' which maps (b, s) onto the longer of b' and $b's$. Since B' is a Schreier system, each $b' \in B'$ except e is the longer of two elements of B', where the shorter element is obtained by deleting the last letter of b'. Therefore

$$o(K \cap (B \times S)) = o(B') - 1 = [F:H] - 1.$$

Let us look at the second presentation of H. Now $eH^* = e$ by (3). Inductively assume that if $a \in B'$ and Len(a) $< n$, then $aH^* \in \langle K \cap (B \times S) \rangle = P$. Let $c = as^i \in B'$ and Len(a) $<$ Len(c) $= n$. Since B' is a Schreier system, $a \in B'$, Therefore by (2),

$$cH^* = (aH^*)(s^i(Ha)^*).$$

By the inductive assumption, $aH^* \in P$. By (6),

$$(s^i(Ha)^*)T = (Ha)'s^i(Has^i)'^{-1} = as^ic^{-1} = e.$$

Hence $(s^i(Ha)^*) \in K$. Moreover, if $i = 1$, then by (1), $s^i(Ha)^* = (Ha, s) \in B \times S$, and if $i = -1$, then by (4),

$$s^i(Ha)^* = (s(Has^{-1})^*)^{-1} = (Has^{-1}, s)^{-1} \in P.$$

Hence, $cH^* \in P$.

It follows from Theorem 8.4.9, and the above paragraph that H has a presentation $(B \times S; V)$, where V is a subset of $B \times S$ of order $[F:H] - 1$. Now $o(B \times S) = [F:H]\operatorname{rank}(F)$. By Exercise 8.4.24, H is free and its rank satisfies the equation

$$\operatorname{rank}(H) + [F:H] - 1 = [F:H]\operatorname{rank}(F).$$

If $[F:H]$ is finite, then

$$\operatorname{rank}(H) = 1 + [F:H](\operatorname{rank}(F) - 1). \parallel$$

If we recall that the presentation $(B \times S; V)$ of H given above has associated homomorphism T, and use (6), we obtain a useful rewording of the facts proved above.

8.4.14. *If F is free on S, $H \subset F$, Q is a Schreier system, and $F = \dot{\cup} \{Ha \mid a \in Q\}$, then H is freely generated by*

$$\{as(as)''^{-1} \neq e \mid a \in Q, s \in S\},$$

where $Hx = Hx''$ and $x'' \in Q$. $\parallel$

Theorem 8.4.13 contains, as an incidental result, the fact that a subgroup of finite index in an infinite cyclic group is again infinite cyclic.

Remark. It is obvious* that the equation given in Theorem 8.4.13 is not valid in general if $[F:H]$ is infinite. For let F be freely generated by $\{x, y\}$, and let $H = \langle x \rangle$. Then $\operatorname{rank}(H) = 1$, $\operatorname{rank}(F) = 2$, and $[F:H] = \aleph_0$. However, $1 \neq 1 + (2 - 1)\aleph_0$.

8.4.15. *If F_1 is a free group, $F_n \supset F_{n+1}$ for $n \in \mathcal{N}$, and*
(*) *if S_n freely generates F_n, then $S_n \cap F_{n+1} = \varnothing$,*
then $\cap F_n = E$.

Proof. Suppose that $e \neq y \in \cap F_n$. Let r be the minimum of all lengths of y with respect to all freely generating subsets S_n of F_n for all $n \in \mathcal{N}$. Since each F_n is free (Theorem 8.4.13), *WLOG* $\exists S$ freely generating F_1 such that

$$y = s_1^{i_1} \cdots s_r^{i_r}, \qquad s_j \in S, i_j = \pm 1.$$

* Or should be. See Specht [1, p. 155].

By $(*)$ and a minor generalization of Theorem 8.4.12, there is a Schreier system Q such that $F_1 = \cup \{F_2 a \mid a \in Q]$, and such that $s_1^{i_1} \in Q$. By Theorem 8.4.14, F_2 is freely generated by

$$S_2 = \{as(as)''^{-1} \neq e \mid a \in Q, s \in S\}.$$

Let $y_0 = e$, $y_n = s_1^{i_1} \cdots s_n^{i_n}$, $n \leq r$. Then $y_0'' = e$ and $y_r'' = y'' = e$ since $y \in F_2$. Therefore,

$$y = \pi\{y_{n-1}'' s_n^{i_n} y_n''^{-1} \mid 1 \leq n \leq r\}.$$

If $i_n = 1$, then

$$y_{n-1}'' s_n^{i_n} y_n''^{-1} = y_{n-1}'' s_n (y_{n-1}'' s_n)''^{-1}$$
$$\cdot = y_{n-1}'' s_n (y_{n-1}'' s_n)''^{-1},$$

which is in S_2 or equals e. If $i_n = -1$, then

$$y_{n-1}'' s_n^{i_n} y_n''^{-1} = (y_n s_n)'' s_n^{-1} y_n''^{-1}$$
$$= (y_n'' s_n (y_n s_n)''^{-1})^{-1} = (y_n'' s_n (y_n'' s_n)''^{-1})^{-1}.$$

Hence $y = c_1^{i_1} \cdots c_r^{i_r}$, where $c_j \in S_2$ or $c_j = e$. But

$$y_0'' s_1^{i_1} y_1''^{-1} = e'' s_1^{i_1} (s_1^{i_1})''^{-1} = e,$$

since $s_1^{i_1} \in Q$, so that $c_1 = e$. Hence the length of y with respect to S_2 is less than r, a contradiction. ‖

One could now quickly prove from Theorem 8.4.5 that if F is free, then $\cap F^n = E$. However, it requires only a little more effort to prove that $\cap Z^n(F) = E$.

8.4.16. *The intersection of the terms of the lower central series of a free group is E.*

Proof. Let F be free on S. Let $F_1 = F$ and, inductively, let F_{n+1} be the subgroup generated by all squares of elements of F_n. Let S_n freely generate F_n and let $x \in S_n$. Now any product of squares of elements of F_n has even length (it has even length in raw form, and cancellation always involves a pair of factors). Hence $x \notin F_{n+1}$. By the preceding theorem, $\cap F_n = E$. Moreover, $F_{n+1} \in \text{Char}(F_n)$, hence all $F_n \triangleleft F$.

CASE 1. S is finite. Inductively suppose that F/F_n is a finite 2-group. Then F_n is of finite rank (Theorem 8.4.13), and every element of F_n/F_{n+1} has order 1 or 2. By Exercise 2.4.13, F_n/F_{n+1} is Abelian. Since it is finitely generated, it is a finite 2-group. Therefore F/F_{n+1} is also a finite 2-group. Hence by induction, F/F_r is a finite 2-group, and therefore nilpotent, for all r. Thus for each r, $\exists\, k$ such that $Z^k(F) \subseteq F_r$. Hence $\cap Z^n(F) \subseteq \cap F_n = E$.

CASE 2. S is infinite. Let $x \in \cap Z^n(F)$, and let S' be the set of letters of S appearing in the S-expansion of x. Then $F = G * H$, where G is free on S' and H is free on $S \backslash S'$. The function h which is the identity on S' and maps $S \backslash S'$ onto e has a unique extension to a homomorphism T of F onto G. Also, if $g \in G$, then $gT = g$. Now (see Theorem 6.4.6),

$$Z^n(F) \cap G = (Z^n(F) \cap G)T \subset (Z^n(F))T \cap GT$$
$$= Z^n(G) \cap G = Z^n(G).$$

Therefore by Case 1,

$$x \in \cap (Z^n(F) \cap G) \subset \cap Z^n(G) = E.$$

Hence $\cap Z^n(F) = E$.

8.4.17. *If F is free, then $\cap F^n = E$.*

Proof. For all n, $F^n \subset Z^n(F)$.

8.4.18. *The intersection of the subgroups of finite index in a free group is E.*

Proof. Let S freely generate F. Let $x \in F^\#$. Just as in the proof of Case 2 of Theorem 8.4.16, there is a free subgroup G of finite rank such that $F = G * H$, $x \in G$, and there is a homomorphism T of F onto G such that $xT = x$. As in Case 1 of Theorem 8.4.16, $\exists K \subset G$ such that $[G:K]$ is finite and $x \notin K$. Let $L = KT^{-1}$. Then $[F:L] = [G:K]$ is finite, and $x \notin L$ since $xT = x \notin K$. The theorem follows.

8.4.19. *If G is a finitely generated group, then $Z^n(G)/Z^{n+1}(G)$ is finitely generated for all n.*

Proof. G/Z^1 is finitely generated. Suppose inductively that Z^{n-1}/Z^n is generated by $\{y_1 Z^n, \ldots, y_r Z^n\}$, and let $G = \langle x_1, \ldots, x_s \rangle$. Now Z^n is generated by all elements of the form

$$z = [x_1^{i_1} \cdots x_m^{i_m}, y_1^{j_1} \cdots y_q^{j_q} f], \qquad f \in Z^n.$$

By Theorem 3.4.2, z is a product of conjugates of elements of the forms $[x_i^{\pm 1}, y_j^{\pm 1}]$ and $[x_i^{\pm 1}, f] \in Z^{n+1}$. Since

$$\frac{Z^n(G)}{Z^{n+1}(G)} \subset Z\left(\frac{G}{Z^{n+1}(G)}\right),$$

any conjugate of an element u of Z^n lies in the same coset of Z^{n+1} as u does. Therefore Z^n/Z^{n+1} is generated by the finite set $\{[x_i^{\pm 1}, y_j^{\pm 1}]Z^{n+1}\}$. ‖

If F is free of finite rank, then $Z^n(F)/Z^{n+1}(F)$ is a free Abelian group of known rank. For details, see M. Hall [1, Chapter 11].

8.4.20. *If F is free of finite rank and $F \cong F/H$, then $H = E$. Equivalently, any endomorphism T of F onto F is an automorphism.*

Proof. Let us prove the second version of the theorem. By Theorem 6.4.6, $Z^n T = Z^n$. Hence T induces an endomorphism T_n of Z^n/Z^{n+1} onto itself given by $(xZ^{n+1})T_n = (xT)Z^{n+1}$. By Theorem 8.4.19, Z^n/Z^{n+1} is a finitely generated Abelian group. By Exercise 5.5.16, T_n is an automorphism. If $e \neq x \in \text{Ker}(T)$, then by Theorem 8.4.16, $\exists\, n$ such that $x \in Z^n \backslash Z^{n+1}$. Then

$$(xZ^{n+1})T_n = (xT)Z^{n+1} = Z^{n+1},$$

contradicting the fact that T_n is an automorphism.

EXERCISES

8.4.21. Show that if G is a group, $H \subset G$, X is a subset of G, $X^G \subset H$, and $G = \cup\{Hb \mid b \in B\}$, then

$$X^G = \{bxb^{-1} \mid b \in B, x \in X\}^H.$$

8.4.22. Verify that $(a, c, d; a^3, c^3, d^2, acd)$ is actually a presentation of Alt(4) (see the example following Theorem 8.4.11).

8.4.23. Show that none of the relations in the presentation of Exercise 8.4.22 can be omitted as follows:

    (a) $J_3 * J_2 * J_3$ has presentation $(a, c, d; a^3, d^2, c^3)$. (Use Theorem 8.2.6.)

    (b) $J_3 * J_2$ has presentation $(a, c, d; a^3, d^2, acd)$. (Use 8.2.8.)

    (c)   Etc.

8.4.24. If G has presentation $(S; T)$ where T is a subset of S, then G is isomorphic to a free group on $S \backslash T$.

8.4.25. Show that a free group of infinite rank has an endomorphism onto itself which is not an automorphism. (Compare with Theorem 8.4.20.)

8.4.26. Let F be a free group.

    (a)   Two elements of F commute iff they are contained in a cyclic subgroup. ($\langle x, y \rangle$ is free and also Abelian, hence has rank 1.)

    (b)   If $x \in F^\#$, then $o(C(x)) = \aleph_0$ and, if rank$(F) > 1$, $o(\text{Cl}(x)) = o(F)$.

    (c)   If F has rank at least two, then $Z(F) = E$.

8.4.27. Two free groups are isomorphic iff they have the same rank.

8.4.28. If F is free of rank r (possibly infinite) and $F = \langle S \rangle$, then $o(S) \geq r$. (Use Theorem 8.4.5 and Exercise 5.7.29.)

8.4.29. Find an example of two elements a and b in a free group of rank 2 such that $[a, b] \neq e$, but Len($[a, b]$) < Len(a).

8.4.30. (a) A free Abelian group $\neq E$ contains a subgroup of index n for all $n \in \mathscr{N}$.

(b) A free group $\neq E$ contains a normal subgroup of index n for all $n \in \mathscr{N}$ (look at G/G^1).

8.4.31. (a) If F is free on $\{x, y\}$ and $S = \{y^{-n}xy^n \mid n \in \mathscr{N}\}$, then $\langle S \rangle$ is freely generated by S.

(b) A free group of rank 2 has subgroups of rank $0, 1, \ldots ; \aleph_0$.

8.4.32. If F is free, $F = F_1 > F_2 > \ldots$, and each $F_{n+1} \in \text{Char}(F_n)$, then $\cap F_n = E$ (use Theorem 8.4.15).

8.4.33. (a) If G is a finitely generated group, $H \subset G$, and $[G:H]$ is finite, then H is finitely generated (use Theorem 8.4.13).

(b) Give an example of a finitely generated group G with a subgroup H which is not finitely generated (see Exercise 8.4.31). (Compare with the situation for Abelian groups.)

8.4.34. If $G = \langle S \rangle$, G is nilpotent of class n, and F is free on S, then G is a homomorphic image of $F/Z^n(F)$.

8.4.35. If G is a finitely generated nilpotent group, then G satisfies the maximal condition for subgroups. (Use Exercise 8.4.34, Theorems 8.4.19 and 7.1.3, and Exercise 5.5.15.)

8.4.36. If G is a finitely generated periodic solvable group, then G is finite. ($G^n = E$. Induct on n and use Exercise 8.4.33.)

8.4.37. What is wrong with the following argument? Let G be a finite group. There is a free group F and a homomorphism T of F onto G.

$$(Z^n(F))T = Z^n(G). \quad \cap Z^n(F) = E. \quad \cap Z^n(G) = E.$$

Since G is finite, some $Z^n(G) = E$. Therefore G is nilpotent.

REFERENCES FOR CHAPTER 8

For the entire chapter, M. Hall [1], Kurosh [1], and Specht [1]; Section 8.2, see also Fox [1].

NINE

EXTENSIONS

Let H and F be groups. The main problem to be considered in this chapter is that of finding all groups G (up to isomorphism) such that $H \lhd G$ and $G/H \cong F$. Because the solution is rather complex and difficult to apply, several special cases will also be considered.

9.1 Definitions

A *sequence* is a finite or infinite sequence $\{A_n\}$ of groups and homomorphisms T_n from A_n into A_{n+1}. Such a sequence is *exact* iff $\mathrm{Ker}(T_{n+1}) = A_n T_n$ whenever both T_n and T_{n+1} are defined. Sequences are usually written

$$\ldots \longrightarrow A \longrightarrow B \longrightarrow C \longrightarrow \ldots.$$

If either of the symbols $E \longrightarrow A$ or $A \longrightarrow E$ occurs, the homomorphism is the only one possible. Unless otherwise specified, in $A \longrightarrow A$, the homomorphism is I_A.

9.1.1. *A homomorphism $A \longrightarrow B$ is an isomorphism iff $E \longrightarrow A \longrightarrow B$ is exact.*

209

9.1.2. *A homomorphism $A \longrightarrow B$ is onto B iff $A \longrightarrow B \longrightarrow E$ is exact.*

9.1.3. *A homomorphism $A \longrightarrow B$ is an isomorphism of A onto B iff $E \longrightarrow A \longrightarrow B \longrightarrow E$ is exact.* ‖

An *extension* of H by F is an exact sequence

$$E \longrightarrow H \longrightarrow G \longrightarrow F \longrightarrow E.$$

9.1.4. *If $E \longrightarrow H \overset{T}{\longrightarrow} G \longrightarrow F \longrightarrow E$ is an extension of H by F, then $F \cong G/HT$ and T is an isomorphism of H onto HT.* ‖

One often speaks of G itself as being an extension of H by F. It should perhaps be remarked that this definition of extension gives more flexibility than would the more rigid one requiring that H be a subgroup of G.

A diagram

is *commutative* iff $TU = VW$. Similar definitions are made for other diagrams.

9.1.5. *(Five lemma.) If the diagram*

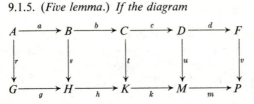

is commutative, has exact rows, and r, s, u, and v are isomorphisms onto, then t is an isomorphism onto.

Proof. We prove only that t is onto. The proof that $\mathrm{Ker}(t) = E$ is similar and is left as an exercise. Let $x \in K$. Then $\exists\, y \in D$ such that $yu = xk$. Since $x(km) = e_P$ (exactness), $y(um) = e_P$. Hence $y(dv) = e_P$ (commutativity). Therefore, $yd = e_F$ (v is an isomorphism). Thus $y \in \mathrm{Ker}(d)$ and $\exists\, z \in C$ such that $zc = y$ (exactness at D). Then

$$(zt)k = zcu = yu = xk.$$

Therefore $x^{-1}(zt) \in \mathrm{Ker}(k)$. Hence (exactness at K) $\exists\, w \in H$ such that $wh = x^{-1}(zt)$. Since s is onto, $\exists\, j \in B$ such that $js = w$. Then

$$(jb)t = jsh = wh = x^{-1}(zt), \qquad x = (z(jb)^{-1})t.$$

Hence t is onto. ‖

An extension $E \longrightarrow H \longrightarrow G \longrightarrow F \longrightarrow E$ is *equivalent* to the extension $E \longrightarrow H \longrightarrow G^* \longrightarrow F \longrightarrow E$ iff there is a homomorphism $G \longrightarrow G^*$ such that the diagram

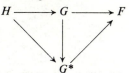

is commutative.

9.1.6. *If the homomorphism $G \longrightarrow G^*$ makes the extension $E \longrightarrow H \longrightarrow G \longrightarrow F \longrightarrow E$ equivalent to $E \longrightarrow H \longrightarrow G^* \longrightarrow F \longrightarrow E$, then $G \longrightarrow G^*$ is an isomorphism onto.*

Proof. The diagram

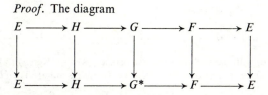

is commutative by assumption and trivialities. By the five lemma, $G \longrightarrow G^*$ is an isomorphism onto.

9.1.7. *Equivalence of extensions is an equivalence relation.*

Proof. This follows readily from Theorem 9.1.6. ‖

The direct product $H \times F$ is an extension of H by F. More precisely

9.1.8. *If $H \longrightarrow H \times F$ is the function $h \longrightarrow (h, e)$ and $H \times F \longrightarrow F$ is the function $(h, f) \longrightarrow f$, then*

$$E \longrightarrow H \longrightarrow H \times F \longrightarrow F \longrightarrow E$$

is an extension of H by F.

EXERCISES

9.1.9. Prove the other half of the five lemma.

9.1.10. If $E \longrightarrow H \longrightarrow G \longrightarrow F \longrightarrow E$ is an extension of H by F, then there is an equivalent extension

$$E \longrightarrow H \overset{I_H}{\longrightarrow} G^* \longrightarrow F \longrightarrow E.$$

9.1.11. Let S be a class of groups closed under the taking of subgroups, homomorphic images, and extensions (if H and G/H are in S, so is G). Let

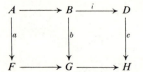

be a commutative diagram with bottom row exact, all images normal subgroups, and a, i, and c onto (mod S) (this means, for example, that $D/Bi \in S$). Then b is onto G (mod S).

9.1.12. Let

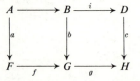

be commutative with second row almost exact ($[\mathrm{Ker}(g): \mathrm{Ker}(g) \cap Ff]$ finite), and i, a, and c almost onto (for example, $[D: Bi]$ is finite). Then b is almost onto G.

9.1.13. (Example of inequivalent but isomorphic extensions.) Let $G = \langle a \rangle$ be cyclic of order 9, and $H = F$ the subgroup of order 3. Let $X = I_H$. Let $a^i Y = a^{3i}$, $a^i Y^* = a^{6i}$.

(a) $E \longrightarrow H \xrightarrow{\ X\ } G \xrightarrow{\ Y\ } F \longrightarrow E$ and

   $E \longrightarrow H \xrightarrow{\ X\ } G \xrightarrow{\ Y^{\bullet}\ } F \longrightarrow E$ are extensions of H by F.

(b) The extensions are not equivalent.

9.1.14. Let $A \cong A'$ and $B \cong B'$. Exhibit a 1–1 function f from the class of all extensions of A by B onto the class of all extensions of A' by B' such that the extensions K_1 and K_2 are equivalent iff $f(K_1)$ and $f(K_2)$ are equivalent.

9.2   Semi-direct products

We shall now construct a general class of groups which will furnish many examples heretofore lacking.

A group G is a *semi-direct product* of its subgroups H and F iff $H \lhd G$, $G = FH$, and $F \cap H = E$ (hence F is a complement of H). One also says that G *splits* over H in this case. For such a group, every element has a unique expression fh with $f \in F$ and $h \in H$. Multiplication is given by

$$(f_1 h_1)(f_2 h_2) = (f_1 f_2)(h_1^{f_2} h_2).$$

This equation can be put in another form. If T_f denotes the inner auto-morphism of G induced by f, then the function T defined by the rule: $fT = T_f \mid H$ is a homomorphism from F into Aut(H). Multiplication in G now becomes

$$(f_1 h_1)(f_2 h_2) = (f_1 f_2)((h_1(f_2 T))h_2). \tag{1}$$

The group G will be called a *semi-direct product* of H by F *with homomorphism T*.

Now consider the converse question. Given H, F, and T, is there a semi-direct product of H by F with homomorphism T? The answer is, technically, no (for example, $F \cap H$ might be empty), but, essentially, yes. First note that if $T \in \text{Hom}(F, \text{Aut}(H))$, U is an isomorphism of H onto H_1, and W is an isomorphism of F onto F_1, then the element $T_1 \in \text{Hom}(F_1, \text{Aut}(H_1))$ corresponding to T in a natural way is defined by the equation

$$(hU)(fWT_1) = (h(fT))U, \qquad h \in H, f \in F. \tag{2}$$

9.2.1. *If H and F are groups, $T \in \text{Hom}(F, \text{Aut}(H))$, $G = F \times H$ as set, but with multiplication*

$$(f_1, h_1)(f_2, h_2) = (f_1 f_2, (h_1(f_2 T))h_2),$$

and U, V, and W are functions defined by the rules:

$$hU = (e, h), (f, h)V = f,$$
$$fW = (f, e), f \in F, h \in H,$$

then

(i) $E \longrightarrow H \overset{U}{\longrightarrow} G \overset{V}{\longrightarrow} F \longrightarrow E$ *is an extension of H by F.*

(ii) *U is an isomorphism of H onto a subgroup H_1 of G.*

(iii) *W is an isomorphism of F onto a subgroup F_1 of G.*

(iv) *G is a semi-direct product of H_1 by F_1 whose homomorphism T_1 satisfies (2).*

(v)   *$WV = I_F$.*

Proof. The given multiplication is an operation on G. We have

$$((f_1, h_1)(f_2, h_2))(f_3, h_3) = (f_1 f_2, (h_1(f_2 T))h_2)(f_3, h_3)$$
$$= (f_1 f_2 f_3, (h_1(f_2 T)(f_3 T))(h_2(f_3 T))h_3),$$
$$(f_1, h_1)((f_2, h_2)(f_3, h_3)) = (f_1, h_1)(f_2 f_3, (h_2(f_3 T))h_3)$$
$$= (f_1 f_2 f_3, (h_1(f_2 T)(f_3 T))(h_2(f_3 T))h_3),$$

so that the multiplication is associative. Also

$$(f, h)(e_F, e_H) = (f, (h(e_F T)e_H) = (f, h),$$

and (e_F, e_H) is a right identity for G. Again

$$(f, h)(f^{-1}, (h(fT)^{-1})^{-1}) = (e, (h(f^{-1}T))(h(f^{-1}T))^{-1}) = (e, e),$$

so that $(f, h)^{-1} = (f^{-1}, (h(fT)^{-1})^{-1})$. Hence G is a group.

It is obvious that U is an isomorphism of H onto a subgroup H_1 of G, and that V is a homomorphism of G onto F. Moreover, $\mathrm{Ker}(V) = \{(e, h)\} = HU$. Therefore the sequence

$$E \longrightarrow H \xrightarrow{U} G \xrightarrow{V} F \longrightarrow E$$

is exact.

Statements (iii) and (v) are obvious. Since $H_1 = HU = \mathrm{Ker}(V)$, $H_1 \lhd G$. Since $(f, h) = (f, e)(e, h)$, $G = F_1 H_1$. Clearly, $F_1 \cap H_1 = E$. It remains to verify (2). We have

$$(hU)(fWT_1) = (fW)^{-1}(hU)(fW) = (f, e)^{-1}(e, h)(f, e)$$
$$= (f^{-1}, e)(f, h(fT)) = (e, h(fT)) = (h(fT))U.$$

9.2.2. *If H and F are groups and $T \in \mathrm{Hom}(F, \mathrm{Aut}(H))$, then there is a semi-direct product K of H by a group F_2 with homomorphism T_2, and an isomorphism X of F onto F_2 such that*

$$h(fXT_2) = h(fT), \qquad h \in H, f \in F. \tag{2$'$}$$

Proof. By Theorem 9.2.1, there is a semi-direct product G of H_1 by F_1 with T_1, U, and W as described there. By Exercise 2.1.36, there are a group $K \supset H$ and an isomorphism R of G onto K such that $UR = I_H$. It follows readily that if $F_2 = F_1 R$, then K is a semi-direct product of H and F_2 with homomorphism T_2, say. Let $X = WR$. Then X is an isomorphism of F onto F_2. Applying (2) twice,

$$h(fXT_2) = (hUR)(fWRT_2) = ((hU)(fWT_1))R$$
$$= (h(fT))UR = h(fT). \;\|$$

If G is a semi-direct product of H by F with homomorphism O, then $G = F + H$.

If $K \subset \mathrm{Aut}(H)$, then a *relative holomorph*, $\mathrm{Hol}(H, K)$, of H by K is a semi-direct product G of H by a group F with associated homomorphism T actually an isomorphism of F onto K. All such groups G are isomorphic (for fixed H and K). If $K = \mathrm{Aut}(H)$, one has the *holomorph* $\mathrm{Hol}(H)$.

9.2.3. *If H is a group, then*

(i) *every automorphism of H is induced by an inner automorphism of $\mathrm{Hol}(H)$, and*

(ii) *if $K \subset H$, then $K \in \mathrm{Char}(H)$ iff $K \lhd \mathrm{Hol}(H)$.*

Example 1. Let $G = \mathrm{Hol}(J_n)$. Then $G/J_n \cong \mathrm{Aut}(J_n)$, which is Abelian (Theorem 5.7.12). Hence G is solvable. For example, $\mathrm{Hol}(J_5)$ has order 20, and is an extension of a cyclic group of order 5 by one of order 4.

Example 2. Let H be Abelian and not an elementary 2-group (or E), and let $F = \langle -I_H \rangle$. Hol(H, F) is called a *generalized dihedral group*, and will be denoted by Dih(H). Thus Dih(H) contains H as a normal subgroup of index 2, and has an element x such that $x^{-1}yx = y^{-1}$ for all $y \in H$. Its order is $2 \cdot o(H)$.

Example 3. Wreath product. Let S be a nonempty set, A a group, $G = \Sigma_E \{A \mid s \in S\}$, and $j \in$ Sym(S). Define j^* by the rule

$$s(gj^*) = (sj^{-1})g, \qquad g \in G \tag{3}$$

(in words, j^* permutes the coordinates of g according to j). Then each $s(gj^*) \in A$, and for all but a finite number of $s \in S$, $s(gj^*) = e$. Hence $gj^* \in G$. Also

$$s((g + g')j^*) = (sj^{-1})(g + g') = ((sj^{-1})g)((sj^{-1})g')$$
$$= (s(gj^*))(s(g'j^*)) = s(gj^* + g'j^*),$$

so that $(g + g')j^* = gj^* + g'j^*$, and j^* is an endomorphism of G. Now

$$s((gj^*)k^*) = (sk^{-1})(gj^*) = (sk^{-1}j^{-1})g = (s(jk)^{-1})g = s(g(jk)^*).$$

Hence $g(j^*k^*) = g(jk)^*$ for all $g \in G$, so that $(jk)^* = j^*k^*$ for all j and k in Sym(S). Since

$$s(ge^*) = (se^{-1})g = sg,$$

$e^* = I_G$. It follows that all endomorphisms j^* have inverses, so that $*$ is a homomorphism of Sym(S) into Aut(G). If $j \neq e$, then $\exists s \in S$ such that $sj^{-1} \neq s$. Next, $\exists g \in G$ such that $sg \neq (sj^{-1})g$. Hence $sg \neq s(gj^*)$, so that $gj^* \neq g$. Therefore $j^* \neq I_G$. Hence $*$ is an isomorphism of Sym(S) into Aut(G).

With this background, we can introduce two versions of the wreath product. If (S, H) is a permutation group, then the *wreath product* $A \wr (S, H)$ is Hol$(\Sigma_E \{A \mid s \in S\}, H^*)$, where $*$ is defined by (3). If H is simply a group, let (H, R_H) be the permutation group given by its Cayley representation (Theorem 3.1.1). Then define the *wreath product* $A \wr H$ to be $A \wr (H, R_H)$.

9.2.4. *If A and B are finite groups, then*

$$o(A \wr B) = o(A)^{o(B)}o(B).$$

Proof. $A \wr B$ has a normal subgroup $G = \Sigma_E \{A \mid b \in B\}$, with factor group $(A \wr B)/G \cong B$.

9.2.5. *If A and B are solvable, then $A \wr B$ is solvable.*

Proof. If $A^n = E$, then $G^n = E$ (Exercise 3.4.13). Since G and

$$(A \wr B)/G \cong B$$

are solvable, so is $A \wr B$.

9.2.6. *If A and B are p-groups, $p \in \mathscr{P}$, then $A \wr B$ is a p-group.*

· *Proof.* G and $(A \wr B)/G$ are p-groups. ‖

Example 4. Let $H = J_2 \wr J$. The normal subgroup $G = \Sigma_E \{J_2 \mid n \in J\}$ contains the subgroup $K = \Sigma_E \{J_2 \mid n \in \mathscr{N}\}$. To the element 1 of J there correspond T_1 in the Cayley representation of J, an automorphism T_1^* of G, and an element t of H. Then for $s \in J$ and $k \in K$,

$$s(t^{-1}kt) = s(kT_1^*) = (sT_1^{-1})k = (s-1)k.$$

Hence if $s \leq 1$, then $s(t^{-1}kt) = 0$. Therefore,

$$t^{-1}kt \in \Sigma_E \{J_2 \mid n > 1\} < K, \, t^{-1}Kt < K.$$

This example furnishes an affirmative answer to the question: "Can conjugation shrink a subgroup?" Note that $H^2 = E$.

Example 5. There is an infinite p-group with center E. Let $o(A) = p$, and let B be a p^∞-group. Then $H = A \wr B$ is a p-group (Theorem 9.2.6). Let $x \in H^{\#}$. If $x \in G = \Sigma_E \{A \mid b \in B\}$, then x has only a finite number n of coordinates different from e. Now $\exists \, y \in B$ such that $o(y) = m > n$. Then the permutation y^* moves all letters of B and is a (formal) product of m-cycles. If z is the element of H corresponding to y^*, then conjugation of x by z must move some non-e component of x onto an e component, since x has too few non-e components to make up an m-cycle. Hence $x \notin Z(H)$. If $x \notin G$, then $x = yz$ where $y \in G$ and z corresponds to some element $z^* \neq I$. But there is an element $u \in G$ with which z does not commute. Then $u^{yz} = u^z \neq u$. Hence $Z(H) = E$.

Example 6. Let H be Abelian and not an elementary 2-group. Let $F = \langle f \rangle$ be infinite cyclic with $H \cap F = E$, and let T be the (unique) homomorphism of F onto $\langle -I_H \rangle$. Since $o(-I_H) = 2$, T is a proper homomorphism. There is then a semi-direct product G of H by F with homomorphism T. G is infinite, and $f^2 \in Z(G)$. ‖

Nilpotent groups share some properties of Abelian groups. For example, a periodic nilpotent group is the direct sum of its Sylow subgroups (Theorem 6.4.13). For finitely generated Abelian groups G, the torsion subgroup P is a direct summand (Exercise 5.4.8). One might ask whether the same statement is true for nilpotent groups. The following example shows that it is not.

Example 7. Let $H = L + M$ where L and M are of order 2, and let t be the automorphism of H interchanging L and M. Let $F = \langle f \rangle$ be infinite cyclic, and let T be the homomorphism of F into $\text{Aut}(H)$ mapping f onto t. Let G be a semi-direct product of H by F with homomorphism T. Then $f^2 \in Z(G)$. Now $G/\langle f^2 \rangle$ has order 8 since it is generated (mod $\langle f^2 \rangle$) by a normal subgroup H of order 4 and an element f of order 2. Hence $G/\langle f^2 \rangle$ is nilpotent, and since $\langle f^2 \rangle \subset Z(G)$, G is nilpotent. (Actually, G/Z has order 4.) Now $H \lhd G$, $o(H) = 4$, and G/H is torsion-free (in fact, infinite cyclic). Hence H is the torsion subgroup of G. If $G = H + X$ for some $X \subset G$, then $X \cong G/H \cong F$, so X is Abelian. Therefore G is Abelian, a contradiction since $f \notin Z(G)$. The group G is generated by $\{a, b, f\}$ with relations

$$a^2 = b^2 = e, \quad ab = ba, \quad f^{-1}af = b, \quad f^{-1}bf = a.$$

Example 8. Let p be prime and $H = J_p \times J_p$. By Exercise 6.3.14, $\exists\, t \in \mathrm{Aut}(H)$ with $o(t) = p$. $\mathrm{Hol}(H, \langle t \rangle)$ is of order p^3, and is nilpotent but not Abelian. ‖

We now have enough information to solve the following problems: for what $n \in \mathcal{N}$ are all groups of order n cyclic (Abelian) (nilpotent)?

9.2.7. Let $1 < n = \pi\,\{p_i^{e_i} \mid 1 \leq i \leq s\}$, $p_i \in \mathscr{P}$, $p_i \neq p_j$ if $i \neq j$, $e_i > 0$, and let $f(p^k) = \pi\,\{p^r - 1 \mid 1 \leq r \leq k\}$. Then

(i)   *all groups of order n are cyclic iff all $e_i = 1$ and*

$$p_j \nmid p_i - 1, \quad 1 \leq i \leq s, 1 \leq j \leq s,$$

(ii) *all groups of order n are Abelian iff all $e_i = 1$ or 2, and*

$$p_j \nmid f(p_i^{e_i}), \quad 1 \leq i \leq s, 1 \leq j \leq s,$$

(iii) *all groups of order n are nilpotent iff*

$$p_j \nmid f(p_i^{e_i}), \quad 1 \leq i \leq s, 1 \leq j \leq s.$$

Proof. If some $p_j \mid f(p_i^{e_i})$, then the direct product H of e_i groups of order p_i has an automorphism t of order p_j by Theorems 5.7.20 and 5.7.22. Hence $\mathrm{Hol}(H, \langle t \rangle)$ is a nonnilpotent group of order n. This establishes the necessity of the condition $p_j \nmid f(p_i^{e_i})$ in all three parts of the theorem. In (i), if some $e_i > 1$, then there is a noncyclic group of order p_i^2, hence one of order n. In (ii), if some $e_i > 2$, then by Example 8, there is a non-Abelian group of order p_i^3, hence one of order n. It therefore remains only to establish the sufficiency of the conditions.

(i) and (ii). Suppose that all $e_i = 1$ or 2, no $p_j \mid f(p_i^{e_i})$, and that G is a group of order n. Let $P_i \in \mathrm{Syl}_{p_i}(G)$. By the N/C theorem and Theorem 7.3.11, $N(P_i) = C(P_i)$. By Burnside, each P_i has a normal complement Q_i. Since $G/Q_i \cong P_i$ is Abelian, $Q_i \supset G^1$. Hence $E = \cap\, Q_i \supset G^1$, and G is Abelian. In case all $e_i = 1$, G is the direct sum of its cyclic Sylow subgroups, hence is cyclic.

(iii). Suppose that $p_j \nmid f(p_i^{e_i})$ for all i and j, $o(G) = n$, and that the theorem is true for all integers less than n. Then all proper subgroups of G are nilpotent. By Theorem 6.5.7, either G is nilpotent or $G = PQ$, $Q \lhd G$, $Q \in \mathrm{Syl}(G)$, and $P \in \mathrm{Syl}(G)$. By Theorem 7.3.11, every element of P centralizes Q. Hence $G = P + Q$, so that G is nilpotent.

EXERCISES

9.2.8. Let $G = \text{Dih}(H)$, H Abelian, and let $x \in G\backslash H$. Then $o(x) = 2$.

9.2.9. Show that $\text{Dih}(J_4)$ has a subnormal subgroup which is not normal.

9.2.10. (a) If D is a division ring and
$$G = \{(a, b) \mid 0 \neq a \in D, b \in D\}$$
with product
$$(a, b)(c, d) = (ac, bc + d),$$
then G is a group.

(b) Relate this example to the relative holomorph.

9.2.11. Let $G = \text{Dih}(J_3 \times J_3)$ and $P \in \text{Syl}_2(G)$. Then $N(P) = P$ and $N(P)$ is neither G nor a maximal proper subgroup of G.

9.2.12. Let H be the group, K the subgroup, and t the element of Example 4. Prove (see Theorem 3.3.10) that the number of right cosets of K in KtK is greater than 1 while the number of left cosets of K in KtK equals 1.

9.2.13. Prove that the converse of the Sylow tower theorem, 7.2.19, is false as follows. Let $H = \langle a \rangle + \langle b \rangle$, with $o(a) = o(b) = 3$.

(a) There is an automorphism t of H of order 4 such that $at = b$, $bt = a^{-1}$.

(b) $G = \text{Hol}(H, \langle t \rangle)$ is of order 36, has a normal Sylow subgroup of order 9, but no normal subgroup of order 3.

(c) G is not supersolvable.

9.2.14. Let G be a group of order $84 = 2^2 \cdot 3 \cdot 7$ with $n_3 = 28$. Determine G as follows.

(a) Let Q be a Sylow 3-subgroup. Then $N(Q) = C(Q) = Q$.

(b) There is a normal subgroup H of order 28. (Burnside.)

(c) There is a normal Sylow 7-subgroup K. (Sylow.)

(d) $o(C(K)) = 14$ or 28. [(a) and N/C.]

(e) $C(K) = H$. (Otherwise, $\exists\, L \in \text{Char}(H)$, $o(L) = 2$, $L \triangleleft G$, $C(L) \supset Q$, contradicting (a).)

(f) A Sylow 2-subgroup P is normal in G.

(g) P is a 4-group. [(f) and (a).]

(h) $H = J_2 \times J_2 \times J_7$. [(c), (f), (g).]

(i) G is the relative holomorph of H by an automorphism of order 3 which acts as an automorphism of order 3 on P and on K.

(j) G has generators $\{x, y, z, u\}$ and relations
$$x^2 = y^2 = z^7 = u^3 = e, \; xy = yx, \; xz = zx,$$
$$yz = zy, \; u^{-1}xu = y, \; u^{-1}yu = xy, \; u^{-1}zu = z^2.$$

9.2.15. (Baer [4].) Example of a group G whose transfinite upper central series reaches G, but whose lower central series does not reach E.

    (a) Let A be a p^∞-group. Then there is an automorphism T of A such that $xT = x^{1+p}$ for $x \in A$.

    (b) Let $G = \mathrm{Hol}(A, \langle T \rangle)$.

    (c) If A_n is the subgroup of A of order p^n, then
$$[G, A_n] = A_{n-1} \text{ for } n \geq 1.$$

    (d) $Z_n(G) = A_n$ for $n \in \mathcal{N}$, $Z_\omega(G) = A$ where ω is the first infinite ordinal, and $Z_{\omega+1}(G) = G$.

    (e) $G^1 = A$, $[G, A] = A$.

    (f) The lower central series terminates at A.

9.2.16. (a) Let G be a non-Abelian group of order p^3 (for existence, see Example 8). Then $Z(G) = G^1$ has order p.

    (b) If G is a finite p-group, $G \cong H$, and
$$G \cap H = A = Z(G) = G^1 = Z(H),$$
then
$$K = (G \times H)_A \text{ has } Z(K) = K^1 = A.$$

    (c) For all $n \in \mathcal{N}$, there is a finite p-group G of order p^m, $m > n$, with $Z(G) = G^1$ of order p.

9.2.17. (See Exercise 3.2.24.) (a) If H is an infinite maximal Abelian normal subgroup of G and K is a normal Abelian subgroup of G, then $o(K) \leq 2^{o(H)}$.

    (b) [Part (a) is best possible.] Let A be an infinite cardinal, S a set of order A, H_i a 4-group for $i \in S$, $x_i \in \mathrm{Aut}(H_i)$ with $o(x_i) = 2$, and $H = \Sigma_E H_i$. $L = \pi \langle x_i \rangle$ may be considered as a subgroup of $\mathrm{Aut}(H)$. Let $G = \mathrm{Hol}(H, L)$. Each H_i has an element $y_i \neq e$ fixed by x_i. Let $M = \Sigma_E \langle y_i \rangle$ and let $K = L'M$, where $L' \subset G$ and L' corresponds to L. Verify that the hypotheses of (a) are satisfied and that $o(K) = 2^{o(H)}$.

9.2.18. Give an example of a finite non-Abelian group G which is the product AB of two normal Abelian subgroups.

9.2.19. (Huppert [2]) Let
$$x = \begin{bmatrix} 2 & 0 \\ 0 & 3 \end{bmatrix}, \quad y = \begin{bmatrix} 0 & 1 \\ -1 & 0 \end{bmatrix}$$
be matrices over J_5.

    (a) Show that $K = \langle x, y \rangle$ is the quaternion group. Let $H = J_5 \times J_5$ and let $G = \mathrm{Hol}(H, K)$ (where K is interpreted as a group of automorphisms of H). (See end of Section 5.7.) Let $A = \langle H, x \rangle$ and $B = \langle H, y \rangle$.

    (b) A and B are supersolvable normal subgroups of G.

    (c) $G = AB$.

(d) G has a normal subgroup of order 5^2 but none of order 5.

(e)   The product of normal supersolvable subgroups is not necessarily supersolvable.

(Compare with Theorem 7.4.1 and Exercises 2.6.9, 7.2.23, and 9.2.18.)

9.2.20. Let p be an odd prime and $G = (U * V) + (W * X)$, where U, V, W, and X are of order p.

(a)   There is an automorphism T of G of order 4 such that $UT = W$, $WT = V$, $VT = X$, $XT = U$. Let $H = \text{Hol}(G, \langle T \rangle)$.

(b)   All Sylow p-subgroups of H are conjugate.

(c) $U + W \in \text{Syl}_p(H)$.

(d) $N(U + W) = U + W$.

(e)   (Compare with Theorems 6.2.5, 6.2.6, and 6.2.7.) U and W are normal subgroups of $U + W$, conjugate in H, but not conjugate in $N(U + W)$.

(f) $G \lhd H$, G contains all Sylow p-subgroups of H, but they are not all conjugate in G.

9.2.21. Let H be free Abelian with basis $\{x, y\}$. Let k be the automorphism of H of order 2 such that $xk = y$ and $yk = x$. Let $\langle U \rangle$ be infinite cyclic, and let T be the homomorphism of $\langle U \rangle$ onto $\langle k \rangle$ such that $UT = k$. Let G be a semi-direct product of H by $\langle U \rangle$ with homomorphism T.

(a) G is torsion-free and $n_2(G) = 1$.

Let $K = \langle x^3, y^3, U^2 \rangle$.

(b) $K \lhd G$, $o(G/K) = 18$.

(c) $n_2(G/K) = 3 > n_2(G)$. (Compare with Theorem 6.1.17.)

9.2.22. Let $G = \text{Hol}(J_p)$ for $p \in \mathscr{P}$. Then

(a) $o(G) = p(p - 1)$,

(b)   if $q \mid p - 1$, $q \in \mathscr{P}$, then $n_q = p$.

(c)   Let $G = HF$, $o(H) = p$, $o(F) = p - 1$. Then $N(F) = F$ and $F \cap F^x = E$ if $x \in G \backslash F$.

9.2.23. (See Exercise 7.3.31.) Let $G = \text{Hol}(J_5)$.

(a) $\text{Fr}(G) = E$. [Use Exercise 9.2.22 (c).]

(b) $\text{Fr}(G/J_5) \neq E$.

9.2.24. The holomorph of the four-group is isomorphic to $\text{Sym}(4)$.

9.2.25. (See Exercise 7.4.22.) Define the *nilpotent length* of a finite solvable group G to be n provided

$$E = A_0 < A_1 < \ldots < A_n = G \quad \text{and} \quad A_{i+1}/A_i = \text{Fit}(G/A_i).$$

Prove that if $n \in \mathscr{N}$, then there is a group of nilpotent length n as follows.

(a)   There is a group of nilpotent length 1.

Let H be solvable of nilpotent length n, and let $G = J_p \wr H$, where $p \in \mathscr{P}$, $p \nmid o(H)$. Let $K = \Sigma_E\{J_p \mid h \in H\}$.

(b) $C_G(K) = K$.

(c) $\text{Fit}(G) = K$.

(d) G has nilpotent length $n + 1$.

9.2.26. (Compare with Theorem 6.1.19.) Let $A = \Sigma_E \{J_3 \mid n \in \mathcal{N}\}$. Then $B = \pi \{\text{Aut}(J_3) \mid n \in \mathcal{N}\}$ can be considered as a subgroup of $\text{Aut}(A)$ (Exercise 5.7.31). Let $G = \text{Hol}(A, B)$, and let B^* be the image of B in G.

(a) $B^* \in \text{Syl}_2(G)$, and B^* is Abelian.

(b)   Any two (in fact, any finite number of) conjugates of B^* have a non-trivial intersection.

(c)   The intersection of all conjugates of B^* is E.

9.2.27. (P. Hall [7].) (Compare with Theorem 7.3.13.) There is a group G with nonnilpotent Frattini subgroup. Let H be a 5^∞-group. Let H_n be the subgroup of H of order 5^n.

(a) H_1 has an automorphism T_1 of order 4.

(b)   If, inductively, H_n has an automorphism T_n of order 4, then $\exists\, T_{n+1} \in \text{Aut}(H_{n+1})$ with $o(T_{n+1}) = 4$ and $T_{n+1} \mid H_n = T_n$.

(c) H has an automorphism T of order 4.

Now let $G = \text{Hol}(H, \langle T \rangle)$.

(d)   If $M \subseteq G$ and $M \cap H = H_n$, then $M < MH_{n+1} < G$.

(e) $\text{Fr}(G) \supset H$.

(f) $\text{Fr}(G) = \langle H, T^2 \rangle$.

(g) $\text{Fr}(G)$ is not nilpotent.

9.2.28. An automorphism of a normal subgroup H of G is not always extendible to an automorphism of G. Let $G = A + B$, $A \cong J_4$, $B \cong J_2$, and $H = 2A + B$.

(a)   There is an automorphism T of H exchanging $2A$ and B.

(b) T cannot be extended to an automorphism of G (or even to an endomorphism of G).

9.2.29. (Compare with $(H \times K)_L$ in Section 8.1.) The following theorem is false. If L is a normal subgroup of both H and K, and $H \cap K = L$, then there is a group G containing H and K as subgroups and L as normal subgroup such that $G/L = H/L + K/L$. Let $H = A + B$, $A \cong J_4$, $B \cong J_2$, and $L = 2A + B$. Let T be the automorphism of L exchanging $2A$ and B, and let $K = \text{Hol}(L, \langle T \rangle)$. Assume G exists.

(a)   The element T^* of K corresponding to T induces an automorphism of H whose restriction to L is T.

(b)   This is impossible (see Exercise 9.2.28).

9.2.30. There is an irregular p-group. Let $G = A \wr B$,
$o(A) = o(B) = p \in \mathscr{P}$, $A = \langle a \rangle$, $B = \langle b \rangle$.
   (a) G^1 is an elementary Abelian p-group.
   (b) $(ab)^p \neq e$.
   (c) G is an irregular p-group.

9.2.31. Let $G = \text{Dih}(H)$ where H is Abelian.
   (a)   If T is the transfer of G into H, then $T = O$.
   (b)   If $[H:K]$ is finite, and U is the transfer of G into K, then $U = O$.

9.2.32. (a)   If H_n is a nilpotent group of class n, then $G = \Sigma_E H_n$ is not (finitely) nilpotent.
   (b)   The union of a chain of normal nilpotent subgroups need not be nilpotent. [Use (a) and Exercise 6.4.27.]

9.2.33. (Compare with Exercise 3.2.24.) There is a finite group G containing maximal normal Abelian subgroups H and K of different orders. Let H be elementary Abelian of order 2^4 with basis (w, x, y, z).
   (a)   There is an automorphism T of H of order 2 such that $wT = wy$, $xT = xz$, $yT = y$, and $zT = z$.
   Let $G = \text{Hol}(H, \langle T \rangle)$, with v corresponding to T.
   (b) H is a maximal normal Abelian subgroup of G.
   (c) $[w, v] = y$, $[x, v] = z$.
   (d)   If $K = \langle v, y, z \rangle$, then K is a maximal normal Abelian subgroup of G.
   (e) $o(H) \neq o(K)$.

9.3   Hall subgroups

If P is a set of primes, a *P-number* is an $n \in \mathcal{N}$ such that every prime divisor of n is in P. The notation P' will be used for the complement $\mathscr{P} \backslash P$ of P. A group G is a *P-group* iff G is periodic and $g \in G$ implies that $o(g)$ is a P-number. Thus a finite group G is a P-group iff $o(G)$ is a P-number. If $P = \{p\}$, the terms p-number and p-group will be used, and the latter term is consistent with our previous usage. A subgroup H of a finite group G is a *Hall P-subgroup* ($H \in \text{Hall}_P(G)$) iff H is a P-group and $[G:H]$ is a P'-number. H is a *Hall subgroup* iff H is a Hall P-subgroup for some P, hence iff $(o(H), [G:H]) = 1$.

9.3.1. *Subgroups and factor groups of P-groups are P-groups (P a set of primes).*

9.3.2. *If H and G/H are P-groups, so is G.*

9.3.3. *If $H \in \text{Hall}(G)$ and $A \lhd G$, then $A \cap H \in \text{Hall}(A)$.*

Proof. Let P be the set of primes dividing $o(H)$. Then H is a Hall P-subgroup of G. By Exercise 2.3.7, $[A:H \cap A] = [HA:H]$, which is a P'-number. Also, $H \cap A$ is a P-group, so $H \cap A$ is a Hall P-subgroup of A.

9.3.4. *If $H \in \mathrm{Hall}(G)$ and $A \lhd \lhd G$, then $A \cap H \in \mathrm{Hall}(A)$.*

Proof. $\exists\ A_0, \ldots, A_n$ such that

$$A = A_0 \lhd A_1 \lhd \ldots \lhd A_n = G.$$

If $n = 0$, the statement is obvious. Induct on n. By the inductive hypothesis, $A_1 \cap H \in \mathrm{Hall}(A_1)$. By Theorem 9.3.3, $A_0 \cap H = A_0 \cap (A_1 \cap H)$ is a Hall subgroup of $A_0 = A$.

9.3.5. *If G is a finite group, A is an Abelian normal subgroup of G, $A \subset H \subset G$, $([G:H], o(A)) = 1$, and A has a complement B in H, then A has a complement in G.*

Proof. There is a subset S of G/A such that $e \in S$ and

$$G/A = \dot\cup\ \{(H/A)s \mid s \in S\}.$$

Then

$$G = \dot\cup\ \{(Ab)s \mid b \in B, s \in S\}.$$

For each $i \in S$, let $x_i \in i$ with $x_e = e$. For $i = Ab \in H/A$ with $b \in B$, let $x_i = b$. Finally, for $i = Ab, j \in S$, let $x_{ij} = x_i x_j$. It then follows that $x_i \in i$ for all $i \in G/A$, and

$$x_i x_j = x_{ij} \quad \text{if} \quad i \in H/A, j \in G/A.$$

We have for all i and j in G/A,

$$x_i x_j \in ij, \ x_i x_j = x_{ij} c_{i,j} \quad \text{with} \quad c_{i,j} \in A,$$

while

$$c_{i,j} = e \quad \text{if} \quad i \in H/A, j \in G/A.$$

Let T_k be the automorphism of A induced by x_k. Then

$$x_i(x_j x_k) = x_i x_{jk} c_{j,k} = x_{ijk} c_{i,jk} c_{j,k},$$
$$(x_i x_j)x_k = x_{ij} c_{i,j} x_k = x_{ij} x_k (c_{i,j} T_k) = x_{ijk} c_{ij,k}(c_{i,j} T_k),$$
$$c_{ij,k}(c_{i,j} T_k) = c_{i,jk} c_{j,k}. \tag{1}$$

In (1), let $i \in H/A$. Since $c_{i,j} = e$ then,

$$c_{ij,k} = c_{j,k} \quad \text{if} \quad i \in H/A. \tag{2}$$

Let $s_1 \in S, s_2 \in S, j \in G/A$. Then $s_i j = r_i u_i$, $r_i \in H/A$, $u_i \in S$, $i = 1, 2$. If $u_1 = u_2$, then

$$s_1 s_2^{-1} = (s_1 j)(s_2 j)^{-1} = (r_1 u_1)(r_2 u_1)^{-1}$$
$$= r_1 r_2^{-1} \in H/A,$$

and $s_1 = s_2$. Hence if $j \in G/A$ is fixed, and for all $s \in S$, $sj = r_s u_s$, $r_s \in H/A$, $u_s \in S$, then the function $s \to u_s$ is a permutation of S. But by (2), $c_{sj,k} = c_{r_s u_s, k} = c_{u_s, k}$. Hence

$$\pi \{c_{ij,k} \mid i \in S\} = \pi \{c_{i,k} \mid i \in S\}. \tag{3}$$

Let $f(j) = \pi \{c_{i,j} \mid i \in S\}$. Then, taking products on both sides of (1) over all $i \in S$, letting $[G:H] = n$, and applying (3), we get

$$f(k)(f(j)T_k) = f(jk)c_{j,k}^n. \tag{4}$$

Since $(n, o(A)) = 1$, $\exists\, r \in J$ such that $rn \equiv -1 \pmod{o(A)}$. Taking rth powers in (4), we get

$$f(k)^r(f(j)T_k)^r f(jk)^{-r} c_{j,k} = e. \tag{5}$$

Let $y_j = x_j f(j)^r$. Then $y_j \in j$ and y_j also induces T_j. Moreover

$$y_i y_j = x_i f(i)^r x_j f(j)^r = x_i x_j (f(i)^r T_j) f(j)^r = x_{ij} c_{i,j} (f(i)T_j)^r f(j)^r$$
$$= y_{ij} f(ij)^{-r} c_{i,j} (f(i)T_j)^r f(j)^r$$
$$= y_{ij}$$

by (5) and the commutativity of A. Hence the set $K = \{y_i \mid i \in G/A\}$ is a subgroup of G. Since $y_i \in i$ for each $i \in G/A$, K contains exactly one element from each coset of A. Therefore, K is a complement of A in G. ‖

The following theorem was credited to Schur by Zassenhaus [4].

9.3.6. (*Schur's splitting theorem.*) *If H is a normal Hall subgroup of a finite group G, then H has a complement.*

Proof. Induct on $o(G)$. Let $E \neq Q \in \mathrm{Syl}(H)$. By Theorem 6.2.4, $G = N(Q)H$. Suppose $N(Q) < G$. Then

$$G/H \cong N(Q)/(N(Q) \cap H),$$

hence $N(Q) \cap H$ is a normal Hall subgroup of $N(Q)$. By induction, there is a complement K of $N(Q) \cap H$ in $N(Q)$, and, since $o(K) = o(G/H)$, K is a complement of H in G.

Next, suppose that $Q \lhd G$. Then $A = Z(Q)$ is an Abelian normal non-E subgroup of G contained in H. By induction H/A has a complement K/A in G/A. Now $G/H \cong (G/A)/(H/A) \cong K/A$, hence A is a normal Hall subgroup of K, and $o(K/A) = o(G/H)$. Since A has complement E in A, by Theorem 9.3.5, A has a complement L in K. Since $o(L) = o(K/A) = o(G/H)$, $H \cap L = E$, and L is a complement of H in G.

9.3.7. (*Gaschütz* [1].) *If A is an Abelian normal subgroup of a finite group G, then the following properties are equivalent:*

(i) *A has a complement in G,*

(ii) *If $P \in \mathrm{Syl}(G)$, then $A \cap P$ has a complement in P.*

Proof. (i) implies (ii). By Theorem 6.1.15, $P \cap A \in \mathrm{Syl}(A)$. Since A is Abelian, $\exists B \subset A$ such that $A = (P \cap A) + B$. By assumption, $\exists D \subset G$ such that $G = AD$, $A \cap D = E$. Then

$$G = AD = (P \cap A)BD.$$

Since B is a Hall subgroup of A it is normal in G. Hence $BD \subset G$. Since $A \cap (BD) = B$ (see Exercise 1.6.15),

$$P \cap A \cap (BD) = P \cap A \cap B = E.$$

Therefore,

$$P = (P \cap A)(P \cap (BD)),$$
$$(P \cap A) \cap (BD \cap P) = P \cap A \cap (BD) = E.$$

(ii) implies (i). If $p \mid o(A)$, $p \in \mathscr{P}$, then $A_p \lhd G$, and $A_p \subset P_p \in \mathrm{Syl}_p(G)$. By assumption, A_p has a complement in P_p. By Theorem 9.3.5, A_p has a complement B_p in G. Let $B = \cap B_p$. By Exercise 1.7.15 and induction, $[G:B] = \pi \, o(A_p) = o(A)$. Let $e \neq x \in A \cap B$. Then

$$x = x_1 \cdots x_r, \qquad x_i \in A_{p_i}^{\#}, \, p_i \neq p_j \text{ if } i \neq j.$$

Raising to a suitable power, one gets

$$x^j = x_1 \in A_{p_1} \cap B \subset A_{p_1} \cap B_{p_1} = E,$$

a contradiction. Hence $A \cap B = E$. Therefore,

$$o(AB) = o(A)o(B) = o(G),$$

so that $G = AB$. ‖

Example. Gaschütz's theorem becomes false if it is not assumed that A is Abelian, as the following example of Baer shows. (The first such example was due to Zassenhaus.)

Let Q be the quaternion group, and H a cyclic group of order 4 such that $Q \cap H = A$ has order 2. Let $L = (Q \times H)_A$. There is an automorphism T of Q of order 3 (Exercise 8.2.16), and T fixes all elements of A. Extend T

to an automorphism T' of L such that $hT' = h$ for all $h \in H$ (see Exercise 8.1.26). Let $G = \text{Hol}(L, \langle T' \rangle)$. Then $o(G) = 48$, $o(L) = 16$, and L and Q are normal subgroups of G.

It will first be shown that Q has a complement in L. Let $H = \langle h \rangle$ and $x \in Q \backslash A$. Then $hx \in L \backslash Q$, and

$$(hx)^2 = h^2 x^2 = a^2 = e \qquad \text{where } A = \langle a \rangle.$$

Hence $\langle hx \rangle$ is a complement of Q in the Sylow 2-subgroup L of G.

Next, it will be shown that Q has no complement in G. Suppose that K is a complement of Q in G. Conjugating, if necessary, it may be assumed (by Sylow) that $K \supset \langle t \rangle$, where t induces T' in L. Since any element of H commutes with t, G/Q is Abelian, hence K is Abelian (therefore cyclic) of order 6. Let $u \in K$, $o(u) = 2$. Then $u \in L$ since L is a normal Sylow 2-subgroup of G. Hence $u = xy$, $x \in Q$, $y \in H$. If $u \in H$, then since $o(u) = 2$, $u \in A \subset Q$, and $Q \cap K \neq E$, a contradiction. Hence $u \notin H$, so $x \notin H$. Therefore

$$u^t = uT' = (xy)T' = (xT)(yT') = (xT)y.$$

Since $x \notin H$, $x \notin A$, so that (Exercise 8.2.16) $xT \neq x$. Therefore $u^t \neq u$, whereas K is Abelian, a contradiction. ‖

As an application of Gaschütz's theorem, we shall prove the converse of Theorem 7.2.8 for finite groups.

9.3.8. (*Huppert* [4].) *If G is a finite group all of whose maximal proper subgroups are of prime index, then G is supersolvable.*

Proof. Deny and induct. Then all proper factor groups of G are supersolvable, so that there is no normal subgroup of prime order (Theorem 7.2.14). Let p be the largest prime dividing $o(G)$, and $P \in \text{Syl}_p(G)$. If $P \ntriangleleft G$, then (Theorem 6.2.3) a maximal proper subgroup X containing $N(P)$ is of prime index $q \equiv 1 \pmod{p}$, an impossibility. Hence $P \triangleleft G$.

Suppose that P is not Abelian. Now $Z(P) \triangleleft G$, so (by the supersolvability of $G/Z(P)$), $\exists H \triangleleft G$ with $[H : Z(P)] = p$, and $H = \langle Z(P), a \rangle$, say. Again $C(H) \cap P \triangleleft G$ and $C(H) \cap P < P$. Hence $\exists K \triangleleft G$ such that

$$[K : C(H) \cap P] = p \quad \text{and} \quad K = \langle C(H) \cap P, b \rangle,$$

say. Since $H/Z(P)$ is a normal subgroup of order p of $P/Z(P)$, $H \subset Z_2(P)$ (Theorem 6.3.3). By Theorem 3.4.4, if $z \in Z(P)$ and $u \in C(H) \cap P$, then

$$[a^i z, b^j u] = [a^i, b^j] = [a, b]^{ij}, \; [a, b]^p = [a^p, b] = e.$$

Therefore $o([H, K]) = p$, and, by Theorem 3.4.5, $[H, K] \triangleleft G$, a contradiction.

Hence P is Abelian. Let M be a minimal normal non-E subgroup of G contained in P. By Theorem 4.4.4, M is an elementary Abelian p-group. First suppose that P is not elementary. Then there is a subgroup H of G which is minimal with respect to being normal, nonelementary, containing M, and being contained in P. The supersolvability of G/M implies that

$$H \cong J_{p^2} + J_p + \ldots + J_p.$$

The subgroup K of pth powers of elements of H is characteristic in H, hence normal in G, and of order p, a contradiction. Therefore P is elementary. Hence $P = M + Q$. By Gaschütz (Theorem 9.3.7), $\exists\, R \subset G$ such that $G = MR$ and $M \cap R = E$. Now $[G:R] = o(M) > p$, hence R is not maximal. Thus $\exists\, L \subset G$ such that $R < L < G$. Now $M \cap L \lhd M$ since M is Abelian, and $M \cap L \lhd L$. Therefore $M \cap L \lhd G$, and, since $L = R\,(M \cap L)$, $E < M \cap L < M$, a contradiction.

9.3.9. (*Zassenhaus.*) *If H is a normal Hall subgroup of a finite group G, and H or G/H is solvable, then any two complements of H are conjugate.*

Proof. Induct on $o(G)$. Let K and L be complements of H.

CASE 1. H is Abelian. For $i \in G/H$, $\exists\,|\, x_i \in K \cap i$ and $y_i \in L \cap i$. Then

$$x_i x_j = x_{ij}, y_i y_j = y_{ij}, y_i = a_i x_i, a_i \in H.$$

Therefore, if T_i is the automorphism of H induced by x_i,

$$a_i x_i a_j x_j = a_{ij} x_{ij}, \quad a_i (a_j T_i^{-1}) x_i x_j = a_{ij} x_{ij},$$
$$a_i (a_j T_i^{-1}) = a_{ij}. \tag{6}$$

Taking the product over all $j \in G/H$ in (6), and setting $b = \pi a_j$, we get

$$a_i^n (b T_i^{-1}) = b, \qquad n = [G:H].$$

If r is such that $nr \equiv 1 \pmod{o(H)}$ and $b^r = c$, then

$$a_i (c T_i^{-1}) = c, \quad a_i (x_i c x_i^{-1}) = c, \quad y_i = c x_i c^{-1}.$$

Hence $L = cKc^{-1}$.

CASE 2. H is solvable but not Abelian. Now KH^1/H^1 and LH^1/H^1 are complements of the normal Hall subgroup H/H^1 of G/H^1. By the inductive hypothesis they are conjugate, and therefore KH^1 and LH^1 are conjugate in G, say $KH^1 = (LH^1)^x = L^x H^1$. Since $KH^1 < G$ and H^1 is a normal Hall subgroup of KH^1, by the inductive hypothesis K and L^x are conjugate, and therefore K and L are conjugate.

CASE 3. G/H is solvable. Now $K \cong G/H \cong L$ and there is an isomorphism T of K onto L such that for $k \in K$, $kT = hk$ with $h \in H$. Let M be a minimal normal non-E subgroup of K. By Theorem 4.4.5, M is a p-group for some prime p. If $M = K$, then $K \in \mathrm{Syl}(G)$, $L = KT \in \mathrm{Syl}(G)$, and K and L are conjugate.

Suppose $M < K$. Then $HM = H(MT)$, so by the inductive hypothesis $\exists\, x$ such that $MT = M^{x^{-1}}$. Since $MT \lhd L$, $M = (MT)^x \lhd L^x$, so K and L^x are contained in $N(M)$. Therefore (Exercise 1.6.15), $N(M) = K(H \cap N(M))$ and $H \cap N(M) \lhd N(M)$. Thus $H \cap N(M)$ is a normal Hall subgroup of $N(M)$. Therefore $(H \cap N(M))M/M$ is a normal Hall subgroup of $N(M)/M$ with complements K/M and L^x/M. By the inductive hypothesis, K/M and L^x/M are conjugate, and therefore K and L^x are conjugate. Thus K and L are conjugate. ∥

It has recently been proved by Feit and Thompson [1] that a group of odd order is solvable. Since one of H and G/H is of odd order, it follows that Theorem 9.3.9 can be improved to read: If H is a normal Hall subgroup of a finite group G, then any two complements of H are conjugate.

9.3.10. (*P. Hall* [1].) *If G is a finite solvable group, P a set of primes, and A a P-subgroup of G, then*

(i)   *there is a Hall P-subgroup of G containing A,*

(ii) *any two Hall P-subgroups of G are conjugate.*

Proof. Induct on $o(G)$. Since G is solvable, there is a normal p-subgroup $H \neq E$, $p \in \mathscr{P}$. Now AH/H is a P-subgroup of G/H. By the inductive hypothesis, $\exists\, K/H \in \mathrm{Hall}_P(G/H)$ such that $K \supset AH$. If $p \in P$, then $K \in \mathrm{Hall}_P(G)$ and (i) holds. Any other Hall P-subgroup L of G must contain H (for HL is a P-group), so $L/H \in \mathrm{Hall}_P(G/H)$. By the inductive hypothesis, K/H is conjugate to L/H, hence K is conjugate to L.

Suppose that $p \notin P$ and $K < G$. By induction, $\exists\, L \in \mathrm{Hall}_P(K)$ such that $L \supset A$. Since $[G:K]$ is a P'-number, $L \in \mathrm{Hall}_P(G)$, and (i) holds. If also $M \in \mathrm{Hall}_P(G)$, then $MH/H \in \mathrm{Hall}_P(G/H)$, so by induction K/H is conjugate to MH/H. Hence K is conjugate to MH, and M is conjugate to a Hall P-subgroup Q of K. By induction, since $K < G$, L and Q are conjugate. Therefore L and M are conjugate.

Finally, suppose that $p \notin P$ and $K = G$. By Schur's theorem, there is a complement L of H. But a complement of H is just a Hall P-subgroup in the present case, hence by Zassenhaus's theorem, 9.3.9, any two Hall P-subgroups of G are conjugate. Thus (ii) holds. If A is a Hall P-subgroup of G, we are done. If not, then $AH < G = LH$. Hence $AH = H(L \cap AH)$, so that A and $L \cap AH$ are Hall P-subgroups of AH. By induction, $\exists\, x$ such that $A = (L \cap AH)^x$. Therefore $A \subset L^x$, and $L^x \in \mathrm{Hall}_P(G)$. ∥

Remark. Since E is a P-subgroup of G, (i) guarantees that a solvable group has a Hall P-subgroup.

A *Sylow basis* of a finite group G is a set S of Sylow subgroups, one for each prime divisor p of $o(G)$, such that $PQ \subset G$ for $P \in S$, $Q \in S$. It follows (Theorem 1.6.8) that the product $P_1 \cdots P_r$ of distinct members of a Sylow basis is a Hall subgroup of G.

9.3.11. (*Hall* [3].) *If G is a finite solvable group, then G has a Sylow basis.*

Proof. Let $p \mid o(G)$, $p \in \mathscr{P}$. By Theorem 9.3.10, there is a Hall p'-subgroup Q_p. Then

$$S_p = \cap \{Q_q \mid q \in \mathscr{P}, q \neq p, q \mid o(G)\}$$

is a Sylow p-subgroup of G (Exercise 1.7.15). If p and r are distinct primes dividing $o(G)$, then $S_p S_r$ is a subset of Q_q for all $q \neq p$ or r, and has the same order as $\cap \{Q_q \mid q \neq p, q \neq r\}$. Hence $S_p S_r$ is a subgroup of G. Therefore $\{S_q \mid q \mid o(G)\}$ is a Sylow basis of G. $\|$

Two Sylow bases S and S' of G are *conjugate* iff $\exists\, x \in G$ such that if $Q \in S$, then $Q^x \in S'$. Expressions such as the conjugate class of S and $N(S)$ have obvious meanings.

9.3.12. *If G is a finite solvable group, then any two Sylow bases are conjugate.*

Proof. Let S^* and S' be two Sylow bases of G. Let

$$Q_p^* = \langle S_q^* \mid q \neq p, S_q^* \in S^* \rangle,$$

and similarly for S'. Let S be a conjugate of S^* such that the number of primes p dividing $o(G)$ for which $Q_p = Q_p'$ is a maximum.

Suppose some $Q_p \neq Q_p'$. Then by Theorem 9.3.10, $G = Q_p S_p$ and $Q_p^x = Q_p'$ for some $x \in G$. But $x = yz$ with $y \in Q_p$ and $z \in S_p$, hence

$$Q_p' = Q_p^x = Q_p^{yz} = Q_p^z.$$

Since $z \in S_p \subset Q_q$ for $q \neq p$, $Q_q^z = Q_q$. Hence S^z has one more prime r (namely p) such that $Q_r^z = Q_r'$ than there are primes r such that $Q_r = Q_r'$. This contradicts the choice of S.

Therefore all $Q_p = Q_p'$. But then

$$S_p = \cap \{Q_q \mid q \neq p\} = \cap \{Q_q' \mid q \neq p\} = S_p'$$

for all p. Hence $S = S'$, and S^* is conjugate to S'.

9.3.13. *If G is a finite solvable group and S is a Sylow basis, then $N(S)$ is nilpotent.*

Proof. Let $p \in \mathscr{P}$, $S_p \in S$, $x \in N(S)$, and $o(x) = p^r$. Then $S_p^x = S_p$, so (Theorem 6.1.9), $x \in S_p$. Hence any Sylow p-subgroup Q of $N(S)$ is contained in S_p. Suppose $h \in N(S)$. Then $Q^h \subset S_p$, so $\langle Q, Q^h \rangle$ is a p-subgroup of $N(S)$. Hence $Q^h = Q$, and $Q \lhd N(S)$. Therefore $N(S)$ is nilpotent.

9.3.14. (*Wielandt* [4].) *If G is a finite group, P a set of primes, H a nilpotent Hall P-subgroup of G, and K a P-subgroup of G, then $\exists \, x \in G$ such that $K \subset H^x$.*

Proof. Deny, and induct on $o(K)$. Every proper subgroup of K is then in some conjugate of H, hence is nilpotent. By Theorem 6.5.7, K has a normal Sylow p-subgroup $Q \neq E$ for some prime p. By Sylow, some conjugate of Q is contained in H, hence $WLOG$, $Q \subset H$. Since H is nilpotent, Q is centralized by the p-complement L of H. Hence $N(Q) \supset \langle K, L \rangle$. Now LQ/Q is a Hall $(P\backslash\{p\})$-subgroup of $N(Q)/Q$, and K/Q is a $(P\backslash\{p\})$-subgroup of $N(Q)/Q$. By the inductive hypothesis, some conjugate of K/Q is contained in LQ/Q, so that a conjugate of K is contained in LQ, hence in H.

EXERCISES

9.3.15. Let P be a set of primes and G a group (not necessarily finite). A *Sylow P-subgroup* of G is a maximal P-subgroup.
   (a) If H is a P-subgroup of G, then there is a Sylow P-subgroup of G containing H. In particular there is a Sylow P-subgroup of G.
   (b) If $n_P(G)$ denotes the number of Sylow P-subgroups of G and $H \subset G$, then $n_P(H) \leqq n_P(G)$.
   (c) If $Q \in \mathrm{Syl}_P(G)$, then $N(N(Q)) = N(Q)$.
   (d) A subnormal Sylow P-subgroup of G is normal.

9.3.16. Let G be a finite solvable group, $H \lhd G$, P a set of primes, and $Q \in \mathrm{Hall}_P(H)$. Then $G = HN(Q)$.

9.3.17. If $G \neq E$ is a finite solvable group, then $\exists \, p \in \mathscr{P}$ such that if P is a set of primes not containing p and $H \in \mathrm{Hall}_P(G)$, then $N(H) > H$.

9.3.18. A finite group G is supersolvable iff $G/\mathrm{Fr}(G)$ is supersolvable. (Use Theorems 7.2.8 and 9.3.8. Compare with Theorem 7.4.10 and Exercises 7.3.29 and 8.2.14.)

9.3.19. If $p \mid o(G)$, $p \in \mathscr{P}$, then $p \mid o(G/\mathrm{Fr}(G))$. (Deny, and use Theorems 7.3.14 and 9.3.6.)

9.3.20. If G is finite, P a set of primes, and H and K are supersolvable Hall P-subgroups of G, then H and K are conjugate. (Use Exercise 7.2.19, induction, and Sylow.)

9.4   Extensions: General case

The solution to the main problem of this chapter will now be given. It is due to Schreier [1], while the main ideas were already given by Hölder [1]. Notation will be cumulative in the section.

9.4.1. *If* $G \longrightarrow B \longrightarrow E$ *is exact, then there is a function* $T: B \longrightarrow G$ *such that the diagram*

is commutative.

Proof. This follows trivially from the fact that the homomorphism $G \longrightarrow B$ is onto.

9.4.2. *If* $E \longrightarrow A \xrightarrow{X} G \longrightarrow B$ *is exact, and* S *is the natural homomorphism of* G *into* $\mathrm{Aut}(AX)$: $gS = T_g \mid AX$, *where* T_g *is the inner automorphism of* G *induced by* g, *then the map* $P: G \longrightarrow \mathrm{Aut}(A)$ *given by the rule* $gP = X(gS)X^{-1}$ *is a homomorphism. Moreover, if* U *is the natural homomorphism of* A *into* $\mathrm{Aut}(A)$, *then the diagram*

is commutative.

9.4.3. *If* $Q: G \longrightarrow G^*$ *is an equivalence of the extension*

$$E \longrightarrow A \xrightarrow{X} G \longrightarrow B \longrightarrow E$$

with the extension $E \longrightarrow A \xrightarrow{X^*} G^* \longrightarrow B \longrightarrow E$, *then the diagram*

is commutative.

Proof. Let $g \in G$, $a \in A$. Then

$$a(gQP^*) = a(X^*(gQS^*)X^{*-1}) = ((gQ)^{-1}(aX^*)(gQ))Q^{-1}QX^{*-1}$$
$$= (g^{-1}(aX)g)QX^{*-1} = (g^{-1}(aX)g)X^{-1}$$
$$= ((aX)(gS))X^{-1} = a(gP).$$

Therefore $gQP^* = gP$ for all $g \in G$. Hence $QP^* = P$. ∥

A *factor system* of A by B is a pair (V, W) of functions $V: B \longrightarrow \text{Aut}(A)$ and $W: B \times B \longrightarrow A$, such that, letting

$$iV = i', \qquad (i, j)W = c_{i,j}, \tag{1}$$

we have

$$i'j' = (ij)'(c_{i,j}U), \tag{2}$$

$$c_{i,jk}c_{j,k} = c_{ij,k}(c_{i,j}k'). \tag{3}$$

Let $I = I_B$. A factor system (V, W) *belongs* to an extension

$$E \longrightarrow A \xrightarrow{X} G \xrightarrow{Y} B \longrightarrow E$$

and a function $T: B \longrightarrow G$ such that $TY = I$, iff $V = TP$ and

$$(iT)(jT) = ((ij)T)(c_{i,j}X) \tag{4}$$

is satisfied for all i and j in B.

9.4.4. *If $E \longrightarrow A \xrightarrow{X} G \xrightarrow{Y} B \longrightarrow E$ is exact and the map $T: B \longrightarrow G$ is such that $TY = I$, then there is a unique factor system belonging to this extension and T.*

Proof. Since the function $T: B \longrightarrow G$ and the homomorphism $P: G \longrightarrow \text{Aut}(A)$ are given, we may define $V = TP$. Then V is a function from B into $\text{Aut}(A)$ as required. Let $i \in B, j \in B$. Then

$$((iT)(jT))Y = (iTY)(jTY) = ij = ((ij)T)Y.$$

Hence

$$((ij)T)^{-1}(iT)(jT) \in \text{Ker}(Y) = AX$$

by the exactness of the sequence. Therefore $\exists \mid c_{i,j} \in A$ such that (4) holds. Define W by (1). Then W is a function from $B \times B$ into A.

For the convenience of the reader, we list the functions so far defined, together with the commutativity conditions.

Iso	X:	$A \longrightarrow G$,	$TY = I_B$,
Hom onto	Y:	$G \longrightarrow B$,	$TP = V$,
Hom	P:	$G \longrightarrow \text{Aut}(A)$,	$XP = U$,
Hom	S:	$G \longrightarrow \text{Aut}(AX)$,	$gP = X(gS)X^-$
Function	T:	$B \longrightarrow G$,	$iV = i'$,
Hom	U:	$A \longrightarrow \text{Aut}(A)$,	$(i, j)W = c_{i,j}$
Function	V:	$B \longrightarrow \text{Aut}(A)$,	
Function	W:	$B \times B \longrightarrow A$.	

Since P is a homomorphism and $V = TP$,

$$\begin{aligned}
i'j' &= (iV)(jV) = (iTP)(jTP) = [(iT)(jT)]P \\
&= [((ij)T)(c_{i,j}X)]P = ((ij)TP)(c_{i,j}XP) \\
&= ((ij)V)(c_{i,j}U) = (ij)'(c_{i,j}U),
\end{aligned}$$

and (2) holds.

By (4),

$$\begin{aligned}
(ijk)T &= (iT)((jk)T)(c_{i,jk}^{-1}X) \\
&= (iT)(jT)(kT)(c_{j,k}^{-1}X)(c_{i,jk}^{-1}X); \\
(ijk)T &= ((ij)T)(kT)(c_{ij,k}^{-1}X) \\
&= (iT)(jT)(c_{i,j}^{-1}X)(kT)(c_{ij,k}^{-1}X) \\
&= (iT)(jT)(kT)((kT)^{-1}(c_{i,j}^{-1}X)(kT))(c_{ij,k}^{-1}X).
\end{aligned}$$

Hence

$$\begin{aligned}
(c_{j,k}^{-1}X)(c_{i,jk}^{-1}X) &= (kT)^{-1}(c_{i,j}^{-1}X)(kT)(c_{ij,k}^{-1}X) \\
&= [(c_{i,j}^{-1}X)(kTS)](c_{ij,k}^{-1}X).
\end{aligned}$$

Now take inverses:

$$(c_{i,jk}X)(c_{j,k}X) = (c_{ij,k}X)[(c_{i,j}X)(kTS)].$$

Applying X^{-1}, one gets

$$\begin{aligned}
c_{i,jk}c_{j,k} &= c_{ij,k}[(c_{i,j}X)(kTS)X^{-1}] = c_{ij,k}(c_{i,j}(kTP)) \\
&= c_{ij,k}(c_{i,j}(kV)) = c_{ij,k}(c_{i,j}k').
\end{aligned}$$

This proves (3) and the theorem.

9.4.5. *Let A and B be groups and (V, W) a factor system of A by B. Let G be the set $B \times A$ with multiplication*

$$(i, m)(j, n) = (ij, c_{i,j}(mj')n), \tag{5}$$

and let $X: A \longrightarrow G$, $Y: G \longrightarrow B$, and $T: B \longrightarrow G$ be the maps

$$nX = (e, c_{e,e}^{-1}n), \tag{6}$$

$$(i, n)Y = i, \tag{7}$$

$$iT = (i, e). \tag{8}$$

Then G is a group, $E \longrightarrow A \xrightarrow{X} G \xrightarrow{Y} B \longrightarrow E$ is an extension, $TY = I$, and the factor system (V, W) belongs to the extension and T.

Proof. G is closed under multiplication. For three elements of G, we have by (5), (3), and (2),

$$[(i, m)(j, n)](k, p) = (ijk, c_{ij,k}(c_{i,j}k')(mj'k')(nk')p)$$
$$= (ijk, c_{i,jk}c_{j,k}[m(jk)'(c_{j,k}U)](nk')p)$$
$$= (ijk, c_{i,jk}(m(jk)')c_{j,k}(nk')p)$$
$$= (i, m)(jk, c_{j,k}(nk')p)$$
$$= (i, m)[(j, n)(k, p)].$$

Therefore the multiplication is associative.

In (2), let $i = j = e$ to get

$$e' = c_{e,e}U. \tag{9}$$

In (3), let $j = k = e$ to get

$$c_{i,e}c_{e,e} = c_{i,e}(c_{i,e}e'),$$

or, using (9),

$$c_{e,e} = c_{e,e}^{-1}c_{i,e}c_{e,e}.$$

Therefore, for all $i \in B$,

$$c_{i,e} = c_{e,e}. \tag{10}$$

In (3), let $i = j = e$ to get

$$c_{e,k}c_{e,k} = c_{e,k}(c_{e,e}k').$$

Hence

$$c_{e,e}k' = c_{e,k}. \tag{11}$$

We now have

$$(i, m)(e, c_{e,e}^{-1}) = (i, c_{i,e}(me')c_{e,e}^{-1})$$
$$= (i, c_{e,e}c_{e,e}^{-1}mc_{e,e}c_{e,e}^{-1}) = (i, m).$$

Therefore $(e, c_{e,e}^{-1})$ is a right identity for G.

Also

$$(i, m)(i^{-1}, (mi^{-1'})^{-1}c_{i,i^{-1}}^{-1}c_{e,e}^{-1})$$
$$= (e, c_{i,i^{-1}}(mi^{-1'})(mi^{-1'})^{-1}c_{i,i^{-1}}^{-1}c_{e,e}^{-1}) = (e, c_{e,e}^{-1}),$$

which is the right identity. Thus G is a group.

By (6),

$$(mX)(nX) = (e, c_{e,e}^{-1}m)(e, c_{e,e}^{-1}n)$$
$$= (e, c_{e,e}((c_{e,e}^{-1}m)e')c_{e,e}^{-1}n)$$
$$= (e, c_{e,e}^{-1}mn) = (mn)X,$$

so that X is a homomorphism. Moreover, if nX is the identity of G, then $n = e$. Hence the sequence $E \longrightarrow A \overset{X}{\longrightarrow} G$ is exact, and $AX = \{(e, n) \mid n \in A\}$.

By (7) and (5), Y is a homomorphism, and it is clear that Y maps G onto B.

$$\mathrm{Ker}(Y) = \{(e, n) \mid n \in A\} = AX,$$

so that the sequence $E \longrightarrow A \overset{X}{\longrightarrow} G \overset{Y}{\longrightarrow} B \longrightarrow E$ is exact.

By (8) and (7), $TY = I$. Let $n \in A$ and $i \in B$. Then

$$n((iT)P) = nX((iT)S)X^{-1} = (e, c_{e,e}^{-1}n)((iT)S)X^{-1}$$
$$= [(i, e)^{-1}(e, c_{e,e}^{-1}n)(i, e)]X^{-1}.$$

Therefore the following statements are equivalent:

$$n(iV) = n((iT)P),$$

$$(i, e)^{-1}(e, c_{e,e}^{-1}n)(i, e) = (e, c_{e,e}^{-1}(ni')),$$

$$(e, c_{e,e}^{-1}n)(i, e) = (i, e)(e, c_{e,e}^{-1}(ni')),$$

$$(i, c_{e,i}[(c_{e,e}^{-1}n)i']) = (i, c_{i,e}c_{e,e}^{-1}(ni')),$$

$$(c_{e,e}i')(c_{e,e}^{-1}i')(ni') = ni',$$

where (10) and (11) have been used. Since the final equation is true, so is the first. Therefore $TP = V$.

Finally, if $i \in B$ and $j \in B$, then by (8), (6), (5), and (10),

$$((ij)T)(c_{i,j}X) = (ij, e)(e, c_{e,e}^{-1}c_{i,j}) = (ij, c_{ij,e}c_{e,e}^{-1}c_{i,j})$$
$$= (ij, c_{i,j}) = (i, e)(j, e) = (iT)(jT).$$

Hence (4) is satisfied and the theorem is true. ‖

Theorems 9.4.4 and 9.4.5 show that there is a correspondence between extensions and factor systems. The nature of this correspondence is obscured somewhat by the presence of T in both theorems. This defect will now be remedied.

The factor system (V, W) of A by B is *equivalent* to the factor system (V^*, W^*) of A by B iff there is a function $R: B \longrightarrow A$ such that if $i \in B$ and $j \in B$, then

$$i^* = i'(iRU), \tag{12}$$

$$c_{i,j}^* = ((ij)R)^{-1}c_{i,j}((iR)j')(jR). \tag{13}$$

The function R is an *equivalence* (of V, W) with (V^*, W^*). Note that (12) states that i^* is the product of i' and an inner automorphism of A.

9.4.6. *If the factor system (V, W) belongs to the extension $E \longrightarrow A \xrightarrow{X}$ $G \xrightarrow{Y} B \longrightarrow E$ and the map $T: B \longrightarrow G$, and the factor system (V^*, W^*) belongs to the extension $E \longrightarrow A \xrightarrow{X^*} G^* \xrightarrow{Y^*} B \longrightarrow E$ and the map T^*: $B \longrightarrow G^*$, then the extensions are equivalent iff the factor systems are equivalent.*

Proof. (i) Assume that the extensions are equivalent. Then by Theorem 9.1.6, there is an isomorphism Q of G onto G^* such that $XQ = X^*$ and $QY^* = Y$. Let $i \in B$. Then

$$iT^*Q^{-1}Y = iT^*Y^* = i = iTY.$$

Hence $\exists \mid n_i \in A$ such that

$$iT^*Q^{-1} = (iT)(n_iX). \tag{14}$$

Define $R: B \longrightarrow A$ by the rule

$$iR = n_i \tag{15}$$

for all $i \in B$.

If $n \in A$ and $i \in B$, then by (15), (14), and Theorem 9.4.3,

$$
\begin{aligned}
(ni')(iRU) &= (iR)^{-1}(ni')(iR) \\
&= n_i^{-1}(ni')n_i = n_i^{-1}(n(iTP))n_i \\
&= n_i^{-1}\{n[((iT^*Q^{-1})(n_iX)^{-1})P]\}n_i \\
&= n_i^{-1}\{n[(iT^*Q^{-1}P)(n_i^{-1}XP)]\}n_i \\
&= n_i^{-1}\{n[(iT^*Q^{-1}P)(n_i^{-1}U)]\}n_i \\
&= n(iT^*Q^{-1}P) = n(i(T^*P^*)) \\
&= n(iV^*) = ni^*.
\end{aligned}
$$

Thus (12) holds.

Next, let $i \in B$ and $j \in B$. Then by (4) and (14)

$$
\begin{aligned}
c^*_{i,j}X^* &= ((ij)T^*)^{-1}(iT^*)(jT^*); \\
c^*_{i,j}X &= ((ij)T^*Q^{-1})^{-1}(iT^*Q^{-1})(jT^*Q^{-1}) \\
&= (n_{ij}X)^{-1}((ij)T)^{-1}(iT)(n_iX)(jT)(n_jX) \\
&= (n_{ij}X)^{-1}((ij)T)^{-1}(iT)(jT)[(jT)^{-1}(n_iX)(jT)](n_jX) \\
&= (n_{ij}^{-1}X)(c_{i,j}X)[(n_iX)(jTS)](n_jX).
\end{aligned}
$$

Therefore

$$
\begin{aligned}
c^*_{i,j} &= n_{ij}^{-1}c_{i,j}[(n_iX)(jTS)X^{-1}]n_j \\
&= n_{ij}^{-1}c_{i,j}(n_i(jTP))n_j \\
&= n_{ij}^{-1}c_{i,j}(n_ij')n_j \\
&= ((ij)R)^{-1}c_{i,j}((iR)j')(jR).
\end{aligned}
$$

Hence (13) also holds. Therefore the factor systems are equivalent.

(ii) Assume that the factor systems are equivalent. Let R be the equivalence and $g \in G$. Now $\exists \mid i \in B$ and $n \in A$ such that $g = (iT)(nX)$. Define the function $Q: G \longrightarrow G^*$ by the rule:

$$((iT)(nX))Q = (iT^*)[((iR)^{-1}n)X^*]. \tag{16}$$

Now by (4), the definition of P, (13), (12), and (16),

$$(iT)(mX)(jT)(nX) = (iT)(jT)[(mX)((jT)S)](nX)$$
$$= ((ij)T)\{[c_{i,j}(m(jTP))n]X\};$$

$$[(iT)(mX)(jT)(nX)]Q = ((ij)T^*)\{[((ij)R)^{-1}c_{i,j}(mj')n]X^*\}$$
$$= ((ij)T^*)\{[c_{i,j}^*(jR)^{-1}((iR)^{-1}j')(mj')n]X^*\}$$
$$= ((ij)T^*)\{[c_{i,j}^*(((iR)^{-1}m)j^*)(jR)^{-1}n]X^*\}$$
$$= ((ij)T^*)\{[c_{i,j}^*(((iR)^{-1}m)(jT^*P^*))(jR)^{-1}n]X^*\}$$
$$= (iT^*)(jT^*)\{[((iR)^{-1}m)(jT^*P^*)]X^*\}$$
$$\hspace{3cm} \cdot \{[(jR)^{-1}n]X^*\}$$
$$= (iT^*)\{[(iR)^{-1}m]X^*\}(jT^*)\{[(jR)^{-1}n]X^*\}$$
$$= ([(iT)(mX)]Q)([(jT)(nX)]Q).$$

Hence Q is a homomorphism.

Again, $[(iT)(nX)]Y = ((iT)Y)(nXY) = i$. Also, by (16),

$$((iT)(nX))QY^* = \{(iT^*)[((iR)^{-1}n)X^*]\}Y^*$$
$$= iT^*Y^* = i.$$

Hence $QY^* = Y$.

Finally, by (4) with $i = j = e$, (16), (13), and (9),

$$(nX)Q = [(eT)(eT)^{-1}(nX)]Q$$
$$= \{(eT)[(c_{e,e}^{-1}X)(nX)]\}Q$$
$$= (eT^*)\{[(eR)^{-1}c_{e,e}^{-1}n]X^*\}$$
$$= [c_{e,e}^*(eR)^{-1}c_{e,e}^{-1}n]X^*$$
$$= [(eR)^{-1}c_{e,e}((eR)e')(eR)(eR)^{-1}c_{e,e}^{-1}n]X^*$$
$$= [(eR)^{-1}(eR)c_{e,e}(eR)(eR)^{-1}c_{e,e}^{-1}n]X^* = nX^*.$$

Hence $XQ = X^*$. Therefore Q is an equivalence of the two extensions.

9.4.7. *If $E \longrightarrow A \longrightarrow G \overset{Y}{\longrightarrow} B \longrightarrow E$ is exact, $T: B \longrightarrow G$ and $T^*: B \longrightarrow G$ are such that $TY = T^*Y = I$, the factor system (V, W) belongs to the extension and T, and the factor system (V^*, W^*) belongs to the extension and T^*, then (V, W) is equivalent to (V^*, W^*).*

Proof. Since $E \longrightarrow A \longrightarrow G \longrightarrow B \longrightarrow E$ is equivalent to itself, this follows from Theorem 9.4.6.

9.4.8. *Equivalence of factor systems is an equivalence relation.*

Proof. Let A and B be groups. By Theorem 9.4.5, there is a function f from the set of factor systems (V, W) of A by B into the class of extensions of A by B such that (V, W) belongs to $(f(V, W), T)$ for some T. By Theorem 9.4.6, (V, W) is equivalent to (V^*, W^*) iff $f(V, W)$ is equivalent to $f(V^*, W^*)$. Since equivalence of extensions is an equivalence relation (Theorem 9.1.7), it follows easily that equivalence of factor systems is an equivalence relation. ‖

Let (V, W) be a factor system. If (V^*, W^*) is an equivalent factor system, there may be more than one equivalence of (V, W) with (V^*, W^*). It is true, however, that any function $R: B \longrightarrow A$ is an equivalence of (V, W) with some factor system.

9.4.9. *If (V, W) is a factor system of A by B and R is a function from B into A, then there is a unique factor system (V^*, W^*) such that R is an equivalence of (V, W) with (V^*, W^*).*

Proof. If such a factor system (V^*, W^*) exists, then Equations (12) and (13) hold, and determine (V^*, W^*).

Now let V^* and W^* be defined by (12) and (13). It is clear that V^* is a function from B into Aut(A) and W^* is a function from $B \times B$ into A. We have by (12), (2), and (13),

$$i^*j^* = i'(iRU)j'(jRU),$$

$$\begin{aligned}
ni^*j^* &= [(iR)^{-1}(ni')(iR)]j'(jRU)\\
&= (jR)^{-1}[((iR)^{-1})j'](ni'j')((iR)j')(jR)\\
&= (jR)^{-1}[((iR)^{-1})j'][(n(ij)')(c_{i,j}U)]((iR)j')(jR)\\
&= (jR)^{-1}[((iR)^{-1})j'](c_{i,j}^{-1})(n(ij)')c_{i,j}((iR)j')(jR)\\
&= c_{i,j}^{*-1}((ij)R)^{-1}(n(ij)')((ij)R)c_{i,j}^*\\
&= [n(ij)'((ij)RU)](c_{i,j}^*U) = n(ij)^*(c_{i,j}^*U).
\end{aligned}$$

Hence $i^*j^* = (ij)^*(c_{i,j}^*U)$, and (2)* holds.

Again, by (13), (12), (2), and (3),

$$\begin{aligned}
c_{ij,k}^*(c_{i,j}^*k^*) &= ((ijk)R)^{-1}c_{ij,k}[((ij)R)k'](kR)(c_{i,j}^*k'(kRU))\\
&= ((ijk)R)^{-1}c_{ij,k}[((ij)R)k'](c_{i,j}^*k')(kR)\\
&= ((ijk)R)^{-1}c_{ij,k}[((ij)R)k'][((ij)R)^{-1}k'](c_{i,j}k')\\
&\qquad\qquad \cdot [(iR)(j'k')][(jR)k'](kR)\\
&= ((ijk)R)^{-1}c_{ij,k}(c_{i,j}k')c_{j,k}^{-1}((iR)(jk)')c_{j,k}((jR)k')(kR)\\
&= ((ijk)R)^{-1}c_{i,jk}((iR)(jk)')c_{j,k}((jR)k')(kR)\\
&= ((ijk)R)^{-1}c_{i,jk}((iR)(jk)')((jk)R)((jk)R)^{-1}c_{j,k}((jR)k')(kR)\\
&= c_{i,jk}^*c_{j,k}^*.
\end{aligned}$$

Hence $(3)^*$ holds and (V^*, W^*) is a factor system. Also, clearly, R is an equivalence of (V, W) with (V^*, W^*). ‖

Theorem 9.4.9 permits a slight simplification in factor systems as follows.

9.4.10. *If (V, W) is a factor system of A by B, then there is an equivalent factor system (V^*, W^*) such that, for all i and j in B,*

$$c^*_{e,e} = c^*_{i,e} = c^*_{e,j} = e.$$

Proof. Since $e' \in \mathrm{Aut}(A)$, $\exists\, a \in A$ such that $ae' = c^{-1}_{e,e}$. Let R be any function from B into A such that $eR = a$. By Theorem 9.4.9, R is an equivalence of (V, W) with a factor system (V^*, W^*). By (13),

$$c^*_{e,e} = (eR)^{-1} c_{e,e}((eR)e')(eR)$$
$$= a^{-1} c_{e,e} c^{-1}_{e,e} a = e.$$

By (10), $c^*_{i,e} \doteq e$. By (11), $c^*_{e,j} = c^*_{e,e} j^* = e j^* = e$. ‖

A factor system (V, W) such that $c_{e,e} = e$ is called a *normalized* factor system. By (4), if a normalized factor system (V, W) belongs to an extension and T, then $eT = e$. Certain other simplifications occur in the theory, but the discussion will be left to the exercises.

EXERCISES

9.4.11. What changes occur in this section if only normalized factor systems are used?

9.4.12. Discuss the dihedral group of order 8 as an extension of a cyclic group of order 4 by one of order 2, finding a factor set, etc.

9.4.13. Let (V, W) be a factor system.
    (a) V is a homomorphism iff W maps $B \times B$ into $Z(A)$.
    (b) If A is Abelian, then V is a homomorphism.
    (c) If $Z(A) = E$, then V is a homomorphism iff all $c_{i,j} = e$.

9.4.14. Let (V, W) be a factor system of A by B. Let u be the natural homomorphism from $\mathrm{Aut}(A)$ onto $\mathrm{Aut}(A)/\mathrm{Inn}(A)$.
    (a) $Vu \in \mathrm{Hom}(B, \mathrm{Aut}(A)/\mathrm{Inn}(A))$.
    (b) If (V^*, W^*) is equivalent to (V, W), then $V^*u = Vu$.

9.4.15. If (V, W) is a factor system of A by B, then there is an extension $E \longrightarrow$ $A \xrightarrow{I_A} G \longrightarrow B \longrightarrow E$ and a map $T: B \longrightarrow G$ such that (V, W) belongs to the extension and T.

9.5 Split extensions

If certain extra conditions are placed on the extension $E \longrightarrow A \longrightarrow G \longrightarrow B \longrightarrow E$, two things occur. First, there are simplifications in the Schreier theory. Second, there is more structure to the class of extensions. Several special types of extensions will be considered in this and succeeding sections.

An extension $E \longrightarrow A \xrightarrow{X} G \xrightarrow{Y} B \longrightarrow E$ is

(i)   *Abelian* iff G is Abelian (therefore, A and B are Abelian),

(ii)   *central* iff $AX \subset Z(G)$ (hence A is Abelian),

(iii)   *cyclic* iff B is cyclic, and

(iv)   *split* iff there is an isomorphism $T: B \longrightarrow G$ such that $TY = I$.

It is obvious that an extension equivalent to an Abelian (central)(cyclic) extension is also Abelian (central)(cyclic). The corresponding statement for split extensions is also true (Theorem 9.5.2).

In terms of the notation of the preceding section, the following simplifications occur in the theory.

			Extension		*A*	*B*
Map	Abelian	Central	Split	Cyclic	Abelian	Abelian
P	O	O				
S	O	O				
T			Iso			
U	O	O			O	
V	O	O	Hom		Hom	
W	Comm		O	Comm		

Here W is *commutative* iff $c_{i,j} = c_{j,i}$ for all i and j in B, and the entries in the split and cyclic extension columns are correct for some equivalent extension and factor system.

9.5.1. *Let* $E \longrightarrow A \xrightarrow{X} G \xrightarrow{Y} B \longrightarrow E$ *be an extension. Then the following statements are equivalent:*

(i)   *the extension splits,*

(ii) $\exists\ H \subset G$ *such that* $G = (AX)H$, $AX \cap H = E$,

(iii)   *there is an equivalent extension*
$$E \longrightarrow A \xrightarrow{I_A} G^* \xrightarrow{Y^*} B \longrightarrow E \ \text{and}\ K \subset G^* \ \text{such that}\ G^* = AK,$$
$A \cap K = E.$

Proof. (i) $\Rightarrow$ (ii). By assumption, there is an isomorphism $T: B \longrightarrow G$ such that $TY = I$. Let $H = BT$. Then $H \subset G$. If $g \in AX \cap H$, then $gY \in AXY = E$ (exactness at G). Since $g = iT$ for some $i \in B$, $e = gY = iTY = i$. Hence $g = eT = e$. Therefore $AX \cap H = E$.

Let $g \in G$. Then $gY = i \in B$ and $(g(i^{-1}T))Y = (gY)i^{-1} = e$. Hence $g(i^{-1}T) \in AX$, $g \in (AX)(iT)$, and $g \in (AX)H$. Thus $G = (AX)H$.

(ii) $\Rightarrow$ (iii). By Exercise 9.1.10, there is an extension $E \longrightarrow A \overset{I_A}{\longrightarrow} G \overset{Y^\bullet}{\longrightarrow} B \longrightarrow E$ and an equivalence $Q: G \longrightarrow G^*$ of the two extensions. Let $K = HQ$. Then

$$G^* = GQ = ((AX)H)Q = (AXQ)(HQ) = (AI_A)K = AK,$$

$$A \cap K = AI_A \cap K = AXQ \cap HQ = (AX \cap H)Q = EQ = E^*.$$

(iii) $\Rightarrow$ (i). Let $Q: G \longrightarrow G^*$ be an equivalence of the two extensions. Then

$$B = G^*Y^* = (AK)Y^* = ((AI_A)K)Y^* = (AI_A Y^*)(KY^*) = KY^*.$$

Hence $Y^* \mid K$ is a homomorphism of K onto B. If $u \in \mathrm{Ker}(Y^* \mid K)$, then $u \in AI_A = A$, so $u \in A \cap K = E$. Thus $Y^* \mid K$ is an isomorphism of K onto B. Hence $T = (Y^* \mid K)^{-1}Q^{-1}$ is an isomorphism of B into G, and

$$TY = (Y^* \mid K)^{-1}Q^{-1}Y = (Y^* \mid K)^{-1}Y^* = I.$$

9.5.2. *If an extension splits, so does any equivalent extension.*

Proof. Let Q be an equivalence of a split extension $E \longrightarrow A \longrightarrow G \overset{Y}{\longrightarrow} B \longrightarrow E$ and $E \longrightarrow A \longrightarrow G^* \overset{Y^\bullet}{\longrightarrow} B \longrightarrow E$. There is an isomorphism T of B into G such that $TY = I$. Let $T^* = TQ$. Then T^* is an isomorphism of B into G^*, and $T^*Y^* = TQY^* = TY = I$. ‖

Because of Theorems 9.4.6 and 9.5.2, it is legitimate to say that a factor system *splits* iff a corresponding extension splits. Hence by Theorem 9.5.2, a factor system equivalent to one which splits also splits.

9.5.3. *Let (V, W) be a factor system of A by B. Then the following statements are equivalent:*

(i) (V, W) *splits,*

(ii) (V, W) *is equivalent to some factor system (V^*, W^*) such that $V^* \in \mathrm{Hom}(B, \mathrm{Aut}(A))$ and $c^*_{i,j} = e$ for all i and j in B,*

(iii) *there is a function $R: B \longrightarrow A$ such that*
$$c_{i,j} = ((ij)R)(jR)^{-1}((iR)j')^{-1}.$$

Proof. (i) $\Rightarrow$ (ii). (V, W) belongs to some split extension $E \longrightarrow A \overset{x}{\longrightarrow}$ $G \overset{Y}{\longrightarrow} B \longrightarrow E$ and $T: B \longrightarrow G$. Since the extension splits, there is an isomorphism $T^*: B \longrightarrow G$ such that $T^* Y = I$. By Theorem 9.4.4, there is a factor system (V^*, W^*) belonging to the extension and T^*. By Theorem 9.4.7, (V, W) is equivalent to (V^*, W^*). Since T^* is an isomorphism, $(ij)T^* = (iT^*)(jT^*)$. By (4) of Section 9.4, $c_{i,j}^* = e$. By (2) of Section 9.4, $i^*j^* = (ij)^*$, and V^* is a homomorphism.

(ii) $\Rightarrow$ (iii). Let $R: B \longrightarrow A$ be an equivalence of (V, W) with (V^*, W^*). Then, for all i and j in B, by (13) of Section 9.4,

$$e = c_{i,j}^* = ((ij)R)^{-1}c_{i,j}((iR)j')(jR),$$

and (iii) follows.

(iii) $\Rightarrow$ (i). By Theorem 9.4.9, R is an equivalence of (V, W) with some factor system (V^*, W^*). By (13) of Section 9.4, $c_{i,j}^* = e$. Let (V^*, W^*) belong to

$$E \longrightarrow A \longrightarrow G^* \overset{Y^*}{\longrightarrow} B \longrightarrow E \quad \text{and} \quad T^*: B \longrightarrow G^*.$$

Then by (4) of Section 9.4, $(iT^*)(jT^*) = (ij)T^*$. Since T^* is 1–1 (for $T^* Y^* = I$), T^* is an isomorphism and the extension splits. Hence (V^*, W^*) splits, and therefore the equivalent factor system (V, W) splits. $\|$

We have met split extensions before in Section 9.2. In fact,

9.5.4. *Every split extension of A by B is equivalent to a semi-direct product (possibly to several).*

Proof. By Theorem 9.5.3, if K is a split extension, then there is an equivalent extension K^*, map T^*, and factor system (V^*, W^*) belonging thereto, such that $V^* \in \mathrm{Hom}(B, \mathrm{Aut}(A))$ and $c_{i,j}^* = e$ for all i and j. By Theorem 9.4.5, there is an extension K', and map T', such that (V^*, W^*) belongs to (K', T') and K' is the semi-direct product of A by B with homomorphism V^* (see Theorem 9.2.1). By Theorem 9.4.6, K and K' are equivalent. Hence every split extension of A by B is equivalent to at least one semi-direct product of A by B. $\|$

Example. Let $A = B = \mathrm{Sym}(3)$, and let

$$E \longrightarrow A \overset{X}{\longrightarrow} A \times A \overset{Y}{\longrightarrow} A \longrightarrow E$$

be the direct product extension. This is a semi-direct product in an obvious way, with the associated homomorphism O. There are, however, other complements to the subgroup $\{(a, e)\}$ than the subgroup $\{(e, a)\}$, for example, the diagonal $\{(a, a) \mid a \in A\}$. The corresponding functions are as follows: $aT = (a, a)$, $aV = T_a$ (the inner automorphism of A induced by a), $c_{i,j} = e$.

One then obtains a semi-direct product equivalent to the direct product extension, but with associated homomorphism the natural isomorphism of A onto Inn(A).

9.5.5. *If B is a free group, then any extension* $E \longrightarrow A \longrightarrow G \longrightarrow B \longrightarrow E$ *is a split extension.*

Proof. This is a rewording of Exercise 8.1.30.

EXERCISES

9.5.6. If (V, W) is a factor system for an extension $E \to A \to G \to B \to E$, where $o(A) = 2$, G is the quaternion group, and B is a 4-group, then W is not commutative. (Thus the table at the beginning of this section cannot be improved by putting Comm at the intersection of row W and column "B Abelian.")

9.5.7. Let $A \lhd G$. Zassenhaus [4] defined a *splitting group* of G over A to be a group $H \supset G$ which contains a normal subgroup B and a subgroup F such that

$$H = GB,\ G \cap B = A,\ H = BF,\ B \cap F = E.$$

Let T be a homomorphism of G onto some group F with Ker(T) $= A$. Let $H = G \times F$, $G_1 = \{(g,f)\,|\,gT = f\}$, $A_1 = A \times E$, $B = G \times E$, and $F_1 = E \times F$.

(a)   There is an isomorphism of G onto G_1 sending A onto A_1.

(b) $B \lhd H$, $H = G_1 B$, $G_1 \cap B = A_1$, $H = BF_1$, $B \cap F_1 = E$.

(c) H is a splitting group of G_1 over A_1.

(d)   There is a splitting group H' of G over A which is isomorphic to H.

(e)   If G is finite, so is the splitting group H' defined in (a)–(d).

9.6   Extensions of Abelian groups

Let A be an Abelian group, B a group, and (V, W) a factor system of A by B. As was noted in Section 9.5 and Exercise 9.4.13, V is then a homomorphism from B into Aut(A). Moreover, if (V^*, W^*) is equivalent to (V, W), then $V^* = V$ by (12) of Section 9.4. It turns out that the set of equivalence classes of factor systems with fixed V forms a group Ext($B, A; V$) in a natural way.

9.6.1. *If A is an Abelian group, B a group, and $V \in$ Hom(B, Aut(A)), then*

(i)   *the set* Fact($B, A; V$) *of functions* $W: B \times B \longrightarrow A$ *such that* (V, W) *is a factor system is an Abelian group (under the usual addition of functions),*

(ii)   *the set* Trans$(B, A; V)$ *of* W *such that* (V, W) *is a split factor system is a subgroup of* Fact$(B, A; V)$,

(iii) *Two factor systems* (V, W) *and* (V, W^*) *are equivalent iff* W *and* W^* *are in the same coset of* Trans$(B, A; V)$,

(iv)   *the group* Ext$(B, A; V) =$ Fact$(B, A; V)/$Trans$(B, A; V)$ *is in* 1–1 *correspondence with the set of equivalence classes of extensions to which the homomorphism* V *belongs.*

Proof. A semi-direct product of A by B with homomorphism V (see Theorem 9.2.1) has factor system (V, W_0), where $(i, j)W_0 = e$ for all i and j. Hence Trans$(B, A; V)$ and Fact$(B, A; V)$ are not empty.

(i) Let W and W^* be in Fact$(B, A; V)$. Then (Section 9.4 (3))

$$c_{i, jk}c_{j, k} = c_{ij, k}(c_{i, j}k'), \quad c^*_{i, jk}c^*_{j, k} = c^*_{ij, k}(c^*_{i, j}k').$$

Since A is Abelian and $k' \in$ Aut(A), it follows that

$$(c_{i, jk}c^{*-1}_{i, jk})(c_{j, k}c^{*-1}_{j, k}) = (c_{ij, k}c^{*-1}_{ij, k})((c_{i, j}c^{*-1}_{i, j})k').$$

Hence $W - W^* \in$ Fact$(B, A; V)$. Therefore Fact$(B, A; V)$ is a group. Since A is Abelian, Fact$(B, A; V)$ is also Abelian.

(ii) Let W and W^* be in Trans$(B, A; V)$. By Theorem 9.5.3, $\exists\ R: B \longrightarrow A$ and $R^*: B \longrightarrow A$ such that

$$c_{i, j} = ((ij)R)(jR)^{-1}((iR)j')^{-1},$$

$$c^*_{i, j} = ((ij)R^*)(jR^*)^{-1}((iR^*)j')^{-1}.$$

Let $R' = R - R^*$. Then

$$c_{i, j}c^{*-1}_{i, j} = ((ij)R)((ij)R^{*-1})(jR)^{-1}(jR^*)\{[(iR)^{-1}(iR^*)]j'\}$$
$$= ((ij)R')(jR')^{-1}((iR')^{-1}j') = ((ij)R')(jR')^{-1}((iR')j')^{-1}.$$

Hence Trans$(B, A; V)$ is a subgroup of Fact$(B, A; V)$.

(iii) (V, W) is equivalent to (V, W^*) iff $\exists\ R: B \longrightarrow A$ such that

$$c^*_{i, j} = c_{i, j}((ij)R)^{-1}((iR)j')(jR).$$

By Theorem 9.5.3, this is true iff $W^* = W - W'$, where

$(V, W') \in$ Trans$(B, A; V)$.

(iv) This follows immediately from (iii) and Theorem 9.4.6. ‖

Thus, in case A is Abelian, the equivalence classes of extensions with fixed V may be considered as a group. Since extensions equivalent to central extensions are central, we may refer to equivalence classes of central extensions and to central factor systems.

9.6.2. *If A is an Abelian group, B a group, and (V, W) a factor system of A by B, then (V, W) is central iff $V = O$.*

Proof. By Theorem 9.4.5, (V, W) belongs to an extension $E \longrightarrow A \longrightarrow G \longrightarrow B \longrightarrow E$, where G has the multiplication (5). The extension is central iff for all i, m, and n,

$$(i, m)(e, n) = (e, n)(i, m).$$

Using (5), (9), (10), and (11), this condition is equivalent to

$$c_{i,e}(me')n = c_{e,i}(ni')m,$$

$$c_{e,e}n = (c_{e,e}n)i'$$

for all $n \in A$ and $i \in B$. This is true iff $i' = I_A$ for all $i \in B$, i.e., iff $V = O$.

9.6.3. *The classes of central extensions of an Abelian group A by a group B form a group* Cext(B, A).

Proof. By Theorems 9.6.1 and 9.6.2, they form a group Ext$(B, A; O)$. In the theorem, the notation Cext(B, A) has been introduced for this group. ‖

By Example 8 of Section 9.2, there is a non-Abelian group G of order p^3. Then $A = Z(G)$ is Abelian of order p, and $B = G/A$ is Abelian of order p^2. Thus G is a central extension of A by B which is non-Abelian.

A factor system is *commutative* iff $c_{i,j} = c_{j,i}$ for all i and j.

9.6.4. *Let A and B be Abelian groups and (V, W) a central factor system of A by B. Then (V, W) belongs to an Abelian extension iff it is commutative.*

Proof. Let (V, W) be a factor system of an Abelian extension $E \longrightarrow A \overset{X}{\longrightarrow} G \overset{Y}{\longrightarrow} B \longrightarrow E$. By Theorem 9.6.2, $V = O$. Now (V, W) belongs to the extension and T for some T. By (4) of Section 9.4 and the fact that G and B are Abelian,

$$c_{i,j}X = (iT)(jT)((ij)T)^{-1}$$

$$= (jT)(iT)((ji)T)^{-1} = c_{j,i}X.$$

Since X is 1–1, $c_{i,j} = c_{j,i}$, so that (V, W) is commutative.

Let (V, W) be a commutative central factor system. Then $V = O$. By Theorem 9.4.5, the extension is equivalent to one in which

$$(i, m)(j, n) = (ij, c_{i,j}mn) = (ji, c_{j,i}nm) = (j, n)(i, m).$$

Hence the extension is Abelian.

9.6.5. *If A and B are Abelian groups, then the classes of Abelian extensions form a subgroup* Ext(B, A) *of* Cext(B, A).

Proof. By Theorem 9.5.3 and the fact that A and B are Abelian, if $W \in \text{Trans}(B, A; O)$, then $\exists R: B \longrightarrow A$ such that

$$c_{i,j} = ((ij)R)(jR)^{-1}(iR)^{-1}$$
$$= ((ji)R)(iR)^{-1}(jR)^{-1} = c_{j,i}.$$

By Theorem 9.6.4, such a factor system (O, W) belongs to an Abelian extension. Hence Trans$(B, A; O)$ is a subgroup of the set of Abelian extensions. If W and W^* are commutative factor systems (with $V = O$) of A by B, then

$$(i, j)(W - W^*) = c_{i,j}c_{i,j}^{*-1} = c_{j,i}c_{j,i}^{*-1}$$
$$= (j, i)(W - W^*).$$

Hence the set of Abelian extensions forms a group H, say. The group Ext$(B, A) = H/\text{Trans}(B, A; O)$ is then a subgroup of Cext(B, A).

9.6.6. *If A is Abelian, B a group, and* $V \in \text{Hom}(B, \text{Aut}(A))$, *then*

(i) *if* Exp(A) *is finite, then* Exp(Fact$(B, A; V))\,|\,$Exp(A), *hence* Exp(Ext$(B, A, V))\,|\,$Exp(A),

(ii) *if A is a P-group for some set of P primes, then any* $W \in \text{Fact}(B, A; V)$ *of finite order is a P-element,*

(iii) *if B is finite, then* Exp(Ext$(B, A; V))\,|\,o(B)$.

Proof. Let $W \in \text{Fact}(B, A; V)$. Then $(i, j)(nW) = c_{i,j}^n$ for $n \in \mathcal{N}$.

(i) Let $n = \text{Exp}(A)$. Then all $c_{i,j}^n = e$, so $nW = O$. Hence $o(W)\,|\,n$ for all W, and (i) follows.

(ii) Let $o(W) = n$. Clearly n is the least common multiple of all $o(c_{i,j})$. Hence n is a P-number since it is finite.

(iii) Let $n = o(B)$. Define the function $R: B \longrightarrow A$ by the rule

$$jR = \pi\{c_{i,j}\,|\,i \in B\}^{-1}.$$

Multiplying the equation $c_{i,jk}c_{j,k} = c_{ij,k}(c_{i,j}k')$ over all $i \in B$, we obtain

$$((jk)R)^{-1}c_{j,k}^n = (kR)^{-1}((jR)k')^{-1}.$$

By Theorem 9.5.3, $nW \in \text{Trans}(B, A; V)$. This proves (iii). ∥

In particular, this theorem says that if A or B is a finite p-group for some prime p, then Ext$(B, A; V)$ is a p-group, finite if both A and B are finite. Moreover,

9.6.7. *If A is a finite Abelian group, B a finite group, $(o(A), o(B)) = 1$, and $V \in \mathrm{Hom}(B, \mathrm{Aut}(A))$, then any extension of A by B with corresponding homomorphism V is equivalent to the semi-direct product of A by B with homomorphism V.*

Proof. By Theorem 9.6.6, $\mathrm{Ext}(B, A; V)$ has order 1. By Theorem 9.6.1 (iv), there is just one equivalence class of extensions with homomorphism V. By Theorem 9.5.4, this class is just the class of a semi-direct product of A by B with homomorphism V (see Theorem 9.2.1). ∥

In case $V = O$, let us write $\mathrm{Fact}(B, A)$ instead of $\mathrm{Fact}(B, A; V)$, etc., and denote the Abelian group of commutative factor systems by $\mathrm{Comm}(B, A)$. A partial reduction of the problem of determining $\mathrm{Cext}(B, A)$ is accomplished by the following theorem.

9.6.8. *Let B be a group, $\pi \{H \mid H \in S\}$ an Abelian group, and p_H the canonical projection of $\pi \{K \mid K \in S\}$ onto H $(H \in S)$. Then the function f defined by the rule: $H(Wf) = Wp_H$ is such that*

(i) *f is an isomorphism of $\mathrm{Fact}(B, \pi H)$ onto $\pi \mathrm{Fact}(B, H)$,*

(ii) *f is an isomorphism of $\mathrm{Trans}(B, \pi H)$ onto $\pi \mathrm{Trans}(B, H)$,*

(iii) *f induces an isomorphism of $\mathrm{Cext}(B, \pi H)$ onto $\pi \mathrm{Cext}(B, H)$.*

If, in addition, B is Abelian, then

(iv) *f is an isomorphism of $\mathrm{Comm}(B, \pi H)$ onto $\pi \mathrm{Comm}(B, H)$,*

(v) *f induces an isomorphism of $\mathrm{Ext}(B, \pi H)$ onto $\pi \mathrm{Ext}(B, H)$.*

Proof. Let $W \in \mathrm{Fact}(B, \pi H)$. Since p_H is a homomorphism, it follows that $Wp_H \in \mathrm{Fact}(B, H)$ [see Section 9.4 (3)]. Hence f is a function from $\mathrm{Fact}(B, \pi H)$ into $\pi \mathrm{Fact}(B, H)$.

(i) Let $i \in B$, $j \in B$, $H \in S$, and let W and W' be in $\mathrm{Fact}(B, \pi H)$. Then

$$(i, j)(H(W + W')f)) = (i, j)((W + W')p_H)$$
$$= [((i, j)W) + ((i, j)W')]p_H$$
$$= [(i, j)Wp_H][(i, j)W'p_H]$$
$$= (i, j)[(Wp_H) + (W'p_H)]$$
$$= (i, j)[(H(Wf)) + (H(W'f))]$$
$$= (i, j)\{H[(Wf)) + (W'f)]\}.$$

Hence $H((W + W')f) = H(Wf + W'f)$ for all $H \in S$. Therefore,

$$(W + W')f = (Wf) + (W'f),$$

and f is a homomorphism.

If $Wf = O$, then $Wp_H = O$ for all H. It follows that $W = O$. Therefore f is an isomorphism.

Let $U \in \pi$ Fact(B, H). For all H, $HU \in$ Fact(B, H). Define W by the rule

$$H((i,j)W) = (i,j)(HU), \quad H \in S, i \in B, j \in B. \tag{1}$$

Then $(i,j)W \in \pi H$. Hence W is a function from $B \times B$ into πH. Moreover,

$$[H((i, jk)W)][H((j, k)W)] = [(i, jk)(HU)][(j, k)(HU)]$$
$$= [(ij, k)(HU)][(i,j)(HU)]$$
$$= [H((ij, k)W)][H((i,j)W)].$$

Therefore $(i, jk)W + (j, k)W = (ij, k)W + (i,j)W$, and W is a factor system. Also,

$$(i,j)(H(Wf)) = (i,j)(Wp_H) = ((i,j)W)p_H = H((i,j)W)$$
$$= (i,j)(HU).$$

Hence $H(Wf) = HU$ for all $H \in S$, so that $Wf = U$. Thus (i) is established.

(ii) Let $W \in$ Trans$(B, \pi H)$. Then $\exists R: B \to \pi H$ such that

$$(i,j)W = ((ij)R) - jR - iR.$$

Thus

$$(i,j)Wp_H = ((ij)Rp_H)(jRp_H)^{-1}(iRp_H)^{-1}.$$

Therefore $Wp_H \in$ Trans(B, H), and, since $H(Wf) = Wp_H$,

$Wf \in \pi$ Trans(B, H).

Let $U \in \pi$ Trans(B, H), and let W be defined by (1). By the proof of (i), $Wf = U$. We must show that $W \in$ Trans$(B, \pi H)$. For all $H \in S$, $HU \in$ Trans(B, H). Hence $\exists R_H: B \longrightarrow H$ such that

$$(i,j)(HU) = ((ij)R_H)(jR_H)^{-1}(iR_H)^{-1}.$$

Let $R: B \longrightarrow \pi H$ be defined by the equation $H(iR) = iR_H$. Then

$$H((i,j)W) = (i,j)(HU)$$
$$= ((ij)R_H)(jR_H)^{-1}(iR_H)^{-1}$$
$$= [H((ij)R)][H(jR)]^{-1}[H(iR)]^{-1}$$
$$= H[((ij)R) - jR - iR].$$

Hence $(i,j)W = (ij)R - jR - iR$, and $W \in$ Trans$(B, \pi H)$.

(iii) This part follows from the definition of Cext, (i), (ii), and elementary properties of the direct product (Exercise 2.2.9).

(iv) Let $W \in$ Comm$(B, \pi H)$. Then, for all $i \in B$ and $j \in B$,

$$(i,j)W = (j,i)W, (i,j)Wp_H = (j,i)Wp_H.$$

Since $H(Wf) = Wp_H$, it follows from (i) that Wp_H is a factor system, and therefore that $Wp_H \in \text{Comm}(B, H)$. Hence $Wf \in \pi \, \text{Comm}(B, H)$.

Let $U \in \pi \, \text{Comm}(B, H)$ and define W by (1). Then we already know that $Wf = U$, and that $W \in \text{Fact}(B, \pi \, H)$. But $HU \in \text{Comm}(B, H)$ for each $H \in S$, hence for all H, i, and j,

$$H((i,j)W) = (i,j)(HU) = (j,i)(HU) = H((j,i)W).$$

Therefore $(i,j)W = (j,i)W$ for all i and j, so that $W \in \text{Comm}(B, \pi \, H)$.

(v) This follows from (ii) and (iv). ∥

This theorem reduces the problem of determining $\text{Cext}(B, A)$ and $\text{Ext}(B, A)$ (for Abelian B) for finitely generated Abelian groups A to the case where A is cyclic of prime power or infinite order.

9.6.9. *If A is an Abelian group and $V \in \text{Hom}(J, \text{Aut}(A))$, then*

$$\text{Ext}(J, A; V) = O, \quad \text{Cext}(J, A) = O, \quad \text{and} \quad \text{Ext}(J, A) = O.$$

Proof. This follows from 9.5.5, 9.6.1, and 9.6.5.

9.6.10. *If B is a free Abelian group, and A is Abelian, then any Abelian extension of A by B is equivalent to the direct product extension. Hence $\text{Ext}(B, A) = O$.*

Proof. Let G be an Abelian group containing A, and let Y be a homomorphism of G onto B with kernel A. Let D be a basis of B, and let T^* be a 1-1 function of D into G such that $T^* Y = I_D$. By Theorem 5.3.1, T^* has an extension T which is a homomorphism of B into G. Let $\pi \, x_i^{n_i} \in B$, $x_i \in D$, $x_i \neq x_j$ if $i \neq j$. Then

$$(\pi \, x_i^{n_i})TY = \pi \, (x_i TY)^{n_i}$$

$$= \pi \, (x_i T^* Y)^{n_i} = \pi \, x_i^{n_i}.$$

Hence $TY = I$, and T is an isomorphism. It follows that G splits over A, i.e., $G = A(BT)$, $A \cap BT = E$. Since G is Abelian, $G = A + BT$, so that G is equivalent to the direct product extension. Since there is only one class of Abelian extensions, $\text{Ext}(B, A) = O$.

EXERCISE

9.6.11. It is false that if B is free Abelian and G is a central extension of A by B, then G is Abelian. (Let G be the set of 3 by 3 matrices over the ring J of the form $I + X$, where X is strictly upper triangular. Then G is nilpotent of class 2, and is a central extension of a free Abelian group of rank 1 by a free Abelian group of rank 2.)

9.7   Cyclic extensions

Consider the problem of determining all extensions of A by B in case B is cyclic. If B is infinite, then any such extension is splitting (Theorem 9.5.5), hence is equivalent to a semi-direct product (Theorem 9.5.4). The semi-direct products were determined in Section 9.2.

The finite case is treated in the next theorem.

9.7.1.   *Let A be a group and B a cyclic group of order n.*

(i) *If $A \lhd G$ and $G/A \cong B$, then $G = \langle A, x \rangle$, $x^n = y \in A$, $yT_x = y$, and $(T_x)^n \mid A = T_y \mid A$.*

(ii) *If $t \in \mathrm{Aut}(A)$, $y \in A$, $yt = y$, and $t^n = T_y$, then there are G and x such that $A \lhd G$, $G = \langle A, x \rangle$, $G/A \cong B$, $x^n = y$, and $T_x \mid A = t$.*

Proof. (i) Since G/A is cyclic, $\exists\, x \in G$ such that $G/A = \langle Ax \rangle$. Then $G = \langle A, x \rangle$ and $x^n \in A$, say $x^n = y$. Thus $x \in C(y)$, so $yT_x = y$. Also $(T_x)^n \mid A = T_{x^n} \mid A = T_y \mid A$.

(ii) Let $B = \langle b \rangle$, and define $V: B \longrightarrow \mathrm{Aut}(A)$ and $W: B \times B \longrightarrow A$ by the rules

$$b^i V = t^i, \quad 0 \leqq i < n,$$

$$(b^i, b^j)W = \begin{cases} e & \text{if } i + j < n, \\ y & \text{if } i + j \geqq n. \end{cases} \quad 0 \leqq i < n,\, 0 \leqq j < n,$$

Then, using the notation of Section 9.4.

$$(b^i V)(b^j V) = t^i t^j$$

$$= \begin{cases} ((b^i b^j)V)I = [(b^i b^j)V][(b^i, b^j)WU] & \text{if } i + j < n, \\ ((b^i b^j)V)T_y = [(b^i b^j)V][(b^i, b^j)WU] & \text{if } i + j \geqq n. \end{cases}$$

so that (2) of Section 9.4 holds. To verify (3) of Section 9.4, note first that $et^i = e$ and $yt^i = y$, hence the automorphism k' in (3) may be omitted. Of the factors

$$(b^i, b^{j+k})W, (b^j, b^k)W, (b^{i+j}, b^k)W, (b^i, b^j)W,$$

all are e if $i + j + k < n$, all are y if $i + j + k \geqq 2n$, while if

$$n \leqq i + j + k < 2n,$$

then one of the first two factors is e and the other y, and one of the last two factors is e and the other y. Hence, in any case, (3) of Section 9.4 is satisfied, and (V, W) is a factor system.

By Theorem 9.4.5, there is an extension $E \longrightarrow A \xrightarrow{X} H \xrightarrow{Y} B \longrightarrow E$ with multiplication in H given by (for $m \in A$ and $n \in A$)

$$(b^i, m)(b^j, n) = (b^{i+j}, ((b^i, b^j)W)(mt^j)n),$$

and with $nX = (e, n), (b^i, n)Y = b^i$. Now

$$(b^i, e)(b^j, e) = (b^{i+j}, (b^i, b^j)W),$$

so that

$$(b, e)^{n-1} = (b^{n-1}, e), (b, e)^n = (e, y) = yX.$$

By Exercise 2.1.36, there are a group G containing A and an isomorphism k of G onto H such that $k \mid A = X$. It follows that $A \lhd G$, and if x is such that $xk = (b, e)$, then $x^n = y$ and $G = \langle A, x \rangle$. Moreover, in H,

$$(b, e)^{-1}(e, n)(b, e) = (b^{n-1}, y^{-1})(e, n)(b, e)$$
$$= (b^{n-1}, y^{-1}n)(b, e) = (e, y((y^{-1}n)t)e)$$
$$= (e, yy^{-1}(nt)) = (e, nt).$$

Hence

$$(x^{-1}nx)k = (xk)^{-1}(nk)(xk) = (b, e)^{-1}(e, n)(b, e)$$
$$= (e, nt),$$
$$x^{-1}nx = nt,$$

and $T_x \mid A = t$. Finally, it is clear that $G/A \cong H/AX \cong B$.

9.7.2. *If A is a group, $y \in Z(A)$, and $n \in \mathcal{N}$, then there is a group G such that $A \lhd G$, $G = \langle A, x \rangle$, G/A is cyclic of order n, $x \in C(A)$, and $x^n = y$.*

Proof. In Theorem 9.7.1 (ii), let $t = I_A$. Then $yt = y$ and $t^n = I_A = T_y$. Since $T_x \mid A = t = I_A$, $x \in C(A)$. ∥

Two remarks are worth making at this point. The first is that it is quite simple to compute in the group G of Theorem 9.7.1 (ii). The elements of G have the form $x^i a$ with $0 \leq i < n$, and $a \in A$. Multiplication is given by

$$(x^i a)(x^j a') = x^{i+j}(at^j)a'$$

where $x^n = y$ is used in case $i + j \geq n$.

The second remark is that Theorem 9.7.1 permits the construction of all finite solvable groups, in theory. For if S is finite and solvable, then there is a normal series

$$E \lhd A_1 \lhd A_2 \lhd \ldots \lhd A_n = S$$

whose factors are cyclic. Thus each A_{i+1} is a cyclic extension of A_i, and is therefore given in terms of A_i and $\mathrm{Aut}(A_i)$ by Theorem 9.7.1 (there being, in general, several cyclic extensions of A_i by the same cyclic group). As a practical matter, however, this program cannot be carried out.

Example. Let A be Abelian, $y \in A$, $o(y) = 2$, $n = 2$, and $t = -I$. Then $yt = y^{-1} = y$, and $t^2 = I_A = T_y$. Therefore there is a group $G = \langle A, x \rangle$ with $[G:A] = 2$, $x^2 = y$, and $x^{-1}ax = a^{-1}$ for $a \in A$. The group G is called a *generalized dicyclic* group and will be denoted by $\mathrm{Dic}(A, y)$. If there is exactly one element of order 2 in A, then we will write $G = \mathrm{Dic}(A)$. If A is cyclic (of even order), then $\mathrm{Dic}(A)$ is called a *dicyclic* group. If A is cyclic of order 2^n, then G is a *generalized quaternion* group. If A is cyclic of order 4 (and $o(G) = 8$), then G is a quaternion group.

9.7.3. *If G is a finite p-group with just one subgroup H of order p, then G is cyclic or generalized quaternion.*

Proof. Deny, and let G be a minimal counterexample. A noncyclic Abelian subgroup would have more than one subgroup of order p (Theorem 5.1.13), hence all Abelian subgroups are cyclic. In particular, G is non-Abelian, and $Z(G) = \langle a \rangle$ is cyclic. Now $\exists\, b \in Z_2(G)\backslash Z(G)$ such that $b^p \in Z$. Since $\langle Z, b \rangle$ is Abelian it is cyclic, so $\langle Z, b \rangle = \langle b \rangle$, and $WLOG$, $b^p = a$. Since $b \notin Z$, $\exists\, c$ such that $\langle c \rangle \cap \langle b \rangle = Z$. Hence $WLOG$, $c^p = a^{-1}$. By Theorem 3.4.4, $[b, c]^p = [b, c^p] = e$. If p is odd, then (Theorem 3.4.4)

$$(cb)^p = c^p b^p [b, c]^{p(p-1)/2} = a^{-1}a = e.$$

Since $cb \notin \langle b \rangle \supset Z$, one has two subgroups of order p, a contradiction. Hence $p = 2$. Again by Theorem 3.4.4,

$$(cb)^4 = c^4 b^4 [b, c]^6 = a^{-2}a^2 = e.$$

Hence there is an element x of order 4 not in Z. By earlier remarks, $o(Z) = 2$.

There is a normal subgroup $K = \langle k \rangle$ of order 4. Since K has just two automorphisms, by the N/C theorem $[G:C(K)] = 2$. Since $o(Z(C(K))) \geq 4$, $C(K)$ cannot be generalized quaternion (Exercise 9.7.6). Hence by induction, $C(K)$ is cyclic, say $C(K) = \langle d \rangle$, $o(d) = 2^n$. Now $\exists\, y \notin C(K)$ with $o(y) = 4$ (for otherwise $o(Z(G)) \geq 4$). Thus, y induces an automorphism T of order 2 of $C(K)$ which moves k. By Theorem 5.7.12, there are just three automorphisms of $C(K)$ of order 2, call them T_1, T_2, T_3, if $n \geq 3$, and one if $n = 2$. If T_1 and T_2 move k, then

$$kT_3 = kT_1T_2 = k^{-1}T_2 = (k^{-1})^{-1} = k,$$

so that only T_1 and T_2 need be considered. It may be verified that $T_1 = -I$ and T_2 given by $dT_2 = d^{2^{n-1}-1}$ have the stated properties. But if $d^y = dT_2$, then

$$(yd)^2 = y(dy)d = y^2(dT_2)d = y^2 d^{2^{n-1}} = e,$$

since both y^2 and $d^{2^{n-1}}$ equal the unique element of order 2. Hence $T = -I$, and $d^y = d^{-1}$. Therefore G is generalized quaternion. ‖

A *Hamiltonian* group is a non-Abelian group in which every subgroup is normal. Such groups are characterized in the following theorem.

9.7.4. (*Baer* [1].) *A group G is Hamiltonian iff $G = A + B + D$, where A is a quaternion group, B is an elementary Abelian 2-group, and D is a periodic Abelian group with all elements of odd order.*

Proof. First assume that $G = A + B + D$ where A, B, and D are as described. If $g \in G$, then $g = abd$, $a \in A$, $b \in B$, and $d \in D$. If $o(a) = 1$ or 2, then $g \in Z(G)$. Suppose $o(a) = 4$. Then g has two conjugates, itself and $a^{-1}bd$. Since $o(d)$ is odd, $2o(d) + 1 \equiv 3 \pmod 4$, and

$$(abd)^{2o(d)+1} = a^{-1}bd.$$

Hence, in any case, $\langle g \rangle \lhd G$. Since this is true for all $g \in G$, every subgroup of G is normal in G. Since G is not Abelian, G is Hamiltonian.

Now let G be Hamiltonian. Let $x \notin Z(G)$, $y \notin C(x)$. Then (Theorem 3.4.5), $[x, y] \in \langle x \rangle \cap \langle y \rangle$. Hence $[x, y] = x^n \in \langle y \rangle$ for some n. Therefore, by Theorem 3.4.4,

$$x^{n^2} = [x, y]^n = [x^n, y] = e.$$

Thus all elements outside $Z(G)$ are of finite order. If $a \in Z(G)$ and $b \notin Z(G)$, then $ab \notin Z(G)$, so that for some m and n, $b^n = e$, $(ab)^m = e$, and $e = (ab)^{mn} = a^{mn}$, hence $o(a)$ is finite. Therefore, G is periodic. Since all Sylow subgroups are normal, G is their direct sum, $G = \Sigma\, G_p$.

Let p be odd and suppose that G_p is non-Abelian. If x and y are non-commuting elements of G_p, then $\langle x, y \rangle = \langle x \rangle \langle y \rangle$ because all subgroups are normal. Hence *WLOG*, G_p is a Hamiltonian p-group of finite minimum order. If M is a subgroup of order p, then G_p/M is a non-Hamiltonian group all of whose subgroups are normal, hence G_p/M is Abelian. Therefore, $M = G_p^1$. Hence there is only one subgroup of order p. By Theorem 9.7.3, G_p is cyclic, a contradiction.

Therefore, G_p is Abelian for all odd primes p. Hence $G = D + P$, where D is a periodic Abelian group, all of whose elements are of odd order, and P is a Hamiltonian 2-group.

First, suppose that there is a quaternion subgroup A. Since $A = \langle x, y \rangle$ where x and y each have two conjugates, $S = C_P(x) \cap C_P(y)$ is of index 4 in P. If $z \in S$ and $o(z) = 4$, then zx is conjugate to zx^{-1}. But $o(zx) = 4$ or 2, $zx \neq zx^{-1}$ and $(zx)^3 = z^{-1}x^{-1} \neq zx^{-1}$, so that $\langle zx \rangle$ is not normal. Hence all elements of S are of order 2. Therefore, S is elementary Abelian (Exercise 2.4.13), so $S = B + \langle x^2 \rangle$ (Theorems 4.4.7 and 5.1.9). Hence $P = B + A$, and we are done.

Next, suppose that there is no quaternion subgroup. Then P has a finite Hamiltonian subgroup, also without quaternion subgroup. Hence $WLOG$, P is of smallest order of this type. P is not a generalized quaternion group by Exercise 9.7.6. Hence, P contains at least two subgroups, L and M, of order 2. One of these, say M, is not P^1, hence P/M is Hamiltonian. By the minimality of P and the preceding paragraph, P/M contains a quaternion subgroup Q/M. Therefore Q is a Hamiltonian group containing no quaternion subgroup. Hence $P = Q$ and $o(P) = 16$. Since L and M are central in P, $Z(P) = L + M$. Therefore $Z(P)/M$ is the unique subgroup of order 2 in P/M. It follows that $Z(P)$ contains all elements of order 2 in P, and that all elements in $P\backslash Z(P)$ are of order 4. Since P is non-Abelian, there are elements x and y of order 4 which do not commute. Therefore $y^{-1}xy = x^{-1}$. If $y^2 \neq x^2$, then $xyx^{-1} = yx^{-2} \notin \langle y \rangle$, a contradiction. Hence $y^2 = x^2$, and $\langle x, y \rangle$ is a quaternion group, a contradiction.

EXERCISES

9.7.5. If A is Abelian, $y \in A$, $o(y) = 2$, and $G = \mathrm{Dic}(A, y)$, then there is no group H such that $H/Z(H) \cong G$.

9.7.6. Let G be a generalized quaternion group with $o(G) \geqq 8$.

    (a) $o(Z(G)) = 2$.

    (b) If $o(G) \geqq 16$, then G has a nonnormal subgroup of order 4.

REFERENCES FOR CHAPTER 9

For the entire chapter, M. Hall [1], Kurosh [1], and Zassenhaus [1]; Theorem 9.2.7, part (ii), Dickson [1], part (iii), Gol'fand, *Mat. Sb.*, 27 (1950) 229–248, the supersolvable case, Pazderski [1]; Section 9.3, P. Hall [1] and [2]; Exercise 9.3.18, Huppert [4]; Exercise 9.3.19, Suzuki [1]; Exercise 9.3.20, P. Hall [4]; Section 9.4, Schreier [1]; Section 9.6, Fuchs [1] (for further results, see Lyndon [1], Nagao [1]), and Zuravskii [1].

TEN

PERMUTATION GROUPS

10.1 Intransitive groups

Let (M, G) be a permutation group. An *orbit* of (M, G) is a subset T of M such that $\exists\, a \in M$ for which $T = aG$.

10.1.1. *If (M, G) is a permutation group, $a \in M$, and $b \in M$, then*

(i) $b \in aG$ *iff* $bG = aG$,

(ii) *M is the disjoint union of the orbits of (M, G).*

Proof. (i) If $b \in aG$, then $b = ag$ for some $g \in G$, and $bG = agG = aG$. If $bG = aG$, then $\exists\, g \in G$ such that $b = be = ag$, so $b \in aG$.

(ii) If $c \in aG \cap bG$, then by (i), $aG = cG = bG$. Hence unequal orbits are disjoint. Since $a = ae \in aG$, M is the disjoint union of the orbits. ‖

A permutation group (M, G) is *transitive* iff it has only one orbit (namely M). Otherwise (M, G) is *intransitive*. If S is a subset of M, the symbol G_S will mean the set

$$G_S = \{x \in G \mid sx = s \text{ for all } s \in S\}.$$

If $S = \{a\}$, then the symbol G_a will be used.

10.1.2. *If (M, G) is a permutation group, S is a subset of M, $a \in M$, and $x \in G$, then*

(i) $G_S \subset G$ and $G_a \subset G$,

(ii) $G_S x = \{g \in G \mid sg = sx \text{ for all } s \in S\}$,
      $G_a x = \{g \in G \mid ag = ax\}$,

(iii) $G_S^x = G_{Sx}, \quad G_a^x = G_{ax}$.

(iv) $N(G_S) \supset \{g \in G \mid Sg = S\}$.

Proof. (i) Clearly, $e \in G_S$. If $g \in G_S$, $h \in G_S$, and $s \in S$, then $sgh^{-1} = sh^{-1} = s$, hence $gh^{-1} \in G_S$. Thus $G_S \subset G$. It follows that $G_a \subset G$.

(ii) The following statements are equivalent: $sg = sx$ for all $s \in S$, $s(gx^{-1}) = s$ for all $s \in S$, $gx^{-1} \in G_S$, $g \in G_S x$. Statement (ii) follows.

(iii) Let $s \in S$. Then $sx(x^{-1}G_S x) = sG_S x = sx$. Hence $G_S^x \subset G_{sx}$. Since $G_{Sx} = \cap \{G_{sx} \mid s \in S\}$, $G_S^x \subset G_{Sx}$. Therefore $G_S \subset G_{Sx}^{x^{-1}}$. Hence

$$G_{Sx} \subset G_{Sxx^{-1}}^{(x^{-1})^{-1}} = G_S^x.$$

Thus $G_S^x = G_{Sx}$. The second statement is a corollary.

(iv) If $Sg = S$, $s \in S$, and $h \in G_S$, then

$$sg^{-1} = s' \in S, \; s(g^{-1}hg) = s'hg = s'g = s.$$

Hence $g^{-1}hg \in G_S$, so that $g^{-1}G_S g \subset G_S$. Similarly $gG_S g^{-1} \subset G_S$, and $g \in N(G_S)$. The proof that $\{g \in G \mid Sg = S\}$ is a subgroup of G is similar to the proof of (i).

10.1.3. *If (M, G) is a transitive permutation group and $a \in M$, then $Cl(G_a) = \{G_b \mid b \in M\}$.*

Proof. This follows from (iii) in the preceding theorem since ax takes on all values of M by the transitivity of G.

10.1.4. *If (M, G) is a permutation group, T is an orbit of G, and $a \in T$, then*
(i) $o(G) = o(G_a)o(T)$,
(ii)   *If G is transitive, then $o(G) = o(G_a)\text{Deg}(G)$.*

Proof. Statement (i) follows from Theorem 10.1.2 (ii). Statement (ii) follows from (i).

10.1.5. *If a permutation group (M, G) has n orbits, then*

$$\sum \{\text{Ch}(g) \mid g \in G\} = n \cdot o(G).$$

If G is finite, then G is transitive iff $\sum \text{Ch}(g) = o(G)$.

Proof. Let T be an orbit of G and $a \in T$. Then $G_a^x = G_{ax}$ (Theorem 10.1.2), hence $o(G_b) = o(G_a)$ for all $b \in T$. Therefore

$$\sum \{o(G_b) \mid b \in T\} = o(G_a)o(T) = o(G)$$

by Theorem 10.1.4. Let

$$S = \{(a, g) \mid a \in M, ag = a\}.$$

Then

$$\begin{aligned}
\sum \mathrm{Ch}(g) = o(S) &= \sum \{ o(G_a) \mid a \in M\} \\
&= \sum \{\sum \{o(G_a) \mid a \in T\} \mid T \text{ is an orbit}\} \\
&= n \cdot o(G).
\end{aligned}$$

Since G is transitive iff $n = 1$, if G is finite then G is transitive iff $\sum \mathrm{Ch}(g) = o(G)$.

10.1.6. *If (M, G) is a transitive permutation group, $a \in M$, and G_a has n orbits, then $\sum (\mathrm{Ch}(g))^2 = n \cdot o(G)$.*

Proof. Let

$$S = \{(b, c, g) \mid bg = b, cg = c, b \in M, c \in M, g \in G\}.$$

By Theorem 10.1.3, each G_b has n orbits. Then by Theorems 10.1.4 and 10.1.5,

$$\begin{aligned}
\sum (\mathrm{Ch}(g))^2 = o(S) &= \sum \{\sum \{\mathrm{Ch}(g) \mid g \in G_b\} \mid b \in M\} \\
&= \mathrm{Deg}(G)n \cdot o(G_b) = n \cdot o(G). \parallel
\end{aligned}$$

If (M, G) is a permutation group and S is a subset of G, let $\mathrm{Ch}(S)$ denote the number of $a \in M$ such that $aS = a$.

10.1.7. *If (M, G) is transitive and $E < H \lhd G$, then $\mathrm{Ch}(H) = 0$.*

Proof. If the theorem is false, then $H \subset G_a$ for some $a \in M$. By Theorem 10.1.3, $H \subset G_b$ for all $b \in M$. Hence $H = E$ contrary to assumption. $\parallel$

Let S be a nonempty subset of M such that $SG = S$ (hence S is a union of orbits). Then each $g \in G$ effects a permutation $g \mid S$ of S. The function U_S: $gU_S = g \mid S$, is a homomorphism of G onto a group of permutations of S. The permutation group (S, GU_S) is the *S-constituent* of (M, G). In case S is an orbit of (M, G), the S-constituent is transitive and is called a *transitive constituent* of (M, G).

Intransitive groups can be described in terms of their transitive constituents.

10.1.8. *If (M, G) is a permutation group, R the set of orbits, U_T the homomorphism of G onto its T-constituent for $T \in R$, and U the function from G into $\pi \, GU_T$ given by $T(gU) = gU_T$, then U is a representation of G as a subdirect product of its transitive constituents.*

Proof.

$$T((gh)U) = (gh)U_T = (gU_T)(hU_T) = (T(gU))(T(hU))$$
$$= T(gU + hU).$$

Hence, $(gh)U = gU + hU$, and U is a homomorphism. If q_T is the canonical projection of $\pi \, GU_t$ onto GU_T, then $Uq_T = U_T$, so Uq_T maps G onto GU_T. The theorem follows. ‖

The converse of Theorem 10.1.8 is also true.

10.1.9. *Let (T, H_T), $T \in R$, be transitive permutation groups, $M = \overset{\cup}{\smile} \{T \mid T \in R\}$, G a group, and U a representation of G as a subdirect product of the H_T. Let gV be the formal product $\pi\{gUp_T \mid T \in R\}$, where p_T is the canonical projection of πH_t onto H_T. Then V is an isomorphism of G onto a group K of permutations of M. The transitive constituents of (M, K) are the T, H_T).*

Proof. The fact that V is a homomorphism follows from the facts that U and p_T are homomorphisms, and the multiplication rule for formal products. If $g \in \text{Ker}(V)$, then $gUp_T = e$ for all $T \in R$. Hence $gU = e$, and since U is an isomorphism, $g = e$. Clearly, the range K of V is a group of permutations of M. Now $(gV) \mid T = gUp_T$, and since G is a subdirect product of the H_T, $GUp_T = H_T$. Since the H_T are transitive, the transitive constituents of (M, K) are the (T, H_T). ‖

An informal version of this theorem reads as follows.

10.1.10. *Any subdirect product of transitive permutation groups is a permutation group having for transitive constituents the given transitive permutation groups.* ‖

Example 1. Let $M = \{1, 2, 3, 4, 5\}$, and let G consist of the six elements: e, $(1, 2, 3)$, $(1, 3, 2)$, $(1, 2)(4, 5)$, $(1, 3)(4, 5)$, and $(2, 3)(4, 5)$. Then the orbits of G are $T_1 = \{1, 2, 3\}$ and $T_2 = \{4, 5\}$. The homomorphism U_1 of G onto the T_1-constituent is an isomorphism, and the image is Sym(3). The homomorphism U_2 of G onto its T_2-constituent is not an isomorphism, but has a subgroup of order 3 as kernel.

Example 2. Let $M = \{1, 2, 3, 4\}$ and $G = \{e, (1, 2), (3, 4), (1, 2)(3, 4)\}$. There are again two orbits, but neither U_1 nor U_2 is an isomorphism.

EXERCISES

10.1.11. (a) Find all transitive groups on $\{1, 2, 3\}$ (there are two, not isomorphic).

(b) Find all transitive groups on $\{1, 2, 3, 4\}$ [one of order 24, one of order 12, three of order 8 all isomorphic (as permutation groups), three cyclic groups of order 4 all isomorphic, and one noncyclic of order 4].

10.1.12. (a) If (M, G) and (M', G') are isomorphic permutation groups, then there is a 1-1 function f from the set of transitive constituents of (M, G) onto the set of transitive constituents of (M', G') such that corresponding constituents are isomorphic.

(b) The converse is false.

10.1.13. Find all intransitive subgroups of Sym(6) with two orbits $T_1 = \{1, 2, 3\}$ and $T_2 = \{4, 5, 6\}$. (There are 13, but only six non-isomorphic ones).

10.1.14. If $g \in$ Sym(M), then $o(g)$ is the least common multiple of the lengths of the cycles in its cyclic decomposition (with an obvious convention about ∞).

10.1.15. If (M, G) is a permutation group containing an element of order 7, (12), (14), (30), then Deg$(G) \geqq 7$, (7), (9), (10). State a general theorem in this connection.

10.1.16. (a) $(1, 2, \ldots, n) = (1, 2)(1, 3) \cdots (1, n)$.

(b) If M is finite, $o(M) > 1$, and $x \in$ Sym(M), then x is a product of 2-cycles.

10.1.17. (a) $(1, \ldots, 2n) = [(n, n + 1)(n - 1, n + 2) \cdots (1, 2n)]$
$$\cdot [(n, n + 2)(n - 1, n + 3) \cdots (2, 2n)]$$

(b) $(1, \ldots, 2n + 1) = [(n, n + 1)(n - 1, n + 2) \cdots (1, 2n)]$
$$\cdot [(n, n + 2)(n - 1, n + 3) \cdots (1, 2n + 1)].$$

(c) $(\ldots, -1, 0, 1, \ldots) = [(0, 1)(-1, 2) \cdots][(0, 2)(-1, 3) \cdots].$

(d) Any permutation (finite or infinite) is the product of two permutations of order 2.

10.1.18. (a) If (M, G) is a permutation group (perhaps infinite), $a \in M$, and $aG = T$ is an orbit, then $[G:G_a] = o(T)$.

(b) If G is transitive, then $[G:G_a] = $ Deg(G).

10.1.19. If $G \neq E$ is a finite transitive group, then $\exists\, g \in G$ such that Ch$(g) = 0$ (use Theorem 10.1.5).

10.2 Transitive groups and representations

A *permutation representation* of a group G is a homomorphism T of G onto a group H of permutations of some set M. The representation is *faithful* iff T is an isomorphism. The representation is transitive, intransitive, etc., iff (M, H) is transitive, intransitive, etc.

A homomorphism of an intransitive permutation group onto a constituent is an example of a permutation representation. The regular representation (Theorem 3.1.1) is a second example. A generalization of the Cayley procedure is given in the following theorem, which gives another important class of permutation representations.

10.2.1. *If $H \subset G$, M is the set of right cosets of H, and U is the function:*

$$m(gU) = mg, m \in M, g \in G,$$

then U is a transitive representation of G of degree $[G:H]$.

Proof. If $m \in M$, then $m = Hx$ for some $x \in G$, hence $m(gU) = mg = Hxg \in M$. If $m(gU) = m'(gU)$, then $mg = m'g$, so that $m = m'$. Also, $(Hxg^{-1})(gU) = Hx = m$. Hence $gU \in \text{Sym}(M)$. Moreover,

$$m(gU)(g'U) = mg(g'U) = mgg' = m((gg')U),$$

so that $(gU)(g'U) = (gg')U$, and U is a homomorphism. Since $H(gU) = Hg$ is an arbitrary element of M, the image group is transitive. Its degree is $o(M) = [G:H]$. ∥

The above representation will be called the *representation of G on H*. The regular representation is (essentially) the representation of G on E.

10.2.2. *In the representation U of G on H, HU is the subgroup of all elements fixing the letter H in GU, and $\text{Ker}(U) = \text{Core}(H)$.*

Proof. The following statements are equivalent: $g \in H$, $Hg = H$, $H(gU) = H$. This establishes the first statement. If $g \in \text{Ker}(U)$, then $gU = e$, hence $H(gU) = H$, and by the first statement, $g \in H$. Hence $\text{Ker}(U) \subset H$, so $\text{Ker}(U) \subset \text{Core}(H)$. Now let $y \in \text{Core}(H)$ and $x \in G$. Then

$$(Hx)(yU) = Hxy = Hxyx^{-1}x = Hy'x = Hx, \qquad y' \in \text{Core}(H).$$

Hence, yU fixes all letters, so that $y \in \text{Ker}(U)$. Therefore,

$$\text{Ker}(U) = \text{Core}(H). ∥$$

Next, we will see that all transitive groups are obtainable from a representation of a group G on a subgroup H.

10.2.3. *If (M, G) is a transitive permutation group, $a \in M$, U is the representation of G on G_a with image group (S, H), then $(M, G) \cong (S, H)$.*

Proof. By Theorem 10.2.2, $\text{Ker}(U) \subset G_a$, hence $\text{Ch}(\text{Ker}(U)) > 0$. By Theorem 10.1.7, $\text{Ker}(U) = E$, so that U is an isomorphism of G onto H. If $m \in M$, let $mV = \{g \in G \mid ag = m\}$. By Theorem 10.1.2 (ii), V is a 1–1 function from M onto the set S of right cosets of G_a in G.

Now let $m \in M$ and $g \in G$. Then $(mg)V = \{x \in G \mid ax = mg\}$. Again by Theorem 10.1.2 (ii),

$$mV = \{y \in G \mid ay = m\} = G_a z, \qquad az = m.$$

By the definition of U, $(mV)(gU) = G_a zg$. Since $azg = mg$, by Theorem 10.1.2 (ii) again, $(mg)V = (mV)(gU)$. Therefore, $(M, G) \cong (S, H)$. ‖

There is a second way in which a transitive representation of G may be obtained.

10.2.4. *If $H \subset G$ and $K(gU) = K^g, K \in \mathrm{Cl}(H), g \in G$, then U is a transitive representation of G of degree $o(\mathrm{Cl}(H))$.*

Proof. For each g, gU is a function from $\mathrm{Cl}(H)$ into $\mathrm{Cl}(H)$. Since conjugation by g is an automorphism of G, gU is onto $\mathrm{Cl}(H)$. If $K_1^g = K_2^g$, then, conjugating by g^{-1}, we have $K_1 = K_2$. Hence, $gU \in \mathrm{Sym}(\mathrm{Cl}(H))$. Moreover,

$$K(gU)(g'U) = K^g(g'U) = (K^g)^{g'} = K^{gg'} = K((gg')U)$$

so that U is a homomorphism. If $K \in \mathrm{Cl}(H)$, then $\exists\, g \in G$ such that $K = H^g = H(gU)$. Thus GU is transitive. ‖

The representation in Theorem 10.2.4 is essentially the same as the representation of G on $N(H)$ in a sense that will be made precise. Let U and V be permutation representations of a group G with images (S, K) and (T, L), respectively. Then U and V are *similar* iff there is an isomorphism (A, B) of (S, K) onto (T, L) such that $UB = V$.

10.2.5. *If $H \subset G$, U is the permutation representation of G on $N(H)$, and V is the representation of G on $\mathrm{Cl}(H)$ given in Theorem 10.2.4, then U and V are similar.*

Proof. Let S be the set of right cosets of $N(H)$ in G, and $T = \mathrm{Cl}(H)$. Define A by the rule: $(N(H)x)A = H^x$. Then A is 1-1 from S onto T. Let B be the relation $\{(gU, gV) \mid g \in G\}$. The following statements are equivalent

$$gU = g'U,$$
$$(N(H)x)(gU) = (N(H)x)(g'U) \quad \text{for all } x \in G,$$
$$N(H)xg = N(H)xg' \quad \text{for all } x \in G,$$
$$xgg'^{-1}x^{-1} \in N(H) \quad \text{for all } x \in G,$$
$$H^{xgg'^{-1}x^{-1}} = H \quad \text{for all } x \in G,$$
$$H^{xg} = H^{xg'} \quad \text{for all } x \in G,$$
$$(H^x)(gV) = (H^x)(g'V) \quad \text{for all } x \in G,$$
$$gV = g'V.$$

Hence B is 1–1 from GU onto GV. Since U and V are homomorphisms, B is an isomorphism of GU onto GV. Also, $UB = V$ by definition. Finally, it must be shown that (A, B) is an isomorphism of (S, GU) onto (T, GV). We have

$$((N(H)x)(gU))A = (N(H)xg)A = H^{xg} = (H^x)(gV)$$
$$= ((N(H)x)A)((gU)B)$$

which is the required relation (following Theorem 2.1.13). ∥

The same sort of considerations apply to the representation of G on the conjugates of an element, or on the conjugates of a subset.

10.2.6. *If $[G:H]$ is finite and U is the representation of G on H, then $[G:\text{Ker}(U)] \mid [G:H]!$. Hence $[G:\text{Core}(H)] \mid [G:H]!$.*

Proof. The first conclusion follows from Theorem 10.2.1, the second from Theorem 10.2.2. ∥

This theorem could have been used to improve a crude estimate made in the proof of Theorem 7.1.6. A numerical illustration of its use will now be given.

10.2.7. *If $o(G) = 144$, then G is not simple.*

Proof. Deny. Then $n_3 = 4$ or 16. If $n_3 = 16$, then by Burnside (Theorem 6.2.9), there is a normal Sylow 2-subgroup. Hence $n_3 = 4$, so $[G:N(P)] = 4$, where $P \in \text{Syl}_3(G)$. Since G is simple, $\text{Core}(P) = E$. Hence by Theorem 10.2.6, $o(G) \mid 4!$, a contradiction.

10.2.8. *If $o(G) = 432$, then G is not simple.*

Proof. Deny. By Theorem 10.2.6, $n_3 \neq 4$. Hence $n_3 = 16$. By Theorem 6.5.4, since $16 \not\equiv 1(\text{mod } 9)$, there are Sylow 3-subgroups H and K such that $o(H \cap K) = 9$. Then $N(H \cap K)$ contains H and K, hence (Sylow) at least four Sylow 3-subgroups. Since G is simple, $N(H \cap K) \neq G$. Therefore $1 < [G:N(H \cap K)] \leq 4$, contradicting Theorem 10.2.6 again.*

10.2.9. *If G is a finite group and $P \in \text{Syl}_p(G)$, then in the representation of G on $N(P)$, $\text{Ch}(P) = 1$.*

---

* This example has some historical interest. In the 1890's, when surveys of simple groups of small orders were begun, Cole [1] was at first unable to handle the case 432. Later [2] he proved Theorem 10.2.8, though not in the above manner.

Proof. By Theorem 10.2.5, we may look at the representation of G on $\text{Cl}(P)$ instead. If $x \in P$, then $P^x = P$, hence $\text{Ch}(P) \geq 1$. If $P^g \neq P$, then, by Theorem 6.1.9, $P \not\subset N(P^g)$, hence $\exists\ x \in P$ such that $P^{gx} \neq P^g$. Therefore $\text{Ch}(P) = 1$. ∥

A special case is worth noting.

10.2.10. *If G is a finite group, $P \in \text{Syl}_p(G)$, and $o(P) = p$, then in the representation of G on $N(P)$ any element of order p is represented by a product of one l-cycle and some p-cycles.*

10.2.11. *If G is finite, $P \in \text{Syl}_p(G)$, $o(P) = p$, $x \in N(P) \setminus C(P)$, and P has k orbits in the representation of G on $N(P)$, then $\text{Ch}(x) \leq k$ in this representation.*

Proof. Deny. Then there are distinct letters a and b in the same orbit of P such that $ax = a$ and $bx = b$. There is a $y \in P$ such that $ay = b$. Then $a(yxy^{-1}x^{-1}) = a$, and $yxy^{-1}x^{-1} \in P$ since $x \in N(P)$. Since $\text{Ch}(P) = 1$ (Theorem 10.2.9), $\text{Ch}(yxy^{-1}x^{-1}) \geq 2$, hence, by Theorem 10.2.10, $yxy^{-1}x^{-1} = e$. But then $x \in C(y) = C(P)$, a contradiction.

EXERCISES

10.2.12. There are no simple groups of orders 36, 72, 108, 216, 300, 324, 540, 600, 648, 728, 900, or 1176.

10.2.13. There are no simple groups of orders 480, 960, or 1200 (see proof of Theorem 10.2.8).

10.2.14. There is no simple group of order 288 (use techniques of this section plus Theorem 6.5.2).

10.2.15. Give an example of a transitive permutation group (M, G) for which $G_a = G_b$ for distinct letters a and b.

10.2.16. It is false that if any two Sylow p-subgroups intersect, then all intersect. (Compare with Theorem 6.1.19.) Let A be elementary Abelian of order 2^6.

   (a) $\exists\ B \subset \text{Aut}(A)$ such that $o(B) = 3^4$.

   Let $G = \text{Hol}(A, B)$.

   (b) If P and Q are Sylow 3-subgroups of G, then $P \cap Q \neq E$ (otherwise $o(PQ) > o(G)$).

   (c) G has no normal 3-subgroup H except E (otherwise, $A + H$ exists, and H induces the identity automorphism, only, on A).

   (d) The intersection of all Sylow 3-subgroups of G is E.

10.2.17. (Compare with Theorem 10.1.7.) Let $G = \text{Sym}(4)$, $H = \langle(1, 2, 3, 4)\rangle$, K the normal subgroup of G of order 4, and $*$ the representation of G on H.

    (a)   The representation $*$ is faithful and transitive.

    (b) $K^* \lhd G^*$, but if $x \in K^*$, then $\text{Ch}(x) > 0$.

10.2.18. Let G and H be finite groups containing A as a subgroup. Prove that there is a finite group K containing isomorphic copies G^*, H^*, and A^* of G, H, and A as subgroups such that $G^* \cap H^* = A^*$, as follows. Let $G = \mathop{\cup} g_i A$, $H = \mathop{\cup} h_j A$, and $M = \{(g_i, h_j, a) \mid a \in A\}$. Let K be the group generated by the following permutations of M:

$$(g_i, h_j, a)x^* = (g_{i'}, h_j, a')$$

if $x \in G$ and $g_i a x = g_{i'} a'$,

$$(g_i, h_j, a)y^* = (g_i, h_{j'}, a'')$$

if $y \in H$ and $h_j a y = h_{j'} a''$. (This construction should be compared with the free product with amalgamated subgroup.)

10.3   Regular permutation groups

A permutation group (M, G) is *regular* iff it is transitive and if $x \in G^{\#}$, then $\text{Ch}(x) = 0$.

10.3.1. *Any group G is isomorphic to a regular permutation group on the elements of G.*

Proof. The isomorphism T given by $x(gT) = xg$ and used in the proof of Cayley's theorem has a regular permutation group as image group.

10.3.2. *If (M, G) is a regular permutation group and (S, H) is its regular representation, then $(M, G) \cong (S, H)$.*

Proof. The regular representation of G is essentially its representation on E. The theorem now follows from Theorem 10.2.3, since $E = G_a$ if G is regular and $a \in M$.

10.3.3. *If (M, G) is a regular permutation group, then $o(G) = \text{Deg}(G)$.*

Proof. By Theorem 10.1.4. ‖

Example. The regular representation of $J_2 \times J_2$ yields a permutation group isomorphic to the 4-group

$$G = \{e, (1, 2)(3, 4), (1, 3)(2, 4), (1, 4)(2, 3)\}.$$

10.3.4. *A transitive Abelian group is regular.*

Proof. Suppose that (M, G) is an irregular transitive Abelian group. Then $\exists\ x \in G^{\#}$, $a \in M$, and $b \in M$ such that $ax = a$ and $bx \neq b$. Since G is transitive, $\exists\ y \in G$ such that $ay = b$. Thus $axy = ay = b$ and $ayx = bx \neq b$. Hence $xy \neq yx$, contradicting the fact that G is Abelian.

10.3.5. *A regular Abelian permutation group (M, G) is its own centralizer in* $\mathrm{Sym}(M)$.

Proof. Suppose that $\exists\ x \in C(G)\backslash G$. Then $\langle G, x \rangle$ is a transitive Abelian group, hence is regular by Theorem 10.3.4. Now $\exists\ a \in M$ such that $ax \neq a$. Since G is transitive, $\exists\ y \in G$ such that $ay = ax$. Then $e \neq xy^{-1} \in \langle G, x \rangle$, and $axy^{-1} = a$, contradicting the regularity of $\langle G, x \rangle$. ∥

A partial generalization of this theorem is true in the non-Abelian case.

10.3.6. *If (M, G) is regular, then $(M, C_{\mathrm{Sym}(M)}(G))$ is regular and isomorphic to (M, G).*

Proof. By Theorem 10.3.2, we may consider the Cayley representation (G, H) of G instead of (M, G). If $g \in G$, $x \in G$, and $y \in G$, let $gL_x = x^{-1}g$ and $gR_y = gy$. Then

$$gL_xR_y = (x^{-1}g)R_y = x^{-1}gy = gR_yL_x,$$

so that $L_xR_y = R_yL_x$, and $L_x \in C(H)$. Conversely, if $T \in C(H)$, then for all $g \in G$,

$$gT = eR_gT = eTR_g = (eT)g,$$

hence $T = L_{(eT)^{-1}}$. Therefore $C(H) = \{L_x \mid x \in G\}$. Now

$$gL_xL_y = y^{-1}x^{-1}g = (xy)^{-1}g = gL_{xy}.$$

Hence, if U is given by $R_xU = L_x$, then

$$(R_xR_y)U = R_{xy}U = L_{xy} = L_xL_y = (R_xU)(R_yU).$$

Thus, U is a homomorphism. But if $x \neq e$, then $eL_x = x^{-1} \neq e$. Hence U is an isomorphism of H onto $C(H)$. The function $-I$ is 1–1 from G onto G. Finally,

$$(gR_x)(-I) = (gx)^{-1} = x^{-1}g^{-1}$$
$$= (g(-I))L_x = (g(-I))(R_xU).$$

Hence $(-I, U)$ is an isomorphism of (G, H) onto $(G, C(H))$.

10.3.7. *If H is a regular normal subgroup of (M, G), $a \in M$, and G_a^* is the group of automorphisms of H induced by elements of G_a, then $(H^{\#}, G_a^*) \cong (M\backslash\{a\}, G_a)$.*

Proof. Let T_x be the automorphism of H induced by x. The function U defined by $hU = ah$ is 1–1 from $H^\#$ onto $M \setminus \{a\}$ because of the regularity of H. If $x \in G_a$ and $T_x = I$, then for all $h \in H^\#$, $ah = ax^{-1}hx = ahx$, hence $bx = b$ for all $b \in M$, and $x = e$. Therefore the function V: $T_xV = x$ is an isomorphism of G_a^* onto G_a. Finally, if $h \in H^\#$ and $x \in G_a$, then

$$(hT_x)U = (x^{-1}hx)U = a(x^{-1}hx) = (ah)x = (hU)(T_xV).$$

Hence, (U, V) is the required isomorphism.

EXERCISE

10.3.8. If G is a cyclic permutation group of degree n whose generator is an n-cycle, then G is regular.

10.3.9. A transitive Hamiltonian group is regular.

10.4 Multiply transitive groups

A permutation group (M, G) is *k-transitive* ($k \leq \mathrm{Deg}(G)$, $k \in \mathcal{N}$) iff whenever T and U are subsets of M with $o(T) = k$ and f is a 1–1 function from T onto U, then $\exists\, g \in G$ such that $g \mid T = f$. This means that if $a_1, \ldots, a_k$ are distinct letters and $b_1, \ldots, b_k$ are distinct letters, then $\exists\, g \in G$ such that $a_i g = b_i$, $i = 1, \ldots, k$.

(M, G) is 1-transitive iff G is transitive. A *k*-transitive group with $k > 1$ is called *multiply transitive*.

If (M, G) is multiply transitive and $a \in M$, then (M, G_a) has two orbits, $\{a\}$ and $M \setminus \{a\}$. The $(M \setminus \{a\})$-constituent of G_a is isomorphic to G_a (as group). Denoting this constituent by $(M \setminus \{a\}, G_a)$, we have

10.4.1. *If* (M, G) *is k-transitive,* $k > 1$, *and* $a \in M$, *then* $(M \setminus \{a\}, G_a)$ *is* $(k - 1)$-*transitive and of degree one less than* (M, G).

Proof. This follows from the definition of k-transitivity.

10.4.2. *If* (M, G) *is transitive,* $a \in M$, *and* $(M \setminus \{a\}, G_a)$ *is* $(k - 1)$-*transitive* $(k > 1)$, *then* G *is k-transitive.*

Proof. Let T and U be subsets of M such that $o(T) = k = o(U)$ and $a \in T$, and let f be a 1–1 function from T onto U. Since G is transitive, $\exists\, g \in G$ such that $ag = af$. Since G_a is $(k - 1)$-transitive, $\exists\, h \in G_a$ such that $bh = bfg^{-1}$ for all $b \in T \setminus \{a\}$ (since $afg^{-1} = a$.) Then $chg = cf$ for all $c \in T$.

If V is a subset of M and f' is a 1–1 function from U onto V, then ff' is a 1–1 function from T onto V. By earlier remarks, $\exists \ x \in G$ and $y \in G$ such that $cx = cf$ and $cy = cff'$ for all $c \in T$. Then if $d \in U$,

$$df' = df^{-1}(ff') = (df^{-1})y = dx^{-1}y.$$

Hence G is k-transitive.

10.4.3. *If (M, G) is k-transitive and of finite degree n, T is a subset of M, and $o(T) = k$, then $[G:G_T] = n!/(n-k)!$*

Proof. This follows by induction on k, Theorems 10.1.4, and 10.4.1.

10.4.4. *If n is finite, then* $\mathrm{Sym}(n)$ *is n-transitive.* ‖

There is a second class of highly transitive finite permutation groups which will be defined shortly.
Since

$$(1, \dots, n) = (1, 2)(1, 3) \dots (1, n),$$

a cycle c is the product of $(o(c) - 1)$ 2-cycles. Hence, if g is a finite permutation with cyclic decomposition $g = c_1 \dots c_n$, then g is the product of $f(g) = \Sigma\,(o(c_i) - 1)$ 2-cycles. We assert that if $g = (a_1, b_1) \cdots (a_m, b_m)$, then $m \equiv f(g)$ (mod 2). To prove this, note that

$$(a, i_1, i_2, \dots, i_r, b, j_1, \dots, j_s)(a, b) = (a, i_1, \dots, i_r)(b, j_1, \dots, j_s),$$

and therefore (since $(a, b)^2 = e$), also

$$(a, i_1, \dots, i_r)(b, j_1, \dots, j_s)(a, b) = (a, i_1, \dots, i_r, b, j_1, \dots, j_s).$$

Hence if a and b occur in the same cycle in the cyclic decomposition of h, then $f(h(ab)) = f(h) - 1$, while if they occur in different cycles, then $f(h(ab)) = f(h) + 1$. Therefore, in any case,

$$f(h(ab)) \equiv f(h) + 1 \ (\mathrm{mod}\ 2).$$

Since $f(e) = 0$, it follows that $f(g) \equiv m$ (mod 2), as asserted.
This justifies the following definition. A finite permutation g is *even* or *odd* according as g is the product of an even or an odd number of 2-cycles.

10.4.5. *If M is a finite set, then the set of even permutations of M is a normal subgroup of index 2 (provided $o(M) > 1$) of* $\mathrm{Sym}(M)$.

Proof. The function T which sends even permutations into [0] and odd permutations into [1] is a homomorphism of $\mathrm{Sym}(M)$ onto J_2. The kernel of T is just the set of even permutations, and has the desired properties. ‖

The group of even permutations of a finite set M is called the *alternating group on* M, and is denoted by $\mathrm{Alt}(M)$. $\mathrm{Alt}(n)$ has an obvious meaning.

10.4.6. $o(\text{Alt}(n)) = n!/2$. $\text{Alt}(n)$ *is* $(n - 2)$-*transitive, but not* $(n - 1)$-*transitive.*

Proof. Since $o(\text{Sym}(n)) = n!$ and $[\text{Sym}(n):\text{Alt}(n)] = 2$, $o(\text{Alt}(n)) = n!/2$. If $\text{Alt}(n)$ is $(n - 1)$-transitive, then, by Theorem 10.4.3, $o(\text{Alt}(n)) = n!$, a contradiction. Let S and T be subsets of $\{1, \ldots, n\}$ of order $n - 2$, and f a 1–1 function from S onto T. Let i and j be the letters omitted from T. Now $\exists\, x \in \text{Sym}(n)$ such that $x \mid S = f$. Also $(x(i,j)) \mid S = f$. One of the permutations x and $x(i,j)$ is even, hence in $\text{Alt}(n)$. Therefore $\text{Alt}(n)$ is $(n - 2)$-transitive. $\|$

Other than the symmetric and alternating groups, no finite 6-transitive group is known. There are two known 5-transitive groups, the Mathieu groups M_{12} and M_{24} of degrees 12 and 24, respectively. There are two known 4-transitive groups, the Mathieu groups M_{11} and M_{23}, which are the subgroups fixing one letter in M_{12} and M_{24}, respectively. These four groups, together with the 3-transitive group M_{22} are all simple. The Mathieu groups will be discussed in Sections 10.6 and 10.8. An infinite class of 3-transitive groups will be discussed in Section 10.6.

The following theorem is quite useful, although trivial.

10.4.7. *If a permutation group* (M, G) *of finite degree contains an odd permutation, then there exists a normal subgroup* H *of index* 2.

Proof. In fact, $H = G \cap \text{Alt}(M)$ is such a group. $\|$

EXERCISES

10.4.8. An n-cycle is an even (odd) permutation iff n is odd (even).

10.4.9. A finite permutation is even iff in its cyclic decomposition there occur an even number of cycles of even length.

10.5   Primitive and imprimitive groups

Let (M, G) be a transitive permutation group. A *block* B of G is a proper subset of M such that (i) $1 < o(B)$, and (ii) if $g \in G$, then either $B = Bg$ or $B \cap Bg = \varnothing$.

It is clear that (ii) is satisfied by the set M itself and by any singleton. This is the reason for the restrictions $1 < o(B)$ and $B < M$.

Condition (ii) is equivalent to the weaker condition $B \cap Bg = B$ or $\varnothing$. For if $B < Bg$, then $B \cap Bg^{-1} < B$ and $B \cap Bg^{-1} \neq B$ or $\varnothing$, so that B does not satisfy the weaker condition.

A *primitive* permutation group is a transitive permutation group without blocks. An *imprimitive* permutation group is a transitive permutation group with blocks.

10.5.1. *If B is a block and $g \in G$, then Bg is a block.*

Proof. $Bg < M$ and $1 < o(Bg)$ trivially. Let $h \in G$. If $Bghg^{-1} = B$, then $(Bg)h = Bg$; if $Bghg^{-1} \cap B = \varnothing$, then $(Bg)h \cap Bg = \varnothing g = \varnothing$. Hence Bg is a block.

10.5.2. *If M is finite and B is a block for the transitive permutation group (M, G), then $o(B) \mid o(M)$.*

Proof. This follows from Theorem 10.5.1 and the transitivity of G.

10.5.3. *If (M, G) is a transitive group of prime degree, then G is primitive.* ‖

A *block system* of an imprimitive group (M, G) is a set S of blocks such that $M = \overset{.}{\cup} \{B \mid B \in S\}$ and such that if $B \in S$ and $g \in G$, then $Bg \in S$.

10.5.4. *Let (M, G) be an imprimitive permutation group. If B is a block, then the set of distinct Bg, $g \in G$, is a block system. Conversely, any block system is of this type.*

Proof. Let B be a block, and let $S = \{Bg \mid g \in G\}$. Since G is imprimitive, it is transitive, and therefore each $a \in M$ is in some Bg. If $Bg \cap Bh \neq \varnothing$, then $Bgh^{-1} \cap B \neq \varnothing$, and, since B is a block, $Bgh^{-1} = B$, $Bg = Bh$. Hence $M = \overset{.}{\cup} \{B' \mid B' \in S\}$. Moreover, if $Bg \in S$ and $h \in G$, then $(Bg)h = B(gh) \in S$. Hence S is a block system.

Conversely, let S be a block system, and let $B \in S$. Then, by definition, $Bg \in S$ for all $g \in G$. Since the set of Bg, $g \in G$, is already a block system by the first half of the proof, and since $M = \overset{.}{\cup} \{B' \mid B' \in S\}$, it follows that $S = \{Bg \mid g \in G\}$. ‖

The structure of imprimitive permutation groups is, to some extent, determined by the following theorem.

10.5.5. *If M is a set, $o(S) > 1$, $M = \overset{.}{\cup} \{B \mid B \in S\}$, and $1 < o(B) = o(B')$ for all B and B' in S, then*

(i) *there is an imprimitive permutation group (M, G) with S as a block system, which contains every other such permutation group as subgroup;*

(ii) $G = \{g \in \operatorname{Sym}(M) \mid \text{if } B \in S \text{ then } Bg \in S\}$;

(iii) *if $o(M) = n$ and $o(B) = k$ are finite for $B \in S$, then $o(G) = (k!)^{n/k}(n/k)!$*

Proof. The set G defined by (ii) is a group. G is transitive since the $B \in S$ all have the same order. It is clear that G has S as a block system and is maximum with this property. Each $g \in G$ induces a permutation gT of S. T is a homomorphism of G onto $\text{Sym}(S)$. $\text{Ker}(T) = \pi\{\text{Sym}(B) \mid B \in S\}$. Statement (iii) follows from these facts. $\|$

The group G in Theorem 10.5.5 is isomorphic to $\text{Sym}(B) \wr \text{Sym}(S)$ for $B \in S$.

10.5.6. *If* (M, G) *is transitive,* $a \in M$, *and* $HT = aH$ *for all H such that* $G_a \subset H \subset G$, *then*

(1) $G_a T = a, GT = M$.

(2) *T is 1-1 from the set of H with $G_a < H < G$ onto the set of blocks of G containing a.*

(3) $(H \cap K)T = (HT) \cap (KT)$.

Proof. (1) $G_a T = aG_a = a$. $GT = aG = M$ since G is transitive.

(2) Let $G_a < H < G$. H is the set union of more than one, but not every, right coset of G_a. Hence, by Theorem 10.1.2 (ii), $HT = aH$ has more than one element, but is not M. If $g \in H$, then $(aH)g = aH$. If $g \in G \backslash H$, then Hg is a set union of right cosets of G_a and $Hg \cap H = \varnothing$. Hence by Theorem 10.1.2 (ii) again, $a(Hg) \cap aH = \varnothing$. Therefore $HT = aH$ is a block. If $H \neq K$, then there is a right coset of G_a in $H \backslash K$ or in $K \backslash H$, hence $aH \neq aK$, by Theorem 10.1.2. Thus T is 1-1. Let B be a block such that $a \in B$, and let $H = \{h \mid ah \in B\}$. Then $e \in H$. If $h \in H$ and $h' \in H$, then $ah' \in B$, $a \in Bh'^{-1} \cap B$, so $Bh'^{-1} = B$, since B is a block. Therefore $ahh'^{-1} \in Bh'^{-1} = B$, and $hh'^{-1} \in H$. Hence, H is a group. By the transitivity of G, $G_a < H < G$ and $HT = aH = B$. Therefore (2) is true.

(3) $(H \cap K)T = a(H \cap K)$ which is a subset of $aH \cap aK = (HT) \cap (KT)$. If $b \in aH \cap aK$, then H and K each contain $G_a x$ where $ax = b$. Hence

$$x \in H \cap K \quad \text{and} \quad b = ax \in a(H \cap K).$$

Therefore,

$$(H \cap K)T = (HT) \cap (KT).$$

10.5.7. *If* (M, G) *is transitive and* $a \in M$, *then G is primitive iff G_a is a maximal proper subgroup of G.*

Proof. It follows from Theorem 10.5.1 (or 10.5.4) that G is primitive iff there is no block containing a. The theorem now follows from Theorem 10.5.6. (2).

10.5.8. *If (M, G) is a 2-transitive group, then G is primitive.*

Proof. Let $B < M$, $1 < o(B)$. Then $\exists \, a \in B$, $b \in B$, and $c \in M \backslash B$ such that $a \neq b$. By 2-transitivity, $\exists \, g \in G$ such that $ag = a$ and $bg = c$. Thus $a \in Bg \cap B$, so that $Bg \cap B \neq \varnothing$. Since $c \in Bg \backslash B$, $Bg \neq B$. Therefore, B is not a block. Hence G is primitive.

10.5.9. *If (M, G) is an imprimitive group with block system S, and*

$$H = \{h \in G \mid Bh = B \quad \text{for all } B \in S\},$$

then H is a normal intransitive subgroup of G.

Proof. If $g \in G$, $h \in H$, and $B \in S$, then $Bg^{-1} \in S$, so that $Bg^{-1}hg = Bg^{-1}g = B$, and $g^{-1}hg \in H$. Thus $H \lhd G$. If $b \in B \in S$, then $B \neq M$, and the orbit bH of H is a subset of B, hence a proper subset of M. Therefore H is intransitive. ‖

A partial converse is true.

10.5.10. *If (M, G) is transitive, $E \neq H \lhd G$, and H is intransitive, then the set S of orbits of H is a block system for G.*

Proof. Since $E \neq H$, $\exists \, B \in S$ such that $o(B) \neq 1$. Since H is intransitive, $B \neq M$. Now $\exists \, b \in B$ such that $B = bH$. Hence if $g \in G$, then $(bg)H = b(Hg) = Bg$, so that $Bg \in S$ for all $g \in G$. Since S is a partition of M, S is a block system for G. ‖

As a corollary, we have the important

10.5.11. *A normal non-E subgroup of a primitive group is transitive.* ‖

A permutation group (M, G) is *k-primitive* iff it is k-transitive, and, if S is a subset of M of order $k - 1$, then $(M \backslash S, G_S)$ is primitive. Thus 1-primitivity is the same as primitivity.

10.5.12. *A $(k + 1)$-transitive group (M, G) is k-primitive.*

Proof. It is certainly k-transitive. If S is a subset of M of order $k - 1$, then G_S is 2-transitive (Theorem 10.4.1 and induction), hence is primitive (Theorem 10.5.8). Therefore G is k-primitive.

10.5.13. *Let G be a group and $A \subset \operatorname{Aut}(G)$.*

(1) *If A is transitive on $G^\#$, then all elements of $G^\#$ have the same order, prime or infinite.*

(2) *If G is finite and A is transitive on $G^\#$, then G is an elementary Abelian p-group for some $p \in \mathscr{P}$.*

(3)   *If A is primitive on $G^\#$ then either $G \cong J_3$ or G is an elementary Abelian 2-group.*

(4)   *If A is 2-primitive on $G^\#$, then $G \cong J_3$ or G is a 4-group.*

(5)   *If A is 3-transitive on $G^\#$, then G is a 4-group.*

   Proof. (1) If $x \in G^\#$ and $y \in G^\#$, then $\exists\ T \in A$ such that $xT = y$. Hence $o(y) = o(xT) = o(x)$. If $o(x)$ is composite, then some power of x has prime order, and not all elements of $G^\#$ have the same order, a contradiction.

   (2) By (1), G is a finite p-group. Since $Z(G)$ is characteristic, $Z(G)^\#$ contains an orbit of A, hence $Z(G) = G$. Therefore, G is elementary Abelian.

   (3) Suppose that G is not an elementary Abelian 2-group. By Exercise 2.4.13, $\exists\ x \in G$ with $o(x) > 2$. Let $B = \{x, x^{-1}\}$. Then $1 < o(B)$. If $B = G^\#$, then $G \cong J_3$. If $B \ne G^\#$ and $T \in A$, then either $BT = B$ or $BT \cap B = \varnothing$, and B is a block, contrary to assumption.

   (4) Let $o(G) > 3$. By (3), G is an elementary Abelian 2-group. There is a subgroup $H = \{e, x, y, xy\}$ of G. The subgroup A_x of A is primitive on $G \backslash \{e, x\}$. If $T \in A_x$, then $HT = H$ or $HT \cap H = \{e, x\}$. Hence, if $H < G$, then $\{y, xy\}$ is a block of A_x, contrary to the primitivity of A_x. Hence $H = G$, and G is a 4-group.

   (5) This is immediate from (4), since $J_3^\#$ has only two elements. ∥

   It should be noted that in (1), the case where all elements of $G^\#$ have infinite order can actually occur, since $\mathrm{Aut}(\mathscr{R})$ is transitive on $\mathscr{R}^\#$.

   The preceding theorem has a partial converse as follows.

   10.5.14. (1) *If G is an elementary Abelian p-group, then* $\mathrm{Aut}(G)$ *is transitive on $G^\#$.*

(2)   *If G is an elementary Abelian 2-group, then* $\mathrm{Aut}(G)$ *is 2-transitive (hence primitive) on $G^\#$.*

(3)   *If $G \cong J_3$, then* $\mathrm{Aut}(G)$ *is 2-primitive (hence 2-transitive) on $G^\#$.*

(4)   *If G is a 4-group, then* $\mathrm{Aut}(G)$ *is 3-transitive (hence 2-primitive) on $G^\#$.*

   Proof. (1) If $x \in G^\#$ and $y \in G^\#$, then

$$G = \langle x \rangle + H = \langle y \rangle + K$$

where $H \cong K$. Then $\exists\ T \in \mathrm{Aut}(G)$ such that $xT = y$ and $HT = K$. Hence $\mathrm{Aut}(G)$ is transitive on $G^\#$.

   (2) If $x_1, x_2, y_1,$ and y_2 are elements of $G^\#$ such that $x_1 \ne x_2$ and $y_1 \ne y_2$, then

$$G = \langle x_1 \rangle + \langle x_2 \rangle + H = \langle y_1 \rangle + \langle y_2 \rangle + K$$

where $H \cong K$. Then $\exists\, T \in \operatorname{Aut}(G)$ such that $x_1 T = y_1, x_2 T = y_2$, and $HT = K$. Hence $\operatorname{Aut}(G)$ is 2-transitive on $G^{\#}$.

(3) There is an automorphism $(-I)$ of J_3 permuting the two elements of $J_3^{\#}$, hence $\operatorname{Aut}(J_3)$ is 2-transitive. The subgroup fixing one element of $J_3^{\#}$ is primitive in a rather trivial fashion.

(4) $\operatorname{Aut}(G) \cong \operatorname{Sym}(3)$, hence $\operatorname{Aut}(G)$ is 3-transitive on $G^{\#}$.

10.5.15. *If (M, G) is a permutation group and H is a regular normal subgroup, then*

(1) *If G is 2-transitive, then all elements of $H^{\#}$ have the same order, prime or infinite.*

(2) *If G is 2-transitive and $\operatorname{Deg}(G)$ is finite, then H is an elementary Abelian p-group, $p \in \mathscr{P}$, and $\operatorname{Deg}(G) = p^n, n \in \mathscr{N}$.*

(3) *If G is 2-primitive, then either $(M, G) \cong \operatorname{Sym}(3)$, or H is an elementary Abelian 2-group. If, moreover, $\operatorname{Deg}(G)$ is finite, then $\operatorname{Deg}(G) = 2^n$ or 3, $n \in \mathscr{N}$.*

(4) *If G is 3-primitive, then $(M, G) \cong \operatorname{Sym}(4)$ or $\operatorname{Sym}(3)$.*

Proof. If A is the group of automorphisms of H induced by G_a, then, by Theorem 10.3.7, $(H^{\#}, A) \cong (M \backslash \{a\}, G_a)$. Therefore, if G is $(k + 1)$-transitive $((k + 1)$-primitive), then A is k-transitive (k-primitive) on $H^{\#}$. Nearly all of the statements of the theorem now follow from Theorem 10.5.13 and the fact that $\operatorname{Deg}(G) = o(H)$. In (3), if $\operatorname{Deg}(G) = 3$, then since G is 2-transitive, $o(G) = 6$ and $(M, G) \cong \operatorname{Sym}(3)$. Similarly in (4), $\operatorname{Deg}(G) = o(H) = 3$ or 4, and G is 3-transitive, hence $(M, G) \cong \operatorname{Sym}(3)$ or $\operatorname{Sym}(4)$.

10.5.16. *If (M, G) is k-primitive, $G \neq \operatorname{Sym}(M)$, and H ($\neq E$) is a non-regular normal subgroup of G, then H is k-transitive.*

Proof. Since G is primitive, H is transitive. For $a \in M$, since H is not regular, $H_a \neq E$. Moreover $H_a = H \cap G_a \lhd G_a$. Now deny the theorem, and take a counterexample with minimum k. If $G_a = \operatorname{Sym}(M \backslash \{a\})$, then by Theorem 10.4.2 and Exercise 10.5.27, $G = \operatorname{Sym}(M)$, contrary to hypothesis. If H_a is not regular, then it is $(k - 1)$-transitive on $M \backslash \{a\}$ by the minimality of k. Hence H is k-transitive by Theorem 10.4.2. Therefore H_a is regular, so that H is 2-transitive and $k > 2$. If $k = 3$, then G_a is 2-primitive, hence by Theorem 10.5.15 (3), H_a has an element x of order 2. Let

$$x = (a)(b, c)(d, f) \cdots .$$

Since G is 3-transitive, $\exists\, g \in G$ such that $ag = d, bg = b$, and $cg = c$. Then $xg^{-1}xg$ is in H, fixes b and c, but maps f onto d. Since the subgroup of H fixing two letters is E (by the regularity of H_a), this is a contradiction. Hence $k > 3$, G_a is 3-primitive, and by Theorem 10.5.15 (4), $G_a = \operatorname{Sym}(M \backslash \{a\})$, a contradiction.

10.5.17. *If G is k-transitive but not symmetric, $k > 3$, and $E \neq H \lhd G$, then H is $(k - 1)$-transitive.*

Proof. By Theorem 10.5.15 (4), H is not regular. By Theorem 10.5.16, H is $(k - 1)$-transitive.

10.5.18. *If (M, G) is k-transitive, $G \neq \mathrm{Sym}(4)$, $k > 3$, and $E \neq H \lhd G$, then H is $(k - 2)$-transitive.*

Proof. Deny, and let G be a counterexample with minimum k. Since G is primitive, H is transitive. By Theorem 10.5.15 (4), H is not regular, hence $E \neq H_a \lhd G_a$. If $k - 1 > 3$, then by the minimality of k, H_a is $(k - 3)$-transitive on $M \backslash a$, hence H is $(k - 2)$-transitive on M by Theorem 10.4.2. Therefore $k = 4$. But G_a is primitive on $M \backslash a$, so that H_a is transitive on $M \backslash a$, hence (Theorem 10.4.2) H is 2-transitive on M.

10.5.19. *If (M, G) is k-transitive for all $k \in \mathcal{N}$, and $E \neq H \lhd G$, then H is k-transitive for all k.*

Proof. Since G is 5-transitive, $G \neq \mathrm{Sym}(4)$. The theorem now follows from Theorem 10.5.18. ‖

It has been noted (Theorem 10.2.15) that there are transitive groups (M, G) in which $G_a = G_b$ with $a \neq b$. In fact, this is true for any regular group. Since a regular group of prime order is primitive, there are primitive groups with $G_a = G_b$ and $a \neq b$. However, according to the next theorem, this is the only type of primitive group with this property.

10.5.20. *If (M, G) is primitive, but not regular of prime degree, $a \in M$, $b \in M$, and $a \neq b$, then $G_a \neq G_b$.*

Proof. Deny the theorem. Since G is transitive. $\exists\, x \in G$ such that $ax = b$. By Theorem 10.1.2,

$$G_a^x = G_{ax} = G_b = G_a.$$

Hence $x \in N(G_a) \backslash G_a$. By Theorem 10.5.7, G_a is a maximal proper subgroup of G. Hence $N(G_a) = G$. Therefore, by the transitivity of G and Theorem 10.1.2 again, $G_a = G_c$ for all $c \in M$. This means that $G_c = E$ for all $c \in M$, and G is regular. Since E is a maximal proper subgroup of G, G is of prime order. Since G is regular, G is of prime degree.

10.5.21. *If (M, G) is a finite primitive solvable group, then $\mathrm{Deg}(G) = p^n$, $p \in \mathscr{P}$, $n \in \mathcal{N}$, and there is a unique minimal, normal non-E subgroup H of G. H is a regular, elementary Abelian group of order p^n.*

Proof. There is a minimal normal non-E subgroup H of G. Since G is solvable, H is an elementary Abelian group of order p^n for some prime p. Since G is primitive, H is transitive (Theorem 10.5.11). Since H is Abelian and transitive, it is regular (Theorem 10.3.4). Hence $\mathrm{Deg}(G) = p^n$. If M is a second minimal, normal non-E subgroup of G, then $H \cap M = E$, hence $M \subset C(H)$, contrary to the fact (Theorem 10.3.5) that a regular Abelian group is its own centralizer.

EXERCISES

10.5.22. Let G be the maximum imprimitive group with block system $S = \{\{1, 2, 3\}, \{4,5,6\}\}$. Show that G has two normal subgroups with S as set of orbits.

10.5.23. An imprimitive group may have blocks of different lengths ($\langle (1, 2, 3, 4, 5, 6) \rangle$).

10.5.24. An imprimitive group may have two different block systems with blocks of the same length (the 4-group).

10.5.25. Compare the order of the maximum imprimitive group of degree 6 with blocks of length 2, with that of the maximum imprimitive group of degree 6 with blocks of length 3.

10.5.26. What is the largest possible order of an imprimitive group of degree 12?

10.5.27. If (M, G) is transitive, $G_a = \mathrm{Sym}(M \setminus a)$, and M is infinite, then $G = \mathrm{Sym}(M)$.

10.5.28. The subgroup H in Theorem 10.5.9 may equal E (let (M, G) be the regular representation of $\mathrm{Sym}(3)$).

10.5.29. (a) A maximal proper subgroup of a finite solvable group has prime power index.

(b) If $2^n - 1$ is prime, then there is a finite solvable group with a maximal proper subgroup of index 2^n.

10.6 Some multiply transitive groups

In this section, several examples of multiply transitive groups will be given.

A permutation group (M, G) is *exactly k-transitive* iff it is k-transitive, and, if S is a subset of M of order k, then $G_S = E$. Thus a permutation group is exactly 1-transitive iff it is regular.

10.6.1. *If a finite permutation group (M, G) is exactly 2-transitive, then* $\mathrm{Deg}(G) = p^j$, $p \in \mathscr{P}$, $j \in \mathscr{N}$, *and a Sylow p-subgroup P of G is an elementary Abelian, regular, normal subgroup of G consisting of e and all $x \in G$ such that* $\mathrm{Ch}(x) = 0$.

Proof. Since G is transitive and finite, it has finite degree n (Theorem 10.1.4). Let $p \mid n$, $p \in \mathscr{P}$, and let $S_i = \{x \in G \mid \mathrm{Ch}(x) = i\}$ for $i = 0$ and 1. Now $o(G_a^\#) = n - 2$ for $a \in M$. Since $G_a \cap G_b = E$ if $a \neq b$, $o(S_1) = (n-2)n$. Since $o(G) = n(n-1)$, $o(S_0) = n - 1$. There is an $x \in G$ such that $o(x) = p$. Since x is a product of p-cycles and at most one 1-cycle, and since $p \mid n$, it follows that $x \in S_0$. Since $G_a \cap G_a^x = G_a \cap G_{ax} = E$, $C(x) \cap G_a = E$. Therefore, $o(\mathrm{Cl}(x)) \geqq o(G_a) = n - 1$. Hence $S_0 = \mathrm{Cl}(x)$. Since this is true for all primes dividing n, n is divisible by just one prime, so that $n = p^j$ for some j. From this it follows that if $P \in \mathrm{Syl}_p(G)$, then $o(P) = p^j$. Since any element y of $P^\#$ is a product of p^i-cycles and at most one 1-cycle, $y \in S_0$. Therefore $P^\# = S_0$. Hence $P \lhd G$ since P is the set union of two conjugate classes. Now P is a minimal, normal non-E subgroup of G since all elements of $P^\#$ are conjugate. By Theorem 4.4.4, P is elementary Abelian. $\parallel$

A *near field* is an algebraic system $(F, +, \cdot)$ such that $+$ and $\cdot$ are associative binary operations on F, $(F, +)$ is a group with identity 0 (say), $(F \backslash 0, \cdot)$ is a group, and

$$(a + b) \cdot c = a \cdot c + b \cdot c$$

for all a, b, and c in F. As usual $a \cdot c$ will be written ac, and the multiplicative identity denoted by 1.

10.6.2. *If F is a near field and G the set of all functions $T_{a,b}$, $a \in F^\#$, $b \in F$, where $xT_{a,b} = xa + b$ for all $x \in F$, then G is an exactly 2-transitive group of permutations of F.*

Proof.

$$xT_{a,b}T_{c,d} = (xa + b)T_{c,d} = x(ac) + (bc + d).$$

Hence $T_{a,b}T_{c,d} = T_{ac,bc+d}$. The function $T_{1,0}$ is the identity I_F, and $T_{a^{-1},-ba^{-1}}$ is an inverse of $T_{a,b}$. Thus G is a group of permutations of F. One checks that the set $\{T_{1,b} \mid b \in F\}$ is a (normal, regular) transitive subgroup isomorphic to $(F, +)$. The subgroup $G_0 = \{T_{a,0} \mid a \in F^\#\}$ (Exercise 10.6.23) is transitive on $F^\#$, since if $a \in F^\#$ and $b \in F^\#$, then $aT_{a^{-1}b,0} = b$ and $0T_{a^{-1}b,0} = 0$. Hence G is 2-transitive. Now let $U = T_{a,b}$ fix 0 and 1. Then

$$0 = 0U = 0a + b = b,$$
$$1 = 1U = 1a + 0 = a,$$
$$U = T_{1,0} = I.$$

Therefore, G is exactly 2-transitive. $\parallel$

For every prime power p^n, there is a field of order p^n, hence by Theorem 10.6.2 an exactly 2-transitive group of degree p^n.

The finite near fields were determined by Zassenhaus [2]. Although this determination will not be included here, certain properties of finite near fields are given in the next corollary, since they follow immediately from Theorem 10.6.1.

10.6.3. *If F is a finite near field, then* $o(F) \in \mathscr{P}^\infty$ *and* $(F, +)$ *is an elementary Abelian group.*

Proof. The group G of Theorem 10.6.2 is exactly 2-transitive of degree $o(F)$. By Theorem 10.6.1, $o(F) = p^j$, $p \in \mathscr{P}$, $j \in \mathscr{N}$. Now the subgroup $\{T_{1,b} \mid b \in F\} \cong (F, +)$ has the right order to be a Sylow p-subgroup of G. By Theorem 10.6.1, it is elementary Abelian. ∥

The converse of Theorem 10.6.2 is true in the finite case.

10.6.4. *If* (M, H) *is a finite exactly 2-transitive group, then there is a near field* $(M, +, \cdot)$ *such that* $H = \{T_{a,b} \mid a \in M^\#, b \in M\}$, *where* $xT_{a,b} = xa + b$ *for* $x \in M$.

Proof. Name two of the elements of M 0 and 1. Define multiplication in M as follows. Let $a0 = 0$ for all $a \in M$. For $b \neq 0$, $b \in M$, $\exists \mid u \in H_0$ such that $1u = b$ since H is exactly 2-transitive. Then let $ab = au$ for $a \in M$. Thus $0b = 0$. If $1u = b$ and $1v = c$ for u and v in H_0, then $1(uv) = bv$, and

$$(ab)c = (au)c = (au)v = a(uv) = a(bv) = a(bc).$$

Also $1e = 1$ and $0e = 0$, so $a1 = ae = a$, and 1 is a multiplicative identity of M. If $1u = a$ and $1u^{-1} = d$ for $u \in H_0$, then

$$ad = au^{-1} = 1uu^{-1} = 1e = 1, \; d = a^{-1}.$$

Hence $M \backslash 0$ is a group under multiplication.

Next, define addition in M as follows. Let (see Theorem 10.6.1) P be the (regular, elementary, primary Abelian) normal subgroup of H such that $x \in P^\#$ iff $Ch(x) = 0$. By the regularity, if $a \in M$ then $\exists \mid x_a \in P$ such that $0x_a = a$. Define $a + b = ax_b$. Since $0x_a x_b = a + b = 0x_{a+b}, x_a x_b = x_{a+b}$ and the mapping $x_a \longrightarrow a$ is an isomorphism of P onto $(M, +)$. Hence $(M, +)$ is a group. Since $0x_0 = 0$ and P is regular, $x_0 = e$. Therefore $a + 0 = ax_0 = a$, and 0 is the identity of $(M, +)$.

Let a, b, and c be in M. If $c = 0$, then $(a + b)c = 0 = ac + bc$. Now assume that $c \neq 0$. Then $\exists \mid u \in G_0$ such that $1u = c$. Now

$$0u^{-1}x_b u = 0x_b u = bu,$$

and $u^{-1}x_b u \in P$ by the normality of P. Therefore, $u^{-1}x_b u = x_{bu}$ and $x_b u = ux_{bu}$. Therefore,

$$(a + b)c = (a + b)u = ax_b u = (au)x_{bu} = (ac)x_{bc} = ac + bc.$$

It follows that $(M, +, \cdot)$ is a near field.

Let $G = \{T_{a,b}\}$. Since (M, H) is exactly 2-transitive by assumption, and (M, G) is exactly 2-transitive by Theorem 10.6.2, $o(G) = o(H)$, and it is sufficient to show that $H \subset G$. Let $u \in P$. Then $u = x_b$ for some $b \in M$. Therefore, if $a \in M$,

$$au = ax_b = a + b = a \cdot 1 + b = aT_{1,b}.$$

Hence $u = T_{1,b} \in G$, and $P \subset G$. Next let $v \in H_0$. Then $1v = b \in M^\#$. Therefore,

$$av = ab = ab + 0 = aT_{b,0}$$

for all $a \in M$. Hence $v = T_{b,0} \in G$, and $H_0 \subset G$. Since $H = H_0 P$, $H \subset G$. ‖

For a discussion of the infinite case, see M. Hall [1, p. 382].

If V is a vector space over a division ring D, an *affine transformation* on V is a function U from V into V such that $\exists\, T \in GL(V)$ and $a \in V$ for which $xU = xT + a$ for all $x \in V$. A *translation* is such a function U for which $xU = x + a$ for all $x \in V$.

10.6.5. *If V is a vector space over a division ring D, then*

(1) *the set G of affine transformations forms a group,*

(2) *the translations form a normal subgroup H of G,*

(3) $G = H \cdot GL(V)$, $H \cap GL(V) = E$.

Proof. Let $U_i \in G$, $i = 1, 2$, with associated $T_i \in GL(V)$ and $a_i \in V$. Then for $x \in V$,

$$xU_1U_2 = (xT_1 + a_1)U_2 = xT_1T_2 + (a_1T_2 + a_2),$$

and $T_1T_2 \in GL(V)$, $a_1T_2 + a_2 \in V$. If V^+ denotes the additive group of V, then (see Section 9.2) there is an obvious isomorphism of G onto $\mathrm{Hol}(V^+, GL(V))$. The theorem now follows from Section 9.2. ‖

The group of affine transformations of V will be denoted by $\mathrm{Aff}(V)$. Notations such as $\mathrm{Aff}(n, F)$ are self-explanatory. It follows from Theorem 10.6.2 that if F is a field, then $\mathrm{Aff}(1, F)$ is exactly 2-transitive.

If V is a vector space over a division ring D, a *scalar* transformation is an element T of $GL(V)$ such that $\exists\, c \in D^\#$ for which $xT = cx$ for all $x \in V$. Note that for such a c, $c \in Z(D)$. The set of scalar transformations forms a normal subgroup S of $GL(V)$ (in fact, $S = Z(GL(V))$). The *projective general linear group*, $PGL(V)$, is the factor group $GL(V)/S$. In case D is a field, the *projective special linear group* $PSL(V)$ is the factor group $SL(V)/(S \cap SL(V))$.

10.6.6. *If V is a 2-dimensional vector space over a division ring D, then $PGL(V)$ is faithfully represented as a 3-transitive permutation group by its action on the 1-dimensional subspaces of V.*

Proof. Let $G = GL(V)$, $P = PGL(V)$, and let M be the set of 1-dimensional subspaces of V. Each $T \in G$ effects a permutation TU of M. The function U is a permutation representation of G. It is clear that any scalar transformation is in $\text{Ker}(U)$. Conversely, let $T \in \text{Ker}(U)$, and suppose that T is not a scalar. Let $x \in V^{\#}$. Then $xT = cx$, $c \in D^{\#}$. First suppose that for all $y \in V \backslash Dx$, $yT = cy$. Then if $a \in D^{\#}$ and $y \in V \backslash Dx$, also $ax - y \in V \backslash Dx$, hence

$$(ax)T = (ax - y)T + yT = c(ax - y) + cy = cax.$$

Therefore $zT = cz$ for all $z \in V$, and T is a scalar transformation. Next, suppose that $\exists \, y \in V \backslash Dx$ such that $yT \neq cy$. Then, since $T \in \text{Ker}(U)$, $yT = dy$ with $c \neq d$. Then

$$cx + dy = (x + y)T = b(x + y) = bx + by, \; b \in D,$$

so that $c = b = d$ by the independence of x and y, a contradiction. Hence $\text{Ker}(U)$ is the group of scalar transformations. Therefore $GU \cong P$, so that P is, in this sense, faithfully represented by its action on M.

Let $\{x, y\}$ be a basis of V. The subspaces Dx, Dy, and $D(x + y)$ are distinct. If u and v are independent elements of V, then $\exists \, T \in G$ such that $xT = u$ and $yT = v$. Therefore P is at least 2-transitive. Let G_{12} be the subgroup of G consisting of those T which fix both Dx and Dy. If $D(ax + by)$ is a subspace distinct from both Dx and Dy, then $a \neq 0$, $b \neq 0$, and $\exists \, T \in G$ such that $xT = ax$ and $yT = by$. Then $T \in G_{12}$ and $(D(x + y))(TU) = D(ax + by)$. Therefore, P is 3-transitive.

10.6.7. *If V is a 2-dimensional vector space over a field F, then $PGL(V)$ is faithfully represented as an exactly 3-transitive permutation group by its action on the 1-dimensional subspaces of V.*

Proof. By Theorem 10.6.6, it remains only to prove that the subgroup fixing three letters is E. Let P, G, x, and y, be as in the proof of Theorem 10.6.6, and let $T \in G$ fix Fx, Fy, and $F(x + y)$. Then for some a, b, c in F,

$$xT = ax, \, yT = by, \, (x + y)T = c(x + y) = cx + cy.$$

Therefore, $a = b = c$, and

$$(rx + sy)T = rax + say = a(rx + sy)$$

for all r and s in F. Hence,

$$[F(rx + sy)]T = F(rx + sy)$$

for all r and s in F, so that T induces the identity permutation on the set of 1-dimensional subspaces. Therefore, the subgroup P_{123} of P which fixes 3 letters is the identity. Hence P is exactly 3-transitive.

10.6.8. *PGL(2, q) has a faithful representation as an exactly 3-transitive group of degree $q + 1$ and order $(q + 1)q(q - 1)$.*

Proof. If V is a 2-dimensional vector space over a field F with q elements, then $o(V^\#) = q^2 - 1$, while each 1-dimensional subspace has $q - 1$ nonzero elements. Therefore, there are $q + 1$ 1-dimensional subspaces. The theorem now follows from Theorem 10.6.7. ‖

A *semi-linear transformation* of a vector space V over a division ring D is a 1-1 function T from V onto V such that $\exists\ t \in \text{Aut}(D)$ such that for all $x \in V$, $y \in V$, and $c \in D$,

$$(x + y)T = xT + yT, (cx)T = (ct)(xT). \tag{1}$$

10.6.9. *If V is a vector space over a division ring D, then the set G of semi-linear transformations of V is a group, and $G/GL(V) \cong \text{Aut}(D)$.*

Proof. The fact that G is a group is left as Exercise 10.6.21. If T and t are as in (1), then define $TU = t$. If $T_i U = t_i$, $i = 1, 2$, then for $c \in D$ and $x \in V$,

$$(cx)T_1 T_2 = ((ct_1)(xT_1))T_2 = (ct_1 t_2)(xT_1 T_2).$$

Hence $(T_1 T_2)U = t_1 t_2$, and U is a homomorphism. Now $T \in \text{Ker}(U)$ iff $t = TU = I$, i.e., iff $(cx)T = c(xT)$ for all c and x, hence iff $T \in GL(V)$. Thus $\text{Ker}(U) = GL(V)$. Finally, let $t \in \text{Aut}(D)$ and let B be a basis of V. Define T by the rule

$$(\Sigma \{c_x x \mid x \in B\})T = \Sigma (c_x t)x.$$

Then if $a \in D$,

$$(\Sigma c_x x + \Sigma d_x x)T = (\Sigma (c_x + d_x)x)T = \Sigma ((c_x + d_x)t)x$$
$$= \Sigma (c_x t)x + \Sigma (d_x t)x$$
$$= (\Sigma c_x x)T + (\Sigma d_x x)T,$$

$$(a \Sigma c_x x)T = (\Sigma (ac_x)x)T = \Sigma ((ac_x)t)x$$
$$= \Sigma (at)(c_x t)x$$
$$= (at) \Sigma (c_x t)x = (at)((\Sigma c_x x)T).$$

Therefore, $T \in G$. Since $(ay)T = (at)(yT)$ by the above equations, $TU = t$. Hence $G/GL(V) \cong \text{Aut}(D)$. ‖

The group of semi-linear transformations of V will be denoted by $\Gamma L(V)$.

10.6.10. *If V is a vector space of dimension at least 2 over a division ring D, then $\Gamma L(V)$ induces a group of permutations of the set M of 1-dimensional subspaces of V, and the kernel of the resulting representation U is the set H of all functions of the form T_a, where $a \in D^{\#}$ and $xT_a = ax$ for all $x \in V$.*

Proof. It follows from (1) that a semi-linear transformation sends subspaces onto subspaces, hence induces a permutation of M. Thus there is a permutation representation U of $\Gamma L(V)$ on M. Let $T \in \mathrm{Ker}(U)$, and let $x \in V^{\#}$. Then $xT = ax$, $a \in D$. If $yT = ay$ for all $y \in V \backslash Dx$, then

$$(bx)T = ((bx - y) + y)T = (bx - y)T + yT$$
$$= a(bx - y) + ay = a(bx).$$

Hence $T = T_a \in H$. Next, suppose that $\exists\, y \in V \backslash Dx$ such that $yT \neq ay$. Then $yT = by$, $b \neq a$, and

$$ax + by = (x + y)T = c(x + y) = cx + cy, \ c \in D.$$

Hence $a = c = b$, a contradiction. Therefore $\mathrm{Ker}(U) \subset H$.

Conversely, if $a \in D^{\#}$, then T_a is semi-linear, for

$$(cx)T_a = acx = (aca^{-1})(ax) = (ct)(xT_a)$$

where t is the inner automorphism of D induced by a^{-1}. Also, T_a induces the identity permutation on M. Hence $H \subset \mathrm{Ker}(U)$, so that $H = \mathrm{Ker}(U)$. ‖

The factor group $\Gamma L(V)/H$ is denoted by $P\Gamma L(V)$ and called the *projective semi-linear group.*

10.6.11. *If V is a vector space of dimension 2 over a division ring, then $P\Gamma L(2, D)$ has a faithful representation as a 3-transitive permutation group on the set M of 1-dimensional subspaces of V, in which the subgroup fixing three letters is isomorphic to $\mathrm{Aut}(D)$.*

Proof. Let $\{x, y\}$ be a basis of V, and for $t \in \mathrm{Aut}(D)$, let U_t be the function

$$(rx + sy)U_t = (rt)x + (st)y, \ r \in D, \ s \in D.$$

Then U_t is semi-linear and fixes Dx, Dy, and $D(x + y)$. However, if $t \neq I$, then $\exists\, r \in D^{\#}$ such that $rt \neq r$. Then

$$(D(rx + y))U_t = D((rt)x + y) \neq D(rx + y),$$

so that U_t does not induce the identity permutation. Also, the set of U_t forms a group isomorphic to $\mathrm{Aut}(D)$.

Now let $T \in \Gamma L(V)$ fix Dx, Dy, and $D(x + y)$. Then by a linearity argument already used twice,

$$xT = ax, \quad yT = ay, \quad a \in D^{\#}.$$

Hence $(rx + sy)T = (rt)(ax) + (st)(ay)$ with $t \in \text{Aut}(D)$. Therefore,

$$
\begin{aligned}
(rx + sy)TT_{a^{-1}} &= a^{-1}(rt)ax + a^{-1}(st)ay \\
&= (rtt_a)x + (stt_a)y = (rx + sy)U_{tt_a},
\end{aligned}
$$

where t_a is the inner automorphism of D induced by a. Thus $TT_{a^{-1}} = U_{tt_a}$, so that (Theorem 10.6.10) T and U_{tt_a} induce the same permutation of M. Putting these facts together, we have that the subgroup of $P\Gamma L(V)$ fixing three letters is isomorphic to $\text{Aut}(D)$. Since the subgroup $GL(V)$ of $\Gamma L(V)$ induces a 3-transitive group (Theorem 10.6.6) on M, $P\Gamma L(V)$ is 3-transitive. $\parallel$

In order to determine the size of $P\Gamma L(V)$ in the finite case, a lemma from field theory is needed.

10.6.12. *If F is a field of order p^n, $p \in \mathscr{P}$, then $\text{Aut}(F)$ is cyclic of order n and is generated by $T \colon yT = y^p$ for $y \in F$.*

Proof. Since the binomial coefficient $\binom{p}{n}$ is an integer divisible by p if $0 < n < p$,

$$
\begin{aligned}
(y + z)T = (y + z)^p &= \Sigma \left\{ \binom{p}{n} y^n z^{p-n} \,\Big|\, 0 \leq n \leq p \right\} \\
&= y^p + z^p = yT + zT.
\end{aligned}
$$

Moreover,

$$(yz)T = (yz)^p = y^p z^p = (yT)(zT).$$

If $x \in \text{Ker}(T)$, then $x^p = 0$, hence $x = 0$. Therefore $T \in \text{Aut}(F)$. Now $F^{\#} = \langle x \rangle$ for some $x \in F$ of multiplicative order $p^n - 1$ (Theorem 5.7.8). If $r < n$, then $xT^r = x^{p^r} \neq x$. Hence $o(T) \geq n$. If $y \in F^{\#}$, then

$$yT^n = y^{p^n} = y^{p^n - 1}y = y$$

by Lagrange's theorem. Hence $T^n = I$, and $o(T) = n$.

Suppose that $o(\text{Aut}(F)) = m > n$. Let P be the prime subfield of F and B a basis of F over P. By the theory of linear equations over fields, there are $a_U \in F$, $U \in \text{Aut}(F)$, such that not all a_U are 0, and such that

$$\Sigma \left\{ a_U(bU) \,\big|\, U \in \text{Aut}(F) \right\} = 0 \quad \text{for all } b \in B.$$

By linearity, $\Sigma\, a_U(yU) = 0$ for all $y \in F$. Hence $\Sigma\, a_U U = 0$. Let $c_1, \ldots, c_r$ be elements of $F^\#$ such that $\Sigma\, c_i U_i = 0$ for distinct $U_1, \ldots, U_r$ in $\text{Aut}(F)$, and r minimal. Clearly $r > 1$. Then for any $y \in F^\#$,

$$0 = \Sigma\, c_i(xy)U_i = \Sigma\, c_i(xU_i)(yU_i),$$
$$0 = \Sigma\, c_i(xU_1)(yU_i),$$
$$0 = \Sigma\, c_i(xU_i - xU_1)(yU_i).$$

In the last equation, the coefficient of yU_1 is 0, and that of yU_2 is not 0 since, if $xU_1 = xU_2$, then $U_1 = U_2$ (for x generates F). Thus the minimality of r is contradicted. Therefore $o(\text{Aut}(F)) \leq n$. Hence $\text{Aut}(F) = \langle T \rangle$.

10.6.13. *If F is a field of order p^n, $p \in \mathscr{P}$, then $P\Gamma L(2, F)$ has a faithful 3-transitive representation of degree $p^n + 1$ and order $(p^n + 1)p^n(p^n - 1)n$.*

Proof. This follows from Theorems 10.6.11 and 10.6.12. ‖

There is a second class of exactly 3-transitive finite groups described in the next theorem.

10.6.14. *Let F be a field of order p^{2n}, $p \in \mathscr{P}$, p odd, H the subgroup of index 2 in $F^\#$, t the automorphism $ct = c^{p^n}$ of F, V a vector space over F with basis $\{x, y\}$, and G the subset of $\Gamma L(2, F)$ such that either*

(i) $T \in GL(V)$ *and* $\text{Det}(T) \in H$, *or*

(ii) T *is semi-linear with associated automorphism t and $xT = ax + by$, $yT = cx + dy$ with $ad - bc \notin H$.*

Then G induces an exactly 3-transitive group P of permutations on the set M of 1-dimensional subspaces of V, and $P \not\cong PGL(2, F)$ (as groups).

Proof. First note that $t \in \text{Aut}(F)$ and $o(t) = 2$ by Theorem 10.6.12. Also, since $F^\#$ is cyclic of even order, there is a unique subgroup H of index 2 in $F^\#$.

For T as in (ii), let $\text{Det}(T) = ad - bc$. Now let $R \in G$ and $S \in G$ with associated automorphisms r and s, respectively. One then verifies that

$$\text{Det}(RS) = (\text{Det}(R)s)(\text{Det}(S)), \tag{2}$$

and rs is the automorphism associated with RS. Since $H \in \text{Char}(F^\#)$, it follows from (2) that $\text{Det}(RS) \equiv \text{Det}(R)\text{Det}(S) \pmod{(H)}$. One then checks that in each of the four possible cases, $RS \in G$. Hence G is a group.

Since $SL(V) \subset G$ and $SL(V)$ is 2-transitive (Exercise 10.6.20), so is G. Let $F(ax + by)$ be distinct from Fx, Fy, and $F(x + y)$. If $ab \in H$, let $T \in GL(V)$ be such that $xT = ax$ and $yT = by$. Then T fixes Fx and Fy, $(F(x + y))T = F(ax + by)$, and $\text{Det}(T) = ab \in H$. Hence, $T \in G$. If $ab \notin H$, let T be given by

$$(cx + dy)T = (ct)(ax) + (dt)(by).$$

Then T is semi-linear with associated automorphism t, fixes Fx and Fy, maps $F(x + y)$ onto $F(ax + by)$, and $\mathrm{Det}(T) = ab \notin H$. Hence, $T \in G$. Thus, G is 3-transitive.

The product of elements T and T' of G both satisfying (ii) is in $GL(V)$ by (2). Hence G and $GL(V)$ both have a subgroup of index 2 consisting of those T satisfying (i). Therefore, $o(G) = o(GL(V))$. A scalar transformation $T \in GL(V)$, $xT = ax$, $yT = ay$, has determinant $a^2 \in H$, hence G contains all scalar transformations. It follows that the group P of permutations of M induced by G is 3-transitive and has the same order as the exactly 3-transitive group of permutations induced by $GL(V)$ (Theorem 10.6.8). Therefore (Theorem 10.4.3), P is also exactly 3-transitive.

Let Q be the subgroup of $GL(V)$ consisting of all T whose matrix with respect to the basis (x, y) has the form

$$\begin{bmatrix} 1 & b \\ 0 & 1 \end{bmatrix}.$$

Then Q is a Sylow p-subgroup of both $GL(V)$ and G, and since Q contains no scalar transformations except I, it is faithfully represented in the respective permutation groups by Q^*, say. Now Q^* fixes one letter (namely Fy), hence no element of $\Gamma L(V)$ moving all letters can normalize Q. By Theorem 10.6.1, Q^* is a regular normal subgroup of the subgroup P_1 of P which fixes the appropriate letter. Hence, in P, $N(Q^*)/Q^* \cong P_{12}$, the subgroup fixing both Fx and Fy. Similar remarks apply to the permutation group induced by $GL(V)$. In $GL(V)$, the group of permutations fixing both Fx and Fy is cyclic, being generated by (the image of)

$$\begin{bmatrix} 1 & 0 \\ 0 & b \end{bmatrix}$$

where $F^{\#} = \langle b \rangle$. In G, the corresponding group is non-Abelian. For let

$$(cx + dy)T = (ct)x + (dt)by,$$
$$(cx + dy)U = cx + b^2\, dy.$$

Then $\mathrm{Det}(T) = b \notin H$ and T has associated automorphism t, hence $T \in G$; $U \in GL(V)$ and $\mathrm{Det}(U) = b^2 \in H$, hence $U \in G$. We have

$$xTU = x = xUT, \quad yTU = (by)U = b^3 y,$$
$$yUT = (b^2 y)T = (b^2 t)by = b^{1+2p^n} y.$$

But $3 < 1 + 2p^n < p^{2n} - 1$, hence $yTU \neq yUT$ and $TU \neq UT$. Since $xTU = xUT = x$, $TU \neq UT$ modulo the scalar transformations. Hence, in P, $N(Q^*)/Q^*$ is non-Abelian. Therefore the two permutation groups are not isomorphic. $\parallel$

It was proved by Zassenhaus [1] that the only finite exactly 3-transitive groups are those given in Theorems 10.6.8 and 10.6.14 (see also Huppert [5]).

After some preliminary lemmas, the 5-transitive Mathieu groups will be constructed (see Witt [1]).

10.6.15. *If (M, G) is 2-transitive, $a \in M$, $x \in G$, and $ax \neq a$, then $G = G_a \cup G_a x G_a$.*

Proof. Since $x \notin G_a$, by Theorem 1.7.1, $G_a \cap G_a x G_a = \varnothing$. Let $ax = b$, $c \in M$, $c \neq a$. Then $\exists\, y \in G_a$ such that $by = c$ since G is 2-transitive. Thus $exy \in G_a x G_a$ and $a(exy) = c$. By Theorem 10.1.2, $G_a x G_a$ contains the right coset of G_a which sends a into c. Since c was arbitrary, $G = G_a \cup G_a x G_a$. ‖

If (M, G) is a permutation group and S is a set such that $S \cap M = \varnothing$, then there is a natural way of considering G as a group of permutations on $M \cup S$ by defining $sg = s$ for all $s \in S$ and $g \in G$. It will be convenient to denote the new group obtained in this manner by G also.

10.6.16. *If (M, G) is k-transitive, $k > 1$, $y \in G$, $b \in M$, $by \neq b$, $x \in \mathrm{Sym}(M \cup \{a\})$, $ax \neq a$, $H = \langle G, x \rangle$, $x^2 = y^2 = (xy)^3 = e$, and $xG_bx = G_b$, then H is $(k+1)$-transitive and $H_a = G$.*

Proof. Let $K = G \cup GxG$. Then $K \neq \varnothing$ and K is closed under the taking of inverses (since $x^2 = e$). Now $xyxyxy = e$, hence $xyx = yxy$. Therefore by Theorem 10.6.15

$$xGx = x(G_b \cup G_b y G_b)x = xG_b x \cup xG_b y G_b x$$
$$= G_b \cup xG_b xxyxxG_b x$$
$$= G_b \cup G_b xyx G_b = G_b \cup G_b yxy G_b$$

which is a subset of $G \cup GxG = K$. It follows readily that K is closed under multiplication. Hence K is a group. But $K \subset H \subset K$, so that $H = K$. Since an orbit of K containing a contains a letter of M, and since G is transitive on M, K is transitive. It follows from the form of K that $K_a = G$. By Theorem 10.4.2, K is $(k+1)$-transitive.

10.6.17. *If (M, G) is 2-transitive, $y \in G$, $a \in M$, $ay \neq a$, x_1, x_2, and x_3 are in $\mathrm{Sym}(M \cup \{1, 2, 3\})$,*

$$x_1 = (1, a)(2)(3) \cdots, x_2 = (1, 2)(3)(a) \cdots,$$
$$x_3 = (2, 3)(1)(a) \cdots.$$
$$y^2 = x_1^2 = x_2^2 = x_3^2 = e,$$
$$(x_1 y)^3 = (x_2 x_1)^3 = (x_3 x_2)^3 = e,$$
$$(y x_2)^2 = (y x_3)^2 = (x_1 x_3)^2 = e,$$
$$x_1 G_a x_1 = x_2 G_a x_2 = x_3 G_a x_3 = G_a,$$

then $H = \langle G, x_1, x_2, x_3 \rangle$ is 5-transitive on $M \cup \{1, 2, 3\}$ and $H_{123} = G$.

Proof. By Theorem 10.6.16, $K = \langle G, x_1 \rangle$ is 3-transitive, and $K_1 = G$. To apply Theorem 10.6.16 again, note that since $(yx_2)^2 = e$, $y^2 = e$, and $x_2^2 = e$, $y^{x_2} = y$. Therefore,

$$x_2 K_1 x_2 = G^{x_2} = \langle G_a, y \rangle^{x_2} = \langle G_a, y \rangle = G = K_1.$$

Hence, by Theorem 10.6.16, if $L = \langle K, x_2 \rangle$, then L is 4-transitive and $L_2 = K$. Again,

$$x_3 L_2 x_3 = K^{x_3} = \langle G, x_1 \rangle^{x_3} = \langle G_a, y, x_1 \rangle^{x_3} = \langle G_a, y, x_1 \rangle = L_2.$$

Hence, by Theorem 10.6.16 again, $H = \langle L, x_3 \rangle$ is 5-transitive, and $H_3 = L$. Therefore, $H_{123} = L_{12} = K_1 = G$.

10.6.18. *If* $A = \langle s, t, u, v, w, x_1, x_2, x_3 \rangle$, *where*

$$s = (4, 5, 6)(7, 8, 9)(10, 11, 12),$$
$$t = (4, 7, 10)(5, 8, 11)(6, 9, 12),$$
$$u = (5, 7, 6, 10)(8, 9, 12, 11),$$
$$v = (5, 8, 6, 12)(7, 11, 10, 9),$$
$$w = (5, 11, 6, 9)(7, 12, 10, 8),$$
$$x_1 = (1, 4)(7, 8)(9, 11)(10, 12),$$
$$x_2 = (1, 2)(7, 10)(8, 11)(9, 12),$$
$$x_3 = (2, 3)(7, 12)(8, 10)(9, 11),$$

then A *is an exactly 5-transitive group of degree* 12.

Proof. $H = \langle s, t \rangle$ is a regular elementary Abelian group of degree 9. $Q = \langle u, v, w \rangle = \langle u, v \rangle$ is a regular group of degree 8 which is a quaternion group. Moreover u and v each induce an automorphism of H fixing only the identity, as does $u^2 = v^2 = w^2$. Hence if $G = \langle s, t, u, v, w \rangle$, then $G = HQ$, $H \triangleleft G$, $H \cap Q = E$. G is an exactly 2-transitive group of degree 9, and $G_4 = Q$. Let

$$y = (4, 6)(7, 12)(8, 11)(9, 10).$$

Then $y = s^{-1} u^2 s \in G$. Now let $a = 4$ and refer to Theorem 10.6.17. Then x_1, x_2, and x_3 have the proper form, and the equations

$$y^2 = x_1^2 = x_2^2 = x_3^2 = e,$$
$$(x_1 y)^3 = (x_2 x_1)^3 = (x_3 x_2)^3 = e$$
$$(yx_2)^3 = (yx_3)^3 = (x_1 x_3)^3 = e$$

are satisfied. Further,

$$u^{x_1} = v, \quad v^{x_1} = u, \quad u^{x_2} = u^{-1},$$
$$v^{x_2} = w, \quad u^{x_3} = v^{-1}, \quad v^{x_3} = u^{-1},$$

so that $G_a^{x_1} = G_a^{x_2} = G_a^{x_3} = G_a$. By Theorem 10.6.17, A is 5-transitive of degree 12, and the subgroup fixing 1, 2, and 3 is G. Since G is exactly 2-transitive, A is exactly 5-transitive. ‖

The group A is the *Mathieu group* M_{12} of degree 12 and order $12!/7! = 95,040$. The subgroup M_{11} fixing one letter is the Mathieu group of degree 11 and order 7920. It is exactly 4-transitive.

In the case of the Mathieu group of degree 24, the situation is a little more complex. If V is a vector space of dimension 3 over F with basis $\{r, s, t\}$, let (for the next proof only) $(a, b, c) = F(ar + bs + ct)$ for $(a, b, c) \neq (0, 0, 0)$. Thus $(a, b, c) = (ia, ib, ic)$ for $i \in F^{\#}$.

10.6.19. *Let* $F = \{0, 1, d, d^2\}$ *be a field,* V *a 3-dimensional vector space over* F, M *the set of 1-dimensional subspaces of* V, (M, G) *the permutation group induced by* $PSL(V)$, I, II, *and* III *distinct letters not in* M, *and*

$$x_1 = (\text{I}, (1, 0, 0)) \cdot h_1,$$
$$x_2 = (\text{I}, \text{II}) \cdot h_2,$$
$$x_3 = (\text{II}, \text{III}) \cdot h_3,$$

where

$(a, b, c)h_1 = (a^2 + bc, b^2, c^2).$
$(a, b, c)h_2 = (a^2, b^2, c^2 d).$
$(a, b, c)h_3 = (a^2, b^2, c^2).$

Then the group $H = \langle G, x_1, x_2, x_3 \rangle$ *is 5-transitive of degree 24 on* $M \cup \{\text{I, II, III}\}$, *and of order*

$$24 \cdot 23 \cdot 22 \cdot 21 \cdot 20 \cdot 48 = 244{,}823{,}040.$$

Proof. Note first that each h_i fixes $(1, 0, 0)$. Since the polynomials appearing in the definitions of the h_i are homogeneous of degree 2, each h_i is a function from M into M. Since the function $x \longrightarrow x^2$ is an automorphism of F, each h_i maps M onto M, hence is a permutation of M. Also,

$$o(M) = (4^3 - 1)/(4 - 1) = 21,$$

By Exercise 10.6.20, $PSL(V)$ is 2-transitive on M.

Let y be the function given by the equation

$$(a, b, c)y = (b, a, c).$$

Then y is a permutation of M induced by an element of $GL(V)$ whose determinant is $-1 = 1$ (since F has characteristic 2). Let us now verify the hypotheses of Theorem 10.6.17. The elements x_1, x_2, and x_3 are of the right form.

The first ten of the thirteen equations in Theorem 10.6.17 can be verified directly, or in the following way. Number the elements of M:

$$1 = (1, 0, 0), \quad 2 = (1, 0, 1), \quad 3 = (1, 0, d),$$
$$4 = (1, 0, d^2), \quad 5 = (1, 1, 0), \quad 6 = (1, 1, 1),$$
$$7 = (1, 1, d), \quad 8 = (1, 1, d^2), \quad 9 = (1, d, 0),$$
$$10 = (1, d, 1), \quad 11 = (1, d, d), \quad 12 = (1, d, d^2),$$
$$13 = (1, d^2, 0), \quad 14 = (1, d^2, 1), \quad 15 = (1, d^2, d),$$
$$16 = (1, d^2, d^2), \quad 17 = (0, 0, 1), \quad 18 = (0, 1, 0),$$
$$19 = (0, 1, 1), \quad 20 = (0, 1, d), \quad 21 = (0, 1, d^2).$$

Then

$$y = (1, 18)(2, 19)(3, 20)(4, 21)(9, 13)(10, 16)(11, 14)(12, 15),$$
$$x_1 = (1, I)(3, 4)(6, 19)(7, 10)(8, 14)(9, 13)(12, 21)(15, 20),$$
$$x_2 = (I, II)(2, 3)(6, 7)(9, 13)(10, 15)(11, 14)(12, 16)(19, 20),$$
$$x_3 = (II, III)(3, 4)(7, 8)(9, 13)(10, 14)(11, 16)(12, 15)(20, 21).$$

The equations

$$y^2 = x_1^2 = x_2^2 = x_3^2 = e,$$
$$(yx_1)^3 = (x_1x_2)^3 = (x_2x_3)^3 = e,$$
$$(yx_2)^2 = (yx_3)^2 = (x_1x_3)^2 = e,$$

may now be verified.

Finally, it must be checked that $x_s G_1 x_s = G_1$ for $s = 1, 2, 3$. Let $T \in G_1$. Then there are elements f, g, h, i, j, and k of F such that

$$(a, b, c)T = (a + fb + gc, hb + ic, jb + kc), \quad hk + ij = 1.$$

Each $x_s T x_s$ fixes I, II, III, and $(1, 0, 0)$. Hence $x_s T x_s = h_s T h_s$. We have

$$(a, b, c)h_3 T h_3 = (a^2, b^2, c^2)T h_3$$
$$= (a^2 + fb^2 + gc^2, hb^2 + ic^2, jb^2 + kc^2)h_3$$
$$= (a + f^2 b + g^2 c, h^2 b + i^2 c, j^2 b + k^2 c).$$

Since

$$h^2 k^2 - i^2 j^2 = h^2 k^2 + i^2 j^2$$
$$= (hk + ij)^2 = 1,$$

$$h_3 T h_3 \in G_1.$$

Similarly,

$$(a, b, c)h_2 T h_2 = (a^2, b^2, c^2 d)T h_2$$
$$= (a^2 + fb^2 + gc^2 d, hb^2 + ic^2 d, jb^2 + kc^2 d)h_2$$
$$= (a + f^2 b + g^2 d^2 c, h^2 b + i^2 d^2 c, j^2 db + k^2 c),$$

and $h^2k^2 + i^2d^2j^2d = (hk + ij)^2 = 1$. Hence $h_2Th_2 \in G_1$. Finally,

$$(a, b, c)h_1Th_1 = (a^2 + bc, b^2, c^2)Th_1$$
$$= (a^2 + bc + fb^2 + gc^2, hb^2 + ic^2, jb^2 + kc^2)h_1$$
$$= (a + b^2c^2 + f^2b + g^2c + hjb + hkb^2c^2 + ijb^2c^2 + ikc,$$
$$h^2b + i^2c, j^2b + k^2c)$$
$$= (a + (f^2 + hj)b + (g^2 + ik)c, h^2b + i^2c, j^2b + k^2c),$$

and $h^2k^2 + i^2j^2 = 1$, as before. Hence $h_1Th_1 \in G_1$. Thus all $x_iG_1x_i = G_1$.

By Theorem 10.6.17, H is 5-transitive of degree 24, and has G as the subgroup fixing I, II, and III. Since

$$o(G) = o(PSL(V)) = \frac{63 \cdot 60 \cdot 48}{3 \cdot 3} = 21 \cdot 20 \cdot 48$$

by Theorem 5.7.21 and the fact that all three scalar transformations are in $SL(V)$, the assertion about $o(H)$ follows. ‖

The group H in Theorem 10.6.19 is the *Mathieu group* M_{24} of degree 24. The subgroup fixing one letter is the 4-transitive Mathieu group M_{23}, and the subgroup fixing two letters is the 3-transitive Mathieu group M_{22}. The simplicity of M_{11}, M_{12}, M_{22}, M_{23}, and M_{24} will be shown in Theorem 10.8.9.

EXERCISES

10.6.20. Let F be a field and V a k-dimensional vector space over F, $1 < k < \infty$. Then $SL(V)$ induces a 2-transitive group P of permutations of the set of 1-dimensional subspaces of V, and $P \cong PSL(V)$.

10.6.21. If V is a vector space over a division ring D, then $\Gamma L(V)$ is a group.

10.6.22. Let A be an infinite cardinal number. Use Theorem 10.6.15 to construct a group G, subgroup H, $x \in G$, such that $G = \langle H, x \rangle$ and $[G:H] > A$.

10.6.23. Let F be a near field and $a \in F$.
   (a) $0a = 0$.
   (b)   If $o(F) > 2$, then $a0 = 0$.

10.6.24. There is a near field of order 2 in which $1 \cdot 0 = 1$. (An additional assumption is usually made in the definition of near field which rules out this example.)

10.7   Numerical applications

Some further examples showing application of some of the theorems in this chapter will be given in this section.

10.7.1. *If $o(G) = 420$, then G is not simple.*

Proof. Deny. By Sylow, $n_7 = 15$, hence if $P \in \mathrm{Syl}_7(G)$, then $o(N(P)) = 28$. By the N/C theorem and Burnside, $o(C(P)) = 14$. Thus $\exists\, x \in G$ such that $o(x) = 14$. In the representation of degree 15 on $N(P)$, by 10.2.10, x^2 has cyclic decomposition 7^2–1 (i.e., is the product of two 7-cycles and one 1-cycle). Hence x has cyclic decomposition 14–1. By Exercise 10.4.9, x is an odd permutation. Therefore (Theorem 10.4.7), G has a normal subgroup of index 2. ‖

In the preceding example, use was made of an element in $C(P)$, $P \in \mathrm{Syl}_p(G)$. Sometimes one may make use of the permutation representation of an element in $N(P)$.

10.7.2. *There are no simple groups of order 264.*

Proof. Suppose that G is such a group. By Sylow, $n_{11} = 12$. If $P \in \mathrm{Syl}_{11}(G)$, then $o(N(P)) = 22$ and $o(C(P)) = 11$. Let $x \in N(P) \backslash C(P)$. In the representation of degree 12 on $N(P)$, since $x \in N(P)$, $\mathrm{Ch}(x) > 0$. By Theorem 10.2.11, $\mathrm{Ch}(x) = 2$. Hence x is the product of five 2-cycles, and is therefore an odd permutation. By Theorem 10.4.7, G has a normal subgroup of index 2.

10.7.3. *If G is a finite group, P and Q are distinct Sylow p-subgroups whose intersection is maximal, then $N(P \cap Q)$ has more than one Sylow p-subgroup (of itself), and if $R \in \mathrm{Syl}_p(N(P \cap Q))$, then $o(R) > o(P \cap Q)$.*

Proof. By Theorem 6.3.9, $N(P \cap Q)$ has subgroups P_1 and Q_1 such that $P \cap Q < P_1 \subset P$ and $P \cap Q < Q_1 \subset Q$. This proves the second assertion in the theorem. If $N(P \cap Q)$ has a single Sylow p-subgroup R, then $P_1 \subset R$ and $Q_1 \subset R$. Now $\exists\, S \in \mathrm{Syl}_p(G)$ such that $R \subset S$. Then $S \cap P \supset P_1 > P \cap Q$, hence $S = P$. Similarly $S = Q$, a contradiction. The theorem follows. ‖

There is a counting technique sometimes useful in connection with permutation representations.

10.7.4. *If $x \in H < G$, G is finite, and $f(x) = o(\mathrm{Cl}(x) \cap H)$, then in the representation of G on H,*

$$o(\mathrm{Cl}(x)) = \frac{[G:H]f(x)}{\mathrm{Ch}(x)}.$$

Proof. H is the subgroup of elements of G whose images in the permutation representation fix one letter, hence, somewhat incorrectly, $H = G_a$. By a slight extension of Theorem 10.1.2, each G_b is $y^{-1}G_a y$ for some $y \in G$. Hence $f(x) = o(\mathrm{Cl}(x) \cap G_b)$ for each letter b. Therefore the number of ordered pairs (b, y) with $y \in G_b \cap \text{Cl}(x)$ is $[G:H]f(x)$. But if $y \in \text{Cl}(x)$, then $\text{Ch}(y) = \text{Ch}(x)$, so that there are $\text{Ch}(x)$ letters b for which $y \in G_b$. Hence the number of pairs (b, y) fulfilling the condition is $\text{Ch}(x) \cdot o(\text{Cl}(x))$. Therefore

$$\text{Ch}(x) \cdot o(\text{Cl}(x)) = [G:H] f(x),$$

and the theorem follows.

10.7.5. *If $o(G) = 1008$, then G is not simple.*

Proof. Deny, and note that $1008 = 2^4 \cdot 3^2 \cdot 7$. If $n_7 = 8$, then $o(N(P)) = 2 \cdot 3^2 \cdot 7$, $P \in \text{Syl}_7(G)$, and by the N/C theorem, $3 \mid o(C(P))$. Hence there is an element of order 21. This is impossible in a permutation group of degree 8 (all nontrivial representations of G are faithful, since G is simple). Therefore, by Sylow, $n_7 = 36$, $o(N(P)) = 28$, and $o(C(P)) = 14$. Hence $\exists\, x \in C(P)$ with $o(x) = 14$. By Theorem 10.2.10, in the representation of degree 36 of G on $N(P)$, x^2 has decomposition 7^5–1, so that x has decomposition 14^2–7–1 (otherwise x is odd). Thus $\text{Ch}(x^7) = 8$. On the other hand, if $y \in N(P)\backslash C(P)$, then, by Theorem 10.2.11, $\text{Ch}(y) \leq 6$. Hence x^7 is not conjugate to y. Since $C(P)$ is cyclic, there is only one element of order 2 in $C(P)$. Using the notation of Theorem 10.7.4, $f(x^7) = 1$, hence by that theorem, $o(\text{Cl}(x^7)) = 36 \cdot \frac{1}{8}$, which is not an integer. Therefore the theorem holds.

10.7.6. *If $o(G) = 1080$, then G is not simple.*

Proof. Deny, and note that $1080 = 2^3 \cdot 3^3 \cdot 5$. Now $n_5 \neq 6$ by Theorem 10.2.6, and $n_5 \neq 216$ by Burnside. Hence $n_5 = 36$, $o(N(P)) = 30$ for $P \in \text{Syl}_5(G)$, and by the usual procedure, $o(C(P)) = 15$. The subgroup H of $C(P)$ of order 3 is characteristic in $C(P)$ since $C(P)$ is cyclic. Hence $H \lhd N(P)$. But $3^2 \mid o(N(H))$, hence $2 \cdot 3^2 \cdot 5 \mid o(N(H))$ and $N(H) > N(P)$. By Sylow's theorem applied to $N(H)$, $n_5(N(H)) = 6$ and $o(N(H)) = 2^2 \cdot 3^2 \cdot 5$. Thus, G has a permutation representation of degree 6 on $N(H)$, an impossibility by Theorem 10.2.6.

EXERCISES

10.7.7. There are no simple groups of order 180, 240, 528, 560, 672, 792, 840, or 1056. (Proof like that of Theorem 10.7.1.)

10.7.8. There are no simple groups of order 760 or 1104. (Proof like Theorem 10.7.2.)

10.7.9. There are no simple groups of order 120 or 336.

10.7.10. If G is a simple noncyclic group of order ≤ 200, then $o(G) = 60$ or 168.

10.7.11. If $o(G) = p^n q$, p and q primes, then G is solvable. (Use Theorem 10.7.3.)

10.7.12. There is no simple group of order 612. (Use Theorem 10.7.4.)

10.7.13. There is no simple group of order 2016 or 1440. (Proof similar to that for 1008, but harder.)

10.7.14. (See Theorem 10.7.4.) If $K \subset H < G$, G is finite, and $f(K)$ is the number of conjugates of K (by elements of G) contained in H, then in the representation of G on H, $o(Cl(K)) = [G:H]f(K)/Ch(K)$.

10.8   Some simple groups

The simplicity of the Mathieu groups, the alternating groups of degree greater than 4, and $PSL(n, F)$ (with two exceptions) will be proved in this section.

Let F be a field, $n \in \mathcal{N}$, and V a vector space of dimension n over F. Recall that $GL(n, F) \cong GL(V) = \mathrm{Aut}(V)$, $SL(n, F) \cong SL(V)$ which is the subgroup of $GL(V)$ consisting of all linear transformations of determinant 1; and $PSL(n, F) \cong PSL(V)$, the factor group of $SL(V)$ by the group of scalar transformations.

A *transvection* is a $g \in SL(V)$ such that there is a subspace W of V and an $x \in V$ such that $V = W + Fx$ and $wg = w$ for all $w \in W$. It follows that $xg = x + w_1$, $w_1 \in W$.

10.8.1.   *If $n \geq 3$, then all transvections ($\neq e$) are conjugate in $SL(V)$.*

Proof. Let g and g' be non-e transvections such that $V = W + Fx$, $wg = w$ for all $w \in W$, $xg = x + w_1$, $w_1 \in W$, $V = W' + Fx'$, $w'g' = w'$ for all $w' \in W'$, $x'g' = x' + w_1'$, $w_1' \in W'$. There are bases $(w_1, \ldots, w_{n-1})$ of W and $(w_1', \ldots, w_{n-1}')$ of W'. There is an element t of $SL(V)$ such that

$$xt = x', \quad w_i t = w_i' \quad \text{for} \quad 1 \leq i < n - 1,$$
$$w_{n-1}t = cw_{n-1}', \quad c \in F^{\#}.$$

It is readily verified that $t^{-1}gt = g'$. ∥

For information on the case $n = 2$, see Exercises 10.8.10 and 10.8.11.

10.8.2.   *Let F be a field, $G = SL(n, F)$, $n > 1$, and $o(F) > 3$ if $n = 2$. Then $G = G^1$.*

Proof. Let V be as before, and let $B = (x_1, \ldots, x_n)$ be a basis of V. First suppose that $n \geq 3$. There are a and b in G such that

$$x_1 a = x_1 - x_2, x_i a = x_i \quad \text{for} \quad i > 1,$$
$$x_2 b = x_2 - x_3, x_i b = x_i \quad \text{for} \quad i \neq 2.$$

Then

$$x_1[a, b] = (x_1 + x_2)(b^{-1}ab) = (x_1 + x_2 + x_3)(ab)$$
$$= (x_1 + x_3)b = x_1 + x_3 \neq x_1,$$
$$x_i[a, b] = x_i b^{-1}ab = x_i b^{-1}b = x_i \quad \text{for} \quad i \neq 1.$$

Hence $[a, b]$ is a transvection different from e. By the preceding theorem, G^1 contains all transvections. Hence (Exercise 10.8.12), $G^1 = G$.

Next let $n = 2$. Since $o(F) > 3$, $\exists\, c \in F$ such that $c \neq 0$, 1, or -1. Let $a \in G$, $b \in G$, and $d \in F$ be such that

$$x_1a = c^{-1}x_1, \; x_2a = cx_2, \; x_1b = x_1 + dx_2, \; x_2b = x_2.$$

Then

$$x_1[a, b] = cx_1(b^{-1}ab) = (cx_1 - cdx_2)ab = (x_1 - c^2dx_2)b$$
$$= x_1 + d(1 - c^2)x_2,$$
$$x_2[a, b] = (c^{-1}x_2)(b^{-1}ab) = (c^{-1}x_2)(ab) = x_2b = x_2.$$

Since $1 - c^2 \neq 0$, G^1 contains all transvections t of the type

$$x_1t = x_1 + kx_2, \; x_2t = x_2, \; k \in F.$$

It follows (see Exercise 10.8.11) that G^1 contains all transvections, hence equals G (Exercise 10.8.12).

10.8.3. *If F is a field, $G = SL(n, F)$ where $n \geq 2$ and $o(F) > 3$ if $n = 2$, $H \lhd G$, and H is not contained in the group of scalars, then $H = G$.*

Proof. Let V be an n-dimensional vector space over F. WLOG, $G = SL(V)$. By Exercise 10.6.20, there is a natural homomorphism T of G onto a 2-transitive permutation group P on the set M of 1-dimensional subspaces of V. Ker(T) is the group of scalar transformations in G. Hence HT is a normal subgroup of P different from E, so (Theorems 10.5.8 and 10.5.11) HT is transitive. Let $V = W + Fx$, $x \in V$. Let A be the subgroup of all $g \in G$ such that gT fixes Fx. Since HT is transitive, $GT = (HT)(AT)$. Since $A \supset$ Ker(T), $G = HA$. Each $a \in A$ induces an automorphism at of W such that, for all $w \in W$,

$$wa = w(at) + cx, \, c \in F$$

(c depends on w). One verifies that t is a homomorphism from A into Aut(W) with kernel B, say. If $h \in H$ and $a \in A$, then

$$(HB)^{ha} = (HB)^a = H^aB^a = HB,$$

so that $HB \lhd G$. Let $(y_1, \ldots, y_{n-1})$ be a basis of W. Any transvection u such that $xu = x$, $y_iu = y_i$ if $i \neq 1$, and $y_1u = y_1 + cx$ for some $c \in F$, is in B.

By Theorem 10.8.1, Exercises 10.8.11 and 10.8.12, $HB = G$. Hence $G/H \cong B/B \cap H$. Now if $b_i \in B$ for $i = 1, 2$ and $w \in W$, then

$$wb_1b_2 = (w + c_1x)b_2 = w + c_2x + c_1x$$
$$= w + c_1x + c_2x$$
$$= wb_2b_1,$$

and, since $\text{Det}(b_i) = 1$,

$$xb_1b_2 = x = xb_2b_1.$$

Hence B is Abelian. Therefore $H \supset G^1 = G$ (Theorem 10.8.2). Hence $H = G$.

10.8.4. *If F is a field, $n \geq 2$, and $o(F) > 3$ if $n = 2$, then $PSL(n, F)$ is simple.*

Proof. This follows immediately from the preceding theorem and the isomorphism theorem. ‖

Dieudonné [1] has defined determinants of matrices whose elements are in a division ring, and has proved the theorems corresponding to Theorems 10.8.3 and 10.8.4 in this case. For an exposition of this more general case, see Artin [2]. Rosenberg [1] determined all normal subgroups of $GL(V)$ for an infinite dimensional vector space V over a division ring.

10.8.5

$$o(PSL(n, q)) = \frac{(\pi\{q^n - q^i \mid 0 \leq i \leq n - 1\})}{(q - 1)(q - 1, n)}.$$

Proof. The number of scalar matrices in $SL(n, q)$ equals the number of nth roots of 1 in F, which equals the number of nth roots of e in a cyclic group of order $q - 1$, which is $(q - 1, n)$. The result now follows from Theorem 5.7.21. ‖

The exceptional cases in Theorem 10.8.4 are actually not simple groups. Thus,

$$GL(2, 2) = SL(2, 2) \cong PSL(2, 2) \cong \text{Sym}(3),$$

since $PSL(2, 2)$ is (at least) 2-transitive on the three 1-dimensional subspaces of the corresponding vector space V, and is of order 6. Again, $PSL(2, 3)$ is 2-transitive of degree 4 and of order 12, hence is isomorphic to $\text{Alt}(4)$, and is therefore, not simple.

The orders of the smaller groups $PSL(2, q)$ are listed below:

q	4	5	7	9	8	11	13
$o(PSL(2, q))$	60	60	168	360	504	660	1092

In view of the fact (proved later in this section) that Alt(n) is simple for $n > 4$, it is of some interest what isomorphisms, if any, hold between groups of the systems $\{PSL(n, q)\}$ and $\{$Alt(n)$\}$. It has been shown (Artin [1]) that the only isomorphisms are

$PSL(2, 4) \cong PSL(2, 5) \cong$ Alt(5),

$PSL(2, 7) \cong PSL(3, 2)$,

$PSL(2, 9) \cong$ Alt(6),

$PSL(4, 2) \cong$ Alt(8).

Example. (P. Hall [4]) Theorem 9.3.14 cannot be improved to say that a supersolvable Hall P-subgroup of a finite group contains conjugates of every P-subgroup. In fact, the example will be of a finite group containing a supersolvable Hall P-subgroup and a nonsupersolvable Hall P-subgroup.

Let $G = PSL(2, 11)$. All we need to know about G is that it is simple (Theorem 10.8.4) and of order $660 = 2^2 \cdot 3 \cdot 5 \cdot 11$ (10.8.5). It follows that $n_{11} = 12$, and, if $P \in \mathrm{Syl}_{11}(G)$, then $o(N(P)) = 5 \cdot 11$ and $o(C(P)) = 11$. Let $Q \in \mathrm{Syl}_5(N(P))$. By Exercise 10.7.14 and Theorem 10.2.11, in the representation of G on $N(P)$, $o(\mathrm{Cl}(Q)) = 12 \cdot 11/2 = 66$, $o(N(Q)) = 2 \cdot 5$, and $o(C(Q)) = 5$ (this last by Burnside). Let $R \in \mathrm{Syl}_2(G)$ and $S \in \mathrm{Syl}_3(G)$. It follows from what has been shown that neither 5 nor 11 divides $o(C(R))$ or $o(C(S))$. Therefore, by Burnside and Sylow, $n_2 = n_3 = 55$, $o(N(R)) = 12$, $o(C(R)) = 4$, $o(N(S)) = 12$, $o(C(S)) = 6$. Since $N(S)$ has normal subgroups of orders 6 and 3, it is supersolvable. However, $N(R)$ has a normal subgroup of order 4 but none of order 2 (an element of order 3 in $N(R)$ must move all three elements of R). Hence $N(R)$ is not supersolvable (in fact $N(R) \cong$ Alt(4)). However, $N(R)$ and $N(S)$ are Hall subgroups of the same order.

10.8.6.   *If G is a primitive permutation group containing no regular normal subgroup, and if G_a is simple, then G is simple.*

Proof. Let $E < H \lhd G$. Since H is not regular, $H \cap G_b > E$ for some letter b. By Theorem 10.1.3, G_b is conjugate to G_a, hence is simple. Since $H \cap G_b \lhd G_b$, $H \cap G_b = G_b$, i.e., $G_b \subset H$. Since G is primitive, G_b is a maximal proper subgroup of G (Theorem 10.5.7), and H is transitive (Theorem 10.5.11). Hence, $H = G$. Therefore, G is simple.

10.8.7.   *If $4 < n < \infty$, then* Alt(n) *is simple.*

Proof. First let $n = 5$ and $E < H \lhd G =$ Alt(5). Since G is 3-transitive, it is primitive. Hence H is transitive, so $5 \mid o(H)$. If a Sylow 5-subgroup K were normal in G, then by the N/C theorem, $3 \mid o(C(K))$, and there is an element of order 15 which is impossible in a group of degree 5. Hence

$n_5(G) = 6$, and, since $H \supset K$, $n_5(H) = 6$, $30 \mid o(H)$. If $H \neq G$, then $o(H) = 30$, so, by Burnside (applied with $p = 2$), H has a normal subgroup L of index 2. L must contain all six conjugates of K in H, which it cannot do. Therefore, $H = G$, so that Alt(5) is simple.

Now assume inductively that Alt$(n - 1)$ is simple, $n \geq 6$. Alt(n) is 4-transitive, hence has no regular normal subgroup by Theorem 10.5.15. The subgroup fixing n is just Alt$(n - 1)$, hence is simple. By Theorem 10.8.6, Alt(n) is simple.

10.8.8. *If $4 < n < \infty$, then the only nontrivial, normal subgroup of* Sym(n) *is* Alt(n).

Proof. Let H be a nontrivial, normal subgroup of Sym(n). Then $H \cap$ Alt$(n) \lhd$ Alt(n). By the simplicity of Alt(n), this intersection is E or Alt(n). If it is E, then $o(H) = 2$, whereas H is transitive (as a normal subgroup of a primitive group), hence of order at least n. Hence $H \supset$ Alt(n). Since [Sym(n):Alt(n)] $= 2$, $H =$ Alt(n).

10.8.9. *The Mathieu groups* M_{11}, M_{12}, M_{22}, M_{23}, *and* M_{24} *are simple.*

Proof. Suppose that H is a nontrivial, normal subgroup of M_{11}. Since M_{11} is 4-transitive, it is primitive, and H is transitive. Therefore, there is a Sylow 11-subgroup P of M_{11} contained in H. By Theorem 10.3.5, $C(P) = P$. If $2 \mid o(N(P))$, then a element x of order 2 in $N(P)$ is the product of five 2-cycles (Theorem 10.2.11), hence is an odd permutation. But by Theorem 10.6.18, $M_{11} \subset$ Alt(11). Hence $2 \nmid o(N(P))$. By Theorem 6.2.4, $M_{11} = N(P)H$. Hence $o(N(P)) = 5 \cdot 11$ and $N(P) \cap H = P$. By Burnside (see also Exercise 6.2.19), there is a characteristic subgroup K of H such that $[H:K] = 11$. Then $K \lhd M_{11}$, $E \neq K$, and $11 \nmid o(K)$, contradicting an earlier statement. Hence M_{11} is simple.

By Theorem 10.6.19, the subgroup M_{21} fixing one letter in M_{22} is isomorphic to $PSL(3, 4)$, which is simple (Theorem 10.8.4). Suppose inductively that M_{n-1} is simple ($n = 12, 22, 23, 24$). Now M_n is primitive, being 3-transitive. Since M_n is 3-transitive, it is 2-primitive, hence by Theorem 10.5.15 (3), M_n contains no regular normal subgroup. By Theorem 10.8.6, M_n is simple, $n = 12, 22, 23$, or 24.

EXERCISES

10.8.10. If V is a vector space of dimension 2 over a field F, then all transvections are conjugate in $GL(V)$.

10.8.11. If V is a vector space of dimension 2 over a field F, u is a transvection of V, and (x, y) is a basis of V, then $\exists \, c \in F$ and a transvection t such that u is conjugate in $SL(V)$ to t, where $xt = x + cy$, $yt = y$.

10.8.12. The subgroup of $SL(V)$ generated by the transvections is $SL(V)$. (This is essentially a statement to the effect that a nonsingular matrix can be reduced to the identity by certain row and column transformations.)

10.8.13. (a) Any normal subgroup of Sym(n) is characteristic.

  (b) Any normal subgroup of Alt(n) is characteristic in both Alt(n) and Sym(n).

  (c) If $n > 3$, then $Z(\text{Alt}(n)) = E$. If $n > 2$, then $Z(\text{Sym}(n)) = E$.

10.8.14. Let V be a vector space over a field, S the group of scalar transformations, and $2 \leqq \text{Dim}(V) < \infty$.

  (a) If $T \in GL(V)$ centralizes $PSL(V)$ (this means that if $U \in SL(V)$, then $[T, U] \in S$), then $T \in S$.

  (b) $Z(GL(V)) = S$.

  (c) $C(SL(V)) = S$.

10.8.15. $PSL(2, 13)$ has no subgroup H of prime index. (If there is, then the index is 13, and a Sylow 7-subgroup of H is normal in H, hence in G.)

REFERENCES FOR CHAPTER 10

For the entire chapter, Wielandt [6]; Section 10.5, Wielandt and Huppert [1] (see also Itô [4]); Section 10.6, Baer [6]; Theorem 10.8.4, Iwasawa [1]; Theorem 10.8.9, Witt [1]; Exercise 10.8.15, Parker [1].

ELEVEN

SYMMETRIC AND ALTERNATING GROUPS

This chapter is concerned principally with the infinite symmetric and alternating groups, and certain of their subgroups. Smaller portions deal with some questions about finite symmetric and alternating groups, and with infinite permutation groups.

11.1 Conjugate classes

This section deals with finite groups only. Let $n \in \mathcal{N}$, and let f be a function such that $f(i)$ is a nonnegative integer for $1 \leq i \leq n$. Let $[f]$ or $[f(1), \ldots, f(n)]$ denote the set of $x \in \text{Sym}(n)$ such that the cyclic decomposition of x contains $f(i)$ i-cycles for $1 \leq i \leq n$.

11.1.1. *If f is a function with values nonnegative integers and $\Sigma \{f(i)i \mid 1 \leq i \leq n\} = n$, then $[f]$ is a conjugate class in $\text{Sym}(n)$. Conversely, if $x \in \text{Sym}(n)$, then there is a function f such that $\text{Cl}(x) = [f]$.*

Proof. This is just a restatement of Theorems 1.3.6 and Exercise 1.3.11.

11.1.2. *A conjugate class* [f] *of* Sym(n) *consists of even permutations iff* $\Sigma f(2i)$ *is even, and otherwise consists of odd permutations.*

Proof. This is a restatement of Exercise 10.4.9.

11.1.3. *The number of conjugate classes of* Sym(n) *equals the number of partitions of n.*

11.1.4. *If* [f] *is a conjugate class in* Sym(n), *then*

$$o([f]) = n!/((\pi i^{f(i)})(\pi(f(i)!)).$$

Proof. There are $n!$ ways of writing n letters as the product of $f(1)$ 1-cycles, followed by $f(2)$ 2-cycles, ..., followed by $f(n)$ n-cycles. Many of these yield the same permutation. In fact, any i-cycle may be started in any of i places (e.g., $(1, 2, 3) = (2, 3, 1) = (3, 1, 2)$), so that there are $i^{f(i)}$ ways of writing the $f(i)$ i-cycles of a given permutation, keeping the cycles in the same order. In addition, the i-cycles may be permuted in $f(i)!$ ways. As these are the only changes permitted, the theorem follows. ‖

Example. Determination of the orders of the conjugate classes in Sym(5). There are seven conjugate classes in Sym(5): $[5, 0, 0, 0, 0]$, $[3, 1, 0, 0, 0]$, $[2, 0, 1, 0, 0]$, $[1, 0, 0, 1, 0]$, $[1, 2, 0, 0, 0]$, $[0, 0, 0, 0, 1]$, and $[0, 1, 1, 0, 0]$. By the preceding theorem, these classes have 1, 10, 20, 30, 15, 24, and 20 elements, respectively. The class $[3, 1, 0, 0, 0]$, for example, has $5!/1^3 \cdot 2^1 \cdot 3! = 10$ elements. By Theorem 11.1.2, the classes $[3, 1, 0, 0, 0]$, $[1, 0, 0, 1, 0]$ and $[0, 1, 1, 0, 0]$ are odd, while the others are even. The fact that the sum of the orders of the even (or odd) classes is 60 furnishes a check on the work.

Next, the conjugate classes in Alt(n) will be found.

11.1.5. *If* [f] *is an even class in* Sym(n), *then*

(i)   *if* $f(2i) > 0$ *or* $f(2i + 1) > 1$ *for some i, then* [f] *is a class in* Alt(n);

(ii) *otherwise,* [f] *is the union of two equal sized classes in* Alt(n).

Proof. Let $x \in [f]$. Then

$$o(\text{Cl}(x) \text{ in Alt}(n)) = [\text{Alt}(n):\text{Alt}(n) \cap C(x)].$$

If $C(x) \not\subset \text{Alt}(n)$, then the second number equals

$$[\text{Alt}(n)C(x):C(x)] = [\text{Sym}(n):C(x)] = o(\text{Cl}(x) \text{ in Sym}(n)).$$

If, on the other hand, $C(x) \subset \text{Alt}(n)$, then

$$[\text{Alt}(n):\text{Alt}(n) \cap C(x)] = [\text{Sym}(n):C(x)]/2,$$

$$o(\text{Cl}(x) \text{ in Alt}(n)) = o(\text{Cl}(x) \text{ in Sym}(n))/2.$$

Hence it suffices to show that $C(x) \subset \text{Alt}(n)$ iff all $f(2i) = 0$ and all $f(2i + 1) = 0$ or 1.

If $f(2i) > 0$, then $x = cy$, where c is a $2i$-cycle and y fixes the letters moved by c. Therefore, $cy = yc$, so that $cx = xc$, and c is an odd permutation in $C(x)$, hence $C(x) \not\subset \text{Alt}(n)$. If $f(2i + 1) > 1$, then

$$x = (1, \ldots, 2i + 1)(1', \ldots, (2i + 1)')y = cc'y,$$

where y fixes the letters moved by c or c'. Then

$$(1, 1')(2, 2') \ldots (2i + 1, (2i + 1)')$$

is an odd permutation in $C(x)$. If, finally, all $f(2i) = 0$ and all $f(2i + 1) = 0$ or 1, then, by Theorem 11.1.4, $o(\text{Cl}(x)) = n!/j$, where j is odd, so that $o(C(x))$ is odd. Therefore every element of $C(x)$ is of odd order. Since every odd permutation is of even order, $C(x) \subset \text{Alt}(n)$. ∥

Note that a conjugate class of $\text{Sym}(n)$ splits in $\text{Alt}(n)$ iff each of its elements is the product of odd-length cycles of different lengths.

Example. The conjugate classes of $\text{Alt}(5)$. By the preceding example, there are four even classes in $\text{Sym}(5)$: [2, 0, 1, 0, 0], [1, 2, 0, 0, 0], [5, 0, 0, 0, 0], and [0, 0, 0, 0, 1]. By the preceding theorem, only the last of these splits in $\text{Alt}(5)$. Hence there are five classes in $\text{Alt}(5)$: [2, 0, 1, 0, 0], [1, 2, 0, 0, 0], [5, 0, 0, 0, 0], [0, 0, 0, 0, 1]$_1$, and [0, 0, 0, 0, 1]$_2$, with 20, 15, 1, 12, and 12 elements respectively.

EXERCISES

11.1.6. Determine the conjugate classes and their orders for $\text{Sym}(n)$, $n \leq 10$.

11.1.7. Determine the conjugate classes and their orders for $\text{Alt}(n)$, $n \leq 10$.

11.1.8. Prove that $\text{Alt}(5)$ is simple as follows. Let $E < H \lhd G$.

    (a) H is the set union of conjugate classes, one of which is $\{e\}$.

    (b) $H = \text{Alt}(5)$. [Use (a) and Lagrange.]

11.2 Infinite groups of finite permutations

For any infinite cardinal number A, let A^+ denote the next larger cardinal number. Let M be an infinite set and A an infinite cardinal such that $A \leq o(M)^+$. If $x \in \text{Sym}(M)$, let $S(x) = \{m \in M \mid mx \neq m\}$ be the set of letters moved by x. Let

$$\text{Sym}(M, A) = \{x \in \text{Sym}(M) \mid o(S(x)) < A\}.$$

Thus $\text{Sym}(M, o(M)^+) = \text{Sym}(M)$.

11.2.1. Sym(M, A) *is a normal subgroup of* Sym(M).

Proof. If x and y are in Sym(M, A), then $S(y^{-1}) = S(y)$, hence $S(xy^{-1})$ is a subset of $S(x) \cup S(y)$. Therefore,

$$o(S(xy^{-1})) \leq o(S(x)) + o(S(y)) < A.$$

Since $e \in$ Sym(M, A), Sym(M, A) $\subset G$. Normality follows from Theorem 1.3.6. ‖

A permutation moving only a finite number of letters will be called a *finite* permutation. If x is a finite permutation, then $x \mid S(x)$ is either even or odd. The set of finite even permutations on M is the *alternating group* Alt(M).

11.2.2. Alt(M) *is a normal subgroup of* Sym(M).

Proof. If x and y are in Alt(M), then there is a finite subset T of M containing $S(x) \cup S(y)$. Then $x \mid T \in$ Alt(T) and $y \mid T \in$ Alt(T), hence $xy^{-1} \mid T \in$ Alt(T). Therefore $xy^{-1} \in$ Alt(M). Since $e \in$ Alt(M), Alt(M) $\subset$ Sym(M). Normality follows from Theorems 1.3.6 and Exercise 10.4.9. ‖

Up to isomorphism, Sym(M, A) and Alt(M) do not depend on M, but merely on $o(M)$. Therefore, if A and B are infinite cardinals with $B \leq A^+$, we may legitimately introduce the notation Sym(A, B) for the subgroup of all $x \in$ Sym(A) with $o(S(x)) < B$, and Alt(A) for the subgroup of all finite even permutations in Sym(A). Thus, Alt(A) and Sym(A, B) are normal subgroups of Sym(A). It is the principal result of the next section that these are the only normal subgroups of Sym(A).

11.2.3. Alt(A) *is simple.*

Proof. Let M be the set of letters, and $E \neq H \lhd$ Alt(A). Let $x \in H^{\#}$, $y \in$ Alt(A), and let T be any finite subset of M containing $S(x) \cup S(y)$ with $o(T) > 4$. Then

$$E < H \cap \text{Alt}(T) \lhd \text{Alt}(T).$$

By the simplicity of Alt(T), $H \supset$ Alt(T). Hence $y \in H$ for all $y \in$ Alt(A). Therefore, $H =$ Alt(A). Thus Alt(A) is simple. ‖

Next, a theorem of G. Higman [1] about subgroups of Alt(A) will be proved. First a lemma, stated more generally than needed for the theorem.

11.2.4. *If* (M, G) *is a transitive subgroup of* Sym(M, A), $A \leq o(M)$, *and* H *is an intransitive subgroup of* G, *then* $[G:H] \geq o(M)$.

Proof. Suppose first that there is an orbit T of H with $o(T) < o(M)$. Let $a \in T$. Then $o(a(Hg)) = o(Tg) = o(T)$ for $g \in G$, yet every letter is in $a(Hg)$ for some $g \in G$. Hence if $[G:H] = B$, then $o(M) \leq Bo(T)$, so $o(M) \leq B$.

Now assume that all orbits of H have order $o(M)$. There is a partition $M = T \mathbin{\dot\cup} U$ with $o(T) = (o)U = o(M)$, and U an orbit. Let $a \in T$. Since G is transitive, $\exists\, g \in G$ such that $ag \in U$. Since U is an orbit of H, $agH = U$. Hence U is covered by the sets $a(Hgh) \cap U$ with $h \in H$. We have

$$o(aHgh \cap U) = o(aHg \cap Uh^{-1})$$
$$\leq o(Tg \cap U) = D < o(M),$$

since g moves fewer than $o(M)$ elements and $T \cap U = \varnothing$. Since $o(U) = M$, there are at least $o(M)$ right cosets of H of the form Hgh. Thus $[G:H] \geq o(M)$.

11.2.5. *If A is infinite and $H < G = \mathrm{Alt}(A)$, then $[G:H] = A$.*

Proof. Since $o(G) = A, [G:H] \leq A$. Suppose that $[G:H] < A$. By Theorem 11.2.4, H is transitive. Suppose that H is not n-transitive with n minimal. Let T be a set of letters, $o(T) = n - 1$. Then

$$[G_T : G_T \cap H] \leq [G:H] < A.$$

Since G_T is the alternating group on A letters again, by what has been proved, $G_T \cap H = H_T$ is transitive, contrary to the supposition. Therefore H is n-transitive for all $n \in \mathcal{N}$.

Let $g \in G$, $h \in H$, and $T = S(h)$. Then T is finite. Since H is $o(T)$-transitive, $\exists\, x \in H$ such that $ag = ax$ for all $a \in T$. Hence $g^{-1}hg = x^{-1}hx \in H$. Therefore $H \lhd G$. Since G is simple (Theorem 11.2.3), this is a contradiction. Hence $[G:H] = A$.

11.2.6. *If G is an infinite subgroup of $\mathrm{Sym}(M, \aleph_0)$ and $a \in M$, then $o(G_a) = o(G)$.*

Proof. Let $A = o(G)$, let T be the orbit of G containing a, and deny the theorem. By Theorem 10.1.4, $A = o(G_a)o(T)$, hence by the denial, $o(T) = A$. Now $\exists\, g \in G$ such that $ag \neq a$. If $b \in T$, then $\exists\, h \in G$ such that $ah = b$. Thus $b \in S(h^{-1}gh)$. Therefore, T is contained in $\cup\, S(h^{-1}gh)$. Since each $S(h^{-1}gh)$ has the same finite order n and $o(T) = A$, g has A conjugates. Also, if $b \in T$, then G_b is conjugate to G_a (Theorem 10.1.2), hence $o(G_b) < A$. There are distinct letters $a_1, \ldots, a_{n+1}$ of T. Each conjugate of g moves only n letters, hence is in some G_{a_i}. Thus

$$A \leq \sum o(G_{a_i}) = (n + 1)o(G_a) < A,$$

a contradiction. $\parallel$

This theorem can be generalized as follows.

11.2.7. *If G is an infinite subgroup of* $\mathrm{Sym}(M, \aleph_0)$, Q *a subset of* M, *and* $o(Q) < o(G)$, *then* $o(G_Q) = o(G)$.

Proof. If $x \in G$, then $S(x) \cap Q$ is finite. Let $o(G) = A$ and let A^- be the immediate predecessor of A if it exists.

CASE 1. A^- exists. There are at most A^- finite subsets of Q. Since A^- exists, there is a subset U of G and a finite subset Q' of Q such that $S(x) \cap Q = Q'$ for all $x \in U$, and $o(U) = A$. If $L = \langle U \rangle$, then $S(x) \cap Q$ is contained in Q' for all $x \in L$, and $o(L) = A$. By Theorem 11.2.6 and induction, $o(L_{Q'}) = A$. Then $L_{Q'} \subset G_Q$, and we are done.

CASE 2. A^- does not exist. If $A = \aleph_0$, then Q is finite, so that Theorem 11.2.6 and induction imply the result. Suppose that $A > \aleph_0$. If $o(Q) \leq B < A$, then there is an infinite cardinal D such that D^- exists and $B < D < A$. There is a subgroup H of G of order D. By Case 1, $o(H_Q) = D$. Since $G_Q \supset H_Q$, $o(G_Q) > B$ for all $B < A$. Therefore, $o(G_Q) = A$.

11.2.8. *If G is an infinite subgroup of* $\mathrm{Sym}(M, \aleph_0)$, U *is a subset of* G, *and* $o(U) < o(G)$, *then* $o(C(U)) = o(G)$.

Proof. Let $Q = \cup \{S(x) \mid x \in U\}$. Then either Q is finite or $o(Q) \leq o(U)$. In either case $o(Q) < o(G)$. Hence by Theorem 11.2.7, $o(G_Q) = o(G)$. Since $G_Q \subset C(U)$, $o(C(U)) = o(G)$.

11.2.9. *If G is an infinite subgroup of* $\mathrm{Sym}(M, \aleph_0)$, *then* $\exists\, H < G$ *such that* $o(H) = o(G)$.

Proof. Some letter a is moved by G. Then $G_a < G$, and, by Theorem 11.2.6, $o(G_a) = o(G)$.

11.2.10. *If G is an infinite subgroup of* $\mathrm{Sym}(M, \aleph_0)$ *and H an Abelian subgroup of G, then there is an Abelian subgroup K of G such that $H \subset K$ and* $o(K) = o(G)$.

Proof. By Zorn, there is a maximal Abelian subgroup $K \neq G$ containing H. If $o(K) < o(G)$, then, by Theorem 11.2.8, $o(C(K)) = o(G)$. Hence $\exists\, x \in C(K)\backslash K$, and $\langle K, x \rangle$ is Abelian, a contradiction. Therefore, $o(K) = o(G)$. ∥

This theorem has a much deeper generalization given in Theorem 15.4.3.

The final theorem of the section requires a lemma about Abelian groups, of some interest in itself.

11.2.11. *If G is an uncountable Abelian group, then G has a well-ordered descending chain of subgroups of length* $o(G)$.

Proof. Let T be the torsion subgroup of G, and first suppose that $o(T) < o(G)$. Then G/T is torsion-free, and $o(G/T) = o(G)$, by Lagrange's theorem. Hence (by the lattice theorem) *WLOG*, G is torsion-free, By Zorn, there is a maximal set S of infinite cyclic subgroups of G such that $H = \sum \{B \mid B \in S\}$ exists. If $o(S) = o(G)$, then H is free Abelian of rank $o(G)$, and H has such a well-ordered descending chain of subgroups (drop out one summand at a time). If $o(S) < o(G)$, then for each $g \in G$, $ng \in H$ for some n, and $ng = ng'$ implies $g = g'$ since G is torsion-free. Therefore $o(G) \leqq \aleph_0 \cdot o(S) < o(G)$, a contradiction.

Next suppose that G is a p-group, $p \in \mathscr{P}$. Let P_n be the subgroup of G consisting of all x such that $p^n x = 0$. By Theorem 5.1.11, $P_n = \sum \{C_i \mid i \in R\}$, where each C_i is cyclic. Then $P_1 = \sum \{D_i \mid i \in R\}$, where $D_i \subset C_i$ and $o(D_i) = p$. Therefore,

$$o(P_n) \leqq \aleph_0 \cdot o(R) = \aleph_0 \cdot o(P_1),$$

$$o(G) \leqq \aleph_0 \cdot o(P_1),$$

and since G is uncountable, $o(G) = o(P_1)$. But P_1 is the direct sum of $o(G)$ cyclic groups, hence has a well-ordered descending sequence of length $o(G)$.

Finally, suppose that $o(T) = o(G)$. Then $G = \sum G_p$, and by the preceding paragraph, *WLOG* $o(G_p) < o(G)$ for all p. The order of G is unchanged if all countable summands G_p are omitted, so it may be supposed that each G_p (not zero) is uncountable. By the preceding paragraph (or induction), each G_p has a well-ordered descending chain of subgroups of length $o(G_p)$. Since

$$G = \sum G_p > \sum \{G_p \mid p \neq p_1\} > \sum \{G_p \mid p \neq p_1 \text{ or } p_2\} > \dots, \qquad (1)$$

one can use the series for the G_p to interpolate in (1) and get a well-ordered descending chain of subgroups of G of length $\sum o(G_p) = o(G)$.

11.2.12. *If G is an infinite subgroup of* Sym$(M, \aleph_0)$, *then G has a well-ordered descending chain of subgroups of length $o(G)$.*

Proof. By Theorem 11.2.10 with $H = E$, there is an Abelian subgroup K of G of order $o(G)$. If G is uncountable, then the theorem follows from Theorem 11.2.11. If $o(G) = \aleph_0$, it follows from Theorem 11.2.9.

EXERCISES

11.2.13. If G is a subgroup of Sym$(A, \aleph_0)$ satisfying the minimal condition for subgroups, then G is finite.

11.2.14. If G is an Abelian group, then $G \cong H \subset$ Sym$(A, \aleph_1)$ for some A. (Use Theorems 5.3.7 and 5.2.12.)

11.2.15. (Compare with Theorem 11.2.4.) Let M be an infinite set, and let $M = M_1 \cup M_2$ with $o(M_1) = o(M_2)$.

    (a) $H = \mathrm{Sym}(M_1) + \mathrm{Sym}(M_2)$ is an intransitive group (when considered as a group of permutations of M).

    (b) There is a transitive group (M, G) such that $[G:H] = 2$.

11.2.16. The only subgroup of $\mathrm{Sym}(A, \aleph_0)$ of index $< A$ is $\mathrm{Alt}(A)$.

11.2.17. Give an example of a descending sequence of simple groups whose intersection is not simple.

11.2.18. The union of an increasing chain of simple groups is simple.

11.3 Normal subgroups of symmetric and alternating groups

The normal subgroups of $\mathrm{Sym}(n)$ and $\mathrm{Alt}(n)$ for n finite were determined in the last chapter. If A, B, and D are infinite cardinals and $D < B \leq A^+$, then it was shown in Section 11.2 that $\mathrm{Sym}(A, D)$ and $\mathrm{Alt}(A)$ are normal subgroups of $\mathrm{Sym}(A, B)$, and that $\mathrm{Alt}(A)$ is simple. In this section it will be shown that, conversely, these are the only normal subgroups of $\mathrm{Sym}(A, B)$. The above groups will be called the infinite symmetric and alternating groups.

Slight improvements of two earlier lemmas are needed. If $x \in \mathrm{Sym}(M)$, let $f_n(x)$ be the cardinal number of n-cycles in the (formal) cyclic decomposition of x for $n = 1, 2, \ldots ; \infty$ [hence $f_1(x) = \mathrm{Ch}(x)$].

11.3.1. *If G is an infinite symmetric or alternating group, then two elements x and y of G are conjugate within G iff $f_n(x) = f_n(y)$ for all n.*

Proof. If $s^{-1}xs = y$ with $s \in G$, then, by Theorem 1.3.6, $f_n(x) = f_n(y)$ for all n.

Conversely, suppose that $f_n(x) = f_n(y)$ for all n. This implies that there is a 1–1 function from the set of cycles of x onto the set of cycles of y which preserves length. This function may be used to construct a 1–1 function s of the set M of letters onto itself such that $y = s^{-1}xs$ by Theorem 1.3.5. If $G = \mathrm{Sym}(A, A^+)$, this completes the proof. Suppose that $G = \mathrm{Sym}(A, B)$ with $B \leq A$. Then

$$o(S(x)) = o(S(y)) < B \leq A.$$

Hence there is a subset P of M such that (i) P contains $S(x) \cup S(y)$, (ii) $o(P \backslash S(x)) = o(P \backslash S(y))$, and (iii) $o(P) < B$. Now define a function t from M into M by requiring that t agree with s on $S(x)$, that it map $P \backslash S(x)$ 1–1 onto $P \backslash S(y)$, and that it be the identity on the remainder of M. Then $t^{-1}xt = y$, and, since $S(t)$ is contained in P, $t \in \mathrm{Sym}(A, B)$. If $G = \mathrm{Alt}(A)$, then x and y

are in $\text{Sym}(A, \aleph_0)$, so $t \in \text{Sym}(A, \aleph_0)$. If $t \notin \text{Alt}(A)$, then there are i and $j \in M \backslash S(t)$, $t(i, j) = t' \in \text{Alt}(A)$, and $t'^{-1}xt' = y$.

11.3.2. *If G is an infinite symmetric or alternating group and $z \in G$, then there are x and y in G such that $z = xy$ and $x^2 = y^2 = e$.*

Proof. Decompose z into cycles, $z = \pi c_i$. By the formulas of Exercise 10.1.17, $c_i = a_i b_i$ where $a_i^2 = e = b_i^2$, and a_i and b_i move no letters unmoved by c_i (if c_i is a 1-cycle, let $a_i = b_i = e$). Let $x = \pi a_i$ and $y = \pi b_i$. Then $z = xy$ and $x^2 = e = y^2$. Moreover, x and y are in G except possibly in the case $G = \text{Alt}(A)$. In this case, if $x \notin \text{Alt}(A)$, then $y \notin \text{Alt}(A)$, and, since x and y are finite permutations, $\exists\, i$ and j in $M \backslash (S(x) \cup S(y))$. Let $x' = x(i, j)$, $y' = y(i, j)$. Then $z = x'y'$, x' and y' are in G, and $x'^2 = y'^2 = e$.

11.3.3. *The only nontrivial normal subgroup of $\text{Sym}(A, \aleph_0)$ is $\text{Alt}(A)$.*

Proof. $\text{Alt}(A)$ is of index 2 in $\text{Sym}(A, \aleph_0)$, hence is a proper normal subgroup. Let $E \neq H \lhd \text{Sym}(A, \aleph_0)$, $H \neq \text{Alt}(A)$. Then $K = H \cap \text{Alt}(A) \lhd \text{Alt}(A)$ and $K \neq \text{Alt}(A)$. By the simplicity of $\text{Alt}(A)$, $K = E$. By the isomorphism theorem, $o(H) = 2$. But since $\text{Sym}(A, \aleph_0)$ is 2-transitive it is primitive, hence H is transitive and therefore infinite, a contradiction.

11.3.4. *The nontrivial normal subgroups of $G = \text{Sym}(A, B)$ are the groups $\text{Sym}(A, D)$ with $D < B$ and $\text{Alt}(A)$.*

Proof. The fact that $\text{Sym}(A, D)$ and $\text{Alt}(A)$ are proper normal subgroups follows from their definitions and Theorems 11.2.1 and 11.2.2.

Conversely, let H be a nontrivial normal subgroup of G. If H contains only finite permutations, then, by Theorem 11.3.3, $H = \text{Sym}(A, \aleph_0)$ or $\text{Alt}(A)$. In particular, it may be assumed that $B > \aleph_0$. We assert that if $\exists\, s_1 \in H$ such that $o(S(s_1)) = D$, then $H \supset \text{Sym}(A, D^+)$. This assertion is evidently equivalent to the unproved portion of the theorem. The proof of this assertion will be broken up into four steps.

(i) $\exists\, s \in H$ such that

$$\begin{cases} f_3(s) = D \\ f_1(s) = A \\ f_n(s) = 0 \text{ if } n \neq 1 \text{ or } 3. \end{cases}$$

If

$$\Sigma \{f_n(s_1) \mid 2 < n < \infty\} + \aleph_0 f_\infty(s_1) = D,$$

then form an element s_2 conjugate to s_1 in G as follows: If s_1 contains a 1-cycle or a 2-cycle, let s_2 also contain the cycle. Well-order the finite cycles of length greater than 2. For every other such cycle $(i_1, \ldots, i_n)$ in s_1, let s_2

contain the n-cycle $(i_{n-1}, i_n, i_{n-2}, i_{n-3}, \ldots, i_1)$. For the remaining finite cycles of s_1, let s_2 contain their inverses. For all ∞-cycles $(\ldots, j_{-1}, j_0, j_1, \ldots)$ in s_1, let s_2 contain the ∞-cycle

$$(\ldots, j_4, j_3, j_1, j_2, j_0, j_{-1}, j_{-3}, j_{-2}, \ldots).$$

Then $s = s_1 s_2 \in H$, and the following 3-cycles occur in the cyclic decomposition of s:

$$(i_{n-2}, i_n, i_{n-1}); \ldots, (j_2, j_1, j_0), (j_{-2}, j_{-3}, j_{-4}), \ldots.$$

Furthermore, there are no other n-cycles for $n > 1$. Therefore, $f_3(s) = D$, and, since enough 1-cycles have been saved in case $D = A$, $f_1(s) = A$. Finally, $f_n(s) = 0$ if $n \neq 1$ or 3.

If $\Sigma \{f_n(s_1) \mid 2 < n < \infty\} + \aleph_0 f_\infty(s_1) < D$,

then $f_2(s_1) = D$. For every three 2-cycles $(1, 2)$, $(3, 4)$, and $(5, 6)$ in s_1, let s_2 contain the 2-cycles $(1, 3)$, $(2, 5)$, and $(4, 6)$. Then $s = s_1 s_2$ contains the 3-cycles $(1, 5, 4)$ and $(2, 3, 6)$. Hence $f_3(s) = D$, and since s_2 clearly can be chosen conjugate to s_1 consistent with the above requirements, the first case applies. Hence, in any case, the assertion in (i) is true.

(ii) $\exists \, s \in H$ such that

$$\begin{cases} f_2(s) = D, \\ f_1(s) = A, \\ f_n(s) = 0 \text{ if } n \neq 1 \text{ or } 2. \end{cases}$$

By (i) $\exists \, s_1 \in H$ and $s_2 \in H$ such that the cyclic decomposition of each element consists of D sets of one 3-cycle and one 1-cycle, together with A additional 1-cycles common to s_1 and s_2. It may further be required (conjugating if necessary) that if a typical combination of one 3-cycle and one 1-cycle in s_1 is $(1, 2, 3)$ and (4), then s_2 contains the combination $(1, 2, 4)$ and (3). Then $s = s_1 s_2$ contains D combinations of the type $(1, 4)(2, 3)$, and A additional 1-cycles. Therefore, s has the required properties.

(iii) H contains every element of order 2 in $\mathrm{Sym}(A, D^+)$.

Let F be a cardinal such that $F < D$ (F may be finite or 0). Let s_1 be of the type described in (ii), and let s_2 contain all but F of the 2-cycles in s_1, and no other n-cycles for $n > 1$. Then $s_2 \in H$, and if $s = s_1 s_2$, then $f_2(s) = F$, $f_1(s) = A$, and $f_n(s) = 0$ otherwise. Unless $A = D$, we are done by (ii) and Theorem 11.3.1.

Now suppose that $A = D$, and again let $F < D$. Let s_1 be of the type described in (ii), and let s_2 also be of that type, and such that $S(s_1) \cap S(s_2) = \varnothing$, while $o(M \backslash (S(s_1) \cup S(s_2))) = F$. Then if $s = s_1 s_2$, $f_2(s) = D$, $f_1(s) = F$, and $f_n(s) = 0$ otherwise. This shows that all elements of order 2 in $\mathrm{Sym}(A, D^+)$ are contained in H.

(iv) $H \supset \mathrm{Sym}(A, D^+)$.

This follows from (iii) and Theorem 11.3.2.

11.3.5.   *If* $\mathrm{Sym}(A, B) \subset G \subset \mathrm{Sym}(A)$, $B > \aleph_0$, *and* $E < H \lhd G$, *then* $H = \mathrm{Alt}(A)$, $H = \mathrm{Sym}(A, D)$ *for some* $D \leqq B$, *or* $H > \mathrm{Sym}(A, B)$.

Proof. Deny the theorem. Since $K = H \cap \mathrm{Sym}(A, B) \lhd \mathrm{Sym}(A, B)$, it follows from the preceding theorem that $K = \mathrm{Sym}(A, D)$ for some $D < B$, $\mathrm{Alt}(A)$, or E. Hence, in any case, $K \subset \mathrm{Sym}(A, D)$ for some $D < B$. By the denial, $\exists\, h \in H \backslash K$. Hence $o(S(h)) \geqq B$. In the cyclic decomposition of h, let T be a set of D cycles. Construct an element $y \in \mathrm{Sym}(A, B)$ as follows. Let y fix all letters not in any cycle of T. Further,

(i)   if $(1, \ldots, n) \in T$ and $n > 2$, let $\begin{pmatrix} 1, \ldots, n \\ n, \ldots, 1 \end{pmatrix}$ be part of y;

(ii)   if $(\ldots, -1, 0, 1, \ldots) \in T$, let $(-1, 1), (-2, 2), \ldots$, be cycles of y;

(iii)   split the 2-cycles in T into disjoint pairs (discarding the odd 2-cycle, if any), and for a pair $(1, 2)$, $(3, 4)$ in T, let $(1, 3)$ be a cycle of y.

Now $[h, y] \in [H, \mathrm{Sym}(A, B)] \subset K$. In case (i), all letters $1, \ldots, n$ are moved by $[h, y]$. In (ii),

$$ih^{-1}y^{-1}hy = (i - 1)y^{-1}hy$$
$$= (1 - i)hy = (2 - i)y = i - 2,$$

so, once again, all letters are moved. In case (iii),

$$((1, 2)(3, 4))^{-1}(1, 3)^{-1}(1, 2)(3, 4)(1, 3) = (1, 3)(2, 4),$$

so all letters (except the two in the discarded 2-cycle, if any) are moved. Hence $o(S([h, y])) = D$, and $[h, y] \notin \mathrm{Sym}(A, D)$, whereas $[h, y] \in K$, a contradiction. ∥

It will now be shown that the only homomorphisms of one infinite symmetric or alternating group onto another are isomorphisms with obviously favorable cardinality conditions. The customary lemma is needed first.

11.3.6. $\mathrm{Sym}(A, B^+)/\mathrm{Sym}(A, B)$ *is not periodic.*

Proof. There is an element $x \in \mathrm{Sym}(A, B^+)$ such that $f_\infty(x) = B$ and $f_n(x) = 0$ for $1 < n < \infty$. Then $s^i \notin \mathrm{Sym}(A, B)$ for $i \in \mathcal{N}$.

11.3.7.   *Let* G *be* $\mathrm{Sym}(A, B)$ *or* $\mathrm{Alt}(A)$, $H = \mathrm{Sym}(A', B')$ *or* $\mathrm{Alt}(A')$, *and* T *a homomorphism of* G *onto* H. *Then* T *is an isomorphism, and either*

(i) $G = \mathrm{Alt}(A) = H$,

or

(ii) $G = \mathrm{Sym}(A, B) = H$.

Proof. If T is not an isomorphism then $ET^{-1} \supset \mathrm{Alt}(A)$ by Theorems 11.2.3, 11.3.3, and 11.3.4. By the latter two theorems, $\mathrm{Alt}(A')T^{-1} \supset \mathrm{Sym}(A, \aleph_0)$. Hence (same references),

$$\mathrm{Alt}(A') \cong \frac{\mathrm{Sym}(A, \aleph_0)}{\mathrm{Alt}(A)} \quad \text{or} \quad \frac{\mathrm{Sym}(A, D^+)}{\mathrm{Sym}(A, D)}$$

for some D. But $\mathrm{Alt}(A')$ is an infinite periodic group, while $\mathrm{Sym}(A, \aleph_0)/\mathrm{Alt}(A)$ has order 2, and $\mathrm{Sym}(A, D^+)/\mathrm{Sym}(A, D)$ is not periodic by Theorem 11.3.6. Hence T is an isomorphism.

Since $\mathrm{Alt}(A)$ is simple, $\mathrm{Alt}(A)T = \mathrm{Alt}(A')$. But $o(\mathrm{Alt}(A)) = A$,

$$o(\mathrm{Alt}(A')) = A',$$

hence $A = A'$. If the theorem is false, then *WLOG*, $G = \mathrm{Sym}(A, B)$ and $H = \mathrm{Sym}(A, B')$ with $B > B'$. Choose the smallest B for which this is possible. Then T maps the subgroup $\mathrm{Sym}(A, B')$ of G onto a subgroup $\mathrm{Sym}(A, B'')$ of H with $B' > B''$, contradicting the minimality of B. ‖

A subgroup H of a group G is *completely characteristic* iff every endomorphism T of G onto G maps H onto H. A completely characteristic subgroup is strictly characteristic, hence characteristic.

11.3.8. *All normal subgroups of* $\mathrm{Sym}(A, B)$ *are completely characteristic.*

Proof. Let T be an endomorphism of $G = \mathrm{Sym}(A, B)$ onto itself. Then T maps every normal subgroup of G onto a normal subgroup of G. The only normal subgroups of G are alternating and symmetric groups. Hence T induces a homomorphism of one alternating or symmetric group onto another. The theorem now follows from Theorem 11.3.7.

EXERCISES

11.3.9. (11.3.5 is false for $B = \aleph_0$.) Let A be an infinite cardinal, $S = \mathrm{Sym}(A, \aleph_0)$, and $T = \mathrm{Alt}(A)$.

    (a)   There is a nontrivial subgroup K of $\mathrm{Sym}(A)$ such that $K \cap S = E$. Let $G = SK$ and $H = TK$.

    (b) $H \not\trianglerighteq S$.

    (c) $H \triangleleft G$.

11.3.10. If $\mathrm{Alt}(A) \subset G \subset \mathrm{Sym}(A)$, and $E \neq H \triangleleft G$, then $H \supset \mathrm{Alt}(A)$.

11.4   Automorphism groups

In this section the automorphism group of all (finite and infinite) symmetric and alternating groups will be determined. In fact, if $\mathrm{Alt}(A) \subset G \subset \mathrm{Sym}(A)$, then $\mathrm{Aut}(G)$ will be found (in a sense).

11.4.1. *If* $n > 3$ *is a cardinal number,* $\mathrm{Alt}(n) \subset G \subset \mathrm{Sym}(n)$, *and* $TU = T \,|\, \mathrm{Alt}(n)$, *then* U *is an isomorphism of* $\mathrm{Aut}(G)$ *into* $\mathrm{Aut}(\mathrm{Alt}(n))$.

Proof. By Theorems 10.8.8, 11.3.4, and Exercise 11.3.10, $\mathrm{Alt}(n)$ is the minimum normal non-E subgroup of G in case $n > 4$. Hence $\mathrm{Alt}(n)$ is characteristic in G if $n > 4$; the same is true if $n = 4$, as it is then the only subgroup of order 12 in G. Therefore $TU \in \mathrm{Aut}(\mathrm{Alt}(n))$. It is clear that U is a homomorphism. Let $T \in \mathrm{Ker}(U)$, $x \in \mathrm{Alt}(n)$, $y \in G$, and $yT = yu$ with $u \in G$. Then

$$yux = (yT)(xT) = (yx)T$$
$$= ((yxy^{-1})y)T = yxy^{-1}yu = yxu,$$

since $T \,|\, \mathrm{Alt}(n) = I$ and $\mathrm{Alt}(n) \lhd G$. Hence $u \in C(\mathrm{Alt}(n)) = E$ (Exercise 11.4.10). Therefore, $yT = y$ for all $y \in G$, and $T = I$. Hence U is an isomorphism. ‖

The set of 3-cycles in $\mathrm{Alt}(n)$ will be denoted by X in the rest of this section. If $n > 4$, X is a conjugate class in $\mathrm{Alt}(n)$ by Theorem 11.1.5.

11.4.2. *If* $T \in \mathrm{Aut}(\mathrm{Alt}(n))$, $n > 3$, *and* $n \neq 6$, *then* $XT = X$.

Proof. If $n = 4$ or 5, then X is the set of elements of order 3, so $XT = X$. Let $n > 6$. Then X is a conjugate class in $\mathrm{Alt}(n)$ such that

(i)   If $x \in X$, then $o(x) = 3$,

(ii) $\max \{o(xy) \mid x \in X, y \in X\} = 5$.

Suppose that Y is a class in $\mathrm{Alt}(n)$ satisfying (i) and (ii). By Theorems 11.1.5 and 11.3.1, Y is a class in $\mathrm{Sym}(n)$ also. If $Y \neq X$, then each element of Y is the product of $f_3(Y)$ 3-cycles and $f_1(Y)$ 1-cycles. If $f_1(Y) > 0$, then

$$x = (1, 2, 3)(4, 5, 6)(7) \cdots \in Y,$$
$$y = (1, 2, 4)(3, 5, 7)(6) \cdots \in Y,$$
$$xy = (1, 4, 7, 3, 2, 5, 6) \cdots ,$$

which contradicts (ii). If $f_1(Y) = 0$, then $n \geq 9$ and

$$x = (1, 2, 3)(4, 5, 6)(7, 8, 9) \cdots \in Y,$$
$$y = (1, 2, 4)(3, 5, 7)(6, 8, 9) \cdots \in Y,$$
$$xy = (1, 4, 7, 9, 3, 2, 5, 8, 6) \cdots ,$$

which again contradicts (ii). Therefore $Y = X$. Now T maps X onto a conjugate class with properties (i) and (ii). Therefore, $XT = X$, as the theorem asserts.

11.4.3. *There is a* $T \in \mathrm{Aut}(\mathrm{Sym}(6))$ *such that* XT *is the conjugate class of products of two 3-cycles.*

Proof. Let $H = \text{Sym}(5)$ and $P \in \text{Syl}_5(H)$. By the example after Theorem 11.1.4, $P \ntriangleleft H$. By Sylow, $[H:N(P)] = 6$. The representation U of H on $N(P)$ is transitive, of degree 6, and faithful by Theorem 10.8.8. Hence $\text{Sym}(6)$ has a transitive subgroup $HU = K$ of order 120. Therefore $[\text{Sym}(6):K] = 6$. Any element of order 3 in K is of the form hU, $h \in H$, $o(h) = 3$. Since $3 \nmid 20 = o(N_H(P))$, $\text{Ch}(hU) = 0$, hence hU is the product of two 3-cycles. The representation V of $\text{Sym}(6)$ on K is faithful by Theorem 10.8.8, and of degree 6. Let G be the image group. Then $KV = G_a$ for some letter a. Any 1–1 function from the letters of G onto the letters of $\text{Sym}(6)$ induces an isomorphism W of G onto $\text{Sym}(6)$. Let $G_a W = L$. Then $VW \in \text{Aut}(\text{Sym}(6))$, and VW maps the transitive subgroup K onto a subgroup L fixing one letter. By an earlier remark, $(VW)^{-1}$ maps a 3-cycle in L onto a product of two 3-cycles in K, and the set X of all 3-cycles in $\text{Sym}(6)$ onto the set Y of all products of two 3-cycles.

11.4.4. *If* $H = \{T \in \text{Aut}(\text{Alt}(6)) \mid XT = X\}$, *then* $[\text{Aut}(\text{Alt}(6)):H] = 2$.

Proof. There are just two conjugate classes of elements of order 3 in $\text{Alt}(6)$, X and the class Y of products of two 3-cycles (Theorem 11.1.5). There is a natural homomorphism U from $\text{Aut}(\text{Alt}(6))$ into the group of permutations of $\{X, Y\}$. U has kernel H, and is onto $\text{Sym}(\{X, Y\})$ by Theorems 11.4.1 and 11.4.3. Therefore, $[\text{Aut}(\text{Alt}(6)):H] = 2$.

11.4.5. *If* $n \geq 5$, $T \in \text{Aut}(\text{Alt}(n))$, *and* $XT = X$, *then* $\exists\, U \in \text{Inn}(\text{Sym}(n))$ *such that* $T = (U \mid \text{Alt}(n))$.

Proof. First note that

$$o((1, 2, 3)(4, 5, 6)) = 3, \quad o((1, 2, 3)(1, 4, 5)) = 5,$$
$$o((1, 2, 4)(1, 2, 3)) = 2, \quad o((1, 2, 3)(1, 4, 2)) = 3,$$
$$o((1, 2, 3)(1, 3, 2)) = 1.$$

Let M be the set of letters. For all distinct i and j in M, let

$$X_{ij} = \{(i, j, k) \mid k \in M, k \neq i \text{ or } j\}.$$

It follows from the equations above that X_{ij} is a maximal subset of X such that

(i)   if x and y are distinct elements of X_{ij}, then $o(xy) = 2$.

Conversely, any subset Y of X which is maximal with respect to having property (i) (with X_{ij} replaced by Y) is an X_{i*j*}. It follows that each $X_{ij}T = X_{i*j*}$ for some $i*$ and $j*$ in M.

Next let

$$Y_i = \{(i, j, k) \mid j \in M, k \in M, i, j, k \text{ distinct}\}.$$

Then we assert that

(ii) Y_i is a union of at least eight X_{jk},

(iii) if X_{jk} and X_{j*k*} are disjoint subsets of Y_i, then $\exists\, x \in X_{jk}$ such that

$$x^{-1} \in X_{j*k*}.$$

In fact, $Y_i = \cup \{X_{ik} \cup X_{ki} \mid k \neq i\}$, and, since $n \geq 5$, (ii) is satisfied. As to (iii), note that

(a) $X_{ik} \cap X_{k*i} \neq \varnothing$ if $k \neq k^*$,

(b) $x = (i, k, j) \in X_{ik}$,

$\quad x^{-1} = (i, j, k) \in X_{ki}$,

(c) $x = (i, k, k^*) \in X_{ik}$,

$\quad x^{-1} = (i, k^*, k) \in X_{ik*},\ k \neq k^*$,

(d) $x = (i, k^*, k) \in X_{ki}$,

$\quad x^{-1} = (i, k, k^*) \in X_{k*i},\ k \neq k^*$.

Conversely, let Y be a subset of X such that (ii) and (iii) are satisfied when Y_i is replaced by Y. Then any two X_{jk} contained in Y must have one or two subscripts in common by (iii). If not all have a subscript in common, then, say, Y contains X_{12} or X_{21}, X_{13} or X_{31}, and X_{23} or X_{32}. But no X_{jk} other than these six has a subscript in common with each of the three known to be contained in Y. This contradicts (ii). Therefore all X_{jk} contained in Y have a common subscript, so that $Y \subset Y_i$ for some i. Hence the maximal sets of 3-cycles satisfying (ii) and (iii) are precisely the Y_i.

Now since each $X_{jk}T$ is some X_{j*k*}, it follows that each $Y_iT = Y_{i'}$. The function $i \longrightarrow i'$ thus induced is 1–1 and onto M (since T^{-1} is also an automorphism), hence is an element x of $\mathrm{Sym}(n)$. Now

$$(i, j, k)T \in (Y_i \cap Y_j \cap Y_k)T = Y_{i'} \cap Y_{j'} \cap Y_{k'},$$

hence, for all i, j, k, $(i, j, k)T = (i', j', k')$ or (i', k', j'). If $(i, j, k)T = (i', k', j')$ for some $i, j,$ and k, then

$$X_{ij}T \subset Y_iT \cap Y_jT = Y_{i'} \cap Y_{j'} = X_{i'j'} \cup X_{j'i'}.$$

It follows readily that $X_{ij}T = X_{j'i'}$. Hence $X_{ji}T = X_{i'j'}$ and

$$[(i, k, j)(i, j, m)(i, j, k)]T = (i', j', k')(i', m', j')(i', k', j'),$$

$$(j, k, m)T = (i', k', m'),$$

a contradiction. Therefore, $(i, j, k)T = (i', j', k')$ for all $i, j,$ and k. Since $\mathrm{Alt}(n)$ is generated by X (any conjugate class generates a normal subgroup, and $\mathrm{Alt}(n)$ is simple), $T = (T_x \mid \mathrm{Alt}(n))$, where $ix = i'$ for all $i \in M$, and $x \in \mathrm{Sym}(n)$.

11.4.6. *If $n > 3$ is a cardinal number, $n \neq 6$, Alt$(n) \subset G \subset$ Sym(n), and $T \in$ Aut(G), then $\exists \,|\, x \in$ Sym(n) such that $T_x \,|\, G = T$.*

Proof. If such an x exists it is unique, since $C_{\mathrm{Sym}(n)}(\mathrm{Alt}(n)) = E$ for $n > 3$.

By Theorem 11.4.1, T induces an automorphism of Alt(n). By Theorem 11.4.2, $XT = X$. By Theorem 11.4.5, if $n > 4$ then $\exists \, x \in$ Sym(n) such that $T_x \,|\,$ Alt$(n) = T \,|\,$ Alt(n). Suppose that $n = 4$. If u and v are elements of Alt(4) of orders 2 and 3, respectively, then Alt$(4) = \langle u, v \rangle$ since there is no subgroup of order 6. Hence an automorphism of Alt(4) is determined by its action on u and v. Since Alt(4) has eight elements of order 3 and three of order 2, there are at most 24 automorphisms of Alt(4). But there are at least 24 such automorphisms induced by inner automorphisms of Sym(4). Hence, again $\exists \, x \in$ Sym(n) such that $T_x \,|\,$ Alt$(n) = T \,|\,$ Alt(n).

If n is finite, then $G =$ Alt(n) or Sym(n). If $G =$ Alt(n), the last equation says that $T_x \,|\, G = T$, and the theorem holds. If $G =$ Sym(n), then, by Theorem 11.4.1, $T_x = T$, and the theorem again holds.

Now let n be infinite. Then $U = TT_x^{-1}$ is an isomorphism of G onto a subgroup H of Sym(n), and $U \,|\,$ Alt$(n) = I$. If $U = I_G$, then $T = T_x \,|\, G$. Suppose that $U \neq I_G$. Then $\exists y \in G$ such that $yU \neq y$. Since $(yu)U = (yU)u \neq yu$ for $u \in$ Alt(n), $\exists \, z \in G$ such that $zU \neq z$ and z fixes at least four letters, a, b, c, and d.

CASE 1. There is a letter i such that $iz = i$ and $i(zU) = j \neq i$. Then, say, a, b, i, and j are distinct. We have

$$(i, a, b)z = z(i, a, b),$$
$$(i, a, b)(zU) = ((i, a, b)z)U = (z(i, a, b))U$$
$$= (zU)(i, a, b),$$
$$i[(zU)(i, a, b)] = j = i(zU) \neq a(zU) = i[(i, a, b)(zU)],$$

a contradiction.

CASE 2. zU fixes all letters fixed by z. Since $zU \neq z$, there are letters i, j, k such that $iz = j \neq i$, $i(zU) = k \neq j$. For $m \neq i, j, k, a, b, c,$ or d,

$$i(z(j, i, m)) = i,$$
$$i[(z(j, i, m))U] = k(j, i, m) \neq i,$$

and $z(j, i, m)$ fixes at least five letters (i, a, b, c, and d). Hence Case 2 reduces to Case 1.

11.4.7. *If $n > 3$, $n \neq 6$, and Alt$(n) \subset G \subset$ Sym(n), then*

$$\mathrm{Aut}(G) \cong N_{\mathrm{Sym}(n)}(G).$$

Proof. This follows directly from Theorem 11.4.6.

11.4.8. (1) *If $n = 1$ or 2, then* Aut(Sym(n)) *and* Aut(Alt(n)) *have order* 1.

(2)   Aut(Sym(3)) $\cong$ Sym(3) *and* Aut(Alt(3)) $\cong J_2$.

(3)   Aut(Sym(6)) $\cong$ Aut(Alt(6)), Inn(Sym(6)) $\cong$ Sym(6), *and*
[Aut(Sym(6)):Inn(Sym(6))] $= 2$.

(4)   *If $n > 3$ and $n \neq 6$ (n may be infinite), then*
Aut(Sym(n)) $\cong$ Aut(Alt(n)) $\cong$ Sym(n).

(5)   *If n is infinite, then* Aut(Sym(n, B)) $\cong$ Sym(n).

Proof. (1) is obvious.

(2). Alt(3) $\cong J_3$, hence Theorem 5.7.12 gives the structure of its auto-morphism group. If x and y are elements of Sym(3) of orders 2 and 3, respectively, then an automorphism T is determined by xT and yT. Since there are three elements of order 2 and two elements of order 3, there are at most six automorphisms of Sym(3). But there are six inner automorphisms, so that (2) follows.

(3). Since $Z(\text{Sym}(6)) = E$, Inn(Sym(6)) $\cong$ Sym(6). By Theorem 11.4.3, $o(\text{Aut}(\text{Sym}(6))) \geq 1440$. By Theorems 11.4.4 and 11.4.5, $o(\text{Aut}(\text{Alt}(6))) \leq 1440$. By Theorem 11.4.1, Aut(Sym(6)) $\cong$ Aut(Alt(6)), and both groups have order 1440. Statement (3) follows.

(4) and (5). These follow from Theorems 10.4.5, 11.2.1, 11.2.2, and 11.4.7. ‖

As a corollary of one of the theorems of this section, the existence of two finite nonisomorphic simple groups of the same order may be proved.

11.4.9. *There are nonisomorphic simple groups of order* 20,160.

Proof. This is the order of both Alt(8) and $PSL(3, 4)$. By Theorem 11.4.8, $o(\text{Aut}(\text{Alt}(8))) = 40,320$. Since $SL(3, 4) \lhd GL(3, 4)$, there is a natural homomorphism U from $GL(3, 4)$ into Aut($PSL(3, 4)$). By Exercise 10.8.14, Ker(U) $= S$, the group of scalar transformations. Thus

$$\text{Aut}(PSL(3, 4)) \stackrel{\sim}{\supseteq} GL(3, 4)/S = PGL(3, 4),$$

which has order 60,480. Hence Aut($PSL(3, 4)$) $\ncong$ Aut(Alt(8)), so that $PSL(3, 4) \ncong$ Alt(8).

EXERCISES

11.4.10. If $n > 3$, then $C_{\text{Sym}(n)}(\text{Alt}(n)) = E$.

11.4.11. Let $2 < n < \aleph_0$, $n \neq 6$. A shorter proof of the theorem Aut(Sym(n)) $\cong$ Sym(n) will be sketched.

(a) $x \in \text{Sym}(n)$ is a 2-cycle iff $o(x) = 2$ and $\max(o(xy^{-1}xy)) = 3$.

(b)   Let Y be a set of 2-cycles maximal with respect to the properties (1) no two elements of Y commute, and (2) if x and y are distinct elements of Y, then $x^{-1}yx \notin Y$. Then $Y = Y_i$ for some letter i, where $Y_i = \{(i, j) \mid j \neq i\}$, and conversely.

(c)   If $T \in \text{Aut}(\text{Sym}(n))$, then $Y_i T = Y_{i'}$.

(d)   The function $i \longrightarrow i'$ in (c) is an element x of $\text{Sym}(n)$.

(e) T in (c) equals T_x.

(f)   The theorem is true.

11.5   Imbedding and splitting theorems

11.5.1. $\text{Alt}(A) \stackrel{.}{\supseteq} \text{Sym}(A, \aleph_0)$.

Proof. Let M_1 and M_2 be disjoint sets of order A, and let $M = M_1 \cup M_2$. Let T_1 and T_2 be 1–1 functions from M onto M_1 and M_2, respectively, and let U_i be the induced isomorphism of $\text{Sym}(M, \aleph_0)$ onto $\text{Sym}(M_i, \aleph_0)$, $i = 1, 2$. Let $sU = (sU_1)(sU_2)$ for $s \in \text{Sym}(M, \aleph_0)$. Then U is an isomorphism of $\text{Sym}(M, \aleph_0)$ into $\text{Alt}(M)$. The theorem follows.

11.5.2.   *If $B > \aleph_0$ and H is a proper normal subgroup of* $\text{Sym}(A, B)$, *then*

$$\frac{\text{Sym}(A, B)}{H} \stackrel{.}{\supseteq} \text{Sym}(A, B).$$

Proof. By Theorem 11.3.4, there is an infinite cardinal $D < B$ such that $H \subset \text{Sym}(A, D)$. Let M be the union of D disjoint sets M_i, each of order A. Let T_i be a 1–1 function from M onto M_i and let U_i be the induced isomorphism of $\text{Sym}(M, B)$ onto $\text{Sym}(M_i, B)$. Define U by the rule: $sU = \pi(sU_i)$ for $s \in \text{Sym}(M, B)$. Then $o(S(sU)) = D \cdot o(S(s))$. It follows that U is an isomorphism of $\text{Sym}(M, B)$ onto a subgroup L of $\text{Sym}(M, B)$. Since $o(S(sU)) \geqq D$ if $s \neq e$, we have $L \cap H = E$. Hence,

$$\frac{LH}{H} \cong L \cong \text{Sym}(A, B),$$

and

$$\frac{LH}{H} \subset \frac{\text{Sym}(A, B)}{H}.$$

11.5.3.   *If $B > \aleph_0$ and H is a proper normal subgroup of* $\text{Sym}(A, B^+)$, *then* $\text{Sym}(A, B^+)/H$ *contains isomorphic copies of all groups of order $\leqq B$.*

Proof. Let G be a group of order $\leq B$. The Cayley representation of G is of degree $o(G) \leq B \leq A$. Hence by Theorem 11.5.2,

$$G \cong \mathrm{Sym}(A, B^+) \cong \frac{\mathrm{Sym}(A, B^+)}{H}.$$

11.5.4. *Any group can be imbedded in a simple group.*

Proof. If $o(G) \leq B$, where B is infinite, then, by Theorem 11.5.3, $G \cong \mathrm{Sym}(B, B^+)/\mathrm{Sym}(B, B)$, which is simple by Theorem 11.3.4.

11.5.5. *If $\aleph_0 \leq B < A$ and $p \in \mathscr{P}$, then all elements of $\mathrm{Sym}(A, B^+)/ \mathrm{Sym}(A, B)$ of order p are conjugate. This is false if $B = A$.*

Proof. Let $x \in \mathrm{Sym}(A, B^+)$ be such that $o(x \cdot \mathrm{Sym}(A, B)) = p$. Then $x = uv$, where $S(u) \cap S(v) = \varnothing$, u is a product of p-cycles, and no cycle of v has length p. Then $x^p = v^p \in \mathrm{Sym}(A, B)$, and $S(v^p) = S(v)$. Hence $v \in \mathrm{Sym}(A, B)$. Therefore, $x \cdot \mathrm{Sym}(A, B) = u \cdot \mathrm{Sym}(A, B)$, and u is the product of fewer than B^+ p-cycles. Since $u \notin \mathrm{Sym}(A, B)$, $f_p(u) = B$. Now let $y \in \mathrm{Sym}(A, B^+)$ be such that $o(y \cdot \mathrm{Sym}(A, B)) = p$. Then $\exists w$ such that $y \cdot \mathrm{Sym}(A, B) = w \cdot \mathrm{Sym}(A, B)$ and w is the product of B p-cycles. By Theorem 11.3.1 and the fact that $B < A$, $\exists t \in \mathrm{Sym}(A, B^+)$ such that $t^{-1}ut = w$. It follows that $u \cdot \mathrm{Sym}(A, B)$ and $w \cdot \mathrm{Sym}(A, B)$ are conjugate in $\mathrm{Sym}(A, B^+)/ \mathrm{Sym}(A, B)$.

Now let $B = A$. There are $x \in \mathrm{Sym}(A)$ and $y \in \mathrm{Sym}(A)$ such that $o(x) = o(y) = p$, $o(S(y)) = A$, $\mathrm{Ch}(x) = 0$, and $\mathrm{Ch}(y) = A$. Then the cosets $x \cdot \mathrm{Sym}(A, A)$ and $y \cdot \mathrm{Sym}(A, A)$ have order p. If they are conjugate, then $\exists u \in \mathrm{Sym}(A)$ such that $u^{-1}xu = yv$ with $v \in \mathrm{Sym}(A, A)$. But $\mathrm{Ch}(u^{-1}xu) = \mathrm{Ch}(x) = 0$, while

$$\mathrm{Ch}(yv) \geq \mathrm{Ch}(y) - o(S(v)) = A,$$

since $o(S(v)) < A$. Hence, $x(\mathrm{Sym}(A, A))$ and $y(\mathrm{Sym}(A, A))$ are not conjugate.

11.5.6. *If $\aleph_0 \leq B < A$ and $\aleph_0 \leq D$, then the (simple) groups $\mathrm{Sym}(D, D^+)/ \mathrm{Sym}(D, D)$ and $\mathrm{Sym}(A, B^+)/\mathrm{Sym}(A, B)$ are not isomorphic.*

Proof. This follows from Theorem 11.5.5.

11.5.7. *If $\aleph_0 < D \leq B$, then*

$$\frac{\mathrm{Sym}(A, \aleph_1)}{\mathrm{Sym}(A, \aleph_0)} \not\cong \frac{\mathrm{Sym}(B, D^+)}{\mathrm{Sym}(B, D)}.$$

Proof. If $y \in \mathrm{Sym}(A, \aleph_1)$ and $z \in \mathrm{Sym}(A, \aleph_0)$, then $f_\infty(y^2z)$ is either 0 or at least 2 (Exercise 11.5.11). Hence if x is an ∞-cycle, then the coset $x \cdot \mathrm{Sym}(A, \aleph_0)$ is not a square of any element of $\mathrm{Sym}(A, \aleph_1)/\mathrm{Sym}(A, \aleph_0)$.

On the other hand, let $u \in \mathrm{Sym}(B, D^+)$. In a cyclic decomposition of u, any pair of cycles of the same length is a square of a single cycle on the same letters. Hence $\exists\, v \in \mathrm{Sym}(B, D^+)$ such that v^2 differs from u by at most one n-cycle for each n, $2 \leq n \leq \infty$, i.e., by an element of $\mathrm{Sym}(B, D)$. This means that

$$(v\, \mathrm{Sym}(B, D))^2 = u\, \mathrm{Sym}(B, D).$$

It follows that the groups are not isomorphic. ‖

It seems to be an unsolved problem whether

$$\frac{\mathrm{Sym}(A, B^+)}{\mathrm{Sym}(A, B)} \cong \frac{\mathrm{Sym}(X, Y^+)}{\mathrm{Sym}(X, Y)}$$

implies $A = X$ and $B = Y$.

11.5.8. *If $B > \aleph_0$, then $\mathrm{Sym}(A, \aleph_0)$ has no complement in $\mathrm{Sym}(A, B)$.*

Proof. There is an $x \in \mathrm{Sym}(A, B)$ such that $f_{2^n}(x) = 2^{n-1}, n \in \mathcal{N}, f_i(x) = 0$, otherwise. For example, one may let

$$x = (1, 2)(3, 4, 5, 6)(7, \dots, 10)(11, \dots, 18)(19, \dots, 26) \cdots.$$

Then $\exists\, x_n \in \mathrm{Sym}(A, B)$ such that

$$x_n^{2^n} \equiv x \ (\mathrm{mod}\ \mathrm{Sym}(A, \aleph_0)).$$

For example, with the above x, one could let

$$x_1 = (3, 7, 4, 8, 5, 9, 6, 10)(11, 19, 12, 20, \dots, 18, 26) \cdots.$$

Now suppose that H is a complement of $\mathrm{Sym}(A, \aleph_0)$ in $\mathrm{Sym}(A, B)$. Then there are y and y_n in H for all $n \in \mathcal{N}$, such that

$$y \equiv x(\mathrm{mod}\ \mathrm{Sym}(A, \aleph_0)),$$
$$y_n \equiv x_n(\mathrm{mod}\ \mathrm{Sym}(A, \aleph_0)).$$

Then $y_n^{2^n} = y$ (since $H \cap \mathrm{Sym}(A, \aleph_0) = E$). Since $y \equiv x$, y must contain a cycle of length 2^r for some $r > 0$. Since $y_n^{2^n} = y$, y contains 2^n cycles of length 2^r. Since this is true for all n, y contains an infinite number of 2^r-cycles. This contradicts the fact that $y \equiv x(\mathrm{mod}\ \mathrm{Sym}(A, \aleph_0))$. Therefore there is no such complement H. ‖

The question as to whether $\mathrm{Sym}(A, B)$ splits over $\mathrm{Sym}(A, D)$ if $\aleph_0 < D < B$ is unsolved.

The argument used to prove Theorem 11.5.2 also turns up in the proof of the following generalization of Theorem 11.2.5 due to E. Gaughan.*

* The author wishes to thank Professor Gaughan for permission to use this result before publication of his paper, "The Index Problem for Infinite Symmetric Groups," Proc. Amer. Math. Soc. (to appear).

11.5.9. *If $\aleph_0 < B \leq A^+$ and $H < G = \text{Sym}(A, B)$, then $[G:H] \geq A$.*

Proof. Deny, and let $\aleph_0 \leq D < B$. Let $a \in G$ be the product of D 3-cycles c_i and A 1-cycles. Then $a \notin \text{Sym}(A, D)$. If M is the set of letters, then there is a partition $M = \overset{.}{\cup} M_i$ such that $o(M_i) = A$ and the three letters appearing in c_i are in M_i. There is an isomorphism T_i of $\text{Alt}(M)$ into $\text{Sym}(M_i)$ induced by a 1-1 function from M onto M_i such that $(1, 2, 3)T_i = c_i$. Define the isomorphism U of $\text{Alt}(M)$ into G by the rule:

$$xU = \pi(xT_i), \qquad x \in \text{Alt}(M).$$

If $K = \text{Alt}(M)U$, then since $[G:H] < A$, $[K:K \cap H] < A$. By Theorem 11.2.5, $K \subseteq H$. Hence, $a \in H$. Since a was arbitrary, all conjugates of a by elements of G are contained in H. Therefore $H \supset a^G = \text{Sym}(A, D^+)$ (Theorem 11.3.4). Since this is true for all $D < B$, $H \supset \text{Sym}(A, B) = G$. This contradiction proves the theorem.

EXERCISES

11.5.10. (a)   Alt(A) has a complement in $\text{Sym}(A, \aleph_0)$.

(b)   If $B > \aleph_0$, then Alt(A) does not have a complement in $\text{Sym}(A, B)$.

11.5.11. (a)   Let $x \in \text{Sym}(A)$, $y \in \text{Sym}(A, \aleph_0)$. Then $f_\infty(x) = f_\infty(xy), f_n(x) = f_n(xy)$ for all but a finite number of n, and $f_n(x) - f_n(xy)$ is finite for each n.

(b)   If $x \in \text{Sym}(A)$ and $y \in \text{Sym}(A, B)$, then

$$\sum \{ \, | f_n(x) - f_n(xy)| \mid 1 \leq n \leq \infty \} < B.$$

Hence if $f_n(x) \geq B$, then $f_n(x) = f_n(xy)$.

REFERENCES

For the entire chapter, Scott, Holmes, and Walker [1]; Exercise 11.2.14, Kneser and Swierczkowski [1]; Theorem 11.3.4, Baer [2]; Theorem 11.3.7, Karrass and Solitar [1]; Section 11.4, Schreier and Ulam [1] in part (see also Dinkines [1]); Theorem 11.4.9, Schottenfels [1]; Section 11.5, Karrass and Solitar [1] (see also Bruijn [1]).

REPRESENTATIONS

12.1 Linear groups

A *linear group* G over a field F (or on V) is a subgroup of $GL(V)$, where V is some finite dimensional vector space over F. A linear group is thus isomorphic to a group of nonsingular matrices over F and will often be considered as such. The *degree* of G is the dimension of V. A linear group G on V is *reducible* iff there is a subspace W, $\{0\} < W < V$, invariant under G; otherwise G is *irreducible*. The linear group induced by G on such a subspace W is a *constituent* of G. G is *decomposable* iff $V = W + X$, $W \neq \{0\}$, $X \neq \{0\}$, where W and X are G-invariant; otherwise G is *indecomposable*. A linear group G on V is *completely reducible* iff V is the direct sum of irreducible G-invariant subspaces. Two subgroups of $GL(V)$ are called *equivalent* iff they are conjugate in $GL(V)$.

12.1.1. *If G and H are equivalent linear groups, then G is reducible, irreducible, decomposable, indecomposable, or completely reducible iff H is.*

Proof. Let $G \subset GL(V)$ and $H = T^{-1}GT$ where $T \in GL(V)$. If W is a nontrivial G-invariant subspace of V, then WT is a nontrivial, H-invariant subspace of V since

$$(WT)H = (WT)(T^{-1}GT) = WGT = WT.$$

Hence if G is reducible, then H is, and conversely, since conjugacy is symmetric. If $V = W + X$ where W and X are G-invariant, then $V = WT + XT$ where WT and XT are H-invariant. Hence G is decomposable or indecomposable depending on whether H is decomposable or indecomposable. The same considerations apply to complete reducibility.

12.1.2. (*Maschke.*) *If G is a finite, reducible linear group over a field F and the characteristic of F does not divide $o(G)$, then G is decomposable.*

Proof. Let $G \subset GL(V)$ where V is a vector space over F. Since G is reducible, there is a G-invariant subspace W with $\{0\} < W < V$. By Theorem 5.6.2, there is a subspace X such that $V = W + X$. Then

$$g \mid W = g_1 \in GL(W), \qquad g \mid X = g_2 + g_3,$$
$$g_2 \in \mathrm{Hom}(X, W), g_3 \in GL(X).$$

We have, for $x \in X$, $g \in G$, and $h \in G$,

$$x(gh) = (xg_2 + xg_3)h$$
$$= xg_2h_1 + xg_3h_2 + xg_3h_3,$$
$$(gh)_2 = g_3h_2 + g_2h_1,$$
$$(gh)_3 = g_3h_3.$$

Let $n = o(G)$ and define T by the equation

$$T = (1/n) \sum \{g_3^{-1}g_2 \mid g \in G\}.$$

Then $T \in \mathrm{Hom}(X, W)$. Now

$$Th = \sum g_3^{-1}g_2h_1/n = (\sum g_3^{-1}(gh)_2 - \sum h_2)/n$$
$$= h_3 \sum (g_3h_3)^{-1}(gh)_2/n - h_2$$
$$= (h_3 \sum (gh)_3^{-1}(gh)_2/n) - h_2$$
$$= h_3T - h_2.$$

Hence, on X,

$$(I + T)h = h + h_3T - h_2 = h_3 + h_3T = h_3(I + T),$$
$$X(I + T)h = Xh_3(I + T) = X(I + T).$$

Hence $X(I + T)$ is a G-invariant subspace. If $x \in \mathrm{Ker}(I + T)$, then $x = -xT \in W \cap X = \{0\}$. Therefore $\mathrm{Dim}(X(I + T)) = \mathrm{Dim}(X)$. If $x \in X$, then $-xT \in W$, hence $x \in \langle W, X(I + T) \rangle$. Therefore,

$$\langle W, X(I + T) \rangle \supset \langle W, X \rangle = V, \qquad V = W + X(I + T).$$

12.1.3. (*Maschke.*) *If G is a finite linear group over a field F whose characteristic does not divide $o(G)$, then G is completely reducible.*

Proof. Induct on Dim(V). If V is irreducible, the theorem is valid. If not, then, by Theorem 12.1.2, $V = X + Y$, where X and Y are G-invariant and not $\{0\}$. By the induction hypothesis, each of X and Y is completely reducible, hence V is also. ‖

If G is a finite linear group, then the set of matrices of the form $\Sigma\, c_g g$, $c_g \in F$, forms a ring (in fact, an algebra) R. One may define the terms reducible, decomposable, etc., for rings of linear transformations just as for groups. The following theorem is almost obvious.

12.1.4. *If G is a finite linear group and R the ring of linear transformations it generates (in the above sense) in* End(V), *then G is reducible, irreducible, decomposable, indecomposable, or completely reducible iff R is reducible, etc.*‖

The following form of Schur's lemma is not very elegant, but is quite useful.

12.1.5. (*Schur's lemma.*) *If G and H are irreducible matrix groups of degree m and n, respectively, over a field F, and A is an m by n matrix over F such that $GA = AH$, then either* (i) $A = 0$, *or* (ii) *A is nonsingular and G and H are equivalent.*

Proof. Let V and W be vector spaces associated with G and H. Then $VA \subset W$, and if $h \in H$ then $\exists\, g \in G$ such that $VAh = VgA \subset VA$. Hence VA is an H-invariant subspace of W. By the irreducibility of H, $VA = 0$ or W. If $VA = 0$, then $A = 0$. Suppose $VA = W$. Now Ker(A) $\subset V$, and if $g \in G$, then $gA = Ah$ for some $h \in H$, hence $(\text{Ker}(A)g)A = (\text{Ker}(A))Ah = 0$, so Ker($A$)$g \subset$ Ker(A). Therefore, Ker(A) is G-invariant, hence by the irreducibility of G, Ker(A) $= 0$ or V. If Ker(A) $= V$, then $A = 0$ again. If Ker(A) $= 0$, then A is an isomorphism. Hence A is a square, nonsingular matrix, and $A^{-1}GA = H$, so that (ii) is valid. ‖

Any linear group G on a vector space V is contained in the algebra End(V). By $C(G)$ will be meant $C_{\text{End}(V)}(G)$. Let $\mathscr{C}$ denote the field of complex numbers.

12.1.6. *If G is an irreducible linear group over $\mathscr{C}$, then $C(G)$ is the ring of scalar transformations.*

Proof. Certainly, any scalar transformation is in $C(G)$. Conversely, suppose that $A \in C(G)$. The characteristic equation Det($A - xI$) $= 0$ has a root $c \in \mathscr{C}$ since $\mathscr{C}$ is algebraically closed. Therefore Det($A - cI$) $= 0$, so that $A - cI$ is singular. But $G(A - cI) = (A - cI)G$, so by Schur's lemma, $A - cI = 0$, and $A = cI$ is a scalar.

EXERCISES

12.1.7. Maschke's theorem becomes false if the hypothesis that the characteristic of F does not divide $o(G)$ is omitted. Let $F = J_2$, $G = \left\{ I, \begin{bmatrix} 1 & 1 \\ 0 & 1 \end{bmatrix} \right\}$. Prove that G is a reducible group but is not decomposable.

12.1.8. Maschke's theorem is false for infinite groups. Let F be any field of characteristic 0, and G the group generated by $\begin{bmatrix} 1 & 1 \\ 0 & 1 \end{bmatrix}$. Prove that G is reducible but not decomposable.

12.1.9. Theorem 12.1.6 is false over the field of reals. $\left(\text{Let } G \text{ be generated by } \begin{bmatrix} 0 & 1 \\ -1 & 0 \end{bmatrix} \right).$

12.1.10. State and prove Schur's lemma for irreducible sets of matrices.

12.1.11. A finite linear group of degree 1 over any field is cyclic.

12.2   Representations and characters

A *representation* of a group G is a homomorphism T of G onto a linear group H. *In the rest of this chapter, unless explicit exception is made, the group G will be finite and all representations and matrices will be over the field $\mathscr{C}$ of complex numbers.* A representation T of G is *reducible*, etc., iff GT is reducible, etc. The *character* X_T of a representation T is given by

$$X_T(g) = \mathrm{Tr}(gT), g \in G,$$

where the *trace* $\mathrm{Tr}(A)$ of a matrix A is the sum of its diagonal entries. The character is a function from G into $\mathscr{C}$. For functions f_1 and f_2 from G into $\mathscr{C}$, there is an *inner product* defined by the rule:

$$(f_1, f_2) = \Sigma \{ f_1(g) f_2(g^{-1}) \mid g \in G \} / o(G). \tag{1}$$

12.2.1. *The inner product is bilinear:*

$$(f_1 + f_2, f_3) = (f_1, f_3) + (f_2, f_3),$$
$$(f_1, f_2 + f_3) = (f_1, f_2) + (f_1, f_3),$$
$$(cf_1, f_2) = c(f_1, f_2) = (f_1, cf_2), \qquad c \in \mathscr{C},$$

and symmetric: $(f_1, f_2) = (f_2, f_1).$

Proof. Let us check symmetry.

$$o(G)(f_2, f_1) = \Sigma f_2(g) f_1(g^{-1}) = \Sigma f_2(g^{-1}) f_1(g)$$
$$= \Sigma f_1(g) f_2(g^{-1}) = o(G)(f_1, f_2),$$

hence $(f_2, f_1) = (f_1, f_2)$ for all f_1 and f_2. ∥

Two representations T and U of G over $\mathscr{C}$ are *equivalent* iff they have the same degree and there is a matrix A such that for all $g \in G$, $A^{-1}(gT)A = gU$.

12.2.2. *Equivalent representations have equal characters.*

Proof. If T and U are equivalent representations, then there is a matrix A such that $A^{-1}(gT)A = gU$ for all $g \in G$. Since $\mathrm{Tr}(A^{-1}BA) = \mathrm{Tr}(B)$, it follows that $X_U(g) = X_T(g)$.

12.2.3. *If T is a representation of G and X_T its character, then X_T is constant on each conjugate class of G.*

Proof. If $a \in G$ and $b \in G$, then

$$X_T(a^{-1}ba) = \mathrm{Tr}((a^{-1}ba)T)$$
$$= \mathrm{Tr}[(aT)^{-1}(bT)(aT)] = \mathrm{Tr}(bT)$$
$$= X_T(b).$$

12.2.4. *If T and U are representations of G of degrees m and n, respectively, A is an m by n matrix, $h \in G$, and $B = \Sigma (gT)A(g^{-1}U)$, then $(hT)B = B(hU)$.*

Proof. We have

$$(hT)B = \Sigma \{(hT)(gT)A(g^{-1}U) \mid g \in G\}$$
$$= \Sigma \{((hg)T)A((hg)^{-1}U)(hU) \mid g \in G\}$$
$$= \Sigma \{(kT)A(k^{-1}U) \mid k \in G\}(hU) = B(hU).$$

12.2.5. *If T and U are inequivalent, irreducible representations of G of degrees m and n, respectively, and A is an m by n matrix, then*

$$\Sigma \{(gT)A(g^{-1}U) \mid g \in G\} = 0.$$

Proof. If $B = \Sigma (gT)A(g^{-1}U)$, then, by Theorem 12.2.4, $(hT)B = B(hU)$ for all $h \in G$. By Schur's lemma (Theorem 12.1.5), $B = 0$. ∥

If T is a representation of G by matrices, then denote the (i, j) entry in $T(g)$ by $T_{ij}(g)$. Thus T_{ij} is a function from G into $\mathscr{C}$. Let δ be the Kronecker delta function.

12.2.6. Let G be a finite group and T and U irreducible representations of G.

(i) *If T is inequivalent to U, then $(T_{ij}, U_{rs}) = 0$ for all i, j, r, and s.*

(ii) $(T_{ij}, T_{rs}) = \delta_{is}\,\delta_{jr}/m$, *where $m = \mathrm{Deg}(T)$.*

Proof. Let I_{jr} be the m by n matrix with (j, r) entry 1 and all other entries 0, where $n = \mathrm{Deg}(U)$.

(i) By Theorem 12.2.5,

$$\frac{1}{o(G)} \sum (gT)I_{jr}(g^{-1}U) = 0.$$

The (i, s) entry of the left member is

$$\frac{1}{o(G)} \sum T_{ij}(g)U_{rs}(g^{-1}) = (T_{ij}, U_{rs}).$$

(ii) Let

$$B_{jr} = \frac{1}{o(G)} \sum (gT)I_{jr}(g^{-1}T).$$

By Theorem 12.2.4, $(hT)B_{jr} = B_{jr}(hT)$ for all $h \in G$. By Theorem 12.1.6, $B_{jr} = c_{jr}I$, $c_{jr} \in \mathscr{C}$. Taking the (i, s) entry of the resulting equation, one has

$$\frac{1}{o(G)} \sum T_{ij}(g)T_{rs}(g^{-1}) = c_{jr}\,\delta_{is},$$

or

$$(T_{ij}, T_{rs}) = c_{jr}\,\delta_{is}.$$

By symmetry (Theorem 12.2.1), $(T_{ij}, T_{rs}) = (T_{rs}, T_{ij}) = \delta_{jr}\,c_{si}$. Hence if $j \neq r$ or $i \neq s$, then the conclusion (ii) holds. Moreover, $(T_{ij}, T_{ji}) = c_{jj} = c_{ii}$ by the above equations. Thus,

$$c_{11} = c_{22} = \ldots = c_{mm} = c \in \mathscr{C},$$

$$mc = \sum \{c_{ii} \mid 1 \leq i \leq m\} = \sum (T_{1i}, T_{i1})$$

$$= \frac{1}{o(G)} \sum \{\sum \{T_{1i}(g)T_{i1}(g^{-1}) \mid g \in G\} \mid 1 \leq i \leq m\}$$

$$= \frac{1}{o(G)} \sum \{\sum \{T_{1i}(g)T_{i1}(g^{-1}) \mid 1 \leq i \leq m\} \mid g \in G\}$$

$$= \frac{1}{o(G)} \sum \{T_{11}(e) \mid g \in G\} = \frac{o(G)}{o(G)} \cdot 1 = 1.$$

Therefore $c = 1/m$. Thus $(T_{ij}, T_{ji}) = 1/m$, and (ii) holds in all cases.

12.2.7. (Orthogonality of irreducible characters.) If T and U are irre-

ducible representations of a finite group G with characters X_T and X_U, respectively, then

$$(X_T, X_U) = \begin{cases} 1 & \text{if } T \text{ and } U \text{ are equivalent,} \\ 0 & \text{if } T \text{ and } U \text{ are inequivalent.} \end{cases}$$

In particular, $(X_T, X_T) = 1$.

Proof. Let T and U have degrees m and n, respectively. Then

$$(X_T, X_U) = \frac{1}{o(G)} \Sigma\, X_T(g) X_U(g^{-1})$$

$$= \frac{1}{o(G)} \Sigma\, \{(\Sigma\, \{T_{ii}(g) \mid 1 \leq i \leq m\})$$

$$(\Sigma\, \{U_{jj}(g^{-1}) \mid 1 \leq j \leq n\}) \mid g \in G\}$$

$$= \frac{1}{o(G)} \Sigma\, \{T_{ii}(g) U_{jj}(g^{-1}) \mid g \in G, 1 \leq i \leq m, 1 \leq j \leq n\}$$

$$= \Sigma\, \{(T_{ii}, U_{jj}) \mid 1 \leq i \leq m, 1 \leq j \leq n\}.$$

If T and U are inequivalent, then all of the terms in the last summation are 0 by Theorem 12.2.6, hence $(X_T, X_U) = 0$. If T and U are equivalent, then by Theorems 12.2.2 and 12.2.6,

$$(X_T, X_U) = (X_T, X_T) = \Sigma\, \{(T_{ii}, T_{jj}) \mid 1 \leq i \leq m, 1 \leq j \leq m\}$$

$$= \Sigma\, \{(T_{ii}, T_{ii}) \mid 1 \leq i \leq m\} = m\frac{1}{m} = 1.$$

12.2.8. *Two irreducible representations of a finite group are equivalent iff they have the same character.*

Proof. Let T and U be irreducible representations of a finite group G. If T and U are equivalent, then by Theorem 12.2.2, $X_T = X_U$. Conversely, if $X_T = X_U$, then, by Theorem 12.2.7, $(X_T, X_U) = 1$, hence, by Theorem 12.2.7 again, T and U are equivalent. ‖

Let T be a representation of a finite group G on a vector space V. By Maschke's theorem, $V = W_1 + \ldots + W_n$, where GT is irreducible on W_i. Thus, T induces an irreducible representation T_i of G on W_i: $w(gT_i) = w(gT)$, $w \in W_i$. Define T to be the *direct sum* of the T_i, written

$$T = T_1 + \ldots + T_n = \Sigma\, T_i,$$

provided the above situation obtains.

12.2.9. *Any representation T of a finite group G on a vector space V is the direct sum of irreducible representations T_i, $T = \Sigma\, T_i$. The characters* *satisfy the relation* $X_T = \Sigma\, X_i$, *where* $X_i = X_{T_i}$. *The* T_i *are unique up to equivalence.*

Proof. It has already been noted that $V = \Sigma\, W_i$ and $T = \Sigma\, T_i$, where T_i is irreducible and acts on W_i. If B_i is a basis of W_i, then $B = \dot{\cup}\, B_i$ is a basis of V. If $g \in G$, then gT has matrix

$$\begin{bmatrix} A_1 & & 0 \\ & \ddots & \\ 0 & & A_n \end{bmatrix}$$

with respect to B, where A_i is the matrix of gT_i with respect to B_i. Hence

$$X_T(g) = \operatorname{Tr}(gT) = \Sigma\, \operatorname{Tr}(A_i) = \Sigma\, \operatorname{Tr}(gT_i) = \Sigma\, X_i(g).$$

Therefore $X_T = \Sigma\, X_i$. Let $T = \Sigma\, U_j$ where the U_j are irreducible, let U be any irreducible representation of G, r the number of T_i equivalent to U, and s the number of U_j equivalent to U. Then, by Theorem 12.2.7 and bilinearity of the inner product,

$$r = \Sigma\, (X_U, X_{T_i}) = (X_U, \Sigma\, X_{T_i}) = (X_U, X_T) = (X_U, \Sigma\, X_{U_j})$$
$$= \Sigma\, (X_U, X_{U_j}) = s. \;\|$$

Theorem 12.2.8 can now be generalized to the case of any two representations of G.

12.2.10. *Two representations T and U of a finite group G are equivalent iff they have equal characters.*

Proof. If T is equivalent to U, then $X_T = X_U$ by Theorem 12.2.2. Conversely, suppose that $X_T = X_U$. Then, adding dummy terms if necessary, $T = \Sigma\, m_i T_i$, $U \sim \Sigma\, n_i T_i$, where T_i is irreducible, $\sim$ denotes equivalence, and m_i and n_i are nonnegative integers. Then $\Sigma\, m_i X_i = \Sigma\, n_i X_i$, hence

$$m_j = (m_j X_j, X_j) = (\Sigma\, m_i X_i, X_j) = (\Sigma\, n_i X_i, X_j) = n_j.$$

Hence $T \sim U$. $\|$

Theorem 12.2.10 permits the introduction of terms such as *irreducible character*, etc. (a character is irreducible iff a representation from which it comes is irreducible).

12.2.11. *The set of irreducible characters is linearly independent.*

Proof. Let $X^1, \ldots, X^s$ be distinct irreducible characters of a finite group G. If $\Sigma\, c_i X^i = 0$, $c_i \in \mathscr{C}$, then, by Theorem 12.2.7, $0 = \Sigma\, (c_i X^i, X^j) = c_j$ for each j. Therefore the set of irreducible characters is linearly independent. $\|$

The next lemma is of a preliminary nature only.

12.2.12. *If G has r conjugate classes, then it has at most r irreducible characters.*

Proof. A character is constant on each conjugate class, hence is an element of the r-dimensional vector space of functions from the set S of conjugate classes of G into $\mathscr{C}$. By the preceding theorem, there are at most r irreducible characters of G. ∥

Theorem 12.2.12 says that the number of irreducible characters of G is finite. Let 1_G be the (irreducible) representation $(g)1_G = 1$, and let X^1 be its character: $X^1(g) = 1$ for all $g \in G$. Let $X^1, \ldots, X^s$ be the irreducible characters of G. Let $C_1, \ldots, C_r$ be the conjugate classes of G with $C_1 = \{e\}$, and let $h_i = o(C_i)$. Let X^i_j be the (constant) value of X^i on C_j.

12.2.13. (i) $(X^i, X^j) = \delta_{ij}$.
(ii) *More generally, if $X = \Sigma\, m_i X^i$ and $Y = \Sigma\, n_i X^i$ with m_i and n_i nonnegative integers, then $(X, Y) = \Sigma\, m_i n_i$.*

Proof. (i) is just a restatement of Theorem 12.2.7, and (ii) follows by bilinearity. ∥

It should be noted that if $m_1, \ldots, m_s$ are nonnegative integers, not all 0, then $\Sigma\, m_i X^i$ is a character. For each X^i is a character of some irreducible representation T_i of G on a vector space W_i. There is then a vector space V which is a direct sum $V = \Sigma\, V_i$ where each V_i is a direct sum of m_i copies of W_i. One can then define a representation T of G by requiring that each (gT) restricted to one of the copies of W_i acts as it should. The result is that T is the direct sum of m_1 representations equivalent to T_1, etc., and that the character $X_T = \Sigma\, m_i X^i$.

12.2.14. (i) $\Sigma\, X^1(g) = o(G)$, $\Sigma\,\{X^i(g) \mid g \in G\} = 0$ *if* $i > 1$.
(ii) $\Sigma\,\{h_j X^i_j \mid 1 \leq j \leq r\} = \delta_{1i}\, o(G)$.
Proof. $X^1(g) = 1$ for all $g \in G$, hence $\Sigma\, X^1(g) = o(G)$. For $i > 1$,

$$0 = (X^i, X^1) = \frac{1}{o(G)} \Sigma\, X^i(g) X^1(g^{-1})$$

$$= \frac{1}{o(G)} \Sigma\, X^i(g).$$

Hence $\Sigma\, X^i(g) = 0$. Since characters are constant on classes, (ii) follows from (i). ∥

If (M, G) is a permutation group of degree m, there is a natural way of considering G as a linear group of degree m. Simply take a vector space V

of degree m with basis M, and extend the permutation g linearly to get an automorphism of V (uniquely). Similar remarks apply to permutation representations of G. In particular, the regular (Cayley) representation of G which, up to now, has been a permutation representation, can be made into a representation.

12.2.15. (i) *If R is the regular representation of G, then $X_R = \Sigma\, n_i X^i$* where $n_i = \mathrm{Deg}(X^i) = X_1^i$.

(ii) $\Sigma\, \{X_1^i X_j^i \mid 1 \leq i \leq s\} = \delta_{1j}\, o(G)$.

Proof. In any case, $X_R = \Sigma\, n_i X^i$ for some nonnegative integers n_i. Now $R(g)$, as permutation, moves all letters if $g \neq e$, and none if $g = e$. Hence,

$$X_R(g) = \begin{cases} 0 & \text{if } g \neq e, \\ o(G) & \text{if } g = e. \end{cases}$$

Therefore,

$$\Sigma\, \{n_i X^i(g) \mid 1 \leq i \leq s\} = \begin{cases} 0 & \text{if } g \neq e, \\ o(G) & \text{if } g = e. \end{cases} \tag{2}$$

Therefore, by orthogonality of the irreducible characters,

$$\begin{aligned} n_j o(G) &= \Sigma\, \{n_i (X^i, X^j) o(G) \mid 1 \leq i \leq s\} \\ &= \sum_i \sum_g n_i X^i(g) X^j(g^{-1}) \\ &= \sum_g [X^j(g^{-1}) \sum_i n_i X^i(g)] \\ &= X^j(e) o(G). \end{aligned}$$

Hence $n_j = X^j(e)$, and $X^j(e)$ is just the degree of a representation from which X^j arises.

If, in (2), n_i is replaced by $X^i(e) = X_1^i$ and $X^i(g)$ by X_j^i, (ii) is obtained.

12.2.16. *If M is a square matrix over a field F, $M^n = I$ for some $n \in \mathcal{N}$, and F contains n nth roots of 1, then M is similar to a diagonal matrix.*

Proof. The minimum polynomial f of M is a divisor of $x^n - 1$ which has distinct roots. Hence, f itself has distinct roots. The conclusion follows from Birkhoff and MacLane [1, p. 327].

12.2.17. *If T is a representation of G and $g \in G$, then there is an equivalent representation U such that gU is a diagonal matrix.*

Proof. Since $g^n = e$ for some $n \in \mathcal{N}$, $(gT)^n = I$, and the result follows from Theorem 12.2.16. ‖

An *algebraic integer* is a complex number c which is a root of a polynomial with integral coefficients and leading coefficient 1:

$$c^n + a_{n-1}c^{n-1} + \ldots + a_0 = 0, \, a_i \in J.$$

For unproved facts about algebraic integers, the reader is referred to Birkhoff and MacLane [1].

12.2.18. *If R is a representation of G with character X, and $g \in G$, then $X(g)$ is an algebraic integer.*

Proof. By Theorem 12.2.17, *WLOG* gR is a diagonal matrix. For some $n \in \mathcal{N}$, $(gR)^n = I$. Therefore the diagonal entries of gR are nth roots of 1, hence algebraic integers. Since the algebraic integers form a ring (Birkhoff and MacLane [1, p. 420]), $X(g)$ is an algebraic integer. ∥

If G is any group and F any field, the *group algebra* of G over F is the set A of formal sums $\Sigma \{c_g g \mid g \in G\}$ such that $c_g \in F$ and $c_g = 0$ for all but a finite number of $g \in G$, with operations:

$$\Sigma c_g g + \Sigma d_g g = \Sigma (c_g + d_g)g,$$
$$a(\Sigma c_g g) = \Sigma (ac_g)g \quad \text{for} \quad a \in F,$$
$$(\Sigma c_g g)(\Sigma d_g g) = \Sigma \{\Sigma c_h d_{h^{-1}g} \mid h \in G\}g \mid g \in G\}.$$

The definition of multiplication is the result of assuming linearity and the multiplication in G. Under the above operations, the group algebra A is an algebra over F. The element $1 \cdot g$ of A will be denoted by g. Let

$$\overline{C_i} = \Sigma \{g \mid g \in C_i\}$$

be the element of A corresponding to the conjugate class C_i of G.

12.2.19. *If A is the group algebra of a finite group G over $\mathscr{C}$, then $\overline{C_1}, \ldots, \overline{C_r}$ form a basis for $Z(A)$.*

Proof. If $g \in G$, then $C_i^g = C_i$, hence $\overline{C_i}g = g\overline{C_i}$. Therefore, by linearity, each $\overline{C_i} \in Z(A)$. Since conjugate classes are disjoint, the set $\{\overline{C_1}, \ldots, \overline{C_r}\}$ is linearly independent over $\mathscr{C}$.

Let $u = \Sigma c_g g \in Z(A)$ and $x \in G$. Then $xu = ux$, so that

$$c_g xg = c_{xgx^{-1}}(xgx^{-1})x, \quad c_g = c_{xgx^{-1}}.$$

Hence the coefficients of all conjugates of g are the same in u. Therefore u is a linear combination of $\overline{C_1}, \ldots, \overline{C_r}$. Thus $(\overline{C_1}, \ldots, \overline{C_r})$ is a basis for $Z(A)$ over $\mathscr{C}$. ∥

Any representation T of G can be extended in just one way to a homomorphism of the group algebra A of G into the algebra of n by n matrices

$(n = \text{Deg}(T))$. The resulting homomorphism will also be denoted by T and called a representation.

12.2.20. *If A is the group algebra of G over $\mathscr{C}$, T an irreducible representation, and X^i the character of T, then*

(i) $\overline{C}_j T = d_j I$, $d_j \in \mathscr{C}$.

(ii) $\overline{C}_j \overline{C}_k = \sum_m c_{jkm} \overline{C}_m$, c_{jkm} *a nonnegative integer.*

(iii) $d_j d_k = \sum_m c_{jkm} d_m$.

(iv) $d_j = h_j X^i_j / X^i_1$, *where* $h_j = o(C_j)$.

Proof. (ii) By Theorem 12.2.19, $\overline{C}_j \overline{C}_k = \sum c_{jkm} \overline{C}_m$, $c_{jkm} \in \mathscr{C}$. It is clear, however, that if this equation is written in terms of elements of G, then the coefficients on the left side are nonnegative integers. Hence the c_{jkm} are nonnegative integers.

(i). By Theorem 12.2.19, $\overline{C}_j \in Z(A)$, hence $\overline{C}_j T \in C(GT)$. Since T is irreducible, it follows from Theorem 12.1.6 that $\overline{C}_j T = d_j I$ for some $d_j \in \mathscr{C}$.

(iii). By (i) and (ii),

$$d_j d_k I = (\overline{C}_j T)(\overline{C}_k T) = (\overline{C}_j \overline{C}_k) T$$
$$= \sum_m c_{jkm} (\overline{C}_m T) = \sum_m c_{jkm} d_m I.$$

Equation (iii) follows.

(iv). By (i),

$$h_j X^i_j = \text{Tr}(\overline{C}_j T) = \text{Tr}(d_j I) = d_j \text{Tr}(I) = d_j X^i_1,$$

so that (iv) holds.

12.2.21. *The d_j in Theorem 12.2.20 are algebraic integers.*

Proof. By Theorem 12.2.20, $d_j d_k = \sum c_{jkm} d_m$, where the c_{jkm} are nonnegative integers. Fix k and let B be the square matrix (c_{jkm}) and D the column matrix (d_m). Then the equations become

$$(B - d_k I) D = 0. \tag{3}$$

Since

$$d_1 = \frac{h_1 X^i_1}{X^i_1} = h_1 = 1 \neq 0$$

by Theorem 12.2.20, $D \neq 0$. Hence

$$\text{Det}(B - d_k I) = 0. \tag{4}$$

Since the c_{jkm} are integers, expansion of (4) leads to an equation for d_k with integral coefficients and leading coefficient ± 1. Hence d_k is an algebraic integer. ‖

If C_i is a conjugate class, then $C_{i'} = \{x \in G \mid x^{-1} \in C_i\}$ is also a conjugate class.

12.2.22. *For each j and k,*

$$\Sigma \{X_j^i X_k^i \mid 1 \leq i \leq s\} = \delta_{jk'} \, o(G)/h_j.$$

Proof. By Theorem 12.2.20, $\overline{C_j}\,\overline{C_k} = \Sigma\, c_{jkm}\overline{C_m}$. Now e occurs in the expansion of $\overline{C_j}\,\overline{C_k}$ iff $j = k'$. Hence $c_{jk1} = 0$ if $j \neq k'$ and $c_{jj'1} = h_j$. By (iii) and (iv) of Theorem 12.2.20, for each i,

$$(h_j X_j^i / X_1^i)(h_k X_k^i / X_1^i) = \sum_m c_{jkm} h_m X_m^i / X_1^i,$$

$$h_j h_k X_j^i X_k^i = \sum_m c_{jkm} h_m X_1^i X_m^i.$$

Hence, by Theorem 12.2.15,

$$h_j h_k \sum_i X_j^i X_k^i = \sum_m (c_{jkm} h_m \sum_i X_1^i X_m^i)$$

$$= c_{jk1} h_1 o(G) = \delta_{jk'} \, h_j o(G).$$

Therefore,

$$\Sigma X_j^i X_k^i = \delta_{jk'} \, o(G)/h_k = \delta_{jk'} \, o(G)/h_j.$$

12.2.23. *The number of irreducible characters of G equals the number of conjugate classes.*

Proof. By Theorem 12.2.22, $\Sigma X_j^i X_k^i = 0$ if $k \neq j'$ and is $o(G)/h_j$ if $k = j'$. If the s-tuples $(X_j^1, \ldots, X_j^s)$, $j = 1, \ldots, r$ are linearly dependent, then $\exists\, c_j \in \mathscr{C}$ such that

$$\Sigma \{c_j X_j^i \mid 1 \leq j \leq r\} = 0, \qquad i = 1, \ldots, s.$$

Therefore, for each k,

$$0 = \sum_{i,j} c_j X_j^i X_k^i = \frac{c_{k'} o(G)}{h_k},$$

hence $c_{k'} = 0$ for all k. This means that the s-tuples $(X_j^1, \ldots, X_j^s)$, $j = 1, \ldots, r$ are linearly independent. Since the set of s-tuples forms a vector space of dimension s over $\mathscr{C}$, $r \leq s$. By Theorem 12.2.12, $s \leq r$. Therefore, $r = s$.

12.2.24. *The sum of the squares of the degrees of the irreducible characters of G is $o(G)$.*

Proof. By Theorem 12.2.22,

$$\sum (X_1^i)^2 = \sum X_1^i X_1^i = \frac{o(G)}{h_1} = o(G). \ \|$$

If T is a representation, let T'' be defined by the equation

$$gT'' = (g^{-1}T)^\circ$$

where A° is the transpose of the matrix A ($A_{ij}^\circ = A_{ji}$). The properties of T'' are left as Exercise 12.2.31. By this exercise, i'' may be defined as follows ($1 \leqq i \leqq r$). Let T be a representation with character X^i and let $X^{i''}$ be the character of T''. Then $X^{i''}$ is irreducible, so that $1 \leqq i'' \leqq r$.

12.2.25. $X_j^{i''} = X_{j'}^i$.

Proof. Let T be a representation with character X^i and let $g \in C_j$. Then

$$X_j^{i''} = \mathrm{Tr}(gT'') = \mathrm{Tr}(g^{-1}T)^\circ$$
$$= \mathrm{Tr}(g^{-1}T) = X_{j'}^i.$$

12.2.26. (*Orthogonality.*) $\sum_j h_j X_j^i X_j^k = \delta_{ik'} \, o(G)$.

Proof. By Theorems 12.2.13 and 12.2.25,

$$\sum_j h_j X_j^i X_j^k = \sum_j h_j X_j^i X_{j'}^{k''} = \sum_g X^i(g) X^{k''}(g^{-1})$$
$$= o(G)(X^i, X^{k''})$$
$$= \delta_{ik''} \, o(G).$$

12.2.27. *The degree of an irreducible representation of G divides $o(G)$.*

Proof. By Theorem 12.2.21, for all i and j, $h_j X_j^i / X_1^i$ is an algebraic integer. By Theorem 12.2.18, each X_j^k is an algebraic integer. Therefore, $\sum_{j,k} h_j X_j^i X_j^k / X_1^i$ is an algebraic integer. By Theorem 12.2.26,

$$\sum_{j,k} h_j X_j^i X_j^k / X_1^i = \sum_k \delta_{ik''} \, o(G)/X_1^i = o(G)/X_1^i,$$

which is rational. But an algebraic integer which is rational must be an (ordinary) integer. Therefore $X_1^i \mid o(G)$. Since $X_1^i = \mathrm{Deg}(X^i)$, this proves the theorem.

EXERCISES

12.2.28. If X is a character of G and $g \in G$, then $X(g^{-1})$ is the complex conjugate of $X(g)$.

12.2.29. If T is a representation of G with character X, and $g \in G$, then $g \in \mathrm{Ker}(T)$ iff $X(g) = X(e)$. (Use Theorem 12.2.17.)

12.2.30. If $g \in G$, then g is conjugate to g^{-1} iff all $X^i(g)$ are real.

12.2.31. Let T be a representation of G, and let $gT'' = (g^{-1}T)°$.

    (a) T'' is a representation of G.

    (b)   If T is irreducible, so is T''.

    (c)   If T is similar to U, then T'' is similar to U''.

12.2.32. (a)   If $X = \Sigma\, n_i X^i$, where the n_i are nonnegative integers, then $(X, X) = \Sigma\, n_i^2$.

    (b)   A character X is irreducible iff $(X, X) = 1$.

12.2.33. If X is a character of G, $x \in G$, and $o(x) = 2$, then $X(x) = \mathrm{Deg}(X) - 2n = X(e) - 2n$, $0 \leq n \leq X(e)$, $n \in J$. In particular, $X(x)$ is an integer.

12.3 The $p^i q^j$ theorem

    In this section the theorem of Burnside that all groups of order $p^i q^j$ ($p \in \mathscr{P}, q \in \mathscr{P}$) are solvable will be proved.

    12.3.1. *If G is a finite irreducible linear group over $\mathscr{C}$ with character X, C_i is a conjugate class of G, and $(\mathrm{Deg}(G), o(C_i)) = 1$, then either $X_i = 0$ or C_i is a (one-element) subset of $Z(G)$.*

    Proof. Since $X_1 = \mathrm{Deg}(G)$, there are integers a and b such that $a \cdot o(C_i) + bX_1 = 1$. By Theorem 12.2.18, X_i is an algebraic integer, and by Theorem 12.2.21, $o(C_i)X_i/X_1$ is also an algebraic integer. Therefore,

$$\frac{a \cdot o(C_i)X_i}{X_1} + bX_i = \frac{X_i}{X_1}$$

is an algebraic integer.

    Let $M \in C_i$. By Theorem 12.2.16, there is a nonsingular matrix A such that $A^{-1}MA$ is diagonal. If $A^{-1}MA$ is a scalar matrix, then $M = A^{-1}MA$ is also, and $M \in Z(G)$, so that C_i is a subset of $Z(G)$. Hence $WLOG$, M is a diagonal, but not a scalar, matrix. All diagonal entries of M are roots of 1, but not all are equal. Therefore, it follows from the triangle inequality that $|X_i| < X_1$, hence $|X_i/X_1| < 1$.

    Now let $f = x^n + \ldots + a_0$ be the minimum polynomial of X_i/X_1 over $\mathscr{R}$. Let F be a finite extension field of $\mathscr{R}$ containing all the roots of f and those roots of 1 occurring as diagonal entries of M. There is an automorphism T of F sending X_i/X_1 onto any given conjugate of X_i/X_1 (Birkhoff and MacLane [1, pp. 437–438]). T sends roots of 1 onto roots of 1. It follows that T maps X_i onto a sum of roots of 1 not all the same. Hence the absolute value of any conjugate of X_i/X_1 is less than 1 also. But $\pm a_0$ equals the product of the conjugates of X_i/X_1. Hence $|a_0| < 1$. Since X_i/X_1 is an algebraic integer, a_0 is an integer. Therefore, $a_0 = 0$. Since f is irreducible, $f = x$, so that $X_i/X_1 = 0$. Hence, $X_i = 0$.

12.3.2. (*Burnside.*) *If G is a finite group, $x \in G$, $o(\mathrm{Cl}(x)) = p^n$, $p \in \mathscr{P}$, and $n > 0$, then G is not simple.*

Proof. Deny the theorem. Let $x \in C_j$. By the orthogonality relations of Theorem 12.2.22.

$$\sum_i X_1^i X_j^i = 0. \tag{1}$$

Now $X_1^1 X_j^1 = 1$. If $i \neq 1$, then by the simplicity of G, the ith irreducible representation T_i of G is an isomorphism. Since the center is a normal subgroup, it is E. Hence $xT_i \notin Z(GT_i)$. By Theorem 12.3.1, if $i \neq 1$, then either $X_j^i = 0$ or $p \mid X_1^i$. Since the algebraic integers form a ring, it follows from (1) and Theorem 12.2.18 that there is an algebraic integer a such that $1 + pa = 0$. This is a contradiction since $a = -1/p$ is not an algebraic integer.

12.3.3. (*Burnside.*) *If $o(G) = p^i q^i$, $p \in \mathscr{P}$, $q \in \mathscr{P}$, then G is solvable.*

Proof. If $p = q$, then G is a p-group, hence solvable. Now induct on $o(G)$. If there is a nontrivial normal subgroup H, then H and G/H are solvable by induction, hence G is solvable. Let $P \in \mathrm{Syl}_p(G)$ and $z \in Z(P)^{\#}$. If $z \in Z(G)$, then $\langle z \rangle \lhd G$ and we are through. If $z \notin Z(G)$, then $o(\mathrm{Cl}(z)) = q^k$, $k > 0$. By Theorem 12.3.2, G has a nontrivial normal subgroup H.

12.3.4. *If G is a (perhaps infinite) group, H a nilpotent subgroup of index p^n, $p \in \mathscr{P}$, then G is solvable.*

Proof. $G/\mathrm{Core}(H)$ is finite, and $\mathrm{Core}(H)$ is nilpotent, therefore solvable. Hence it suffices to prove that $G/\mathrm{Core}(H)$ is solvable. But $G/\mathrm{Core}(H)$ has a nilpotent subgroup $H/\mathrm{Core}(H)$ of index p^n. Therefore $WLOG$, G is finite.

Now induct on $o(G)$. Since H is nilpotent, $\exists\, x \in Z(H)^{\#}$. Hence

$$o(\mathrm{Cl}(x)) = p^s, s \geq 0.$$

By Theorem 12.3.2, $\exists\, K \lhd G$, $E < K < G$ (the case $s = 0$ yields either $K = \langle x \rangle$ or a triviality). Now $HK/K \cong H/H \cap K$ is a nilpotent subgroup of G/K, and

$$[G/K : HK/K] = [G:HK] \mid [G:H].$$

Hence $[G/K : HK/K] = p^t$, so, by induction, G/K is solvable. Also $H \cap K$ is a nilpotent subgroup of K and

$$[K : H \cap K] = [HK : H] \mid [G:H].$$

Therefore, K is solvable by induction. Hence G is solvable. ‖

This generalizes the $p^i q^j$ theorem. Another generalization will be given in Theorem 13.2.9.

Theorem 12.3.2 gives a nonsimplicity criterion. A numerical illustration of its use will be given here and others in the exercises.

12.3.5. *There is no simple group of order* 4400.

Proof. Suppose that G is simple of order $4400 = 2^4 \cdot 5^2 \cdot 11$. Let $P \in \text{Syl}_5(G)$. If $n_5 = 16$, then $11 \mid o(N(P))$, hence $11 \mid o(C(P))$ by the N/C theorem, so that $N(P) = C(P)$, contradicting Burnside. Therefore $n_5 = 11$. Since $11 \not\equiv 1 \pmod{5^2}$, $\exists \; Q \in \text{Syl}_5(G)$ such that $o(P \cap Q) = 5$ (Theorem 6.5.4). Now $C(P \cap Q)$ contains at least two Sylow 5-subgroups, hence by Sylow, either 16 or 11 divides $o(C(P \cap Q))$. If $x \in (P \cap Q)^{\#}$, it follows that $o(\text{Cl}(x)) = 11$ or 2^i, so that, by Theorem 12.3.2, G is not simple, a contradiction. ‖

Recall the theorem of Hall, 9.3.11, which states that any finite solvable group has a Sylow basis. It is now possible to prove the converse, also due to P. Hall [2].

12.3.6. *If a finite group G has a Sylow basis, then it is solvable.*

Proof. Let $\{P_1, \dots, P_n\}$ be a Sylow basis of G, with P_i a Sylow p_i-subgroup. The product Q_i of all P_j, $j \neq i$, is a Hall p_i'-subgroup of G. Now induct on $o(G)$.

By the $p^i q^j$ theorem, $P_1 P_2$ is solvable, hence has a normal subgroup M of prime power order. WLOG, $M \subset P_1$. Then $M \subset Q_2$ also, and $P_2 Q_2 = G$. Hence $M^G = M^{P_2 Q_2} = M^{Q_2} \subset Q_2$, so that $H = M^G$ is a proper normal subgroup of G. Since Q_2 has a Sylow basis, it is solvable by the inductive hypothesis. Therefore, H is solvable, being a subgroup of Q_2. Also $P_i H/H \in \text{Syl}(G/H)$ by Theorem 6.1.16, and $(P_i H/H)(P_j H/H) = P_i P_j H/H$, so that $\{P_i H/H \mid 1 \leq i \leq n\}$ is a Sylow basis of G/H. Therefore, by the inductive hypothesis, G/H is solvable. Thus, G is solvable. ‖

The next theorem is also due to P. Hall (see M. Hall [1, p. 161]).

12.3.7. *If every maximal proper subgroup of a finite group G is of prime or prime squared index, then G is solvable.*

Proof. Let p be the largest prime dividing $o(G)$ and $P \in \text{Syl}_p(G)$. If $P \lhd G$, then G/P is solvable by induction, hence G is solvable. Suppose $P \not\lhd G$, and let M be a maximal proper subgroup of G containing $N(P)$. Then $[G:M] = q$ or q^2, $q \in \mathscr{P}$. By Theorem 6.2.3, $[G:M] \equiv 1 \pmod p$. Hence $[G:M] = q^2$, and $p \mid (q-1)(q+1)$. Therefore, $p \mid (q+1)$, so $p = q + 1$, $q = 2$ and $p = 3$. Thus $o(G) = 2^i 3^j$, so G is solvable (Theorem 12.3.3). ‖

Example. Theorem 12.3.7 is best possible in the sense that there is a nonsolvable (in fact simple) group, all of whose maximal proper subgroups H have index p, p^2, or p^3 for some prime p (which depends on H). Let $G = PSL(2, 7)$. Then $o(G) = 2^3 \cdot 3 \cdot 7 = 168$, and G is simple. Hence, $n_7 = 8$, $o(N(P)) = 21$, and $o(C(P)) = 7$ for $P \in \mathrm{Syl}_7(G)$. Let $Q \subset N(P)$, $o(Q) = 3$. By the counting theorem, 10.7.14,

$$o(\mathrm{Cl}(Q)) = 8 \cdot 7/2 = 28.$$

Thus

$$o(N(Q)) = 6 \quad \text{and} \quad o(C(Q)) = 3$$

(by Burnside). Counting elements, there are 48 of order 7, 56 of order 3, and 1 of order 1, leaving 63 of order 2^i, $i > 0$. Therefore $n_2 \neq 7$ (this would yield at most 49 elements of order 2^i), so that $n_2 = 21$. Since $21 \not\equiv 1 \pmod 8$, there are Sylow 2-subgroups R and S such that $o(R \cap S) = 4$ or 2, and the intersection is a maximal one for Sylow 2-subgroups. Since $3 \mid o(N(R \cap S))$, if $o(R \cap S) = 2$, then there is an element of order 6, and $2 \mid o(C(Q))$, a contradiction. Hence, $o(R \cap S) = 4$ and $o(N(R \cap S)) = 24$.

Now let H be a maximal proper subgroup of G. First suppose that $7 \mid o(H)$. *WLOG, $P \subset H$.* If $P \not\vartriangleleft H$, then $8 \mid o(H)$ and G has a subgroup of index 3, an impossibility (Theorem 10.2.6). If $P \vartriangleleft H$, then by its maximality, $H = N(P)$, so $[G:H] = 8$ as asserted. Next suppose that $7 \nmid o(H)$. Since there is a subgroup L of order 24, $3 \mid o(H)$. If $o(H) = 24$, we are done. If $o(H) = 12$, then H has four Sylow 3-subgroups, hence (counting elements) a normal subgroup K of order 4. But then $2^3 \mid o(N(K))$, so that H is not maximal. If $o(H) = 3$, then $H < N(H)$, a contradiction. Therefore $o(H) = 6$, so that *WLOG, $H = N(Q)$.* Also *WLOG, $Q \subset L$.* Since $n_3(L) = 4$, $N(Q) \subset L$, and $H = N(Q)$ is not maximal, the final contradiction. ‖

Baer [7] has given an interesting generalization of Theorem 12.3.2. This generalization depends on a lemma of perhaps more interest than the generalization itself.

12.3.8. *If a finite group G has a nontrivial characteristic subgroup, then it has a nontrivial, fully characteristic subgroup.*

Proof. Deny. Let $A \neq E$ be a homomorphic image of G of least order. Let

$$H = \cap \{\mathrm{Ker}(T) \mid T \in \mathrm{Hom}(G, A)\}.$$

If $U \in \mathrm{End}(G)$, then $UT \in \mathrm{Hom}(G, A)$ for all $T \in \mathrm{Hom}(G, A)$. Hence,

$$(HU)T = H(UT) = E, \ T \in \mathrm{Hom}(G, A).$$

Thus $HU \subset H$, and H is fully characteristic in G. Since $A \neq E$, $H \neq G$. Therefore, by the denial, $H = E$.

Let M be a minimal normal non-E subgroup of G. Since $H = E$, $\exists\, T \in \mathrm{Hom}(G, A)$ such that $MT \neq E$. By the minimality of A, $GT = A$. Again by the minimality of A, A is simple. Since $MT \lhd GT = A$, $MT = A$. Since $M \cap \mathrm{Ker}(T) \lhd G$ and M is minimal normal, $M \cap \mathrm{Ker}(T) = E$. It follows that $G = \mathrm{Ker}(T)M$, hence $G = \mathrm{Ker}(T) + M$. Moreover, $M \cong A$.

Let $S = \mathrm{Soc}(G)$, and let L be a maximal subgroup of S such that $G = L + R$ for some R. If $R \neq E$, then there is a minimal normal non-E subgroup M of R which is necessarily a minimal normal non-E subgroup of G. By the preceding paragraph, $G = M + T$ for some T. Since $R \supset M$, $R = M + U$ for some U (Exercise 4.1.13). Hence

$$G = M + (L + U) = (L + M) + U.$$

Since M is minimal normal, $M \subset S$ and $L + M \subset S$. This contradicts the maximality of L. Therefore, $R = E$, and $G = L = S$.

Thus G is the direct sum of some of its minimal normal non-E subgroups. All such subgroups were proved isomorphic to A, so that

$$G = M_1 + \ldots + M_r, M_i \cong A.$$

Therefore (Exercise 4.4.17), G has no nontrivial characteristic subgroups, a contradiction.

12.3.9. *If G is a finite group, $x \in G$, and $o(\mathrm{Cl}(x)) = p^n \neq 1$ for some prime p, then there is a nontrivial, fully characteristic subgroup of G.*

Proof. Deny. By Theorem 12.3.8, G has no nontrivial, characteristic subgroup. Therefore (Theorem 4.4.2), $G = H_1 + \ldots + H_r$ where the H_i are isomorphic simple groups. Thus $x = \pi x_i$, $x_i \in H_i$, and

$$p^n = o(\mathrm{Cl}(x)) = \pi o(\mathrm{Cl}_{H_i}(x_i)).$$

Therefore, for some i, $o(\mathrm{Cl}_{H_i}(x_i)) = p^j \neq 1$. But then, by Theorem 12.3.2, H_i is not simple, a contradiction.

EXERCISES

12.3.10. There are no simple groups of order 1200, 2240, or 2800.

12.3.11. (Compare with Theorem 12.3.2.) It is false that if H is a cyclic subgroup of a finite group G with $o(\mathrm{Cl}(H)) = p^n$, $p \in \mathscr{P}$, and $n > 0$, then G is not simple. In fact, let $G = \mathrm{PSL}(2, 7)$ and $o(H) = 7$.

12.3.12. If a finite group G contains a Hall p'-subgroup for each prime p dividing $o(G)$, then G is solvable. (Use Theorem 12.3.6.)

12.3.13. The converse of Theorem 12.3.7 is false. There is a solvable group with a maximal proper subgroup of index 8. (See Exercise 10.5.29.)

12.4 Representations of direct sums

Let A and B be matrices of degrees m and n with (i, j) entries A_{ij} and B_{ij}. Define a matrix $A \times B$ of degree mn by the rule

$$(A \times B)(i, j; s, t) = A_{is}B_{jt}, \qquad 1 \leq i \leq m, 1 \leq s \leq m,$$
$$1 \leq j \leq n, 1 \leq t \leq n.$$

12.4.1.

$$(A_1 + A_2) \times B = (A_1 \times B) + (A_2 \times B),$$
$$A \times (B_1 + B_2) = (A \times B_1) + (A \times B_2),$$
$$(A \times B)(A' \times B') = (AA') \times (BB').$$

Proof. Let us verify the last relation and leave the others as exercises.

$$[(A \times B)(A' \times B')](i, j; s, t) = \sum_{u,v} (A \times B)(i, j; u, v)(A' \times B')(u, v; s, t)$$
$$= \sum_{u,v} A_{iu}B_{jv}A'_{us}B'_{vt} = \sum_{u} A_{iu}A'_{us} \sum_{v} B_{jv}B'_{vt}$$
$$= (AA')_{is}(BB')_{jt} = (AA' \times BB')(i, j; s, t).$$

Hence, $(A \times B)(A' \times B') = (AA' \times BB')$.

12.4.2. $\mathrm{Tr}(A \times B) = \mathrm{Tr}(A)\mathrm{Tr}(B)$.

Proof. $\mathrm{Tr}(A \times B) = \sum_{i,j} (A \times B)(i, j; i, j)$
$$= \sum_{i,j} A_{ii}B_{jj}$$
$$= \sum_{i} A_{ii} \sum_{j} B_{jj} = \mathrm{Tr}(A)\mathrm{Tr}(B). \parallel$$

If T and U are (matrix) representations of G, let $T \times U$ be defined by

$$g(T \times U) = (gT) \times (gU), \quad g \in G.$$

12.4.3. *If T and U are representations of G, then*

(i) $T \times U$ *is a representation of G,*

(ii) $X_{T \times U} = X_T X_U$.

Proof. (i) We have, for $g \in G$ and $h \in G$,
$$(gh)(T \times U) = (gh)T \times (gh)U$$
$$= (gT)(hT) \times (gU)(hU)$$
$$= [(gT) \times (gU)][(hT) \times (hU)]$$
$$= (g(T \times U))(h(T \times U)).$$

(ii) By Theorem 12.4.2,

$$X_{T\times U}(g) = \mathrm{Tr}(g(T \times U)) = \mathrm{Tr}((gT) \times (gU))$$
$$= \mathrm{Tr}(gT)\mathrm{Tr}(gU) = X_T(g)X_U(g). \parallel$$

Thus the product of two characters of a group is again a character (the same is true of the sum of two characters by the discussion preceding Theorem 12.2.14).

Now let $G = H + K$. To every representation T of H there is a representation T' of G given by $(hk)T' = hT$ for $h \in H$ and $k \in K$, and similarly for K.

12.4.4. *Let $G = H + K$.*

(i) *If H has r_1 irreducible representations and K has r_2, then G has $r_1 r_2$ irreducible representations.*

(ii) *The irreducible representations of G are precisely the representations $T' \times U'$, where T is an irreducible representation of H and U an irreducible representation of K.*

Proof. (i) Let $h \in H$, $h' \in H$, $k \in K$, and $k' \in K$. Then hk is conjugate to $h'k'$ iff h is conjugate to h' and k is conjugate to k'. By Theorem 12.2.23, H has r_1 conjugate classes and K has r_2. Hence G has $r_1 r_2$ conjugate classes. Therefore, by Theorem 12.2.23 again, G has $r_1 r_2$ irreducible representations.

(ii) If T and U are representations of H and K respectively, then T' and U' are representations of G, hence, by Theorem 12.4.3, $T' \times U'$ is a representation of G.

Let T_1 and T_2 be irreducible representations of H, and U_1 and U_2 irreducible representations of K. Then

$$(X_{T_1' \times U_1'}, X_{T_2' \times U_2'})$$
$$= \frac{1}{o(G)} \sum X_{T_1' \times U_1'}(g) X_{T_2' \times U_2'}(g^{-1})$$
$$= \frac{1}{o(G)} \sum_{h,k} X_{T_1' \times U_1'}(hk) X_{T_2' \times U_2'}(h^{-1}k^{-1})$$
$$= \frac{1}{o(G)} \sum_{h,k} X_{T_1'}(hk) X_{U_1'}(hk) X_{T_2'}(h^{-1}k^{-1}) X_{U_2'}(h^{-1}k^{-1})$$
$$= \frac{1}{o(H)} \frac{1}{o(K)} \sum_{h,k} X_{T_1}(h) X_{T_2}(h^{-1}) X_{U_1}(k) X_{U_2}(k^{-1})$$
$$= \left[\frac{1}{o(H)} \sum X_{T_1}(h) X_{T_2}(h^{-1})\right]\left[\frac{1}{o(K)} \sum X_{U_1}(k) X_{U_2}(k^{-1})\right]$$
$$= \delta_{T_1 T_2} \delta_{U_1 U_2}.$$

It follows that each $T' \times U'$, with T an irreducible representation of H and U

an irreducible representation of K, is an irreducible representation of G (Exercise 12.2.32), and that no two of them are equivalent. By part (i), therefore, they are all of the irreducible representations of G, up to equivalence. ‖

We have enough theory available to enable us to find all representations of Abelian groups.

12.4.5. *An irreducible representation T of G is of degree* 1 *iff* $\mathrm{Ker}(T) \supset G^1$.

Proof. $\mathrm{Ker}(T) \supset G^1$ iff GT is Abelian, which occurs iff all elements of GT are scalar matrices (Theorem 12.1.6), which, because of the irreducibility, is true iff $\mathrm{Deg}(T) = 1$.

12.4.6. *All irreducible representations of an Abelian group are of degree* 1.

Proof. This follows from Theorem 12.4.5. ‖

The irreducible representations of Abelian groups are just homomorphisms of G into the group of complex numbers of absolute value 1 and, as such, have been more or less determined in Theorems 5.8.1 and 5.8.5. Thus a representation of a cyclic group $\langle g \rangle$ of order n is determined by mapping the generator g into some power of a primitive nth root of 1. Irreducible representations of any finite Abelian group G can then be determined by repeated application of Theorem 12.4.4, since G is the direct sum of cyclic groups. Note that irreducible characters and irreducible representations of an Abelian group are the same thing.

Example 1. Let $G = \langle a \rangle$ be cyclic of order 4. The character table for G is given below.

	e	a	a^2	a^3
X^1	1	1	1	1
X^2	1	i	-1	$-i$
X^3	1	-1	1	-1
X^4	1	$-i$	-1	i

Here, of course, the ordering of the last three characters has no significance.

Example 2. Let $G = \{e, a, b, c\}$ be the 4-group. The character table is:

	e	a	b	c
X^1	1	1	1	1
X^2	1	1	-1	-1
X^3	1	-1	1	-1
X^4	1	-1	-1	1

EXERCISES

12.4.7. Verify that if A and B are square matrices of degree m, and D a square matrix, then
$$(A + B) \times D = (A \times D) + (B \times D).$$

12.4.8. (a) If $U \in \text{End}(G)$ and T is a representation of G, then UT is a representation of G.

(b) If $U \in \text{Aut}(G)$ and X is an irreducible character of the representation T, then $X^U = X_{UT}$ is also an irreducible character. $X^U(g) = X(gU)$.

(c) If $U \in \text{Aut}(G)$, then the function $X^i \longrightarrow (X^i)^U$ is a permutation of the set of irreducible characters.

(d) Illustrate (c) with the 4-group.

12.5 Induced representations

Let G be a finite group and let $H \subset G$. If T is a representation of G, then $T \mid H$ is a representation of H. In this section the reverse process, going from a representation of H to one of G, will be studied. The results are due to Frobenius.

Let $G = \cup \, Hx_i$, and let T be a representation of H. Define

$$
gT^* =
\begin{bmatrix}
(x_1 g x_1^{-1})T, & \ldots, & (x_1 g x_n^{-1})T \\
\vdots & & \vdots \\
(x_n g x_1^{-1})T, & \ldots, & (x_n g x_n^{-1})T
\end{bmatrix}
$$

where the entries listed are square submatrices of gT^*, and where $(x_i g x_j^{-1})T = 0$ if $x_i g x_j^{-1} \notin H$.

12.5.1. *If T is a representation of H and $H \subset G$, then T^* is a representation of G. Up to equivalence, T^* is independent of the choice of coset representatives $x_1, \ldots, x_n$. If U is a representation of H equivalent to T, then U^* is equivalent to T^*.*

Proof. The (i, j)th block of $(gT^*)(hT^*)$ is

$$\sum_k (x_i g x_k^{-1})T(x_k h x_j^{-1})T. \tag{1}$$

Now $x_i g \in Hx_{k_0}$ for a unique k_0, hence $\exists \mid k_0$ such that $x_i g x_{k_0}^{-1} \in H$. Hence there is at most one nonzero term in the summation (1). If $x_{k_0} h x_j^{-1} \in H$, then $x_i g h x_j^{-1} \in H$, and the (i, j)th block in $(gh)T^*$ is

$$(x_i g h x_j^{-1})T = \sum_k (x_i g x_k^{-1})T(x_k h x_j^{-1})T.$$

If $x_{k_0} h x_j^{-1} \notin H$, then $x_i g h x_j^{-1} \notin H$ and the (i,j)th block of both $(gh)T^*$ and $(gT^*)(hT^*) = 0$. It follows that $(gh)T^* = (gT^*)(hT^*)$.

If one x_i is replaced by $h x_i$ with $h \in H$, for convenience say $i = 1$, then the only entries in gT^* that are affected are those in the first row and column. Now

$$(h x_1 g x_j^{-1})T = (hT)(x_1 g x_j^{-1})T,$$

$$(x_i g x_1^{-1} h^{-1})T = (x_i g x_1^{-1})T(h^{-1}T),$$

whether or not $x_1 g x_j^{-1} \in H \ (x_i g x_1^{-1} \in H)$. A calculation then shows that the new matrix has the form

$$
\begin{bmatrix}
hT & & & 0 \\
 & I & & \\
 & & \ddots & \\
0 & & & I
\end{bmatrix}
(gT^*)
\begin{bmatrix}
hT & & & 0 \\
 & I & & \\
 & & \ddots & \\
0 & & & I
\end{bmatrix}^{-1}.
$$

Hence, up to equivalence, T^* is unaffected by change of one coset representative. Change of a number of coset representatives can be made one at a time. Finally, change in the order of cosets merely effects a simultaneous permutation on the rows and columns which can also be effected by conjugation by a permutation matrix.†

If U is equivalent to T, then $hU = A^{-1}(hT)A$. Hence, abbreviating notation somewhat,

$$gU^* = ((x_i g x_j^{-1})U) = (A^{-1}((x_i g x_j^{-1})T)A)$$

$$
=
\begin{bmatrix}
A & & 0 \\
 & \ddots & \\
0 & & A
\end{bmatrix}^{-1}
((x_i g x_j^{-1})T)
\begin{bmatrix}
A & & 0 \\
 & \ddots & \\
0 & & A
\end{bmatrix}.
$$

Hence, U^* is equivalent to T^*. ‖

The representation T^* is called the representation of G *induced* by the representation T of H.

12.5.2. *If $H \subset G$, T is a representation of H with character X, and T^* an induced representation of G with character X^*, then X^* is independent of the choice of representatives of H in the definition of T.*

Proof. This follows from Theorem 12.5.1.

† Since a matrix is a function of a certain sort, the rows and columns need not be ordered (for this portion of the theory), so that permutation of the cosets actually has no effect on the matrix.

12.5.3. (*Frobenius reciprocity theorem.*) *If* $H \subset G$, $\{Y^j\}$ *is the set of irreducible characters of* H, *and* $\{X^i\}$ *that for* G, *then*

$$(Y^j)^* = \sum_i c_{ij} X^i, \qquad X^i \,|\, H = \sum_j c_{ij} Y^j,$$

where the c_{ij} *are nonnegative integers.*

Proof. By Theorems 12.5.1 and 12.5.2, the $(Y^j)^*$ are characters of G, hence (by complete reducibility) there are nonnegative integers c_{ij} such that $(Y^j)^* = \sum c_{ij} X^i$. Since $X^i \,|\, H$ is a character of H, there are nonnegative integers d_{ij} such that $X^i \,|\, H = \sum d_{ij} Y^j$. It must be shown that $c_{ij} = d_{ij}$.
Let $G = \overset{.}{\cup} H x_k$. Then

$$c_{ij} = (X^i, (Y^j)^*) = \frac{1}{o(G)} \sum_g X^i(g)(Y^j)^*(g^{-1})$$

$$= \frac{1}{o(G)} \sum_{k,g} X^i(g) Y^j(x_k g^{-1} x_k^{-1}).$$

A term in the sum is 0 unless $x_k g^{-1} x_k^{-1} = h^{-1} \in H$, in which case, since $x_k g x_k^{-1} = h$, it is $X^i(h) Y^j(h^{-1})$. But given $h \in H$ and x_k, there is exactly one $g \in G$ such that $x_k g^{-1} x_k^{-1} = h^{-1}$. Hence, given $h \in H$, there are $[G:H]$ pairs (k, g) such that $x_k g^{-1} x_k^{-1} = h^{-1}$. Therefore,

$$c_{ij} = \frac{[G:H]}{o(G)} \sum \{X^i(h) Y^j(h^{-1}) \,|\, h \in H\}$$

$$= \frac{1}{o(H)} \sum_h X^i(h) Y^j(h^{-1}) = d_{ij}. \;\|$$

One can also obtain an explicit formula for the value of an induced character on the *j*th conjugate class.

12.5.4. *If* $H \subset G$, X *is a character of* H, *and* X^* *the induced character of* G, *then*

$$X_j^* = ([G:H]/h_j) \sum \{X(h) \,|\, h \in C_j \cap H\}.$$

Proof. Let $G = \overset{.}{\cup} H x_i$. Then for $g \in C_j$,

$$X_j^* = \sum_i X(x_i g x_i^{-1})$$

$$= \frac{1}{o(H)} \sum_i \sum \{X(h x_i g x_i^{-1} h^{-1}) \,|\, h \in H\}$$

$$= \frac{1}{o(H)} \sum \{X(y g y^{-1}) \,|\, y \in G\}.$$

There are just $o(C(g)) = o(G)/h_j$ elements y of G such that ygy^{-1} is a given element of C_j. Hence

$$X_j^* = \frac{1}{o(H)} \sum \left\{ \frac{o(G)}{h_j} X(u) \mid u \in C_j \right\}$$

$$= \frac{[G:H]}{h_j} \sum \{X(u) \mid u \in C_j \cap H\}. \; \|$$

The notion of induced character can be generalized to that of induced class function in an obvious way. Let $H \subset G$ and $G = \cup Hx_i$ as before. If f is a function from H to the complex numbers which is constant on all classes of H, first define $f(x)$ to be 0 if $x \in G \backslash H$, and then define $f*$ on G by

$$f^*(y) = \sum f(x_i y x_i^{-1}).$$

Since, for $h \in H$, $(hx_i)y(hx_i)^{-1} \in H$ iff $x_i y x_i^{-1} \in H$,

$$f(hx_i y x_i^{-1} h^{-1}) = f(x_i y x_i^{-1}).$$

Therefore, (i) the definition of $f*$ is independent of the choice of coset representatives, and (ii)

$$f^*(y) = \frac{1}{o(H)} \sum_h \sum_i f(hx_i y x_i^{-1} h^{-1})$$

$$= \frac{1}{o(H)} \sum \{f(gyg^{-1}) \mid g \in G\}.$$

12.5.5. *The operation $*$ is linear:* $(f_1 + f_2)^* = f_1^* + f_2^*$, $(cf)^* = c(f^*)$ *for* $c \in \mathscr{C}$. $\|$

A *generalized character* of a group G is a function X of the form $X = \sum n_i X^i$, $n_i \in J$, where each X^i is an irreducible character as usual.

12.5.6. *The generalized characters form a ring.*

Proof. This follows from Theorem 12.4.3 and complete reducibility.

12.5.7. *If* $X = \sum n_i X^i$ *is a generalized character, then* $(X, X) = \sum n_i^2$.

Proof. This follows from bilinearity of the inner product.

12.5.8. *If Y is a generalized character of $H \subset G$, then $Y*$ is a generalized character of G.*

Proof. This follows from Theorems 12.5.1. and 12.5.6.

12.5.9. *Let $H \subset G$, $H^x \cap H = E$ for all $x \in G \backslash H$, and let f be a function from H into $\mathscr{C}$, constant on classes. Then*

(i) $f^*(h) = f(h)$ *for* $h \in H^\#$,

(ii) *if* $f(e) = 0$, *then* $f^*(e) = 0$ *and* $(f^*, f^*) = (f, f)$.

Proof. (i) If $G = \cup\, Hx_i$ and $h \in H^\#$, then just one of the conjugates $x_i h x_i^{-1}$ is in H, namely for that i for which $x_i \in H$, so that

$$f^*(h) = \Sigma f(x_i h x_i^{-1}) = f(h).$$

(ii) As above, if $f(e) = 0$, then $f^*(e) = \Sigma f(x_i e x_i^{-1}) = 0$. Therefore, using (i) at one place, we have

$$(f^*, f^*) = \frac{1}{o(G)} \Sigma \{ f^*(x) f^*(x^{-1}) \mid x \in G \}$$

$$= \frac{1}{o(G)} \Sigma \{ f^*(x) f^*(x^{-1}) \mid x \in H^{\#y} \quad \text{for some} \quad y \}$$

$$= \frac{[G:H]}{o(G)} \Sigma \{ f^*(x) f^*(x^{-1}) \mid x \in H^\# \}$$

$$= \frac{1}{o(H)} \Sigma \{ f(x) f(x^{-1}) \mid x \in H \}$$

$$= (f, f). \;\|$$

By Exercise 12.2.29, if T is a representation of G with character X, then $\mathrm{Ker}(T)$ can be recovered from a knowledge of X alone. In fact,

$$\mathrm{Ker}(T) = \{ g \in G \mid X(g) = X(e) \}.$$

This justifies the notation:

$$\mathrm{Ker}(X) = \{ g \in G \mid X(g) = X(e) \}.$$

12.5.10. *If $x \in G^\#$, then there is an irreducible character X^i of G such that $x \notin \mathrm{Ker}(X^i)$.*

Proof. Suppose the contrary. Then $H = \cap \,\mathrm{Ker}(X^i) \lhd G$, and $x \in H$. Each representation T of G yields a representation T_1 of G/H where $(gH)T_1 = gT$. It may then be verified that representations T and U of G are equivalent iff T_1 and U_1 are equivalent, and T is irreducible iff T_1 is irreducible. Moreover $\mathrm{Deg}(T) = \mathrm{Deg}(T_1)$. Since $o(G)$ is the sum of the squares of the degrees of the irreducible characters of G, this leads to the contradiction that the sum of the squares of the degrees of the irreducible characters of G/H is greater than $o(G/H)$.

12.5.11. (*Frobenius.*) *If $H \subset G$ (G finite), $H^x \cap H = E$ for all $x \in G \backslash H$, and M is the set of elements of G not in any conjugate of $H^\#$, then M is a normal subgroup of G.*

Proof. Let $\{Y^i\}$ be the set of irreducible characters of H. Let

$$f_i = Y^i(e)Y^1 - Y^i, \qquad i > 1.$$

Then f_i is a generalized character of H, and $f_i(e) = 0$. Now, by the linearity of $*$ (Theorem 12.5.5),

$$f_i^* = Y^i(e)(Y^1)^* - (Y^i)^*,$$

which is a generalized character of G (Theorem 12.5.8). By the Frobenius reciprocity theorem, 12.5.3,

$$((Y^i)^*, X^1) = (Y^i, X^1 \mid H)$$
$$= (Y^i, Y^1) = \delta_{1i}.$$

Hence

$$(f_i^*, X^1) = Y^i(e)((Y^1)^*, X^1) - ((Y^i)^*, X^1)$$
$$= Y^i(e), \qquad i > 1.$$

Therefore, since f_i^* is a generalized character of G,

$$f_i^* = Y^i(e)X^1 + Z^i, \qquad (Z^i, X^1) = 0,$$

where Z^i is a generalized character of G. Now by Theorem 12.5.9,

$$(Y^i(e))^2 + 1 = (f_i, f_i) = (f_i^*, f_i^*)$$
$$= (Y^i(e))^2 + (Z^i, Z^i).$$

Hence $(Z^i, Z^i) = 1$, so (Theorem 12.5.7), $Z^i = \pm X^{j(i)}$. But

$$0 = f_i^*(e) = Y^i(e) + Z^i(e),$$

hence $Z^i(e) < 0$, so that $Z^i = -X^{j(i)}$. Thus,

$$f_i^* = Y^i(e)X^1 - X^{j(i)}.$$

Since $f_i^*(e) = 0$, $X^{j(i)}(e) = Y^i(e)$. If $x \in H^{\#}$, then (see Theorem 12.5.9)

$$Y^i(e) - Y^i(x) = f_i(x) = f_i^*(x)$$
$$= Y^i(e) - X^{j(i)}(x),$$
$$X^{j(i)}(x) = Y^i(x).$$

Therefore $X^{j(i)}(x) = Y^i(x)$ for all $x \in H$.

If $y \in M^{\#}$, then $f_i^*(y) = 0$ by the definition of f_i^*. Thus,

$$0 = f_i^*(y) = Y^i(e) - X^{j(i)}(y)$$
$$= X^{j(i)}(e) - X^{j(i)}(y),$$

so that $X^{j(i)}(y) = X^{j(i)}(e)$, and $y \in \text{Ker}(X^{j(i)})$. Hence,

$$M \subseteq \cap \text{Ker}(X^{j(i)}).$$

Now if $y \in H^{\#}$, then by Theorem 12.5.10, $\exists\, i$ such that $y \notin \mathrm{Ker}(Y^i)$. Hence $Y^i(y) \neq Y^i(e)$, so $X^{j(i)}(y) \neq X^{j(i)}(e)$, and $y \notin \cap\, \mathrm{Ker}(X^{j(i)})$. Since the $X^{j(i)}$ are functions, constant on classes, and a member of each class has been examined, $M = \cap\, \mathrm{Ker}(X^{j(i)})$. Therefore, M is a normal subgroup of G. ‖

Another form of the theorem of Frobenius is the following theorem about permutation groups.

12.5.12. *If G is a finite transitive permutation group such that if $x \in G^{\#}$, then $\mathrm{Ch}(x) = 0$ or 1, then the set $M = \{x \in G \mid \mathrm{Ch}(x) = 0 \text{ or } x = e\}$ is a regular, normal subgroup of G.*

Proof. Let a be any letter and $x \in G \backslash G_a$. Then $G_a^x = G_{ax} = G_b$ with $b \neq a$, and $G_a \cap G_a^x = G_a \cap G_b = E$, since any $y \in G_a \cap G_b$ fixes at least two letters. Any element not in any conjugate of G_a must fix no letters, hence is in M. Conversely, any element of $M^{\#}$ fixes no letters, hence is not in any conjugate of G_a. By Frobenius' theorem, 12.5.11, M is a normal subgroup of G. Now

$$o(G) = (o(G_a) - 1)[G:G_a] + o(M)$$

$$= o(G) - \mathrm{Deg}(G) + o(M).$$

Hence $o(M) = \mathrm{Deg}(G)$. Since $o(M_a) = 1$, it follows (Theorem 10.1.4) that M is transitive, hence regular.

EXERCISES

12.5.13. (Induction is transitive.) Let $K \subset H \subset G$, $G = \cup\, Hx_i$, $H = \cup\, Ky_j$, T a representation of K, T_1 the induced representation of H (using the given coset decomposition), T_1^* the representation of G induced by T_1, and T^* the representation of G induced by T with respect to the decomposition $G = \cup\, Ky_j x_i$. Prove that $T_1^* = T^*$.

12.5.14. Let G be finite, $H \subset G$, $H^x \cap H = E$ for all $x \in G \backslash H$, M the set of elements of G not in any conjugate of $H^{\#}$, and T the permutation representation of G on H. Then T is an isomorphism, GT is transitive, if $x \in (GT)^{\#}$, then $\mathrm{Ch}(x) = 0$ or 1, and $MT = \{xT \in G \mid \mathrm{Ch}(xT) = 0 \text{ or } x = e\}$. (Compare with Theorems 12.5.11 and 12.5.12.)

12.5.15. (Shaw [1].) Prove the following special case of the theorem of Frobenius without the use of the theory of characters. Let H be a solvable subgroup of a finite group G such that $H \cap H^x = E$ if $x \in G \backslash H$. Prove that H has a normal complement in G. (See the proof of Theorem 6.2.9. If the exercise gives difficulty, first try the case where H is Abelian.)

12.5.16. The theorem of Frobenius, 12.5.11, is false for infinite groups G, even if H is finite. (See Exercise 8.3.6.)

12.6 Frobenius groups

The type of group arising in Frobenius' theorem is of rather frequent occurrence. The more elementary properties of such groups will be studied in this section, but the deeper (and more important) theorems lie outside the scope of this book.

A *Frobenius group* is a finite group G containing a nontrivial normal subgroup M such that if $x \in M^{\#}$, then $C(x) \subset M$. The subgroup M is called a *Frobenius kernel* of G. It will be shown later (Theorem 12.6.12) that a Frobenius group has just one Frobenius kernel.

12.6.1. *If G is a Frobenius group with a Frobenius kernel M, then $M \in \mathrm{Hall}(G)$, and $\exists\, H \in \mathrm{Hall}(G)$ such that $G = MH$, $M \cap H = E$. For any such H, $H \cap H^x = E$ for all $x \in G\backslash H$, and M is the set of elements of G not in any conjugate of $H^{\#}$.*

Proof. If M is not a Hall subgroup, then $\exists\, p \in \mathscr{P}$, $P \in \mathrm{Syl}_p(M)$, and $Q \in \mathrm{Syl}_p(G)$ such that $E < P < Q$. There is an element x of order p in $Z(Q)$. If $x \in P$, then $C(x) \supset Q$, so that $C(x) \not\subset M$, a contradiction. If $x \notin P$, then for $y \in P^{\#}$, $x \in C(y)\backslash M$, a contradiction. Hence $M \in \mathrm{Hall}(G)$.

The Schur splitting theorem, 9.3.6, now guarantees the existence of a subgroup H such that $G = MH$ and $M \cap H = E$. Since M is a Hall subgroup and $o(G) = o(M)o(H)$, H is a Hall subgroup. Let $x \in G\backslash H$ and suppose that $H \cap H^x \neq E$. Since $G = HM$, $x = hm$ with $h \in H$ and $m \in M$. Hence $H^x = H^{hm} = H^m$. Since $H \cap H^m \neq E$, $\exists\, y \in H^{\#}$ such that $m^{-1}ym \in H$. Therefore, $y^{-1}m^{-1}ym \in H \cap M = E$ and $y \in C(m)$, a contradiction. Thus, $H \cap H^x = E$ if $x \in G\backslash H$. A count of elements proves the final statement (with the details given in the next proof).

12.6.2. *If $H \subset G$, G is finite, and $H \cap H^x = E$ for all $x \in G\backslash H$, then G is a Frobenius group with a Frobenius kernel M consisting of e and all elements outside all conjugates of H. Moreover, $M \cap H = E$ and $G = MH$.*

Proof. By the theorem of Frobenius, the set M is a normal subgroup of G, and clearly $H \cap M = E$. Let $x \in M^{\#}$ and suppose that $C(x) \not\subset M$. Then there is an element $y \neq e$ in $C(x)$ which is in some conjugate of H. Conjugation leads to elements $z \in M^{\#}$ and $h \in H^{\#}$ such that $h \in C(z)$. But then, $H \cap H^z \neq E$, a contradiction. Hence $C(x) \subset M$ for all $x \in M^{\#}$, and G is a

Frobenius group with a Frobenius kernel M. Finally,

$$o(G) = (o(H) - 1)[G:H] + o(M)$$

$$= o(G) - [G:H] + o(M),$$

so that $o(M) = [G:H]$ and $o(G) = o(M)o(H)$. Since $H \cap M = E$, $G = HM$. ‖

The next lemma (due to Brauer) will be used in proving that the Frobenius kernel of a Frobenius group is unique.

12.6.3. *If D is a nonsingular matrix on $S \times S$, and G is a group of permutations of $S \times S$ such that if $g \in G$ then $\exists\, g_R$ and g_C in $\mathrm{Sym}(S)$ such that for all i and j*

$$D((i, j)g) = D(ig_R, j) = D(i, jg_C), \tag{1}$$

then

(i)   *$G_R = \{g_R \mid g \in G\}$ is a group; $G_C = \{g_C \mid g \in G\}$ is a group.*

(ii) *The number of orbits of G_R equals the number of orbits of G_C.*

(iii) *If G is cyclic, then $\mathrm{Ch}(G_R) = \mathrm{Ch}(G_C)$.*

Proof. (i) and (ii). Because D is nonsingular, no two rows are identical. Hence given $g \in \mathrm{Sym}(S \times S)$, there is at most one $g_R \in \mathrm{Sym}(S)$ such that (1) holds, and similarly for g_C.

Let $A(g)$ be the permutation matrix corresponding to g_R, so that $Dg = A(g)D$. Then

$$A(gh)D = Dgh = (Dg)h = (A(g)D)h = A(h)A(g)D.$$

Since D is nonsingular, $A(gh) = A(h)A(g)$. Since the function which maps a permutation onto the corresponding (left) permutation matrix is 1–1 and reverses products, G_R is a group. Moreover the mapping $g \longrightarrow A^T(g)$ (T means transpose) is a representation of G with character X, say. Similarly one finds that if $Dg = DB(g)$, where $B(g)$ is a permutation matrix, then $g \longrightarrow B(g)$ is a representation of G with character Y, say. Now by assumption, $A(g)D = DB(g)$. Hence,

$$A(g) = DB(g)D^{-1},$$

$$\mathrm{Tr}(A^T(g)) = \mathrm{Tr}(B(g)),$$

$$X(g) = Y(g),$$

for all $g \in G$. By Theorem 10.1.5 and the fact that $X(g) = \mathrm{Ch}(g_R)$,

$$\Sigma \{X(g) \mid g \in G\} = o(G)t,$$

$$\Sigma \{Y(g) \mid g \in G\} = o(G)u,$$

where t and u are the number of orbits of G_R and G_C, respectively. It follows that $t = u$.

(iii) If G is cyclic, say $G = \langle g \rangle$, then

$$\text{Ch}(G_R) = \text{Ch}(g_R) = X(g) = Y(g) = \text{Ch}(G_C).$$

12.6.4. *If G is a Frobenius group with Frobenius kernel M, Y is an irreducible character of M other than the unit character and X is an irreducible character of G, then*

(1) *Y^* is an irreducible character of G.*

(2) *Either $M \subset \text{Ker}(X)$ or $X = U^*$ for some irreducible character U of M.*

Proof. There is a complement H of M as in Theorem 12.6.1. If $x \in H$, let T_x be the induced automorphism of M. Let $\{C_j\}$ be the set of classes of M. Let $\{Y^j\}$ be the set of irreducible characters of M, and D^i an irreducible representation with character Y^i. If $u_j \in C_j$, then

$$(3) \qquad \text{Tr}(u_j^x D^i) = \text{Tr}(u_j(T_x D^i)).$$

By Exercise 12.4.8, there is a permutation xA of $\{1, \ldots, r\}$ such that $Y^{i(xA)}$ is the character of $T_x D^i$. Let xB be the permutation of $\{1, \ldots, r\}$ such that $C_{j(xB)} = C_j^x$. By (3), $Y_j^{i(xA)} = Y_{j(xB)}^i$. Letting $(i, j)x = (i, j(xB))$, one sees that the hypotheses of Theorem 12.6.3 are satisfied. Suppose $x \in H^\#$, $C_j \neq \{e\}$, and $C_j^x = C_j$. Then $\exists \, y \in C_j$ and $z \in M$ such that $x^{-1}yx = z^{-1}yz$, so that $xz^{-1} \in C(y)$, hence $xz^{-1} \in M$, a contradiction. Thus $C_j^x \neq C_j$ for all $x \in H^\#$ and $C_j \neq \{e\}$. Therefore, $\text{Ch}(xB) = 1$ if $x \in H^\#$. By Theorem 12.6.3, $\text{Ch}(xA) = 1$ for all $x \in H^\#$. Changing notation slightly, this means that $Y(xA) \neq Y(yA)$ if x and y are distinct elements of H.

Now $G = \overset{\cup}{} \{Mx \mid x \in H\}$. Hence,

$$Y^*(z) = \sum \{Y(x^{-1}zx \mid x \in H\} = \sum \{(Y(xA))(z) \mid x \in H\}.$$

Therefore, $Y^* \mid M = \sum \{Y(xA) \mid x \in H\}$, and, of course, $Y^*(z) = 0$ if $z \in G \backslash M$. We have

$$(Y^*, Y^*) = \frac{1}{o(G)} \sum \{Y^*(y)Y^*(y^{-1}) \mid y \in G\}$$

$$= \frac{1}{o(G)} \sum \{(Y(xA))(y)(Y(uA))(y^{-1}) \mid y \in M, x \in H, u \in H\}$$

$$= \frac{o(M)}{o(G)} \sum \{(Y(xA), Y(uA)) \mid x \in H, u \in H\}$$

$$= \frac{o(M)}{o(G)} \sum \{(Y(xA), Y(xA)) \mid x \in H\}$$

$$= \frac{o(M)o(H)}{o(G)} = 1,$$

by the orthogonality of the irreducible characters of M. This proves (1) (Exercise 12.2.32).

Let $X \mid M = \Sigma \, a_i Y^i$ with Y^i irreducible. If some $a_i \neq 0$ for $i \neq 1$, then by the reciprocity theorem, 12.5.3, $((Y^i)^*, X) \neq 0$, and by (1), $(Y^i)^* = X$. If all $a_i = 0$ for $i \neq 1$, then $X \mid M = a Y^1$, $X(x) = a$ for all $x \in M$. Hence $M \subset \mathrm{Ker}(X)$.

12.6.5. *If G is a Frobenius group with Frobenius kernel M, H is a complement of M, X is an irreducible character of G such that $\mathrm{Ker}(X) \not\supset M$, and Y is the character of the regular representation of H, then $X \mid H = n Y$ for some $n \in J$.*

Proof. By Theorem 12.6.4, there is an irreducible character U of M such that $X = U^*$. $U^*(y) = 0$ for all $y \in H^\#$. Also

$$U^*(e) = o(H)U(e) = U(e)Y(e).$$

Thus, $X(y) = U^*(y) = U(e)Y(y)$ for all $y \in H$, so that $X \mid H = n Y$ where $n = U(e)$ is a positive integer.

12.6.6. *Let G be a Frobenius group with Frobenius kernel M, and let H be a complement of M.*

(i) *If $E < H_1 < H$, then $H_1 M$ is Frobenius with kernel M.*

(ii) *If $E < K < M$ and $K \lhd G$, then G/K is Frobenius with kernel M/K.*

(iii) *If $E < K \subset M$, $E < H_1 \subset H$, and $H_1 \subset N(K)$, then $H_1 K$ is Frobenius with kernel K.*

Proof. Clearly (iii) implies (i). As to (iii), certainly $H_1 K \subset G$. If $x \in K^\#$ then $C_{H_1 K}(x) \subset H_1 K \cap M = K$. Hence $H_1 K$ is Frobenius with kernel K.

(ii) Let $x \in H^\#$, $y \in M \backslash K$, and $xK \in C_{G/K}(yK)$. WLOG, $o(x) = p \in \mathscr{P}$. Now conjugation by x induces a permutation of yK. By Theorem 12.6.1, the number of elements in yK is not divisible by p. Hence the permutation fixes an element of $M^\#$, contrary to the definition of a Frobenius group. Statement (ii) follows.

12.6.7. *If G is a Frobenius group, then there is a Frobenius subgroup B with an elementary Abelian p-group K for Frobenius kernel ($p \in \mathscr{P}$), and a complement L of K of prime order.*

Proof. WLOG, G has no proper Frobenius subgroup. Let M be a Frobenius kernel of G and H a complement of M. If $E < A < H$, then by Theorem 12.6.6, MA is a proper Frobenius subgroup of G, a contradiction. Hence H is of prime order. Let $P \in \mathrm{Syl}(M)$, $P \neq E$. By Theorem 6.2.4, $G = N(P)M$. Since $H \in \mathrm{Syl}(G)$ (Theorem 12.6.1), $o(H) \mid o(N(P))$ and some conjugate of H, say H, is contained in $N(P)$. By Theorem 12.6.6, PH is Frobenius. Hence $P = M$, and M is a p-group for some prime p. If M has a nontrivial characteristic subgroup Q, then $Q \lhd G$ and QH is Frobenius. Hence, M is characteristically simple, therefore an elementary Abelian p-group.

12.6.8. *If G is a Frobenius group with Frobenius kernel M and $A \lhd G$, then either $A \subset M$ or $M \subset A$.*

Proof. Induct on $o(G)$. Suppose that $M \nsubseteq A$, so that $A \cap M < M$. If $A \cap M = E$, then $AM = A \times M$ and $A \subset C(M)$. Hence $A \subset M$ in this case. If $A \cap M \neq E$, then $G/(A \cap M)$ is a Frobenius group (12.6.6) with Kernel $M/(A \cap M) \nsubseteq A/(A \cap M)$. By the induction hypothesis, $A \subset M$.*

If D is a representation of a finite group G over a field F, then D is also a representation of G over any field K containing F. A representation of G over F is *absolutely irreducible* iff it is irreducible over every field K of finite degree over F.

12.6.9. *If D is a representation of a finite group G over a field F and the characteristic of F does not divide $o(G)$, then there is a field K of finite degree over F such that $D = \Sigma\, D_i$ where each D_i is an absolutely irreducible representation of G over K.*

Proof. Induct on $\mathrm{Deg}(D)$. If D is absolutely irreducible, we are done. If not, then there is a finite extension F_1 of F over which D is reducible. By Maschke, $D = D' + D''$ over F_1 with $\mathrm{Deg}(D') < \mathrm{Deg}(D)$. By the inductive hypothesis, there is a finite extension F_2 of F_1 such that $D' = \Sigma\, D_i$ over F_2, and the D_i are absolutely irreducible. By the inductive hypothesis again, there is a finite extension K of F_2 such that $D'' = \Sigma\, D_j''$ over K, and the D_j'' are absolutely irreducible. Thus, K is a finite extension of F, and over K, D is the direct sum of absolutely irreducible representations.

12.6.10. *If D is an absolutely irreducible representation of a finite Abelian group G over a field F whose characteristic does not divide $o(G)$, then $\mathrm{Deg}(D) = 1$.*

Proof. Induct on $o(G)$. $G = A + B$ where $A = \langle x \rangle$ is cyclic and $x \neq e$. By inductive hypothesis, complete reducibility, and Theorem 12.6.9, there are a finite extension F_1 of F and a representation D_1 of G over F_1 which is equivalent to D such that the matrices BD_1 are diagonal. If all the matrices of BD_1 are scalars, then a further extension of the field and a change to an equivalent

---

*The author wishes to thank B. Wehrfritz for pointing out the error in the original proof and furnishing this proof.

representation D_2 makes xD_2 diagonal (Theorem 12.2.16). Since a scalar matrix is unchanged by conjugation, all matrices of GD_2 are diagonal. Since D was absolutely irreducible, D_2 must be irreducible. Therefore, $\text{Deg}(D_2) = \text{Deg}(D) = 1$.

Suppose that not all matrices of BD_1 are scalars, and let V be the vector space being acted on. Then $V = V_1 + \ldots + V_n$, where $\text{Dim}(V_i) = 1$, and each V_i is invariant under BD_1. By combining several V_i, one gets $V = W_1 + \ldots + W_r$, where B acts as a group of scalars on each W_i, and W_i is maximal with this property. If $b \in B$ and $v \in W_i$, then $vb = f_i(b)v$, and

$$vbx = f_i(b)vx, \qquad f_i(b) \in F_1,$$
$$vx = c_1v_1 + \ldots + c_rv_r, \qquad c_i \in F_1, v_i \in W_i,$$
$$(vx)b = f_1(b)c_1v_1 + \ldots + f_r(b)c_rv_r.$$

Thus, $f_i(b)c_j = f_j(b)c_j$ for all j. Since $i \neq j$ implies that $\exists \, b \in B$ such that $f_i(b) \neq f_j(b)$, $c_j = 0$. Therefore $vx = c_iv_i$ and $W_ix = W_i$. Thus W_i is a G-subspace. Since V was irreducible this is a contradiction.

12.6.11. *If G is a Frobenius group with Frobenius kernel M, and if H is a complement of M, then H does not contain a Frobenius group.*

Proof. Deny, and let G be a counterexample of least order. By Theorems 12.6.6 and 12.6.7, H is itself Frobenius with elementary Abelian kernel Q and a complement K of Q (in H) of prime order. Let $E < P \in \text{Syl}(M)$. By Theorem 6.2.4, $o(H) \mid o(N(P))$. Now $N(P) \cap M$ is a normal Hall subgroup of $N(P)$. By Schur's splitting theorem 9.3.6, $N(P) = (N(P) \cap M)S$ where $o(S) = o(H)$. Hence, PS exists and $MS = G$. Therefore, $S \cong G/M \cong H$. Moreover, PS is a Frobenius group, since if $x \in P$, then $C_{PS}(x) \in PS \cap M = P$. By the minimality of M and the fact that S is a Frobenius group (being isomorphic to H) which is a complement to P in PS, $M = P$. Hence M is a p-group for some $p \in \mathscr{P}$. By Theorem 12.6.6, M is characteristically simple. Therefore, M is an elementary Abelian p-group. Moreover, M is a minimal normal non-E subgroup of G.

Now H has a faithful representation A as an irreducible group of linear transformations on M considered as a vector space over J_p. By Theorem 12.6.9, there is a finite field $F \supset J_p$ such that A is the direct sum of absolutely irreducible representations of H over F. Let B be one of these absolutely irreducible representations. By a further extension of the field, it may be assumed that, if V is the vector space acted on, then $V = V_1 + \ldots + V_n$, where each V_i is an absolutely irreducible Q-space. By Theorem 12.6.10, $\text{Dim}(V_i) = 1$, so that xB is a scalar on each V_i for $x \in Q$. As in the proof of Theorem 12.6.10, $V = W_1 + \ldots + W_r$, each W_i is a maximal subspace such that xB is a scalar for all $x \in Q$. If $K = \langle y \rangle$, $v \in W_i$, and $x \in Q$, then

$$vyx = vyxy^{-1}y = cvy, \qquad c \in F,$$

where c is independent of v, since $yxy^{-1} \in Q$. It follows that each $W_i y \subset W_j$ for some j. If any $W_i y < W_j$, then $Vy < V$, an impossibility since all linear transformations in a representation are nonsingular. Hence $W_i y = W_j$. Since $H = \langle Q, y \rangle$ and V is irreducible, it follows that y permutes the W_i cyclically. If $w \in W_1^\#$, then

$$v = w + wy + \ldots + wy^{r-1}$$

where $o(y) = r$, is invariant under y and not 0. Therefore, yB has 1 as one of its characteristic values. But characteristic values of matrices are unchanged by conjugation, and B is (equivalent to) a direct summand of A. Hence, 1 is a characteristic value of yA. Therefore there is a characteristic vector $z \in M$ for y for the value 1. Changing to multiplicative notation again, this means that $y^{-1}zy = z$, contrary to the definition of a Frobenius group.

12.6.12. *A Frobenius group has a unique Frobenius kernel.*

Proof. Let G be a Frobenius group with Frobenius kernels M and M_1 and with H a complement of M. By Theorem 12.6.8, *WLOG*, $M \subseteq M_1$. Hence $M_1 = MK$ where $K = H \cap M_1 \lhd H$. If $x \in K$, then by Theorem 12.6.1, $C(x) \subseteq H \cap M_1 = K$. Therefore, if $M < M_1$, then H is a Frobenius group with Frobenius kernel K. This contradicts Theorem 12.6.11. Hence, $M = M_1$.

12.6.13. *If G is a Frobenius group with solvable Frobenius kernel M, then M is nilpotent.*

Proof. Deny, and let G be a counterexample of least order. By Theorem 12.6.6, a complement H of M is of prime order, so $H = \langle x \rangle$, say. Since M is not nilpotent, its order is divisible by at least two distinct primes. Since M and H are solvable, G is solvable.

CASE 1. $o(M) = p^i q^j$ with p and q distinct primes. Let P_1 be a minimal normal non-E subgroup of G contained in M. *WLOG*, P_1 is an elementary Abelian p-group. Since G/P_1 is a Frobenius group with Frobenius kernel M/P_1 (Theorem 12.6.6), M/P_1 is nilpotent by assumption. Suppose that $P_1 < P \in \mathrm{Syl}_p(G)$ and $Q \in \mathrm{Syl}_q(G)$. Then $QP_1/P_1 \in \mathrm{Syl}_q(M/P_1)$; hence by the nilpotence of M/P_1, $QP_1 \lhd M$. Any conjugate of Q by an element of H is in M, hence in QP_1. Therefore $QP_1 \lhd G$. By Theorem 12.6.6, $H(QP_1)$ is Frobenius with kernel QP_1. Therefore, by induction, QP_1 is nilpotent. Hence $Q \lhd QP_1$, so $Q \in \mathrm{Char}(QP_1)$ and $Q \lhd M$. Since M/P_1 is nilpotent, the Sylow p-subgroup P/P_1 is normal, hence $P \lhd M$. Therefore, M is nilpotent.

Now assume that $P_1 = P$, so that P itself is an elementary Abelian minimal normal non-E subgroup of G. By Hall's theorem, 9.3.11, $\exists Q \in \mathrm{Syl}_q(M)$ such that $HQ \subseteq G$. Since $HQ \cap M = Q$, $Q \lhd HQ$. Hence

$Q^1 \lhd HQ$. If $Q^1 \not\subseteq C(P)$, then HQ^1P is Frobenius, with Frobenius kernel Q^1P which is not nilpotent, contradicting the minimality of G. Hence, $Q^1 \subset C(P)$, but $Q \not\subseteq C(P)$. There is therefore an irreducible representation T of HQ on P. Here, HQ is a Frobenius group with kernel Q and complement H. $\mathrm{Ker}(T) \lhd HQ$, and $\mathrm{Ker}(T) \not\supseteq Q$. By Theorem 12.6.8, $\mathrm{Ker}(T) \subset Q$, and by an earlier remark, $Q^1 \subset \mathrm{Ker}(T)$. By Theorem 12.6.6, $S = HQ/\mathrm{Ker}(T)$ is a Frobenius group, with Abelian Frobenius kernel $Q/\mathrm{Ker}(T)$ and complement H, and is represented faithfully by T. By Theorem 12.6.5, $T/H \sim nY$ for some n, where Y is the regular representation of H. By Theorem 12.2.15, $T/H \sim nl_H + \cdots$. But this means that there are fixed vectors under H in P, contrary to the definition of a Frobenius group.

CASE 2. $o(M)$ is divisible by three or more distinct primes. Let p and q be two such primes. By Hall's theorem, $\exists\, P \in \mathrm{Syl}_p(M)$ and $Q \in \mathrm{Syl}_q(M)$ such that HP, HQ, PQ, and HPQ are subgroups of G. Then HPQ is a Frobenius group (Theorem 12.6.2) with kernel PQ. By Case 1, PQ is nilpotent, so $Q \subset N(P)$. Since q is arbitrary, $P \lhd M$. Since p is arbitrary, M is nilpotent. $\|$

A much deeper theorem of Thompson [1] says that the Frobenius kernel M of a Frobenius group G is always nilpotent.

The complement H of a Frobenius kernel also has some interesting properties.

12.6.14. *If G is a Frobenius group with kernel M and complement H, and p and q are primes, then any subgroup of H of order p^2 or pq is cyclic.*

Proof. Let G be a smallest counterexample. Then $o(H) = p^2$ or pq, and M is a minimal normal subgroup of G. If $o(H) = pq$, then, since H is not cyclic, it is Frobenius (Exercise 12.6.22), but this contradicts Theorem 12.6.11. Hence, H is an Abelian noncyclic group of order p^2. If $E < P \in \mathrm{Syl}_r(M)$, then $o(H) \,|\, o(N(P))$, so some conjugate of H is in $N(P)$, say H itself. But then PH is Frobenius, so that by the minimality of M, $M = P$. Since M is minimal normal, M is an elementary Abelian primary group.

There is an irreducible representation A of H on M considered as a vector space over a field J_r. By Theorems 12.6.9 and 12.6.10, there is a finite extension field of J_r over which A is the direct sum of one-dimensional representations. Hence, each hA, $h \in H$, is diagonal. The entries of hA on the diagonal are pth roots of 1 in F. Since there are (at most) p such pth roots and $o(H) = p^2$, there are distinct elements x and y of H with the upper left entry in xA and yA the same. Then $(xy^{-1})A$ is a diagonal matrix with 1 in the upper left corner, hence it has 1 as a characteristic value. Therefore, the matrix $(xy^{-1})A$ over J_r has characteristic value 1, i.e., some element of $H^{\#}$ centralizes an element of $M^{\#}$, a contradiction.

12.6.15. *If G is a Frobenius group with kernel M and complement H, and* $P \in \mathrm{Syl}_p(H)$, *then*

(i) *if* $p \neq 2$, *then P is cyclic*,

(ii) *if* $p = 2$, *then P is cyclic or generalized quaternion.*

Proof. By Theorem 12.6.14. P contains no noncyclic subgroup of order p^2. Now $\exists \, K \subset Z(P)$, with $o(K) = p$. If P contains another subgroup L of order p, then $KL = K + L$ is noncyclic, a contradiction. Hence, P has just one subgroup of order p. By Theorem 9.7.3, P is cyclic or generalized quaternion. $\|$

The finite groups all of whose Sylow subgroups are cyclic will now be determined.

12.6.16. *If G is a (possibly infinite) group and* G^{i-1}/G^i *and* G^i/G^{i+1} *are cyclic for some* $i \geq 2$, *then* $G^i/G^{i+1} = E$.

Proof. Let $H = G^{i-2}/G^{i+1}$. Then H^1/H^2 and H^2 are cyclic, and $H^3 = E$. By the N/C theorem, $H/C(H^2) \cong \mathrm{Aut}(H^2)$ which is Abelian, hence $C(H^2) \supset H^1$. Therefore, $H^2 \subset Z(H^1)$, so $H^1/Z(H^1)$ is cyclic. Hence (Theorem 3.2.8) H^1 is Abelian, so that $E = H^2 = G^i/G^{i+1}$. $\|$

The preceding theorem is false for $i = 1$, as the example $G = \mathrm{Sym}(3)$ shows.

A group G is *metacyclic* iff G/G^1 and G^1 are cyclic.

12.6.17. *Let G be a finite group.*

(1) *The following conditions are equivalent.*

    (a) *All Sylow subgroups of G are cyclic.*

    (b) $G = HK$, $H \cap K = E$, *H is a cyclic normal Hall subgroup, K is a cyclic Hall subgroup, and* $[H, K] = H$.

    (c) *G has generators and relations:*
$$G = \langle x, y \rangle, \ x^m = y^n = e, \ y^{-1}xy = x^r,$$
and $(m, n) = 1 = (m, r - 1), \ r^n \equiv 1 \pmod m$.

(2) *If all Sylow subgroups of G are cyclic, then G is metacyclic.*

Proof. (2) Any subgroup or factor group of G has all Sylow subgroups cyclic, hence may be inductively assumed to be solvable. By Theorem 6.2.11, if p is the smallest prime dividing $o(G)$ and $P \in \mathrm{Syl}_p(G)$ then P has a normal complement M. Thus G/M and, by induction, M are solvable, so that G is solvable. The groups $G/G^1, G^1/G^2, \ldots,$ are Abelian with all Sylow subgroups cyclic, hence are cyclic. By Theorem 12.6.16 and the solvability of G, $G^2 = E$. This proves (2).

(a) implies (c). Let $G^1 = \langle x \rangle$, and $G/G^1 = \langle yG^1 \rangle$ with $y \in G$. Also let $o(x) = m$, and $o(G/G^1) = n$. Then $x^m = e$, $y^n = x^s$ for some s, and $y^{-1}xy = x^r$ for some r. It follows that $x = y^{-n}xy^n = x^{r^n}$, and $r^n \equiv 1 \pmod{m}$.

Since every element of G is of the form x^iy^j, Theorem 3.4.2 implies that G^1 is normally generated by $[x, y] = x^{r-1}$. Since all subgroups of $G^1 = \langle x \rangle$ are normal in G, $\langle x^{r-1} \rangle = \langle x \rangle$, so that $(r - 1, m) = 1$. Also

$$x^s = y^{-1}x^sy = x^{rs}, \qquad x^{(r-1)s} = e.$$

But $\langle x^{r-1} \rangle = \langle x \rangle$, hence $x^s = e$ and $y^n = e$. If a prime p divides (m, n), then there is a subgroup A of G^1 of order p and a subgroup B of $\langle y \rangle$ of order p. Since $B \subset N(A)$, AB is a noncyclic subgroup of G of order p^2, so that a Sylow p-subgroup of G is not cyclic. Hence $(m, n) = 1$.

Now let L be a group with generators $\{x, y\}$ and relations $x^m = y^n = e$ and $y^{-1}xy = x^r$. Then $\langle x \rangle \lhd L$, and

$$o(L) \leq o(x)o(y) \leq mn.$$

Since $o(G) = mn$, it follows from Theorem 8.2.4 that $G \cong L$, i.e., that (c) is true.

(c) implies (b). Let $H = \langle x \rangle$, and $K = \langle y \rangle$. The statements in (b) are clearly true with the possible exception of the last one. We have

$$x^{r-1} = [x, y] \in [H, K],$$

$$[H, K] \supset \langle x^{r-1} \rangle = \langle x \rangle = H,$$

since $(r - 1, m) = 1$.

(b) implies (a). Any Sylow subgroup of G is either in H or in a conjugate of K, hence is cyclic. ‖

The converse of (2) in Theorem 12.6.17 is false (see Exercises 12.6.24 and 12.6.25).

12.6.18. *If a finite group G has an automorphism T of order 2 without fixed points (except e), then $gT = g^{-1}$ for all $g \in G$, and G is an Abelian group of odd order.*

Proof. Let $gU = g^{-1}(gT)$ for all $g \in G$. If $gU = hU$, then $g^{-1}(gT) = h^{-1}(hT)$, $hg^{-1} = (hg^{-1})T$, $hg^{-1} = e$, and $h = g$. Hence U is 1-1. Since G is finite, U is a permutation of G. Now let $x \in G$. Then $x = gU$ for some $g \in G$. Therefore,

$$xT = (gU)T = (g^{-1}(gT))T$$
$$= (gT)^{-1}g = (g^{-1}(gT))^{-1} = x^{-1}.$$

If also $y \in G$, then

$$(xT)(yT) = x^{-1}y^{-1} = (yx)^{-1}$$
$$= (yx)T = (yT)(xT).$$

Therefore, G is Abelian. Since T, as a permutation, is a product of disjoint 2-cycles and one 1-cycle, G is of odd order.

12.6.19. *If G is a Frobenius group, M its Frobenius kernel, and H a complement of M of even order, then M is Abelian.*

Proof. H has an element x of order 2 which induces an automorphism of order 2 in M without fixed points. By the preceding theorem, M is Abelian.

EXERCISES

12.6.20. Let H be a subgroup of a finite group G such that if $x \in H^\#$, then $C(x) \subset H$. Prove that H is a Hall subgroup of G. (Compare with Theorem 12.6.1.)

12.6.21. Let G be a Frobenius group with kernel M and complement H.

    (a) If $E < K \subset H$, then $N(K) \subset H$.

    (b) If $E < L \subset G$ and $o(L) \mid o(H)$, then $L \subset H^x$ for some x.

12.6.22. If G is a noncyclic group of order pq, p and q primes with $p < q$, then G is Frobenius with kernel $M \in \mathrm{Syl}_q(G)$ and complement $H \in \mathrm{Syl}_p(G)$.

12.6.23. There is a Frobenius group G with non-Abelian kernel M.

    (a) There is a non-Abelian group $M = \langle x, y, z \rangle$ of order 7^3, exponent 7, such that $Z(M) = \langle x \rangle$ and $[y, z] = x$.

    (b) M has an automorphism T such that $xT = x^4$, $yT = xy^2$, and $zT = z^2$.

    (c) $o(T) = 3$.

    (d) T moves all elements of M except e.

    (e) $G = \mathrm{Hol}(M, \langle T \rangle)$ is Frobenius with non-Abelian kernel M.

12.6.24. (Compare with, and use the proof of, Theorem 12.6.17.) A finite group is metacyclic iff it has generators x, y and relations $x^m = y^n = e$, $y^{-1}xy = x^r$, where $(m, r - 1) = 1$ and $r^n \equiv 1 \pmod{m}$.

12.6.25. A metacyclic group need not have Abelian Sylow subgroups. Let G have generators x, y and relations $x^9 = y^6 = e$, $y^{-1}xy = x^5$.

    (a) G is metacyclic of order 54 (see Exercise 12.6.24).

    (b) A Sylow 3-subgroup of G is not Abelian.

12.6.26. There is a Frobenius group G with kernel M and complement H of M such that all Sylow subgroups of H are cyclic but H is not Abelian. Over J_5, let

$$x = \begin{bmatrix} 3 & 0 \\ 3 & 2 \end{bmatrix}, \, y = \begin{bmatrix} 1 & -1 \\ 1 & 0 \end{bmatrix}.$$

(a) $H = \langle x, y \rangle$ is dicyclic of order 12:

   $o(y) = 6$, $x^{-1}yx = y^{-1}$, and $x^2 = y^3$.

(b) No element of $H^{\#}$ has characteristic value 1. (Only the unique element x^2 of order 2 and an element y^2 of order 3 need be examined.)

(c) If M is elementary Abelian of order 25, then $\mathrm{Hol}(M, H)$ is Frobenius with kernel M and complement H with the required properties.

12.6.27. The Frobenius kernel of a Frobenius group G is fully characteristic in G.

12.6.28. If G is a Frobenius group, M its kernel, and H a complement of M of odd order, then H is metacyclic, hence solvable.

12.7 Representations of transitive groups

If G is a permutation group on M, then G has a natural representation as a linear group on a vector space with basis M.

12.7.1. *Let G be a permutation group, X its character, and $\{X^1, \ldots, X^r\}$ the irreducible characters of G with X^1 the unit character.*

(i) $X = \Sigma \, n_i X^i$, *where n_1 is the number of orbits of G.*

(ii) *G is transitive iff $n_1 = 1$.*

(iii) *If G is transitive, G_a the subgroup fixing a, and m the number of orbits of G_a, then $\Sigma \, n_i^2 = m$.*

(iv) *G is 2-transitive iff $X = X^1 + X^2$ (with possible change of notation).*

(v) *If $G \neq E$ is transitive but not 2-transitive, then $X = X^1 + n_2 X^2 + n_3 X^3 + \ldots$, where $n_2 \neq 0$ and $n_3 \neq 0$ (with possible change of notation).*

Proof. (i) and (ii). By complete reducibility, $X = \Sigma \, n_i X^i$. If k is the number of orbits of G, then, by Theorem 10.1.5,

$$n_1 = (X, X^1) = \frac{1}{o(G)} \Sigma \, X(g) X^1(g^{-1})$$

$$= \frac{1}{o(G)} \Sigma \, X(g) = k.$$

(iii) By Theorem 10.1.6, (i), and orthogonality (Theorem 12.2.13),

$$mo(G) = \sum (X(g))^2 = \sum X(g)X(g^{-1})$$
$$= \sum_i \sum_j \sum_g n_i X^i(g) n_j X^j(g^{-1})$$
$$= o(G) \sum_{i,j} n_i n_j (X^i, X^j)$$
$$= o(G) \sum n_i^2.$$

Therefore, $\sum n_i^2 = m$.

(iv) G is 2-transitive iff $m = 2$. By (ii) and (iii), this happens iff n_1 and exactly one other n_i equal 1, all others 0.

(v) By assumption, $m \geq 3$. If (v) is false, then by (i) and (ii), $X = X^1 + n_2 X^2$, where by (iii), $n_2 > 1$. By Exercise 10.1.19, $\exists\ g \in G$ such that $X(g) = 0$. Thus

$$0 = X(g) = X^1(g) + n_2 X^2(g),$$
$$X^2(g) = -1/n_2 \notin J.$$

Since $X^2(g)$ is an algebraic integer (Theorem 12.2.18) which is rational, this is a contradiction. Hence (v) holds. ‖

The usefulness of the preceding theorem in computing the character table of some groups will now be illustrated.

Example 1. Character table for Sym(3). Any finite symmetric group Sym(n) ($n > 1$) has exactly two one-dimensional, irreducible characters since its commutator subgroup is Alt(n), which is of index 2 (see Theorem 12.4.5). These characters are X^1 and X^2, where $X^2(g) = 1$ if g is an even permutation and $X^2(g) = -1$ if g is odd. By Theorem 12.2.23, $G = \text{Sym}(3)$ has three irreducible characters. Since Sym(3) is 2-transitive, its natural character X (of degree 3) is the sum of X^1 and an irreducible character of degree 2. This other character must be X^3, so that $X = X^1 + X^3$. Now $X(e) = 3$, $X((1, 2)) = 1$, and $X((1, 2, 3)) = 0$. Thus the entire table of characters is determined to be as follows.

Character Table of Sym(3)

	e	$(1, 2)$	$(1, 2, 3)$
h_j	1	3	2
X^1	1	1	1
X^2	1	-1	1
X^3	2	0	-1

The size h_j of the jth conjugate class is not part of the character table proper but is convenient for checking orthogonality. If x is a complex number, let

$\bar{x}$ denote its complex conjugate. The orthogonality relations are as follows (see Exercise 12.2.28).

Rows:

$$(X^1, X^1) = \frac{1}{o(G)} \Sigma\, h_j X_j^1 \overline{X_j^1} = \frac{1+3+2}{6} = 1,$$

$$(X^1, X^2) = \frac{1-3+2}{6} = 0,$$

$$(X^1, X^3) = \frac{2+0-2}{6} = 0,$$

$$(X^2, X^2) = \frac{1+3+2}{6} = 1, \text{ etc.}$$

Columns:

$$\frac{h_1}{o(G)} \Sigma\, X^i(e)\overline{X^i(e)} = \frac{1+1+4}{6} = 1,$$

$$\Sigma\, X^i(e)\overline{X^i((1, 2))} = 1 - 1 + 0 = 0, \text{ etc.}$$

Example 2. Character table for $G = \text{Alt}(4)$. There are three one-dimensional characters X^1, X^2, and X^3, arising from the characters of G/H where H is the 4-group. There is one remaining irreducible character X^4 of degree 3. If X is the natural character of G, then $X = X^1 + X^4$ by Theorem 12.7.1. This gives the complete table of characters,

Character Table for Alt(4)

	e	$(1, 2)(3, 4)$	$(1, 2, 3)$	$(1, 3, 2)$
h_j	1	3	4	4
X^1	1	1	1	1
X^2	1	1	u	u^2
X^3	1	1	u^2	u
X^4	3	-1	0	0

where u is a primitive cube root of 1. This table illustrates some other results. There is an automorphism of $\mathscr{C}$ which maps u onto u^2, thereby inducing a permutation of the characters which interchanges X^2 and X^3 (and fixes X^1 and X^4). All of the entries in the table are algebraic integers, and, if rational, are integers. The character X^3 is the square of X^2 (see Exercise 12.7.2). Since $\text{Deg}(X^4) = X_1^4 = 3$ is relatively prime to $h_3 = 4$, $X_3^4 = 0$ by Theorem 12.3.1. Similarly $X_4^4 = 0$.

Example 3. Character table for Alt(5). By the example after Theorem 11.1.5, there are five conjugate classes in $G = \text{Alt}(5)$, hence five irreducible characters. Representatives for the classes are e, $(1, 2, 3)$, $(1, 2)(3, 4)$,

(1, 2, 3, 4, 5), and (1, 2, 3, 5, 4). Since G is simple, only the character X^1 can be obtained from characters of factor groups as before. Ordering the conjugate classes as above, $X^1 = (1, 1, 1, 1, 1)$. From the natural character $X = (5, 2, 1, 0, 0)$, one gets $X^2 = (4, 1, 0, -1, -1)$. In G, $n_5 = 6$, hence G has a transitive permutation representation of degree 6 on $N(P)$, where $P \in \mathrm{Syl}_5(G)$. In this representation, which is necessarily faithful, an element of order 5 is represented by a 5-cycle, hence the representation is 2-transitive. Since elements outside $N(P)$ have character 0, its character Y is $(6, 0, 2, 1, 1)$ (see Theorem 10.2.11). By Theorem 12.7.1,

$$X^3 = Y - X^1 = (5, -1, 1, 0, 0)$$

is an irreducible character.

Since every element of G is conjugate to its inverse, all characters are real (Exercise 12.2.30). The last two entries in column 1 are 3 by Theorem 12.2.24. The last two entries in column 2 are 0, since $(3, 20) = 1$ (Theorem 12.3.1), or by the orthogonality relations. The last two entries in column 3 are integers (Exercise 12.2.33), hence are both -1 from the orthogonality relations for columns 1 and 3. Let $X_4^4 = x$ and $X_5^4 = y$. Then

$$0 = (X^4, X^2) = 12 - 12(x + y), \qquad y = 1 - x.$$

Orthogonality of columns 4 and 3 gives $X_4^5 = 1 - x$. Similarly, $X_5^5 = x$. From $1 = (X^4, X^4)$, one gets

$$x^2 + (1 - x)^2 = 3, \qquad x = \frac{1 \pm \sqrt{5}}{2}.$$

Thus WLOG, $x = \dfrac{1 + \sqrt{5}}{2}.$

Character Table for Alt(5)

	e	(1, 2, 3)	(1, 2)(3, 4)	(1, 2, 3, 4, 5)	(1, 2, 3, 5, 4)
h_j	1	20	15	12	12
X^1	1	1	1	1	1
X^2	4	1	0	-1	-1
X^3	5	-1	1	0	0
X^4	3	0	-1	$\dfrac{1 + \sqrt{5}}{2}$	$\dfrac{1 - \sqrt{5}}{2}$
X^5	3	0	-1	$\dfrac{1 - \sqrt{5}}{2}$	$\dfrac{1 + \sqrt{5}}{2}$

There is a general method for computing the character tables of symmetric and alternating groups (see Boerner [1]). There is even a general method for obtaining the character table of any finite group, but it involves factorization of polynomials in several variables and cannot always be carried out. Both these methods will be omitted.

EXERCISES

12.7.2. If X and Y are irreducible characters of G with X of degree 1, then XY is an irreducible character of G.

12.7.3. Compute the character table of Sym(4). [There are five irreducible characters. Find three of them as in the examples. A fourth comes from a representation of a factor group of Sym(4), and the fifth from Exercise 12.7.2 (or orthogonality).]

12.8   Monomial representations

This section contains a brief discussion of a type of representation of a finite group which is a generalization of permutation representations and a specialization of linear representations.

A *monomial matrix* is a square matrix over $\mathscr{C}$ with just one nonzero entry in each row and each column. A monomial matrix is automatically nonsingular. A *monomial group* is a group of monomial matrices.

12.8.1. *If M is the group of all monomial matrices, D the group of all nonsingular diagonal matrices, and P the group of permutation matrices, then $M = DP$, $D \cap P = E$, and $D \lhd M$.*

12.8.2. *Using the notation of Theorem 12.8.1, let T be the function such that if $u = vw$, $u \in M$, $v \in D$, and $w \in P$, then $uT = w$. Then T is a homomorphism of M onto P with kernel D.*

Proof. This follows immediately from Theorem 12.8.1. ‖

Another way of describing T is to say that it maps each monomial matrix A onto the matrix B where all nonzero entries of A have been replaced by 1's.

A *monomial representation* of G is a representation T of G such that GT is a monomial group.

The terminology of permutation groups will sometimes be applied to monomial groups. For example, a monomial group G is *transitive* iff the corresponding permutation group (see Theorem 12.8.2) is transitive on the relevant basis of the vector space being acted upon.

12.8.3. *A monomial representation is the direct sum of transitive monomial representations.*

Proof. If T is a monomial representation of G, then T gives a transitive monomial representation on each orbit of GT, and T is the direct sum of these transitive constituents. ‖

Because of this theorem, it is enough to study transitive monomial representations. They are described, up to equivalence, by the following theorem.

12.8.4. *A representation T of a finite group G is equivalent to some transitive monomial representation iff T is equivalent to the induced representation of a representation of degree 1 of a subgroup H of G.*

Proof. Suppose that $T = U^*$, where U is a representation of $H \subset G$, and $\mathrm{Deg}(U) = 1$. Then (Section 12.5), with $G = \overset{.}{\cup}\, Hx_i$,

$$(gT)_{ij} = (x_i g x_j^{-1})U,$$
$$(x_i g x_j^{-1})U = 0 \quad \text{if} \quad x_i g x_j^{-1} \notin H.$$

For any i, $x_i g \in Hx_j$ for a unique j, hence $(gT)_{ij} \neq 0$ for exactly one j. Hence there is just one nonzero entry in each row of gT. If $x_i g x_j^{-1} \in H$ and $x_{i'} g x_j^{-1} \in H$, then $x_i x_{i'}^{-1} \in H$ and $i = i'$. Hence there is just one nonzero entry in each column of gT. Therefore, T is a monomial representation. Since

$$((x_1^{-1}x_j)T)_{1j} = (x_1 x_1^{-1}x_j x_j^{-1})U = 1,$$

T is transitive.

Conversely, let T be a transitive monomial representation of G. Let $H = \{g \in G \mid (gT)_{11} \neq 0\}$. Then $H \subset G$. Let $hU = (hT)_{11}$. Then U is a representation of degree 1 of H with character Y, say. Since T is transitive, $\exists\, x_i \in G$ such that $(x_i T)_{1i} \neq 0$. One checks that $G = \overset{.}{\cup}\, Hx_i$. Let X be the character of T and let $g \in G$. Then

$$(gT)_{ii} = (x_i T)_{1i}(gT)_{ii}((x_i T)_{1i})^{-1}$$
$$= (x_i T)_{1i}(gT)_{ii}((x_i T)^{-1})_{i1}$$
$$= ((x_i g x_i^{-1})T)_{11},$$
$$X(g) = \sum (gT)_{ii} = \sum ((x_i g x_i^{-1})T)_{11}$$
$$= \sum (x_i g x_i^{-1})U = Y^*(g).$$

Hence, $X = Y^*$. Since two representations are equivalent iff they have the same character (Theorem 12.2.10), T is equivalent to U^*. ‖

Note that a representation U of H of degree 1 has kernel K such that H/K is cyclic (Exercise 12.1.11). Hence its induced representation U^* has all entries 0 or an $[H:K]$th root of 1. Any such representation U^* will be called a *monomial representation of G on (H, K)*.

The following theorem of Huppert [2], itself a generalization of earlier theorems of Zassenhaus and Itô, gives a wide class of groups whose representations are all monomial.

   12.8.5. *If G is a finite group, S a solvable subgroup with all Sylow subgroups Abelian, and G/S supersolvable, then all representations of G are equivalent to monomial representations.*

   Proof. Induct on $o(G)$, and note that G is solvable. If $H < G$, then $H \cap S$ is a normal solvable subgroup of H with all Sylow subgroups Abelian, and $H/(S \cap H) \cong SH/S$ is supersolvable. If $E < H \lhd G$, then $SH/H \cong S/(S \cap H)$ is a normal solvable subgroup of G/H with all Sylow subgroups Abelian, and

$$\frac{G/H}{SH/H} \cong \frac{G}{SH} \cong \frac{G/S}{SH/S}$$

is supersolvable. Hence, by the inductive hypothesis, all representations of proper subgroups or factor groups of G are equivalent to monomial representations.

   By complete reducibility, it is sufficient to prove the theorem for irreducible representations. Let T be an irreducible representation of G acting on a vector space V. If $\mathrm{Ker}(T) \neq E$, then T is (essentially) an irreducible representation of $G/\mathrm{Ker}(T)$, hence is equivalent to a monomial representation. Therefore *WLOG*, T is faithful. Now $Z(G)S$ is a normal subgroup of G and $G/Z(G)S$ is supersolvable. If $p \in \mathscr{P}$, $P \in \mathrm{Syl}_p(Z(G))$, $Q \in \mathrm{Syl}_p(S)$, then PQ is an Abelian Sylow p-subgroup of $Z(G)S$. Hence *WLOG* $S \supset Z(G)$.

   First, suppose that there is a normal Abelian subgroup H of G such that $H > Z(G)$. Then as in the proof of Theorem 12.6.10, $V = V_1 + \ldots + V_r$, where H acts as a group of scalars on each V_i, and V_i is maximal with this property. If $r = 1$, then H is a group of scalars, hence $H \subset Z(G)$, a contradiction. Hence $r > 1$. If $g \in G$, $v \in V_i$, and $h \in H$, then

$$(vg)h = v(ghg^{-1})g = c_h vg,$$

where $c_h \in \mathscr{C}$ depends on h and g but not on $v \in V_i$. It follows that $V_i g \subset V_j$ for some j. Since $V = \Sigma V_i$ and $Vg = V$, each $g \in G$ permutes the subspaces V_i. Since V is irreducible, G acts transitively on the V_i. Let Q be the subgroup fixing V_1. Then $[G:Q] = r$, so $G = \cup Qu_j$ where $V_1 u_j = V_j$. Since $Q < G$, the representation of Q on V_1 is monomial. Let B_1 be a monomial Q-basis of V_1. Then $B = \cup B_1 u_j$ is a basis of V. If $b \in B_1$ and $g \in G$, then

$$(bu_j)g = bqu_k = cb'u_k$$

for some k, $q \in Q$, $c \in \mathscr{C}$, and $b' \in B_1$. Therefore, B is a monomial G-basis of V, and the theorem is true in this case.

Next suppose that $Z(G)$ is a maximal Abelian normal subgroup of G. Since $Z(G) \subset S$, also $Z(G) \subset Z(S) \subset \text{Fit}(S) = F$. But F is a nilpotent group with all Sylow subgroups Abelian, hence Abelian itself. Since F is characteristic in S, it is normal in G. Therefore, by the maximality of $Z(G)$, $Z(G) = Z(S) = F$. By Theorem 7.4.7, $F \supset C_S(F) = S$. Hence $F = S = Z(G)$. If G is Abelian, then T is of degree 1, hence monomial. If G is not Abelian, then since $G/S = G/Z(G)$ is supersolvable, $\exists\ A \lhd G$ such that $[A:Z(G)] \in \mathscr{P}$. But then A is an Abelian normal subgroup of G, contradicting the maximality property of $Z(G)$. ‖

In particular, if G is supersolvable, then all representations of G are equivalent to monomial ones.

In the other direction, one has the following theorem.

12.8.6. *If G is a finite group all of whose representations are equivalent to monomial representations, then G is solvable.*

Proof. Suppose that G is not solvable. Then $\exists\ H \lhd G$ such that $E < H$ and $H = H^1$. There is a representation (for example, the regular representation) of G whose kernel does not contain H. Hence there is an irreducible monomial representation T of G of smallest possible degree such that $H \nsubseteq \text{Ker}(T)$. If $\text{Deg}(T) = 1$, then $G/\text{Ker}(T) \cong GT$ is Abelian, so $H/(H \cap \text{Ker}(T))$ is Abelian and $H^1 < H$, a contradiction. Hence $\text{Deg}(T) > 1$. Let U be the homomorphism mapping each gT onto the corresponding permutation (Theorem 12.8.2). Since T is irreducible, it is a transitive monomial representation, hence TU is a transitive permutation representation of G. By Theorem 12.7.1, TU is reducible:

$$TU = D_1 + D_2 + \dots,$$

where D_i is irreducible. Thus $\text{Deg}(D_i) < \text{Deg}(TU) = \text{Deg}(T)$. Therefore, by the minimality of $\text{Deg}(T)$, $\text{Ker}(D_i) \supset H$ for all i, hence $\text{Ker}(TU) \supset H$. Now $\text{Ker}(U)$ is a group of diagonal matrices, hence is Abelian. This means that

$$\frac{\text{Ker}(TU)}{\text{Ker}(T)} \cong \text{Ker}(U)$$

is Abelian. Hence

$$H = H^1 \subset (\text{Ker}(TU))^1 \subset \text{Ker}(T),$$

a contradiction. Therefore the theorem is true.

EXERCISES

12.8.7. An irreducible representation T of a group G on a vector space V is *imprimitive* iff $V = \Sigma\ \{V_i \mid 1 \leq i \leq r\}$ where the V_i are subspaces of V, $r > 1$, and each $V_i\ (gT)$ is some V_j. An irreducible representation is *primitive* iff it is not imprimitive.

(a)   If T is an irreducible representation of G on a vector space V over $\mathscr{C}$ and $H \lhd G$, then either $T \mid H$ is irreducible, or G is imprimitive, $V = \Sigma\, V_i$, the V_i are irreducible H-modules all of the same dimension, and each $g \in G$ permutes the V_i.

(b)   If T is a faithful, irreducible, primitive representation of G over $\mathscr{C}$ and A is an Abelian normal subgroup of G, then A is a cyclic, central subgroup of G.

12.8.8.   There is a solvable group not all of whose representations are equivalent to monomial representations (compare with Theorem 12.8.6). Let G be the relative holomorph of the quaternion group by an automorphism of order 3.

(a) $o(G/G^1) = 3$. G is solvable.

(b) G has exactly three irreducible characters of degree 1.

(c) $\exists\, H \lhd G$ such that $G/H \cong \mathrm{Alt}(4)$.

(d) G/H has irreducible characters of degrees 1, 1, 1, and 3.

(e) G has irreducible characters of degrees 1, 1, 1, 3, 2, 2, and 2.

(f) G has no subgroup of index 2 [see (a)].

(g)   An irreducible representation of degree 2 cannot be monomial (see Theorem 12.8.4).

[Itô [1] has shown that there is an overgroup K of G all of whose representations are equivalent to monomial ones, so that this property is not inherited by subgroups.]

12.9   Transitive groups of prime degree

The object of this section is to present a useful theorem of Burnside to the effect that a transitive permutation group of prime degree is either 2-transitive or solvable of known type. It will be necessary to assume certain facts about cyclotomic fields.

12.9.1.   *If S is a nonempty subset of a group G such that for x and y in G, $Sx \cap Sy = Sx$ or $\varnothing$, then there is a subgroup H of G such that $S = Hx$ for some x.*

Proof. Let $x \in S$ and let $H = Sx^{-1}$. Then $e \in H$, and the same hypotheses hold for H as for S. If $a \in H$ and $b \in H$, then $e \in H \cap Hb^{-1}$, hence $Hb^{-1} = H$ by assumption. Therefore, $ab^{-1} \in H$, and H is a subgroup of G. Since $S = Hx$, this proves the lemma.

12.9.2.   *If G is a transitive group of prime degree p, then G is 2-transitive or solvable.*

Proof. Let $\{x_1, \ldots, x_p\}$ be the set of letters. There is a natural representation T of G on a vector space V with basis $(x_1, \ldots, x_p)$. Assume that G is not 2-transitive. By Theorem 12.7.1,

$$T = 1_G + T_1 + \ldots + T_t, \qquad t \geq 2, \tag{1}$$

where the representations T_i are irreducible and inequivalent to the one representation 1_G. Let $V = V_0 + V_1 + \ldots + V_t$ be the corresponding decomposition of V, so that T_i acts irreducibly on V_i and 1_G on V_0.

Since G is transitive, $p \mid o(G)$, and there are elements of order p. Let $c \in G$ have order p. Then c is a p-cycle, say $c = (x_1, \ldots, x_p)$. Let r be a primitive pth root of 1. Then

$$y_i = x_1 + r^{-i}x_2 + \ldots + r^{-(p-1)i}x_p \tag{2}$$

is a characteristic vector of cT with characteristic value r^i for $i = 1, \ldots, p$. Therefore $(y_1, \ldots, y_p)$ is a basis of V. Also, the subspaces $\mathscr{C}y_i$ are the only one-dimensional subspaces invariant under cT. Since (Theorem 12.2.16) each cT_i is diagonal on V_i with suitable choice of the basis, a basis of V_i may be chosen to consist of some of the y_j. The set

$$Q = \{r^i \mid 0 \leq i < p\}$$

of characteristic roots of cT is partitioned

$$Q = \overset{.}{\cup} \{Q_i \mid 0 \leq i \leq t\},$$

where $Q_0 = \{1\}$ is the characteristic root of $(c)1_G$ and Q_i the set of characteristic roots of cT_i, $i > 0$. Thus, $o(Q_i) = \mathrm{Dim}(V_i)$, and $T_1, \ldots, T_t$ are inequivalent.

Let F be the field $\mathscr{R}(r)$. Since both $(x_1, \ldots, x_p)$ and $(y_1, \ldots, y_p)$ are bases of V, one can solve (2) over F for the x_i in terms of the y_j. Thus if $g \in G$, then

$$y_i(gT_n) = \sum_j a_{ij}x_j(gT_n) = \sum_j \sum_k a_{ij}b_{jk}x_k$$

$$= \sum_j \sum_k \sum_m a_{ij}b_{jk}c_{km}y_m,$$

where all coefficients are in F, since the b_{jk} are 0 or 1. Therefore the representations T_n have entries in F.

If $k \in J_p^\#$, then there is an automorphism U_k of F such that $rU_k = r^k$ (van der Waerden [1, paragraph 53]). U_k induces a permutation of the irreducible representations of G which are over F, but it fixes both T and 1_G since they are integral valued. Hence, U_k permutes the T_i, $i > 0$, and therefore the set $\{Q_1, \ldots, Q_t\}$. Moreover, the group of all U_i is transitive on $\{Q_1, \ldots, Q_t\}$. Now $Q_1 = \{r^i \mid i \in S\}$ where S is a subset of $J_p^\#$. Therefore,

$$Q_1 U_k = \{r^{ik} \mid i \in S\} = \{r^j \mid j \in Sk\}.$$

Since the Q_i are disjoint, $Si \cap Sj = Si$ or $\varnothing$ for all i and j in $J_p^\#$. By Theorem 12.9.1, $\exists \ H \subseteq J_p^\#$ such that the Si are the cosets of H. Since $c^k T_1$ has

$$\{r^{ik} \mid i \in S\} = Q_1 U_k$$

as its set of characteristic roots, *WLOG*

$$Q_1 = \{r^i \mid i \in H\}.$$

Moreover, each $c^k T_1$ has some Q_j as its set of characteristic roots. Now H is the unique subgroup of $J_p^\#$ of order $s = (p-1)/t$. Since c was arbitrary and t is independent of c, it follows that if $o(g) = p$, then $g T_j$ has some Q_k as its set of characteristic roots. Now $c^j T_1$ and $c^k T_1$ have the same set of characteristic roots iff $j \in Hk$. Therefore the number of elements g of $\langle c \rangle^\#$ such that the set of characteristic roots of $g T_1$ is Q_i is the same as the number such that this set is Q_m for any i and m. Let N be the number of elements g of order p such that the set of characteristic roots of $g T_1$ is Q_1. Since c is arbitrary, N is also the number of elements g of order p such that the set of characteristic roots of $g T_1$ is Q_i for any (fixed) $i > 1$.

Let $q \in \mathcal{N}, p \nmid q$. We assert that the only qth roots of 1 in F are ± 1. If there are others, then $q > 2$, and F contains a primitive (pq)th root of 1, namely a product of a primitive pth root and a primitive qth root. If $\phi(m)$ denotes the number of natural numbers $i \leq m$ which are relatively prime to m, then (van der Waerden [1, par. 53])

$$[F : \mathscr{R}] \geq \phi(pq) = \phi(p)\phi(q) > \phi(p) = [F : \mathscr{R}],$$

a contradiction. This proves the assertion.

Let X be the character of T and X^i the character of T_i.* Let $g \in G$ and $p \nmid o(g) = q$. Since $X^i(g) \in F$ and $X^i(g)$ is a sum of qth roots of 1, $X^i(g)$ is an integer. Since the group of all U_j permutes the X^i transitively but fixes all rationals,

$$X^1(g) = \ldots = X^t(g), \qquad p \nmid o(g). \tag{3}$$

Therefore, by (1),

$$X(g) = 1 + tX^1(g), \qquad p \nmid o(g).$$

Hence $0 \leq X^1(g) \leq s$ and $X(g) \equiv 1 \pmod{t}$. In particular, $X(g) \neq 0$ if $p \nmid o(g)$.

Let N_i be the number of $g \in G$ such that $X(g) = 1 + it$ for $0 \leq i \leq s$ (hence $p \nmid o(g)$ and $X^1(g) = i$). Since T_1 and 1_G are inequivalent, their characters are orthogonal, and

$$0 = \Sigma \{X^1(h) \mid h \in G\}$$
$$= N \Sigma \{r^i \mid 0 < i < p\} + N_1 + 2N_2 + \ldots + sN_s.$$

---

* X^1 is not the unit character.

Since $\Sigma \{ r^i \mid 0 \leq i < p \} = 0$ (geometric progression), it follows that

$$N = N_1 + 2N_2 + \ldots + sN_s. \tag{4}$$

Let $X^{j''}$ denote the irreducible character given by $X^{j''}(g) = X^j(g^{-1})$ (see after Theorem 12.2.24), and let $j'' \neq 1$. Now $X^j = X^1 U_k$ for some k. By the orthogonality of X^1 and $X^{j''}$,

$$
\begin{aligned}
0 &= \Sigma \{ X^1(g) X^{j''}(g^{-1}) \mid g \in G \} \tag{5} \\
&= \Sigma \{ X^1(g) X^j(g) \mid g \in G \} \\
&= N \sum_{m \in R} (\Sigma \{ r^{im} \mid i \in H \})(\Sigma \{ r^{ikm} \mid i \in H \}) + N_1 + 2^2 N_2 + \ldots + s^2 N_s,
\end{aligned}
$$

where $J_p^{\#} = \cup \{ Hm \mid m \in R \}$. The next job is to evaluate the first term on the right side of (5). First, suppose that $v \in H$ and $kv = -1$ (in J_p). Let $o(g) = p$, and let Q_h be the set of characteristic roots of gT_1. Then for some m,

$$
\begin{aligned}
Q_h &= Q_1 U_m = \{ r^{im} \mid i \in H \}, \\
X^j(g) &= X^1(g^k) = \Sigma \{ r^{imk} \mid i \in H \} \\
&= \Sigma \{ r^{imkv} \mid i \in H \} \\
&= \Sigma \{ r^{-im} \mid i \in H \} = X^1(g^{-1}) \\
&= X^{1''}(g).
\end{aligned}
$$

Since $X^j(g) = X^{1''}(g)$ for $p \nmid o(g)$ by (3), this implies that $X^j(g) = X^{1''}(g)$ for all $g \in G$. Therefore $j = 1''$, a contradiction. Hence $1 + kv \neq 0$ if $v \in H$. We have

$$
\begin{aligned}
\sum_{m \in R} (\Sigma \{ r^{im} \mid i \in H \})(\Sigma \{ r^{hkm} \mid h \in H \}) \\
&= \sum_{m \in R} \Sigma \{ r^{mi(1 + ki^{-1}h)} \mid i \in H, h \in H \} \\
&= \sum_{m \in R} \sum_{i \in H} \sum_{v \in H} r^{mi(1 + kv)} \\
&= \sum_{v \in H} \Sigma \{ r^{(1 + kv)mi} \mid i \in H, m \in R \} \\
&= \Sigma \{ -1 \mid v \in H \} = -s,
\end{aligned}
$$

since $r^{1 + kv}$ is a primitive pth root of 1. Therefore, (5) becomes

$$Ns = N_1 + 2^2 N_2 + \ldots + s^2 N_s.$$

By (4),

$$N_1 + 2^2 N_2 + \ldots + s^2 N_s = sN_1 + 2sN_2 + \ldots + s^2 N_s. \tag{6}$$

Since all terms in (6) are nonnegative and $i^2 < is$ for $i < s$, it follows that $N_1 = \ldots = N_{s-1} = 0$. This means that all elements of $G^{\#}$ fix 0 or 1 letters. Hence (Theorem 12.5.12), G is a Frobenius group with a regular normal subgroup P of order p. Since (Theorem 10.3.5) $C(P) = P$, it follows that $G/P \stackrel{\sim}{\subseteq} \text{Aut}(P)$, hence G/P is cyclic. Therefore, G is solvable. In fact,

12.9.3. *If G is a transitive but not 2-transitive group of prime degree p and $P \in \text{Syl}_p(G)$, then $P \lhd G \subseteq \text{Hol}(P)$.* ‖

Here is an example of a numerical application of the degree p theorem.

12.9.4. *There are no simple groups of order* 4960.

Proof. Let G be a simple group of order $4960 = 2^5 \cdot 5 \cdot 31$, and let $P \in \text{Syl}_2(G)$. If P is not a maximal proper subgroup, then there is a subgroup H of index 31, hence a faithful, transitive permutation representation of G of degree 31. Since G is simple, this representation is 2-transitive by Theorem 12.9.2, and therefore $30 \mid 4960$, which it doesn't. Hence P is maximal and $n_2 = 155$. Since $n_2 \not\equiv 1 \pmod 4$, it follows (Theorem 6.5.3) that $\exists \, Q \in \text{Syl}_2(G)$ such that $o(P \cap Q) = 2^4$. But then $N(P \cap Q) > P$, contradicting either the maximality of P or the simplicity of G.

REFERENCES FOR CHAPTER 12

For the entire chapter, Curtis and Reiner [1], M. Hall [1], and Speiser [1]; Theorem 12.6.17, Zassenhaus [4]; Section 12.9, Burnside [1].

THIRTEEN

PRODUCTS OF SUBGROUPS

Suppose that a group G is the product AB of subgroups A and B. This chapter is primarily concerned with the following question: What conclusions can be made about G if A and B are suitably restricted? The most striking result of this type known so far is that if A and B are finite nilpotent groups, then G is solvable (Theorem 13.2.9).

13.1 Factorizable groups

If A and B are subgroups of G, at least one of which is normal, then AB is a subgroup of G. If neither is normal, then AB need not be a subgroup. For example, if $G = \mathrm{Sym}(3)$ and A and B are different subgroups of order 2, then AB contains four elements and cannot be a subgroup. On the other hand, AB may be a subgroup even though neither A nor B is normal. Examples illustrating this fact will be given after some preliminary lemmas.

13.1.1. *If A and B (but possibly not AB) are subgroups of G, then $[AB:A] = [B:A \cap B]$, where $[AB:A]$ is the number of right cosets of A in AB.*

Proof. Let $R = \{((A \cap B)b, Ab) \mid b \in B\}$. If $b_1 \in B$ and $b_2 \in B$, then $(A \cap B)b_1 = (A \cap B)b_2$ iff $b_1 b_2^{-1} \in A \cap B$ iff $Ab_1 = Ab_2$. Hence the relation

R is actually a 1–1 function from the set S of right cosets of $A \cap B$ in B onto the set T of right cosets of A in AB. The conclusion follows. ‖

This lemma yields a useful corollary.

13.1.2. *If A and B are finite subgroups of a group, then*

$$o(AB) = \frac{o(A)o(B)}{o(A \cap B)}.$$

Proof. For $o(AB) = o(A)[AB:A] = o(A)o(B)/o(A \cap B)$.

13.1.3. *If $o(G) = mn$; $(m, n) = 1$, $A \subset G$, $B \subset G$, $o(A) = m$, and $o(B) = n$, then $G = AB$.*

Proof By the preceding corollary, $o(AB) = o(A)o(B) = o(G)$, and surely AB is contained in G. Hence $AB = G$. ‖

This remark shows that Sym(4) is the product of a Sylow 2-subgroup and a Sylow 3-subgroup, and these subgroups are not normal. Similarly $G =$ Alt(5) is the product of a Hall subgroup A fixing one letter ($A \cong$ Alt(4)) of order 12 and a Sylow 5-subgroup B. Since Alt(5) is simple, neither A nor B is normal.

It is obvious that if $G = AB$, $A \subset H \subset G$, and $B \subset K \subset G$, then $G = HK$ also. This remark could be applied to the above examples to give further examples of products.

A group G is *factorizable* iff $G = AB$ with $A < G$ and $B < G$, and, in this case, A and B furnish a *factorization* of G. The general problem of determining which groups are factorizable has not yet been solved. The following theorems will treat some of the cases which are easy to handle.

13.1.4. *If G/H is factorizable, so is G.*

Proof. Let $G/H = A^*B^*$ with $A^* < G/H$ and $B^* < G/H$. Then there are subgroups A and B of G such that $A/H = A^*$ and $B/H = B^*$, hence $A < G$ and $B < G$. Since $(A/H)(B/H) = G/H$, if $g \in G$, then $\exists\, a \in A$ and $b \in B$ such that $gH = (aH)(bH)$, so $g = abh$ for some $h \in H$. But $h \in B$, so $g = a(bh) \in AB$. Hence $G = AB$. (Similar arguments below will be much abbreviated.)

13.1.5. *A finite solvable group G is factorizable iff it is not a cyclic p-group, $p \in \mathscr{P}$.*

Proof. A group of order 1 is trivially not factorizable. If G is cyclic of order p^n, $p \in \mathscr{P}$, $n > 0$, then there is a maximum proper subgroup M. If $A < G$ and $B < G$, then $A \subset M$ and $B \subset M$, so that $AB \subset M$ and $G \neq AB$. Hence, G is not factorizable.

Now, suppose that G is not a cyclic p-group. If G is a p-group, then by the Burnside basis theorem, 7.3.10, there are distinct maximal proper subgroups A and B. Each is normal by Theorem 6.3.9. Hence, $A < AB = G$. If G is not a p-group, then by Hall's theorem $G = PH$ where $P \in \mathrm{Syl}_p(G)$ for some prime p dividing $o(G)$, H is a product of Sylow subgroups, and $H < G$.

13.1.6. *An Abelian group G is factorizable iff G is neither a cyclic p-group nor a p^∞-group for some prime p.*

Proof. Cyclic p-groups are handled in the preceding theorem. If G is a p^∞-group, then the product of any two proper subgroups is the larger of the two and therefore not G.

Now let G be Abelian but not a cyclic p-group or a p^∞-group. If G is decomposable, it is factorizable. Assume that G is not decomposable. By Theorem 5.2.10, G is torsion free. Let $x \in G^{\#}$. Then $G/\langle 6x \rangle$ contains an element of order 6 and is therefore factorizable by what has already been said. By Theorem 13.1.4, G is factorizable.

13.1.7. *If G is an uncountable, solvable group, then G is factorizable.*

Proof. If G/G^1 is factorizable, so is G. If G/G^1 is not factorizable, then, by the preceding theorem, it is countable. Let S be a set of representatives of the cosets of G^1. Then $\langle S \rangle$ is countable, and $G = G^1\langle S \rangle$ is a factorization of G. ‖

Before stating the next theorem, it should perhaps be remarked that factorizability is invariant under isomorphism.

13.1.8. *If G is a finite, insolvable group such that every non-Abelian composition factor group of G is factorizable, then G is factorizable.*

Proof. Let M be a maximal proper normal subgroup of G. If G/M is non-Abelian, then it is factorizable by assumption, so G is also. If G/M is Abelian, it is of prime order p. Let $H \in \mathrm{Syl}_p(G)$. Then $H \nsubseteq M$, so $G = MH$ with $H < G$.

13.1.9. *If G is a permutation group (perhaps infinite) and $H < G$ is transitive, then G is factorizable.*

Proof. Let G_a be the subgroup fixing a letter a. Then $G_a < G$ since G is transitive. If $g \in G$, then $\exists\, h \in H$ such that $ah = ag$. Thus, $agh^{-1} = a$, $gh^{-1} \in G_a$, and $g \in G_a H$. Therefore, $G = G_a H$.

13.1.10. *If (M, G) is a permutation group containing $\mathrm{Alt}(M)$, $o(M) > 3$, then G is factorizable.*

Proof. By the preceding theorem, it suffices to prove the existence of a proper transitive subgroup H. If $n = o(M)$ is odd, then there is an n-cycle $x \in G$, $H = \langle x \rangle$ is transitive, and $o(H) = n < n!/2 \leqq o(G)$.

Let n be even or infinite. Then there is a partition $M = \overset{.}{\cup} A_i$ with $o(A_i) = 2$. Let H be the set of $x \in G$ such that each $A_i x$ is some A_j. Let $A_i = \{a_i, b_i\}$. If $i \neq j$, then $(a_i, b_i)(a_j, b_j) \in H$, $(a_i, a_j)(b_i, b_j) \in H$, but $(a_i, b_i, a_j) \in G \backslash H$. It follows that H is a transitive proper subgroup of G. ‖

Other facts about factorizability are known. In particular, Itô [2] has determined all factorizations of the simple groups $PSL(2, q)$. Here it will only be shown that there are nonfactorizable, finite, simple non-Abelian groups.

13.1.11. $G = PSL(2, 13)$ *is not factorizable.*

Proof. By Theorem 10.8.4, G is simple. Its order is $14 \cdot 13 \cdot 12 / 2 = 1092 = 2^2 \cdot 3 \cdot 7 \cdot 13$. Now suppose that $G = AB$ with $A < G$ and $B < G$. Then $13 \mid o(A)$, say. If $7 \mid o(A)$, then there is a permutation representation of G of degree less than 13, an impossibility. Thus $7 \nmid o(A)$, so $7 \mid o(B)$ (Theorem 13.1.2) and $13 \nmid o(B)$. Let $P \in \mathrm{Syl}_{13}(A)$ and $Q \in \mathrm{Syl}_7(B)$. By Sylow, $n_{13}(G) = 14$, $n_{13}(A) = 1$, $A \subset N(P)$, $o(N(P)) = 78$, $n_7(G) = 78$, $n_7(B) = 1$, $B \subset N(Q)$, $o(N(Q)) = 14$. Since $14 \cdot 78 = 1092$, $A = N(P)$, $B = N(Q)$, and $A \cap B = E$. By Burnside, if $R \in \mathrm{Syl}_2(G)$, then $o(N(R)/C(R)) = 3$, hence, by Sylow, all elements of order 2 are conjugate. There are elements x in A and y in B of order 2. Then $\exists\, g = ab$, $a \in A$, $b \in B$, such that $y = x^g = (x^a)^b$. Hence $x^a \in A$, $y^{b^{-1}} \in B$, and $x^a = y^{b^{-1}}$, a contradiction. ‖

The following theorem of Huppert [4] is concerned with factorizations of a special type. A class S of groups will be called *hereditary* iff

(a) If $H \subset G \in S$, then $H \in S$,

(b) If $G \in S$ and $T \in \mathrm{Hom}(G)$, then $GT \in S$.

13.1.12. *Let S be a hereditary class of finite groups such that if G is finite and $G/Fr(G) \in S$, then $G \in S$. If G is a finite group, $G/H \in S$, and U is a minimal subgroup such that $G = HU$, then $U \in S$.*

Proof. Let $F = \mathrm{Fr}(G)$. First suppose that $H \subset F$. Then $G/F \cong (G/H)/(F/H) \in S$. By assumption, $G \in S$. Therefore, $U \in S$.

Now suppose that $H \not\subset F$. Then $H \not\subset K$ for some maximal proper subgroup K of G. Hence $HK = G$, so that $U \neq G$. Now $U/(U \cap H) \cong HU/H = G/H \in S$. By induction, $\exists\, T \in S$ such that $U = (U \cap H)T$. Then

$$G = HU = H(U \cap H)T = HT, \qquad T \subset U.$$

By the minimality of U, $T = U$. Therefore $U \in S$.

13.1.13. *If G is a finite group, G/H is solvable (supersolvable) (nilpotent) (cyclic), and U is a minimal subgroup such that $G = HU$, then U is solvable (supersolvable) (nilpotent) (cyclic).*

Proof. The class S of solvable (supersolvable) (nilpotent) (cyclic) finite groups is hereditary. If G is a finite group and $G/\text{Fr}(G) \in S$, then $G \in S$ by Theorem 7.4.10 and Exercises 7.3.29 and 9.3.18. The assertions now follow from Theorem 13.1.12.

EXERCISES

13.1.14. If A and B are (possibly infinite) subgroups of G, then
$$o(AB)o(A \cap B) = o(A)o(B).$$

13.1.15. It is false that if A and B are subgroups of a finite primary group G, then AB is also a subgroup.

13.1.16. If F is a field with at least three elements and $n > 1$, then $GL(n, F)$ is factorizable.

13.1.17. If $G = AB$ and $A \cap B = E$, then in the permutation representation of G on B, A is regularly represented.

13.2 Nilpotent times nilpotent

The principal theorem of this section is that the product of two nilpotent groups is solvable if one of the factors is finite.

If $H \subset G$ the *focal* subgroup $\text{Foc}_G(H) = \text{Foc}(H)$ is the subgroup generated by all $[h, g] \in H$, with $h \in H$ and $g \in G$. Thus, $\text{Foc}(H) \supset H^1$. The *focal series* of H is $H = H_0, H_1, \ldots$, where $H_{n+1} = \text{Foc}(H_n)$ for all n. H is *hyperfocal* (in G) iff its focal series reaches E.

13.2.1. *A subgroup of a hyperfocal group is hyperfocal.*

Proof. If $K \subset H \subset G$, then $\text{Foc}(K) \subset \text{Foc}(H)$ from the definition. Hence if H is hyperfocal, so is K.

13.2.2. *If H is a hyperfocal Hall subgroup of G, then H has a normal complement in G.*

Proof. Induct on $o(G)$. Let P be the set of primes dividing $o(H)$, and let Q be the subgroup generated by all P'-elements of G. Then $Q \lhd G$ and $G = HQ$. If $Q \cap H = E$, we are done. Suppose that $Q \cap H \neq E$. Then $Q \cap H$

is a Hall P-subgroup of Q (Theorem 9.3.3), which is hyperfocal by Theorem 13.2.1. If $Q < G$, then, by induction, $Q \cap H$ has a normal complement R in Q. Now R must be the subgroup generated by all P'-elements of Q, hence is Q itself. Therefore, $Q \cap H = E$, a contradiction. It follows that $Q = G$. Therefore $\nexists \, M < G$ such that G/M is a P-group, for if so, any $g \in G \backslash M$ would have order divisible by some $p \in P$, hence $M \supset Q = G$, a contradiction.

Let T be the transfer of G into H. By Theorem 3.5.6, if $x \in G$ then $\exists \, t_i \in G$ and $f_i \in \mathcal{N}$ such that

$$xT = H^1 \pi t_i^{-1} x^{f_i} t_i, \qquad t_i^{-1} x^{f_i} t_i \in H, \qquad \sum f_i = [G : H].$$

If $x \in H$, then $[x^{f_i}, t_i] \in H$, so $[x^{f_i}, t_i] \in \mathrm{Foc}[H]$. Since H is hyperfocal, $\mathrm{Foc}(H) < H$. If $x \in H \backslash \mathrm{Foc}(H)$, then for some $a \in \mathrm{Foc}(H)$,

$$xT = H^1 \pi x^{f_i} [x^{f_i}, t_i] = H^1 a \pi x^{f_i} = H^1 a x^{[G : H]} \neq e.$$

Therefore, $G/\mathrm{Ker}(T)$ is a nontrivial P-group, contradicting an earlier statement.

13.2.3. *If H is a subgroup but not a Sylow subgroup of G such that $N(P) = H$ for all $P \in \mathrm{Syl}(H)$, then H has a normal complement in G.*

Proof. The hypotheses imply that H is a nilpotent Hall subgroup. Let $P \in \mathrm{Syl}(H)$, $x \in P$, $y \in P$, and $y = x^z$, with $z \in G$. Since H is not a Sylow subgroup, $\exists \, Q \in \mathrm{Syl}(G)$, $Q \neq E$, $Q \neq P$, such that $x \in C(Q)$ and $y \in C(Q)$. By Theorem 6.2.8, $y = x^u$ with $u \in N(Q) = H$. Since H is nilpotent, $y = x^v$, with $v \in P$.

If $x \in H$, $y \in H$, and $y = x^z$, with $z \in G$, then $x = \pi x_i$, $y = \pi y_i$, x_i and y_i in $P_i \in \mathrm{Syl}(H)$, and $P_i \neq P_j$ if $i \neq j$. Now $\exists \, k_i \in J$ such that $x^{k_i} = x_i$ for all i (Theorem 5.1.5). Hence,

$$x_i^z = (x^{k_i})^z = (x^z)^{k_i} = y^{k_i} \in H.$$

Since $o(y^{k_i}) = o(x_i)$ and H is nilpotent, $y^{k_i} \in P_i$. Thus,

$$\pi y_i = y = x^z = \pi x_i^z = \pi y^{k_i}, \qquad y_i = y^{k_i} = x_i^z.$$

By the first paragraph, $\exists \, u_i \in P_i$ such that $x_i^{u_i} = y_i$. If $u = \pi u_i$, then by the nilpotence of H, $x^u = y$, $u \in H$.

If $h \in H$, $g \in G$, and $[h, g] \in H$, then, by the preceding remarks,

$$[h, g] = h^{-1} h^g = h^{-1} h^u = [h, u]$$

for some $u \in H$. Therefore, if $K \subset H$, then $\mathrm{Foc}(K) \subset [K, H]$. If $[K, H] \supset K$ for some K with $E < K \subset H$, then inductively

$$Z^{n+1}(H) = [Z^n(H), H] \supset [K, H] \supset K,$$

a contradiction since H is nilpotent. Therefore, $[K, H] \not\supset K$, and $\mathrm{Foc}(K) < K$ for all K such that $E < K \subset H$. Thus, H is hyperfocal. By Theorem 13.2.2, H has a normal complement.

13.2.4. *If $G = HK$, H is conjugate to A, and K is conjugate to B, then $G = AB$ and $\exists\, g \in G$ such that $A = H^g$ and $B = K^g$.*

Proof. By assumption, $K^x = B$, where $x = kh$, $k \in K$, and $h \in H$. Hence $K^h = K^{kh} = K^x = B$. Thus $G = (HK)^h = HK^h = HB$. Again, $\exists\, y = h'b$ such that $H^y = A$, $h' \in H$, and $b \in B$. Then

$$H^{hb} = H^b = A, \quad K^{hb} = B^b = B, \quad G = G^{hb} = (HK)^{hb} = AB. \parallel$$

Because of Theorem 13.2.4, the factorizations $G = HK$ and $G = H^g K^g$ are considered to be (essentially) the same.

13.2.5. *If $G = HK$, $p \mid o(G)$, and $p \in \mathscr{P}$, then $\exists\, P \in \mathrm{Syl}_p(H)$ and $Q \in \mathrm{Syl}_p(K)$ such that $PQ \in \mathrm{Syl}_p(G)$.*

Proof. Let $P_1 \in \mathrm{Syl}_p(H)$, $Q_1 \in \mathrm{Syl}_p(K)$, and $R \in \mathrm{Syl}_p(G)$. By Sylow (Theorem 6.1.12), there are x and y in G such that $P_1^x \subset R$ and $Q_1^y \subset R$. By Theorem 13.2.4, $G = H^x K^y$ and $\exists\, g \in G$ such that $H^{xg} = H$, $K^{yg} = K$. Thus,

$$P = P_1^{xg} \in \mathrm{Syl}_p(H),$$
$$Q = Q_1^{yg} \in \mathrm{Syl}_p(K), \quad R^g \in \mathrm{Syl}_p(G), \quad P \subset R^g, \quad Q \subset R^g.$$

Let $o(H) = p^i m$, $o(K) = p^j n$, and $o(H \cap K) = p^k r$, where m, n, and r are not divisible by p. Then,

$$o(G) = \frac{o(H)o(K)}{o(H \cap K)} = p^{i+j-k}s, \qquad (s, p) = 1.$$

Since

$$o(PQ) = \frac{o(P)o(Q)}{o(P \cap Q)} \geq p^{i+j-k} = o(R)$$

(Theorem 13.1.2), and $PQ \subset R^g$, we have $PQ = R^g$. Hence, $PQ \in \mathrm{Syl}_p(G)$.

13.2.6. *If $G = HK$, $A \lhd H$, and $B \lhd K$, then*

(i)   *Either $A^x B^y \subset G$ for all x and y in G, or $\not\exists\, x$ and y in G such that $A^x B^y \subset G$.*

(ii) *If G is solvable and A and B are Hall subgroups of G, then $A^x B^y \subset G$ for all x and y in G.*

Proof. (i) If x and y are in G, then $yx^{-1} = kh^{-1}$ for some $h \in H$ and $k \in K$. Hence $A^x B^y \subset G$ iff $AB^{yx^{-1}} \subset G$ iff $AB^{kh^{-1}} \subset G$ iff $A^h B^k \subset G$ iff $AB \subset G$ This implies (i).

(ii) By Hall's theorems, 9.3.10 and 9.3.11, there are conjugates A^x and B^y such that $A^x B^y \subset G$. The conclusion now follows from (i).

13.2.7. *If $G = HK$ is finite, $A \lhd H$, $B \lhd K$, and $L = N(\langle A, B \rangle)$, then $L = (L \cap H)(L \cap K)$.*

Proof. Let $g \in L$. Then $g = hk^{-1}$ with $h \in H$ and $k \in K$. If $R = \langle A, B \rangle$, $R^{hk^{-1}} = R$, so $R^h = R^k$. Now

$$R^h \supset A^h = A, \qquad R^h = R^k \supset B^k = B.$$

Hence $R^h \supset \langle A, B \rangle = R$. Since G is finite, $R^h = R$. Therefore, $h \in L \cap H$, $k = g^{-1}h \in L \cap K$, and $g \in (L \cap H)(L \cap K)$. Thus, $L = (L \cap H)(L \cap K)$. ‖

The main theorem can now be proved. The most difficult case is when the two subgroups have relatively prime orders. This case and proof are due to Wielandt [5] and will be given first.

13.2.8. *If a group $G = AB$, where A and B are finite nilpotent subgroups of relatively prime orders, then* (1) *G is solvable, and* (2) *if $P \in \mathrm{Syl}(A)$ and $Q \in \mathrm{Syl}(B)$, then $PQ \subset G$.*

Proof. By Theorem 13.2.6, it is sufficient to prove (1). Deny the theorem and let G be a counterexample of smallest order.

(i) *G is simple.*

Let $E < M \lhd G$. G/M is solvable by the inductive hypothesis. Hence M, and therefore AM, is insolvable. If $AM < G$, then $AM = A(M \cap B)$ is solvable by induction. Therefore, $AM = G$, and similarly, $BM = G$. From $AM = G$ it follows that $o(B) \mid o(M)$; similarly $o(A) \mid o(M)$. Hence, $G = M$ and G is simple.

(ii) *$A \notin \mathrm{Syl}(G)$ and $B \notin \mathrm{Syl}(G)$.*

If, for example, $A \in \mathrm{Syl}(G)$, then B is a nilpotent subgroup of G of prime power index, hence G is solvable (Theorem 12.3.4).

(iii) *If $a \in Z(A)^{\#}$ and $b \in Z(B)^{\#}$, then $\langle a, b \rangle = G$.*

Deny. There are P, Q, R, and S with $E \neq R \lhd P \in \mathrm{Syl}(A)$, $E \neq S \lhd Q \in \mathrm{Syl}(B)$, $H = \langle R, S \rangle < G$, and as many of R and S in $\mathrm{Syl}(G)$ as possible. By Theorem 13.2.7, $N(H) = (N(H) \cap A)(N(H) \cap B)$, so $N(H)$ satisfies (1) and (2) by (i) and induction. If P is a p-group, then $N(H) \cap A$ contains a Sylow p-subgroup of $N(H)$, hence by the nilpotence of A, $N(H) \cap P \in \mathrm{Syl}_p(N(H))$. By a similar argument on Q and (2),

$$L = (N(H) \cap P)(N(H) \cap Q) \subset G.$$

Let M be a minimal normal non-E subgroup of L. Since L is solvable, either $M \subset P$ or $M \subset Q$, say $M \subset P$. If $R \neq P$, then by (ii), $G > N(M) \supset \langle U, S \rangle$ where $E \neq U \in \mathrm{Syl}(A)$, $U \neq P$. This contradicts the assumptions on R and S. Hence $R = P$, $M \lhd P$, and

$$N(M) \supset \langle A, S \rangle \supset S^A = S^{BA} = S^G = G$$

by (i), and then contradicting (i).

(iv) *If $e \neq a \in Z(A)$, then $C(a) = A$.*

Deny. Then $C(a) = AB_1$ with $E \neq B_1 \subset B$. Hence $\exists\, b \in (C(a) \cap B)^{\#}$. But then $C(b) \supset \langle a, Z(B) \rangle = G$ by (iii), and G is not simple.

(v) If $A \cap Z(A^g) > E$, then $g \in N(A)$.

Let $a \in A \cap Z(A^g)$, $e \neq a \in P \in \mathrm{Syl}(A)$. By (ii), $\exists\, Q \in \mathrm{Syl}(A)$ with $Q \neq P$ and $\exists\, b \in Z(Q)^{\#}$. By (iv),

$$\langle Q, A^g \rangle \subset C(a) = A^g, \qquad Q \subset A^g.$$

By (iv) again, $\langle A, A^g \rangle \subset C(b) = A$. Hence $A^g = A$ and $g \in N(A)$.

(vi) If $Z(A) \subset H$ and $P \in \mathrm{Syl}(A)$, then $P \cap H \in \mathrm{Syl}(H)$.

Otherwise there is a Sylow subgroup Q of H with $Q > P \cap H$, hence two distinct Sylow subgroups P and P^* of G containing $Z(P)$. This contradicts (v).

(vii) If $Z(A) \subset H < G$ and $N_H(H \cap P) \subset A$ for all $P \in \mathrm{Syl}(A)$, then $H \subset A$.

By (vi), $H \cap A$ is a nilpotent Hall subgroup of H, and by assumption and (vi), $N_H(Q) = H \cap A$ for all $Q \in \mathrm{Syl}(H \cap A)$. Since $Z(A) \subset H \cap A$, $H \cap A$ is not a Sylow subgroup of H by (ii). By Theorem 13.2.3, $\exists\, K \lhd H$ such that $H = (H \cap A)K$, $H \cap A \cap K = E$. Thus, $o(K) \mid o(B)$. By Theorem 9.3.14, $\exists\, g \in G$ such that $K^g \subset B$. Now $g = ab$ with $a \in A$ and $b \in B$, so $K^a \subset B$. By (iii),

$$N(K^a) \supset \langle Z(B), H^a \rangle \supset \langle Z(B), Z(A) \rangle = G.$$

Hence $K = E$, $H = H \cap A$, and $H \subset A$.

(viii) If $p \in \mathscr{P}$, then $A \cap N(B)$ contains at most one subgroup of order p.

By (iv) applied to B, $A \cap N(B)$ is isomorphic to a regular automorphism group of $Z(B)$. Thus $(A \cap N(B))Z(B)$ is a Frobenius group with kernel $Z(B)$ and complement $A \cap N(B)$. By Theorem 12.6.15, $A \cap N(B)$ has at most one subgroup of order p.

(ix) If $A \cap N(B) \cap A^g > E$, then $g \in N(A)$.

Let $D \subset A \cap N(B) \cap A^g$ with $o(D) = p \in \mathscr{P}$. If $ab^{-1} \in N(D)$ with $a \in A$, $b \in B$, then $D^a \subset A$, $D^a = D^b \subset N(B)$, so $D^a \subset A \cap N(B)$. By (viii), $D^a = D$ and $a \in N(D) \cap A$. Hence, $N(D) = (N(D) \cap A)(N(D) \cap B)$. By (ii), $\exists\, Q \in \mathrm{Syl}(A)$ such that $Q \not\supset D$. Then $Q \subset N(D)$, and, by induction, QP is a solvable group if P is a Sylow subgroup of $N(D) \cap B$ ($P \neq E$ if possible). Hence, $\exists\, K \lhd QP$, $K \neq E$ such that $K \subset Q$ or $K \subset P$. If $K \subset P$, $R \in \mathrm{Syl}(B)$, $E \neq R \not\supset P$, then $N(K) \supset \langle Q, R \rangle = G$, by (iii). Hence, $K \subset Q$ and $\exists\, x \in K \cap Z(Q)$. Now $\langle x \rangle^P \subset K \subset Q \subset A$. By (v), $P \subset N(A)$. Since this is true for all $P \in \mathrm{Syl}(N(D) \cap B)$, $N(D) \cap B \subset N(A)$. Therefore,

$$N(D) = (N(D) \cap A)(N(D) \cap B) \subset N(A).$$

Since $D \subset A^g$, $Z(A^g) \subset N(D) \subset N(A)$. But A is a normal Hall subgroup of $N(A)$ and $o(Z(A^g)) \mid o(A)$, so that $Z(A^g) \subset A$. By (v), $g \in N(A)$.

(x) (1) is true.

WLOG $o(A) > o(B)$. Since G is simple, $\exists\, g \notin N(A), g \in G$. If $A \cap A^g = E$, then $o(AA^g) = o(A)^2 > o(G)$. Hence, $A \cap A^g > E$. By (v), $Z(A^g) \cap A = E$, so that $H = C(A \cap A^g) \nsubseteq A$. Also, $Z(A) \subset H$, hence by (vii), $\exists\, P \in \mathrm{Syl}(A)$ such that $N_H(H \cap P) \nsubseteq A$. By (vi), $H \cap A \in \mathrm{Hall}\,(H)$. Therefore (since $N_H(H \cap P) > H \cap A$) $\exists\, x \in N_H(P \cap H)$ such that $o(x) \mid o(B)$, $x \neq e$. Now $Z(P) \subset H \cap P = (H \cap P)^x \subset A^x$. By (v), $x \in N(A)$. By Theorem 9.3.14, $\exists\, y \in G$ such that $x^y \in B$. Since $G = AB$, *WLOG*, $y \in A$. Therefore, $x^y \in N(A) \cap H^y \cap B$ or

$$E < N(A) \cap B \cap C(A \cap A^{gy}).$$

This last group equals $(N(A) \cap B \cap C(A \cap A^{gy}))^a$ for all $a \in A \cap A^{gy}$. Hence, $E < B \cap N(A) \cap B^a$ for all $a \in A \cap A^{gy}$. By (ix), $A \cap A^{gy} \subset N(B)$, so that $E < A \cap A^{gy} \cap N(B)$. By (ix) again, $gy \in N(A)$, hence $g \in N(A)$, a contradiction. ‖

The proof of the theorem in the general case is due to Kegel [1].

13.2.9. *If a finite group is the product of two nilpotent subgroups, then it is solvable.*

Proof. Deny the theorem, and let $G = AB$, where A and B are nilpotent subgroups, be a counterexample of minimum order. By the preceding theorem, $\exists\, p \in \mathscr{P}$ such that $p \mid o(A)$ and $p \mid o(B)$. Let $P \in \mathrm{Syl}_p(A)$ and $Q \in \mathrm{Syl}_p(B)$.

Let $H = P^G$. If $AH < G$, then $AH = AB_1$ with $B_1 < B$, so that AH, and therefore H, is solvable by the minimality of G. Moreover, $G/H = (AH/H)(BH/H)$ is solvable, hence G is solvable, a contradiction. Therefore, $AH = G$, and similarly $BH = G$. Thus $G/H = AH/H = BH/H$ is nilpotent with Sylow p-subgroup $PH/H = QH/H$. Since $H \supset P$, this implies that $H = PH = QH$, and $Q^G \subset H = P^G$. By symmetry, $P^G \subset Q^G$, so that $P^G = Q^G = H$.

By Theorem 13.2.5, $PQ \subset G$, hence, by Theorem 13.2.6, $PQ^x \subset G$ for all $x \in G$. Let P_1 be a maximal p-subgroup of H containing P such that $P_1 Q^x \subset G$ for all $x \in G$. Then $P_1^G = H$. Moreover $N(P_1) \nsupseteq H$ since otherwise H is generated by p-subgroups normal in H (namely conjugates of P_1), hence is a p-group, therefore solvable, and G/H is solvable. Since $Q^G = H$, $\exists\, x$ and y such that $y \in Q^x \backslash N(P_1)$. Now $P_1^y Q^{xt} \subset G$ for all $t \in G$, so that $\langle P_1, P_1^y \rangle Q^{xt} \subset G$ for all $t \in G$. But $H \supset \langle P_1, P_1^y \rangle > P_1$, so that, by the defining property of P_1, $\langle P_1, P_1^y \rangle$ is not a p-group. However,

$$\langle P_1, P_1^y \rangle Q^x = \langle P_1, Q^x \rangle = P_1 Q^x$$

is a p-group, a contradiction.

13.2.10. *If a finite group $G = AB$, where A and B are nilpotent subgroups at least one of which is proper, then there is a proper normal subgroup H of G containing A or B.*

Proof. Deny and induct. G is solvable by the preceding theorem. Let M be a minimal normal non-E subgroup of G. M is a p-group for some prime p. If $AM < G$, say, then since $G/M = (AM/M)(BM/M)$, by the inductive hypothesis there is a proper normal subgroup K/M of G/M containing AM/M or BM/M. But then K is a proper normal subgroup of G containing A or B, a contradiction. Hence $AM = BM = G$.

If G is a p-group and $A < G$, then any maximal proper subgroup $K \supset A$ is normal. Hence $\exists q \in \mathscr{P}$, $q \neq p$, such that $q \mid o(G)$. Let $Q \in \mathrm{Syl}_q(A)$ and $R \in \mathrm{Syl}_q(B)$. Since $AM = BM = G$ and $q \nmid o(M)$, $Q \in \mathrm{Syl}_q(G)$ and $R \in \mathrm{Syl}_q(G)$. By Theorem 13.2.5, $QR \in \mathrm{Syl}_q(G)$. Hence, $Q = R$. Therefore, $N(Q) \supset AB = G$. By the first paragraph of the proof, applied to a minimal normal subgroup of G contained in Q, $A = AQ = G$, $B = BQ = G$, a contradiction.

13.2.11. *Let $\mathscr{A}$ be a hereditary class of solvable groups and $\mathscr{B}$ a hereditary class of groups such that if $L = AB$ is a finite group with $A \in \mathscr{A}$ and $B \in \mathscr{B}$, then L is solvable. If $G = HK$ is a group, $H \in \mathscr{A}$, $K \in \mathscr{B}$, and K finite, then G is solvable.*

Proof. $[G:H]$ is finite. Hence G/M is finite with $M = \mathrm{Core}(H)$. Now $G/M = (H/M)(KM/M)$, $H/M \in \mathscr{A}$, and $KM/M \in \mathscr{B}$. By assumption, G/M is solvable. Since $M \subset H$, $M \in \mathscr{A}$, so M is solvable. Therefore, G is solvable.

13.2.12. *If a group G is the product of two nilpotent subgroups, at least one of which is finite, then G is solvable.*

Proof. Let $\mathscr{A} = \mathscr{B}$ be the class of nilpotent groups. This is hereditary (Theorems 6.4.5 and 6.4.6). By Kegel's theorem, 13.2.9, the hypotheses of Theorem 13.2.11 are satisfied, hence G is solvable. ‖

It is an unsolved problem as to whether a group which is the product of two infinite nilpotent subgroups is always solvable.

EXERCISES

13.2.13. (Kegel [1].) If a finite group G contains subgroups $A \neq E$ and $B \neq E$ such that AB^x is a proper subgroup of G for all $x \in G$, then there is a proper normal subgroup of G containing A or B. (See the last part of the proof of Theorem 13.2.9.)

13.2.14. Give an example of a finite simple group G with nontrivial subgroups A and B such that $\langle A, B^x \rangle < G$ for all $x \in G$. [Take $G = \mathrm{Alt}(5)$ and $A = B$ of order 2.] This shows that the statement obtained from Exercise 13.2.13 by replacing AB^x by $\langle A, B^x \rangle$ is false.

13.2.15. Show that $\mathrm{Sym}(4) = HK$ where $o(H) = 4$, $o(K) = 6$, and neither H nor K is contained in a proper normal subgroup of G. Compare with Theorem 13.2.10.

13.2.16. The product of two nilpotent groups is not necessarily supersolvable.

13.2.17. Theorem 13.2.4 becomes false for three factors. Let $G = \mathrm{Alt}(5)$, H be the 4-group on $\{1, 2, 3, 4\}$, $K = \langle (1, 2, 3) \rangle$, $L = \langle (1, 2, 3, 4, 5) \rangle$, $H_1 = H$, $K_1 = \langle (3, 4, 5) \rangle$, and $L_1 = L$.

 (a) $HK = \mathrm{Alt}(4)$, $G = HKL$.

 (b) $H_1 K_1 \cap L_1 \neq E$, $G \neq H_1 K_1 L_1$.

 (c) $H_1 \sim H$, $K_1 \sim K$, $L_1 \sim L$ ($\sim$ denotes conjugacy).

 (d) $\nexists\, g \in G$ such that $H_1 = H^g$, $K_1 = K^g$, and $L_1 = L^g$.

13.2.18. Theorem 13.2.10 is valid for infinite groups provided B, say, is finite. (Use Theorem 6.4.10.)

13.2.19. If $G = AB$, A is a finite nilpotent group, and B is a solvable p-group, then G is solvable.

13.3 Special cases of nilpotent times nilpotent

Theorems relating to the product of two cyclic groups, two Abelian groups, or an Abelian group and a nilpotent group of class 2 are proved in this section.

13.3.1. *If $G = AB$ where A and B are cyclic groups and B is finite, then G is supersolvable.*

Proof. If A is infinite, then $\mathrm{Core}(A)$ is a normal cyclic subgroup and $G/\mathrm{Core}(A)$ is the product of two finite cyclic groups. By Theorem 7.2.14, it suffices to consider the case where A is finite.

Let $A = \langle a \rangle$, $B = \langle b \rangle$, and induct on $o(G)$. *WLOG*, $o(A) \geq o(B)$. If $A^b \cap A = E$, then $o(AA^b) = o(A)^2 \geq o(G)$, hence $G = AA^b$. By 13.2.4, $G = A$, so that G is supersolvable. If $A^b \cap A \neq E$, then $\exists H \subset A^b \cap A$ with $o(H) = $

$p \in \mathscr{P}$. Hence, $H^{b-1} \subset A$, and since A as a cyclic group has only one subgroup of order p, $H^{b-1} = H$. Therefore, $N(H) \supset \langle A, b^{-1} \rangle = G$. By the inductive hypothesis, $G/H = (A/H)(BH/H)$ is supersolvable, hence (Theorem 7.2.14) G is also. ‖

This theorem is also true if both A and B are infinite (see Redei [1] and Cohn [1]).

13.3.2. (*Itô* [3].) *If a group* $G = AB$ *where* A *and* B *are* (*possibly infinite*) *Abelian subgroups, then* $G^2 = E$.

Proof. By Theorem 3.4.6, $[A, B] \triangleleft G$. Now G^1 is generated by commutators $[a_1 b_1, a_2 b_2]$ where $a_i \in A$ and $b_i \in B$, hence, by Theorem 3.4.2, $G^1 \subset [A, B]$. Therefore, $G^1 = [A, B]$.

Let $a_i \in A$ and $b_i \in B$ for $i = 1, 2$, and let $b_1^{a_2} = a_3 b_3$, $a_1^{b_2} = b_4 a_4$. Then, by Theorem 3.4.1,

$$[a_1, b_1]^{a_2 b_2} = [a_1, b_1^{a_2}]^{b_2}$$

$$= [a_1, a_3 b_3]^{b_2} = [a_1, b_3]^{b_2}$$

$$= [a_1^{b_2}, b_3] = [b_4 a_4, b_3]$$

$$= [a_4, b_3];$$

$$[a_1, b_1]^{b_2 a_2} = [b_4 a_4, b_1]^{a_2} = [a_4, b_1]^{a_2}$$

$$= [a_4, a_3 b_3] = [a_4, b_3].$$

Hence,

$$[a_1, b_1]^{a_2 b_2 a_2^{-1} b_2^{-1}} = [a_1, b_1].$$

Therefore, $[A, B] = G^1$ is commutative and $G^2 = E$. ‖

Both Sym(3) and Alt(4) are the product of two Abelian subgroups. The first example shows that such a product is not always Abelian, and the second that it need not be supersolvable.

13.3.3. *If* $G = AB \neq E$ *is finite and* A *and* B *are Abelian subgroups, then* $\exists H \triangleleft G$ *such that* $E \neq H$ *and* $H \subset A$ *or* $H \subset B$.

Proof. If $A = G$ or $B = G$, the result is trivial. Hence *WLOG*, $E < A < G$ and $E < B < G$. Induct on $o(G)$. We assert:

(∗) $\exists$ $U \triangleleft G$ such that $Z(U) \neq E$ and $U \supset A$ or $U \supset B$.

If $A \cap G^1 \neq E$, then, since G^1 is Abelian by Theorem 13.3.2,

$$C(A \cap G^1) \supset AG^1.$$

Since $AG^1 \supset G^1$, $AG^1 \lhd G$, and $Z(AG^1) \supset A \cap G^1 \neq E$. Hence $(*)$ is true in this case. *WLOG*, therefore, $A \cap G^1 = B \cap G^1 = E$. By Theorem 13.2.10, there is a proper normal subgroup U of G containing A, say. Then $U = AB_1$ with $B_1 < B$. By the inductive hypothesis, $\exists L \lhd U$ such that $E \neq L$ and $L \subset A$ or $L \subset B_1$. Now

$$A \cap U^1 = B \cap U^1 = E, \qquad L \cap U^1 = E.$$

But $[U, L] \subset L \cap U^1 = E$. Therefore, $[U, L] = E$, that is $L \subset Z(U)$. This proves $(*)$.

The subgroup U guaranteed by $(*)$ contains A, say, so that $U = AB_2$, $B_2 \subset B$. If $ab \in Z(U)$, $a \in A$, $b \in B_2$, and $b' \in B_2$, then

$$(b'a)b = abb' = (ab')b, \qquad b'a = ab',$$

hence, $C(a) \supset AB_2 = U$ and $a \in Z(U)$. Therefore, $b \in Z(U)$ also, and $Z(U) = A_1 + B_3$, where $A_1 \subset A$ and $B_3 \subset B_2$. If $B_3 = E$, then $Z(U) = A_1 \subset A$, and since $Z(U)$ is characteristic in U, $Z(U) \lhd G$, and the theorem is true. If $B_3 \neq E$, then $B_3 \subset Z(U)$, so

$$C(B_3) \supset UB \supset AB = G, \qquad B_3 \subset Z(G),$$

so B_3 is the required normal subgroup. ‖

The last theorem is trivially true if A is infinite and B is finite [let $H = \mathrm{Core}(A)$]. However it appears to be an unsolved problem as to whether Theorem 13.3.3 or 13.2.10 holds for the case of two infinite Abelian groups A and B. The same doubt holds as to Theorem 13.3.3 for the case where A and B are finite nilpotent subgroups, or even finite p-groups.

In connection with Kegel's theorem, it has been conjectured that if $G = AB$ where A and B are finite nilpotent subgroups of classes m and n, respectively, then $G^{m+n} = E$. By Itô's theorem, this is true if A and B are both Abelian. No further case is known in which the conjecture is certainly valid. The next theorem gives a rather poor estimate in case A is Abelian and B is nilpotent of class 2.

13.3.4. *If a finite group $G = AB$ where A is an Abelian subgroup, B a nilpotent subgroup of class 2, and $o(B^1) = \pi p_j^{i_j}$, where the p_j are distinct primes, then $G^n = E$ where $n = 2(1 + \Sigma\, i_j)$.*

Proof. Let $(B^1)^G = Q$. Then $G/Q = (AQ/Q)(BQ/Q)$, AQ/Q is Abelian, and $BQ/Q \cong B/(B \cap Q)$ is Abelian since $B \cap Q \supset B^1$. Therefore, by Itô's theorem, 13.3.2, $G^2 \subset Q$.

Let L be any subgroup of B^1 such that $L^G \cap B^1 = L$ (B^1 is such a subgroup). Since $L \subset B^1 \subset Z(B)$, $L \lhd B$. Hence, $L^G = L^A$. For $a_i \in A$ and $c_i \in L$, we have

$$[a_1^{-1}c_1a_1,\ a_2^{-1}c_2a_2] \in [A, L]^G$$

by Theorem 3.4.2. By Theorem 3.4.1, if $a \in A$, $a' \in A$, $b \in B$, and $c \in L$, then

$$[a, c]^{a'} = [aa', c][a', c]^{-1} \in [A, L],$$
$$[a, c]^{b} = [ab, c][b, c]^{-1} = [b'a'', c]$$
$$= [a'', c] \in [A, L],$$

where $a'' \in A$ and $b' \in B$. Hence $[A, L]^{G} = [A, L]$. Therefore, by earlier remarks,

$$(L^{G})^{1} = (L^{A})^{1} \subset [A, L]^{G} = [A, L] \subset L^{G}.$$

By Kegel's theorem, G is solvable, hence L^{G} is solvable. Therefore,

$$(L^{G})^{2} \subset [A, L]^{1} < [A, L] \subset L^{G}. \tag{1}$$

We assert that $\exists \, M \subset B^{1}$ such that

$$[A, L]^{1} = M^{G}, \, M^{G} \cap B^{1} = M. \tag{2}$$

Granting this for the moment, it follows from (1) that $M^{G} < L^{G}$ while

$$M = M^{G} \cap B^{1} \subset L^{G} \cap B^{1} = L.$$

Therefore, $M < L$. Moreover, $(L^{G})^{2} \subset [A, L]^{1} = M^{G}$. Therefore, by an easy induction, $((B^{1})^{G})^{n-2} = E$. Hence $G^{n} = E$, as asserted. It remains only to prove the existence of $M \subset B^{1}$ such that (2) holds. In fact, only the first equation in (2) need be verified, for M may be replaced by $M^{G} \cap B^{1}$ if necessary.

As in the proof of Itô's theorem, let $a_{i} \in A$, $b_{i} \in L$, $i = 1, 2$, $b_{1}^{a_{2}} = a_{3}b_{3}$, $a_{1}^{b_{2}} = b_{4}a_{4}$, where $a_{i} \in A$ and $b_{i} \in B$ for $i = 3, 4$. Then by Theorem 3.4.1,

$$[a_{1}, b_{1}]^{b_{2}a_{2}} = [b_{4}a_{4}, b_{1}]^{a_{2}} = [a_{4}, b_{1}]^{a_{2}}$$
$$= [a_{4}, a_{3}b_{3}] = [a_{4}, b_{3}],$$
$$[a_{1}, b_{1}]^{a_{2}b_{2}} = [a_{1}, a_{3}b_{3}]^{b_{2}} = [a_{1}, b_{3}]^{b_{2}}$$
$$= [b_{4}a_{4}, b_{3}] = [b_{4}, b_{3}]^{a_{4}}[a_{4}, b_{3}].$$

Therefore, using $\sim$ to denote conjugacy in G,

$$[[a_{1}, b_{1}], [b_{2}^{-1}, a_{2}^{-1}]] \sim ([a_{1}, b_{1}]^{-1})^{a_{2}b_{2}}[a_{1}, b_{1}]^{b_{2}a_{2}}$$
$$= [a_{4}, b_{3}]^{-1}([b_{4}, b_{3}]^{-1})^{a_{4}}[a_{4}, b_{3}]$$
$$\sim [b_{4}, b_{3}]^{-1}.$$

Therefore the group $[A, L]^{1}$ is normally generated by a certain subset S of B^{1}. Hence $[A, L]^{1} = M^{G}$ for some subgroup M of B^{1}, and the theorem follows. ‖

Some corollaries of this theorem are indicated in the exercises.

EXERCISES

13.3.5. If $G = AB$ is finite, A Abelian, and B Hamiltonian, then $G^4 = E$.

13.3.6. Let A be elementary Abelian of order 9, B the (Hamiltonian) quaternion group. Prove that there is a Frobenius group $G = AB$ with $G^2 \neq E$ and $G^3 = E$. (The gap between this exercise and the last represents an unsolved problem.)

13.3.7. State the corollaries of Theorem 13.3.4 for the case where A is infinite and B finite; for the case where A is finite and B infinite. (See Theorem 13.2.11.)

13.4 Nilpotent maximal subgroups

If a finite group G has a nilpotent, maximal proper subgroup whose Sylow 2-subgroup is Abelian, then G is solvable. This theorem, due essentially to Thompson [1] (see Janko [1]) is beyond the scope of this book. However, a useful special case, due to Huppert [4], will be presented.

13.4.1. *If G is a transitive monomial group of degree p which is a regular p-group, $y \in G$ is diagonal, and $o(y) = p \in \mathscr{P}$, then $\mathrm{Det}(y) = 1$.*

Proof. Since G is transitive, $\exists\, x \in G$, whose induced permutation is a p-cycle. Then $H = \langle x, y \rangle$ satisfies the hypotheses of the theorem. The subgroup D of all diagonal matrices in H is normal in H. The subgroup D_1 of all $z \in D$ such that $z^p = e$ is characteristic in D, hence normal in H. By Theorem 3.4.2, H^1 is normally generated by $[x, y]$. Since $y \in D_1$, $[x, y] \in D_1$, so that $H^1 \subset D_1$. Therefore, by Theorem 7.5.6, $\exists\, u \in H^1$ such that

$$(xy)^p = x^p y^p u^p = x^p.$$

Now

$$x = \begin{bmatrix} 0, & a_1 & & & \\ & & \cdot & \cdot & \\ & & & \cdot & \cdot \\ & & & & \cdot \;\; a_{p-1} \\ a_p & & & & 0 \end{bmatrix}, \quad y = \begin{bmatrix} c_1 & & & \\ & \cdot & & \\ & & \cdot & \\ & & & \cdot \\ & & & & c_p \end{bmatrix},$$

so that $(xy)^p = a_1 \dots a_p c_1 \dots c_p I = \mathrm{Det}(y) x^p$. Hence $\mathrm{Det}(y) = 1$.

13.4.2. *If G is a finite group, P is a regular Sylow p-subgroup, $N(P) = P$, and P/Q is Abelian, then $\exists\, H \lhd G$ such that $H \cap P = Q$ and $G/H \cong P/Q$.*

Proof. CASE 1. P/Q is cyclic. There is a monomial representation of G on (P, Q) of degree $[G: P]$ with all non-zero entries $(o(P/Q)$)th roots of 1, and with kernel contained in Q. Using induction, it may be assumed that the representation is faithful, hence *WLOG, G* is itself the monomial group. Let $(x_1, \ldots, x_n)$ be the basis of the vector space in use, and X_i the subspace spanned by x_i. By Theorem 10.2.9 and the definition of the monomial representation on (P, Q), P has just one orbit of length 1, say $\{X_1\}$.

Let K be a transitive constituent of P of degree greater than 1. The subgroup R of P, consisting of all elements whose K-constituent is diagonal, is normal in P. Let uR be an element of order p in $Z(P/R)$. As a normal subgroup of a transitive group, $\langle uR \rangle$ moves all letters of K (Theorem 10.1.7), hence u does also. Therefore the K-constituent of u acts on the corresponding X_j's as a product of p-cycles. Let S be one of these cycles. If $y \in P$ fixes a letter X_j of S, then for all k,

$$X_j u^k y = X_j y u^k = X_j u^k,$$

since $uR \in Z(P/R)$. Hence, y fixes all letters of S. Applying these remarks to all transitive constituents of P, one obtains a partition

$$\{X_2, \ldots, X_n\} = \dot\cup \, S_j, \qquad o(S_j) = p,$$

such that if $y \in P$ fixes one letter of S_j, then it fixes all of them, and such that each S_j is an orbit for some $u_j \in P$.

Now let $y \in P$. Then $x_1 y = c_1 x_1, c_1 \in \mathscr{C}$. To save time, the term "diagonal entry" will mean "nonzero diagonal entry."

(i)   If all diagonal entries of y are pth roots of 1, then the product of the diagonal entries is c_1.

For suppose that y fixes all X_j in S_i. Then the corresponding constituent of $\langle y, u_i \rangle$ is a transitive monomial group of degree p which is a regular p-group, hence $\mathrm{Det}(y \mid S_i) = 1$, by Theorem 13.4.1. Taking the product over all relevant S_i, one has (i).

(ii)   If all diagonal entries of y are pth roots of 1, then they are all equal.

Let $x_i y = c_i x_i$. By the transitivity of G, $\exists g \in G$ such that $x_1 g = d x_i$. Hence

$$x_1 g y g^{-1} = c_i x_1,$$

and by a similar argument, the diagonal entries of $g y g^{-1}$ are those of y rearranged. Since $g y g^{-1}$ fixes X_1, $g y g^{-1} \in P$, so the product of the diagonal entries of $g y g^{-1}$ equals c_i by (i). Therefore, $c_1 = c_i$, and (ii) holds.

(iii)   If y has one diagonal entry 1, than all diagonal entries are 1.

For if not, then some power y^{p^m} of y has all diagonal entries pth roots of 1 but not all equal to 1, contradicting (ii).

Let $f(y)$ be the product of the nonzero entries of y [thus $f(y) = \pm \text{Det}(y)$].

(iv)   If $c_1^p = 1$, then $f(y) = c_1$.

By (iii), all diagonal entries of y^p equal 1. Hence all diagonal entries of y are pth roots of 1. By (i), the product of the diagonal entries is c_1. Let z be any cycle of y,

$$x_i z = c_i x_{i+1}, \ldots, x_{i+p^j-1} z = c_{i+p^j-1} x_i, \qquad j > 0.$$

Then $x_i z^{p^j} = (\pi c_k) x_i$. Thus, y^{p^j} has some diagonal entries πc_k, but all diagonal entries 1 by (iii). Therefore, $\pi c_k = 1$. Using this fact for all cycles z of y, one gets $f(y) = c_1$ as asserted.

(v) $f(y) = 1$ iff $c_1 = 1$.

If $c_1 = 1$, then $f(y) = 1$ by (iv). If $c_1 \neq 1$, then some y^m satisfies (iv) with $c_1^m \neq 1$, so that $f(y^m) \neq 1$, hence $f(y) \neq 1$.

Thus f is a homomorphism of G with kernel H, say, and $G/H \overset{\sim}{\subset} P/Q$ by the construction of the monomial representation. But $P \cap H = Q$ since $f(y) = 1$ iff $c_1 = 1$. Therefore, $G/H \cong P/Q$.

CASE 2. P/Q is not cyclic.

In any case, P/Q is Abelian, hence the direct sum of cyclic p-groups. There are normal subgroups $Q_1, \ldots, Q_r$ of P such that P/Q_i is cyclic and $\cap Q_i = Q$. By Case 1, $\exists H_i \lhd G$ such that $H_i \cap P = Q_i$ and $G/H_i \cong P/Q_i$. Let $H = \cap H_i$. Then $H \lhd G$ and

$$P \cap H = P \cap (\cap H_i) = \cap Q_i = Q.$$

Since each G/H_i is a p-group, so is G/H; hence H contains all Sylow q-subgroups for $q \neq p$. Therefore, $G = HP$, and

$$\frac{G}{H} = \frac{HP}{H} \cong \frac{P}{H \cap P} = \frac{P}{Q}.$$

13.4.3. *If G is a finite group, H is a nilpotent maximal proper subgroup of G, and all Sylow subgroups of H are regular, then G is solvable.*

Proof. Induct on $o(G)$. If $\exists P \in \text{Syl}(H)$ such that $P \neq E$ and $N(P) > H$, then $N(P) = G$, G/P is solvable by inductive hypothesis, so G is solvable. Now suppose that $N(P) = H$ for all $P \in \text{Syl}(H)$. Therefore H is a Hall subgroup of G. If H is not a Sylow subgroup of G, then, by Theorem 13.2.3, there is a normal complement L of H in G.

Suppose that $H \in \text{Syl}(G)$. Let K be a minimal normal non-E subgroup of H such that $\exists M \lhd G$ with $M \cap H = K$ (note that $H \lhd H$, $G \lhd G$, and $G \cap H = H$, hence K exists). If $N(K) = G$, then G/K is solvable by Theorem 7.5.2 and the inductive hypothesis. If $N(K) < G$, then $N(K) = H$ by the maximality of H, hence $N_M(K) = K$. By Theorem 13.4.2, $\exists L \lhd M$ such that

$L \cap K = K^1$ and $M/L \cong K/K^1$. Since L is the smallest normal subgroup of M whose factor group M/L is an Abelian p-group, L is characteristic in M, hence normal in G, and $L \cap H = K^1$. By the minimality of K, $K^1 = E$. Hence, again, H has a normal complement L.

Let $Q \in \text{Syl}(L)$. By Theorem 6.2.4, $G = N(Q)L$, so that $o(H) \,|\, o(N(Q))$. Since $N(Q) \cap L$ is a normal Hall subgroup of $N(Q)$ [all Sylow q-subgroups of $N(Q)$ for $q \,|\, o(L)$ are contained in L], by the Schur splitting theorem, there is a complement R of $N(Q) \cap L$ in $N(Q)$. Since $o(R) = o(H)$, R is a conjugate of H (Theorem 9.3.14). Therefore, R is also a maximal proper subgroup of G, and $N(Q) = G$ since it properly contains R. Since all Sylow subgroups of L are normal, L is nilpotent. Since $G/L \cong H$, G is solvable. $\|$

The theorem becomes false if the hypothesis of regularity is omitted entirely, as the following theorem shows.

13.4.4. *If* $p = 2^n - 1$ *is prime and* $n \geq 5$, *then a Sylow 2-subgroup* P *of* $G = PSL(2, p)$ *is a maximal proper subgroup of* G.

Proof.

$$o(G) = \frac{(p + 1)p(p - 1)}{2} = 2^n(2^n - 1)(2^{n-1} - 1).$$

Hence $o(P) = 2^n$. Let F be a field of order p^2 containing J_p as a subfield. There is an automorphism T of F given by $xT = x^p$, and T fixes all elements of J_p and no others (Theorem 10.6.12). Since $F^\#$ is cyclic of order $p^2 - 1$, $\exists a \in F^\#$ of multiplicative order $p + 1 = 2^n$. Since $4 \nmid o(J_p^\#)$, $a \notin J_p$. Now $a + a^p$ is fixed by T, so $a + a^p = u \in J_p$. Hence

$$a^2 - ua + 1 = a^2 - a(a + a^p) + 1$$
$$= 1 - a^{p+1} = 0.$$

Since $a \notin J_p$, $x^2 - ux + 1$ is the minimum polynomial of a over J_p. Since there are $(p + 1)/2$ squares in J_p and only $(p - 1)/2$ nonsquares, $\exists b \in J_p$ such that $u^2 - 4 - 4b^2$ is a square. Then $\exists c \in J_p$ such that

$$c^2 - uc + (1 + b^2) = 0.$$

Let $d = u - c$ and $y = \begin{bmatrix} c & b \\ b & d \end{bmatrix}$. Then y has characteristic polynomial

$$x^2 - (c + d)x + cd - b^2 = x^2 - ux + cu - c^2 - b^2$$
$$= x^2 - ux + 1.$$

But there are just two roots a and a^{-1} of $x^2 - ux + 1 = 0$ in F. Hence the characteristic roots of y^{p+1} are 1 and 1. Therefore, $y^{p+1} = I$. By Theorem

12.2.16, some conjugate z of y in $GL(2, p^2)$ is diagonal. Now z has the same characteristic equation and characteristic roots as y. Hence, by a further conjugation if necessary,

$$z = \begin{bmatrix} a & 0 \\ 0 & a^{-1} \end{bmatrix},$$

and $o(y) = o(z) = 2^n$.

If $A \in SL(2, p)$, let A' be the corresponding element of G. Now $y \in SL(2, p)$, so $y' \in G$ and $o(y') = 2^{n-1}$ since $z^{2^{n-1}} = -I$. Let

$$v = \begin{bmatrix} 0 & -1 \\ 1 & 0 \end{bmatrix}.$$

Then $v' \in G$, $o(v') = 2$, and since $v^{-1}yv = y^{-1}$, $v'^{-1}y'v' = y'^{-1}$. It follows that P is a dihedral group.

A direct computation shows that if

$$w = \begin{bmatrix} c & 0 \\ 0 & c^{-1} \end{bmatrix}', \qquad c \neq \pm 1, 0, \qquad c \in J_p,$$

then $C_G(w)$ is the group D' where D is the group of all diagonal matrices in $SL(2, p)$. If q is a prime dividing $2^{n-1} - 1$, then D' contains a Sylow q-subgroup. Therefore there are no elements of order $2q$.

Assume that $P < H < G$. If $p \mid o(H)$, then G has a permutation representation of degree at most $2^{n-1} - 1$ which must be faithful since G is simple. This is impossible since G contains elements of the prime order $p = 2^n - 1$. If $P \lhd H$, then since the cyclic subgroup $\langle y' \rangle$ is characteristic in P, it is normal in H. Since $o(\mathrm{Aut}(\langle y' \rangle))$ is a power of 2 (Theorem 5.7.12), by the N/C theorem, $C(\langle y' \rangle)$ is divisible by some prime q dividing $2^{n-1} - 1$. Hence there is an element of order $2q$, a contradiction. Thus, in particular, $N(P) = P$. Hence,

$$n_2(H) = [H{:}P] \mid 2^{n-1} - 1.$$

Now (since $n > 3$), no divisor of $2^{n-1} - 1$ (except 1) is congruent to $1 \pmod{2^{n-2}}$. Hence (Theorem 6.5.3) there is a conjugate P^* of P in H with $P^* \neq P$ and $8 \mid o(P \cap P^*)$. WLOG, $P \cap P^*$ is a maximal such intersection. It is then either cyclic or dihedral and, in either case, contains a characteristic cyclic subgroup K of order at least 4. Then $N(K) \supset N(P \cap P^*)$, which contains more than one Sylow 2-subgroup. Therefore an odd prime divides $o(N(K))$, a contradiction as before, since $P \subset N(K)$. It follows that P is a maximal proper subgroup of G.

13.4.5. *If a non-Abelian group G has an Abelian maximal proper subgroup A, $x \in G$, and $A^x \neq A$, then $A \cap A^x = Z(G)$.*

Proof. If $A^x < A$, then $A < A^{x^{-1}}$, so $A^{x^{-1}} = G$, an impossibility, Therefore $A^x \nsubseteq A$. Now

$$C(A \cap A^x) \supset \langle A, A^x \rangle = G, A \cap A^x \subset Z(G).$$

If $y \in Z(G)$, then $\langle A, y \rangle$ is Abelian, and $\langle A, y \rangle \neq G$ since G is not Abelian. so $y \in A$. Thus $Z(G) \subset A$ and $Z(G) = Z(G)^x \subset A^x$. Therefore, $Z(G) \subset A \cap A^x$, and finally, $Z(G) = A \cap A^x$.

13.4.6. *If a finite group G has an Abelian maximal proper subgroup A, then $G^3 = E$.*

Proof. If $A \lhd G$, then G/A is of prime order, so $G^1 \subset A$ and $G^2 = E$. Suppose that $A \ntriangleleft G$. Then $N(A) = A$. By Theorem 13.4.5, any two different conjugates of A intersect in $Z(G)$. Thus, $H = G/Z(G)$ has an Abelian maximal proper subgroup $B = A/Z(G)$ such that $N(B) = B$ and $B \cap B^x = E$ if $x \in H \backslash B$. Therefore, H is a Frobenius group with Frobenius kernel M, say: $H = MB$, $M \cap B = E$ (Theorem 12.5.11). Since an Abelian p-group is regular, H is solvable by Theorem 13.4.3. Let K be a minimal normal non-E subgroup of H contained in M. Since H is solvable, K is Abelian. Then $BK > B$, so $BK = H$. Therefore $H/K \cong B$, $H^1 \subset K$, and $H^2 = E$. Hence, $G^2 \subset Z(G)$, $G^3 = E$.

EXERCISES

13.4.7. (Yonaha [1].) If H is a nilpotent, maximal proper subgroup of a finite solvable group G, then there are normal subgroups A and B such that $A \subset B$, B/A is Abelian, and A and G/B are nilpotent. (Use Theorem 10.5.21.)

13.4.8. Theorem 13.4.6 is best possible. Let Q be the quaternion group, T an automorphism of Q of order 3, and $G = \text{Hol}(Q, \langle T \rangle)$. Show that G has an Abelian maximal proper subgroup of order 6, but $G^2 \neq E$.

13.4.9. There is no simple group of order 756. (Use Section 7.5 and Theorem 13.4.3.)

13.4.10. If a finite group G has a nilpotent, maximal subgroup H which is not a Hall subgroup, then G has a normal p-subgroup $K \neq E$ for some prime p.

13.5   Transfer theorems

In this section a theorem of Grün, similar to, but stronger than, Burnside's theorem 6.2.9 about the existence of normal complements, will be proved.

13.5.1. *If G is a finite group and $p \in \mathscr{P}$, then there is a subgroup H such that G/H is a p-group and such that H is minimum with this property. There is a minimum subgroup K such that G/K is an Abelian p-group.*

Proof. If G/A and G/B are p-groups and $g \in G$, then $\exists\, m$ such that $g^{p^m} \in A$. Then $\exists\, n$ such that $(g^{p^m})^{p^n} \in B$. Therefore the element $g(A \cap B)$ of $G/(A \cap B)$ has order dividing p^{m+n}. Thus, $G/(A \cap B)$ is also a p-group. The first statement now follows by induction. The second statement is proved similarly. ‖

The (characteristic) minimum subgroup K of G such that G/K is an Abelian p-group will be denoted by $G^1(p)$.

13.5.2. *If G is finite, $p \in \mathscr{P}$, $P \in \mathrm{Syl}_p(G)$, T is the transfer of G into P, and*

$$L = \langle P \cap N(P)^1, \{P \cap (P^1)^x \mid x \in G\}\rangle,$$

then $P \cap \mathrm{Ker}(T) = L = P \cap G^1$, $\mathrm{Ker}(T) = G^1(p)$, and

$$GT \cong \frac{G}{G^1(p)} \cong \frac{P}{P \cap G^1}.$$

Proof. We have

$$G = PG^1(p), \qquad \frac{G}{G^1(p)} \cong \frac{P}{P \cap G^1(p)} = \frac{P}{P \cap G^1},$$

where the fact has been used that $[G^1(p):G^1]$ is prime to p.

Next, the equation $G^1(p) = \mathrm{Ker}(T)$ follows from the other equations; for $G/\mathrm{Ker}(T)$ is an Abelian p-group, hence $\mathrm{Ker}(T) \supset G^1(p)$. Also,

$$G/G^1(p) \cong P/(P \cap G^1) = P/(P \cap \mathrm{Ker}(T))$$
$$\cong P\,\mathrm{Ker}(T)/\mathrm{Ker}(T) \subset G/\mathrm{Ker}(T).$$

Hence $o(\mathrm{Ker}(T)) \leqq o(G^1(p))$, so that $\mathrm{Ker}(T) = G^1(p)$.

The other isomorphism now follows, since

$$GT \cong G/\mathrm{Ker}(T) = G/G^1(p).$$

Now $P \cap N(P)^1 \subset P \cap G^1$, and by normality of G^1, $P \cap (P^1)^x \subset P \cap G^1$ for all $x \in G$. Hence $L \subset P \cap G^1$. Since $G/\mathrm{Ker}(T)$ is an Abelian p-group, $\mathrm{Ker}(T) \supset G^1$, so

$$P \cap G^1 \subset P \cap \mathrm{Ker}(T).$$

It thus remains only to show that $P \cap \mathrm{Ker}(T) \subset L$. Deny this and let y be of smallest order in $(P \cap \mathrm{Ker}(T)) \backslash L$. We wish to calculate $yT \pmod{L}$, so it should be noted that $P^1 = P \cap (P^1)^e \subset L$, and $L \lhd P$.

Let $G = \bigcup \{x_i P \mid x_i \in M\}$, and let U be the permutation representation of P on M associated with the transfer T. Recall (Theorem 3.5.6 and its proof) that the x_i can be so chosen that

$$yT = \pi\{(x_i^{-1} y^{n_i} x_i)P^1 \mid x_i \in R\},$$

where R contains just one x_i from each cycle of yU, and n_i is the smallest natural number such that $x_i^{-1} y^{n_i} x_i \in P$. Therefore, n_i is a power of p. If $n_i > 1$, then $x_i^{-1} y^{n_i} x_i \in P \cap \mathrm{Ker}(T)$ and has order less than that of y, hence is in L.

Let S be an orbit of P of length greater than 1, and therefore a power of p. The number of 1-cycles of yU in S is a multiple of p. Let xP be one such 1-cycle (assuming one exists), and zxP any other, with $z \in P$. Then

$$y^x \in P, \; y^{zx} \in P,$$
$$(zx)^{-1} y(zx) = x^{-1}[z, y^{-1}]x(x^{-1}yx) \in P.$$

Since $x^{-1}[z, y^{-1}]x \in P \cap (P^1)^x \subset L$, all of the factors y^{zx} are $\equiv x^{-1}yx \pmod{L}$. Hence the contribution of the orbit S to yT is $(x^{-1}yx)^{ep} \equiv e \pmod{L}$ by the minimality property of y again.

There remain the factors of yT arising from orbits of P of length 1. But an orbit of P is of length 1 iff it is contained in $N(P)$. Therefore, from what has been said,

$$yT \equiv \pi\{x_i^{-1} y x_i \mid x_i \in N(P) \cap M\}.$$

But $x_i^{-1} y x_i = y[y, x_i]$, and $[y, x_i] \in P \cap N(P)^1 \subset L$, so $x_i^{-1} y x_i \equiv y \pmod{L}$. Therefore,

$$yT \equiv y^{[N(P):P]} \pmod{L}.$$

Since $p \nmid [N(P):P]$ and $y \notin L$, $yT \not\equiv e \pmod{L}$. But $yT = P^1 \subset L$, so we have reached a contradiction. ‖

A finite group is *p-normal* iff $P \in \mathrm{Syl}_p(G)$, $P^* \in \mathrm{Syl}_p(G)$, and $Z(P) \subset P^*$ imply that $Z(P) \subset Z(P^*)$. It follows that $Z(P) = Z(P^*)$ for such P and P^*. Any finite group with an Abelian Sylow p-subgroup is p-normal. Sym(4) is an example of a group which is not 2-normal.

13.5.3. *If P and P^* are Sylow p-subgroups of a finite group G and $Z(P) \triangleleft P^*$, then $Z(P) = Z(P^*)$.*

Proof. Both P and P^* are Sylow p-subgroups of $N(Z(P))$. Hence $\exists \, x \in N(Z(P))$ such that $P^* = P^x$. Now,

$$Z(P) = (Z(P))^x = Z(P^x) = Z(P^*).$$

13.5.4. *(Grün.) If G is p-normal and $P \in \mathrm{Syl}_p(G)$, then the maximum Abelian factor p-group of G is isomorphic to that of $N(Z(P))$.*

Proof. By Theorem 13.5.2,

$$G/G^1(p) \cong P/(P \cap G^1),$$

$$N(Z(P))/N(Z(P))^1(p) \cong P/(P \cap N(Z(P))^1) = P/P_1,$$

$$P_1 = P \cap N(Z(P))^1 \subseteq P \cap G^1.$$

It remains only to show that $P \cap G^1 \subseteq P_1$.

Let T and T_1 be the transfers of G and $N(Z(P))$ into P. By Theorem 13.5.2,

$$P \cap G^1 = P \cap \mathrm{Ker}(T) = L = \langle P \cap N(P)^1, \{P \cap (P^1)^x\}\rangle,$$

$$P_1 = P \cap \mathrm{Ker}(T_1).$$

Now $Z(P)$ is characteristic in P, hence normal in $N(P)$, so $N(Z(P)) \supset N(P)$, $N(Z(P))^1 \supset N(P)^1$, and

$$P \cap N(P)^1 \subseteq P_1. \tag{1}$$

Let $x \in G$ and $Q = P \cap (P^1)^x$. Then $Z(P) \subseteq N(Q)$ and $Z(P)^x = Z(P^x) \subseteq N(Q)$. There are Sylow p-subgroups R and R_1 of $N(Q)$ such that $Z(P) \subseteq R$ and $Z(P)^x \subseteq R_1$. Then $\exists y \in N(Q)$ such that $R_1^y = R$. Thus,

$$Z(P)^{xy} \subseteq R \subseteq S \in \mathrm{Syl}_p(G), \qquad Z(P) \subseteq R \subseteq S.$$

By p-normality, $Z(P) = Z(S) = Z(P)^{xy}$, hence $xy \in N(Z(P))$. Then

$$Q = Q^y = P^y \cap (P^1)^{xy}, \qquad Q \subseteq (P^1)^{xy} \subseteq N(Z(P))^1,$$

$$Q \subseteq P \cap N(Z(P))^1 = P_1.$$

Thus,

$$P \cap (P^1)^x \subseteq P_1, \qquad x \in G. \tag{2}$$

It follows from (1) and (2) that $P \cap G^1 = L \subseteq P_1$, and we are done.

13.5.5. *If P is an Abelian Sylow p-subgroup of a finite group G, and T is the transfer of G into P, then $GT \cong P \cap Z(N(P))$.*

Proof. Let U be the transfer of $N(P)$ into P, and let $N(P) = \dot\cup\, x_i P$. If $y \in P$, then $yU = \pi x_i^{-1} y x_i$. If also $u \in N(P)$, then $N(P) = \dot\cup\, x_i u P$, and

$$u^{-1}(yU)u = \pi(x_i u)^{-1} y (x_i u) = yU.$$

Therefore, $PU \subseteq Z(N(P)) \cap P$. If $e \neq y \in Z(N(P)) \cap P$, then

$$yU = y^{[N(P):P]} \neq e.$$

Therefore, U is an isomorphism on $Z(N(P)) \cap P$, $PU = Z(N(P)) \cap P$, and

$$P = (P \cap Z(N(P))) + (P \cap \mathrm{Ker}(U)). \tag{3}$$

Since P is Abelian, G is p-normal. By Grün's theorem 13.5.4, Theorem 13.5.2, the fact that $Z(P) = P$, and (3),

$$GT \cong G/G^1(p) \cong N(P)/N(P)^1(p) \cong P/(P \cap N(P)^1(p))$$
$$= P/(P \cap \mathrm{Ker}(U)) \cong P \cap Z(N(P)). \parallel$$

There is a theorem of Burnside that gives information about the case when G is not p-normal (see Theorem 13.5.7), which will be given shortly.

Let H be a subgroup of G. The *quasi-centralizer* $\mathrm{Qua}(H)$ of H in G is the set of all $g \in G$ such that $K^g = K$ for all subgroups K of H. $\mathrm{Qua}(H)$ is a subgroup of $N(H)$, and, if G is finite, it is the set of all $g \in G$ such that if $h \in H$, then $h^g = h^n$ for some integer n. Since the elements of $N(H)$ induce permutations of $\mathrm{Lat}(H)$, there is a natural permutation representation T of $N(H)$ on $\mathrm{Lat}(H)$. Then $\mathrm{Qua}(H) = \mathrm{Ker}(T)$, hence $\mathrm{Qua}(H) \lhd N(H)$ and $N(H)/\mathrm{Qua}(H)$ is isomorphic to this group of permutations.

13.5.6. *If G is a finite group and H is a normal subgroup of one Sylow p-subgroup P and a nonnormal subgroup of a second Sylow subgroup Q, then there is a p-subgroup K such that $N(K)/\mathrm{Qua}(K)$ is not a p-group.*

Proof. WLOG, Q is such that $K = N_Q(H)$ is as large as possible. Then $H < K < Q$ and $K < N_Q(K)$. Let $L = H^{N(K)}$. Since $N_Q(H) = K < N_Q(K)$, $H < L$. Also, $L \lhd N(K)$, so $N(K) \subset N(L)$. Since $H < L = H^{N(K)}$, $N(K) \nsubseteq N(H)$. Therefore, $N(L) \nsubseteq N(H)$, and $r = [N(L):N(H) \cap N(L)] \neq 1$. Since $L = H^{N(K)} \subset K^{N(K)} = K$, L is a p-group.

We have

$$K \subset N(K) \cap N(H) \subset N(L) \cap N(H) \subset N(L) \subset G.$$

Hence there are Sylow p-subgroups R, S, T, and U of $N(K) \cap N(H)$, $N(L) \cap N(H)$, $N(L)$, and G, respectively, such that

$$K \subset R \subset S \subset T \subset U.$$

$N(H)$ contains a Sylow p-subgroup P of G by assumption, while K is a p-subgroup of $N(H)$ which is not Sylow. Hence there is a p-subgroup of $N(H)$ larger than K in which K is normal, that is $K < R$. Again,

$$N_U(H) = U \cap N(H) \supset R \cap N(H) = R > K.$$

By the maximality property of K, $H \lhd U$. Hence $T \subset U \subset N(H)$, so $T \subset N(H) \cap N(L)$. Therefore, $T = S$, that is, the same power of p which divides $o(N(L))$ also divides $o(N(L) \cap N(H))$. Therefore, $(r, p) = 1$.

Now, certainly, $\mathrm{Qua}(L) \subset N(H)$, since $H \subset L$ and the automorphisms induced by $\mathrm{Qua}(L)$ fix all subgroups of L. Hence

$$\mathrm{Qua}(L) \subset N(H) \cap N(L).$$

Therefore, $r \mid o(N(L)/\mathrm{Qua}(L))$ and, as asserted, $N(L)/\mathrm{Qua}(L)$ is not a p-group.

13.5.7. *If a finite group G is not p-normal, then there is a p-subgroup $H \neq E$ such that $N(H)/\mathrm{Qua}(H)$ is not a p-group.*

Proof. There are Sylow p-subgroups P and Q such that $Z(P) \subset Q$ but $Z(P) \neq Z(Q)$. By Theorem 13.5.3, $Z(P) \ntrianglelefteq Q$. Therefore, by Theorem 13.5.6, there is a p-subgroup H with the described property. ‖

Some applications of the preceding theorems will now be given.

13.5.8. *If $o(G) = 1040$, then G is not simple.*

Proof. Deny. We have $o(G) = 2^4 \cdot 5 \cdot 13$. If there is a permutation representation of G of degree 13, then by Theorem 12.9.2, it is 2-transitive, hence $12 \mid o(G)$, a contradiction. Therefore a Sylow 2-subgroup P is a maximal proper subgroup of G. Hence, $N(Z(P)) = P$. If G is 2-normal, then by Grün's theorem 13.5.4, $G/G^1(2) \cong P/P^1(2) \neq E$, a contradiction. Hence, G is not 2-normal. By Theorem 13.5.7, there is a 2-subgroup H such that $N(H)/\mathrm{Qua}(H)$ is not a 2-group. WLOG, $H \subset P$. If $o(H) = 2^3$ or 2^4, then $N(H) > P$, a contradiction. Now there is an element $x \in N(H)$ which induces a permutation of prime order $q = 5$ or 13 on the set of subgroups of H. Since $H^x = H$ and $E^x = E$, H must have at least five nontrivial subgroups. This is impossible if $o(H) = 2$ or 2^2.

13.5.9. (*Huppert* [4].) *If G is a finite group, all of whose proper subgroups are supersolvable, then G is solvable.*

Proof. Let G be a counterexample of smallest order. If $E < H < G$ and $H \triangleleft G$, then all proper subgroups of G/H are supersolvable (Theorem 7.2.4), hence G/H is solvable by assumption, H is supersolvable, and therefore G is solvable, a contradiction. Hence G is simple.

Let p be the smallest prime divisor of $o(G)$, and let $P \in \mathrm{Syl}_p(G)$. Then $N(Z(P)) < G$ by the simplicity of G, hence $N(Z(P))$ is supersolvable. By the Sylow tower theorem 7.2.19, $N(Z(P))$ has a normal subgroup of index p. If G is p-normal, then by Grün's theorem, G has a normal subgroup of index p, a contradiction.

Therefore, G is not p-normal. By Theorem 13.5.7, there is a p-subgroup H of G and an $x \in N(H)\backslash C(H)$ with $o(x) = q$, where $p \nmid q$. Since $N(H) < G$, $N(H)$ is supersolvable. By the Sylow tower theorem again, there is a normal p-complement K in $N(H)$. Therefore, $H + K$ exists. Now $x \in K$, since K contains all elements of $N(H)$ of order prime to p, hence $x \in C(H)$, a contradiction. ‖

This theorem has a nice corollary. A *chain* of subgroups of G: $H_0 < H_1 < \ldots < H_n$ is of *length* n, and is maximal iff no further terms can be added, i.e., iff $H_0 = E$, $H_n = G$, and each H_i is maximal proper in H_{i+1}.

13.5.10. (*Iwasawa* [3].) *If G is a finite group, then all of its maximal chains of subgroups have the same length iff G is supersolvable.*

Proof. If G is supersolvable and $(H_0, \ldots, H_n)$ is a maximal chain, then each H_i is supersolvable and each index $[H_{i+1}:H_i]$ is a prime (Theorem 7.2.8). Hence all such maximal chains have the same length (the largest possible length).

Suppose that all maximal chains have the same length, and induct. If $H < G$, then any maximal chain of H is an initial segment of one for G, hence all maximal chains of H have the same length. By the inductive hypothesis, all proper subgroups H of G are supersolvable. By Theorem 13.5.9, G is solvable. Thus, G has a normal chain $(H_0, \ldots, H_n)$, with all H_{i+1}/H_i of prime order. Therefore, by the assumption, all maximal chains $(K_0, \ldots, K_n)$ have each index $[K_{i+1}:K_i]$ prime. In particular, all maximal subgroups of G are of prime index, hence (Theorem 9.3.8) G is supersolvable.

13.5.11. *If G is a Frobenius group with Frobenius kernel M and complement H, and M is p-normal for all primes p dividing $o(M)$, then M is nilpotent.*

Proof. Let G be a counterexample of smallest order. Then $o(H) = q \in \mathscr{P}$ (Theorem 12.6.6), and $H \in \mathrm{Syl}_q(G)$ (Theorem 12.6.1). Let $P \in \mathrm{Syl}_p(M)$. Then (Theorem 6.2.4), $N(P)M = G$, so $q \mid o(N(P))$ and $WLOG$, $H \subset N(P)$.

Suppose $Z(P) \lhd M$. By the p-normality of M, $Z(P)$ is the common center of all Sylow p-subgroups of M, hence is characteristic in M and normal in G. By Theorem 12.6.6, $G/Z(P)$ is Frobenius with Frobenius kernel $M/Z(P)$. By induction, $M/Z(P)$ is nilpotent, hence M is solvable. Therefore (Theorem 12.6.13), M is nilpotent, a contradiction.

Hence $Z(P) \ntrianglelefteq M$. Therefore,

$$G > N(Z(P)) = H(N(Z(P)) \cap M),$$

$$N(Z(P)) \cap M \lhd N(Z(P)).$$

By Theorem 12.6.6, $N(Z(P))$ is a Frobenius group with kernel $N(Z(P)) \cap M$. By induction, $N(Z(P)) \cap M$ is nilpotent. Therefore, $N(Z(P)) \cap M$ has a nontrivial Abelian factor p-group. By Grün's theorem, $M > M^1(p)$. Now $M^1(p)$ is characteristic in M, hence normal in G, so that $HM^1(p)$ is Frobenius. Therefore, by induction, $M^1(p)$ is nilpotent, so that M is solvable. By Theorem 12.6.13, M is nilpotent, a contradiction.

13.5.12. *If G is a finite group, $E \neq P \in \mathrm{Syl}_p(G)$, $Q \in \mathrm{Syl}_p(G)$, $H \lhd P$, $H \subset Q$, and G has a normal p-complement K, then $H \lhd Q$.*

Proof. Deny. By Theorem 13.5.6, there is a p-subgroup L of P such that $N(L)/C(L)$ is not a p-group. Hence $\exists\, g \in N(L)\backslash C(L)$ such that $p \nmid o(g)$. Therefore, $g \in K$. Thus $[L, g] \subset L \cap K = E$, so $g \in C(L)$, a contradiction.

13.5.13. *If a finite group G has a normal p-complement $(p \in \mathscr{P})$, then G is p-normal.*

Proof. This follows from Theorems 13.5.3 and 13.5.12 and the definition of p-normality.

13.5.14. *If $o(G) = p^n m, p \in \mathscr{P}, p \nmid m, k = \pi\{(p^i - 1) \mid 1 \leq i \leq n\}$, and $(k, m) = 1$, then G has a normal p-complement.*

Proof. Induct. If any $x \in G$ induces an automorphism T of prime order $q \neq p$ in some p-subgroup H, then, since T is a product of q-cycles and 1-cycles, it follows that $q \mid (o(H) - o(C_H(x)))$, hence $q \mid k$, a contradiction. Therefore (Theorem 13.5.7), G is p-normal.

Let $P \in \mathrm{Syl}_p(G)$. Suppose first that $Z(P) \lhd G$. Then by induction, $G/Z(P)$ has a normal p-complement, $R/Z(P)$. By the first part of the proof, $Z(P) \subset Z(R)$. By Schur's splitting theorem, there is a complement S of $Z(P)$ in R. But then, $R = Z(P) + S$, so S is characteristic in R, hence normal in G. S is then the desired normal p-complement of G.

Now suppose that $N(Z(P)) < G$. By induction, $N(Z(P))$ has a normal p-complement, hence a nontrivial Abelian factor p-group. By Grün's theorem, G has a nontrivial factor p-group, G/T. Then, by induction, T has a normal p-complement U. By the usual argument, $U \lhd G$, and U is a normal p-complement of G.

13.5.15. *If G is a simple group of even order (not 2), then 12, 16, or 56 divides $o(G)$.*

Proof. Let $o(G) = 2^n m$ where m is odd. Suppose $16 \nmid o(G)$. If $n = 1$, then G has a normal 2-complement by Burnside's theorem. If $n \neq 1$, then in the notation of the preceding theorem, either $n = 2$ and $k = 3$, or $n = 3$ and $k = 21$. Since G does not have a normal 2-complement, Theorem 13.5.14 implies that either $2^2 \cdot 3 = 12$ or $2^3 \cdot 7 = 56$ divides $o(G)$. ∥

By the theorem of Feit and Thompson mentioned earlier, this can be improved to read: the order of a nonsolvable finite group is divisible by 12, 16, or 56. It was only in 1960 that the first simple (noncyclic) groups of order not divisible by 12 were discovered by Suzuki [3]. In fact, he gave an infinite set of such groups. Some of them have orders not divisible by 56.

EXERCISES

13.5.16. There is no simple group of order 2268. (Apply Grün's theorem, Theorem 13.5.7 for $p = 3$, Exercise 10.7.14, and Theorem 6.5.5).

13.5.17. Show that Theorem 13.5.7 implies that if G is not p-normal, then there are a p-subgroup H and an $x \in G$ such that $p \nmid o(x)$, and x induces an automorphism of H of prime order $q \neq p$.

13.5.18. (a) If $H \subset K \subset G$, then $\mathrm{Qua}(K) \subset \mathrm{Qua}(H)$.

   (b) If H is Abelian and $G = \mathrm{Dih}(H)$, then $\mathrm{Qua}(H) = G$.

   (c) If H is Abelian, $x \in H$ has order 2, and $G = \mathrm{Dic}(H, x)$, then $\mathrm{Qua}(H) = G$.

   (d) A *Dedekind* group is one which is Abelian or Hamiltonian. Show that if $H \subset G$ and $\mathrm{Qua}(H) \supset H$, then H is a Dedekind group.

13.5.19. Prove Burnside's theorem, 6.2.9, from Grün's theorem and Schur's splitting theorem.

13.5.20. If G is p-normal and $P \in \mathrm{Syl}_p(G)$, then

$$N(Z(P)) \cap G^1(p) = N(Z(P))^1(p).$$

13.6   Generalized dihedral times odd-order nilpotent

In this section, two product theorems will be proved, the first of which is (somewhat inaccurately) indicated by the title of the section.

13.6.1. If $G = AB$, G is finite, A is a nilpotent subgroup of odd order, and B has a subgroup H of index 2 such that $B \subset \mathrm{Qua}(H)$, then G is solvable.

Proof. One verifies that any subgroup or factor group of B is either Dedekind (hence nilpotent) or of the same type as B. Moreover, B is solvable. Now let G be a smallest counterexample.

If $E \neq M \lhd G$, then by induction and Theorem 13.2.9, G/M is solvable. Therefore, G has no normal solvable subgroup except E.

Let D be a maximal proper subgroup of G containing A. If $D \cap H \neq E$, then

$$(D \cap H)^G = (D \cap H)^{BA} = (D \cap H)^A \subset D.$$

Since $D = A(D \cap B)$ is solvable by induction, this yields a normal solvable subgroup, a contradiction. Hence $D \cap H = E$.

If $K = D \cap B \neq E$, then $B = KH$, $G = AB = AKH = DH$, and $o(K) = 2$. Represent G on H. The representation is faithful, since otherwise its kernel is a subgroup of H, hence is solvable. Since $D \cap H = E$ and $G = DH$, D is regularly represented (Exercise 13.1.17). Since $o(D)$ is twice an odd number, an element x of order 2 in D is represented as a product of an odd number of 2-cycles. Hence there is an odd permutation in the representation. Therefore (Theorem 10.4.7), G has a normal subgroup M of index 2. Now $M = A(B \cap M)$ is solvable by induction, a contradiction.

Therefore, $D \cap B = E$. Since $A \subset D$ and $G = AB$, $A = D$ and A is a maximal proper subgroup of G. Moreover, $A \cap B = E$. If $E \neq P \in \mathrm{Syl}_p(A)$, then $N(P) = A$ by the maximality of A and the absence of a normal solvable subgroup. Therefore such a P is a Sylow subgroup of G. Since the product of P and some Sylow p-subgroup of B is a Sylow p-subgroup of G (Theorem 13.2.5), it follows that $(o(A), o(B)) = 1$. Therefore, A and B are Hall subgroups of G.

Let $E < M \lhd G$. If $AM < G$, then $AM = A(B \cap AM)$ is solvable by induction, hence M is solvable. Therefore, $AM = G$, so $o(B) \mid o(M)$. Similarly $o(A) \mid o(M)$, and $M = G$. Therefore, G is simple.

If $A \notin \mathrm{Syl}(G)$, then by Theorem 13.2.3, there is a normal complement of A in G, and G is not simple. Hence $A \in \mathrm{Syl}_p(G)$ for some p. If $2 \nmid o(H)$, then $2 \mid o(G)$ but $2^2 \nmid o(G)$. By $\nmid$ Theorem 6.2.11, $\exists K \lhd G$ with $[G:K] = 2$, so G is not simple. Therefore, $2 \mid o(H)$, and $\exists x \in H$ such that $o(x) = 2$. By the assumptions on H and B, $\langle x \rangle \lhd B$, hence $x \in Z(B)$. Therefore, $o(Cl(x)) = p^i$ for some i. By Theorem 12.3.2, G is not simple, a contradiction.

13.6.2. *If $G = AB$ is a finite group, A is a Dedekind group, $[B:H] = 2$, and $B \subset \mathrm{Qua}(H)$, then G is solvable.*

Proof. Let G be a counterexample of least order. By induction, G has no normal solvable subgroup. If $B \subset D < G$ and $D \cap A \neq E$, then D is solvable by induction, and

$$(D \cap A)^G = (D \cap A)^B \subset D,$$

since all subgroups of A are normal in A. Therefore, $(D \cap A)^G$ is a normal solvable subgroup, a contradiction. Hence, B is a maximal proper subgroup, and $A \cap B = E$.

If A is a maximal proper subgroup, then $N(A) = A$. If $x \notin A$ and $A \cap A^x \neq E$, then $N(A \cap A^x) = G$, since all subgroups of A and A^x are normal. Hence $A \cap A^x = E$ for all $x \notin A$. By the theorem of Frobenius, 12.5.11, $B \lhd G$ (see Theorem 13.2.4), a contradiction. Hence, $\exists M$ with $A < M < G$. If $H \cap M \neq E$, then $(H \cap M)^G \subset M$, M is solvable by induction, hence $(H \cap M)^G$ is a normal solvable subgroup. Therefore, $H \cap M = E$. Thus, $M = AK$, where $o(K) = 2$ and $K \subset B$. Also, $G = MH$. If also $A < M_1 < G$, then by the above remarks, $[M_1:A] = 2$ also, so $N(A) \supset \langle M, M_1 \rangle > M$, hence by the above remarks again, $N(A) = G$, a contradiction. Therefore, M is the only subgroup properly containing A, and $M = N(A)$.

We assert

$$A \cap A^x = E \text{ if } x \notin M. \tag{1}$$

In fact, since $M = N(A)$, $A^x \neq A$. If (1) is false, then $N(A \cap A^x) \supset \langle A, A^x \rangle > A$, so $N(A \cap A^x) = M$ and $A^x \subset M$. Since $[M:A^x] = 2$, $N(A^x) \supset \langle M, M^x \rangle$ which is greater than M since $N(M) = M$. Hence, $N(A^x) = G$, a contradiction.

$$H \cap H^x = E \text{ if } x \notin B. \tag{2}$$

If (2) is false, then $N(H \cap H^x) = B$ by the maximality of B and the assumptions on H and B. Hence $H^x \subset B$, and since $[B:H^x] = [B:H] = 2$, $H^x \lhd B$. Therefore, $N(H^x) \supset \langle B, B^x \rangle = G$, a contradiction.

Now let $a = o(A)$ and $h = o(H)$, so that $o(G) = 2ah$. For some subset S of $G \backslash B$,

$$G = B \, \dot{\cup} \, (\dot{\cup} \, \{HxH \mid x \in S\}).$$

By Theorem 3.3.10, there are $[H:H^x \cap H] = h$ right cosets of H in HxH for $x \in S$. Hence, counting elements, we have

$$2ah = 2h + sh^2, \qquad s \in \mathcal{N}.$$

Similarly from (1), we get $2ah = 2a + ta^2$, $t \in \mathcal{N}$. Therefore,

$$a = \frac{4 + 2s}{4 - st}, \qquad h = \frac{4 + 2t}{4 - st},$$

so that $st < 4$. If $(s, t) = (1, 1)$ or $(1, 2)$, then $a = 2$ or 3, hence $[G:B] \leq 3$, so that, since $\text{Core}(B) = E$, G has a faithful permutation representation of degree ≤ 3, hence is solvable. If $(s, t) = (1, 3)$, then $a = 6$ and $h = 10$. Since H is a Dedekind group of order 10, it is cyclic. Thus, G has a faithful permutation representation of degree 6, and an element of order 10, an impossibility. The cases $(s, t) = (2, 1)$ and $(s, t) = (3, 1)$ are similar.

EXERCISES

13.6.3. If $G = AB$, $E < M \lhd A$, and $M \subset A \cap B$, then $\exists \, H \lhd G$ such that $M \subseteq H \subset B$.

13.6.4. If $G = AB$ is finite, $A \cap B = E$, and A is a Dedekind maximal proper subgroup, then $\exists \, H \lhd G$ such that $H \neq E$ and $H \subset A$ or $H = B$.

13.7   S-rings

The next few sections will be devoted to the study of Schur rings or S-rings, first studied by Schur [1] and Wielandt [2]. Only finite groups will be considered.

In the study of products of subgroups, the following situation sometimes arises. A finite group $G = AB$, where $A \cap B = E$, $\text{Core}(A) = E$, and A is a maximal proper subgroup. The permutation representation of G on A then has the following properties: the representation is faithful, primitive, and B is regularly represented. Moreover, each coset of A is of the form Ab, so the set of letters might as well be B itself. Thus, essentially, G is a primitive group acting on the set B, and B is a subgroup of G acting on B by the regular representation. The primary goal of the S-ring theory is to show that, with some additional hypotheses on B (but not on A or G), the representation is 2-transitive. This, of course, gives information about the order of G.

Let (M, G) be a permutation group containing a regular subgroup (M, H). Then (Theorem 10.3.2), $(M, H) \cong (H, H)$ where (H, H) means the regular representation of H. By changing the names of the letters of M, we then have a group isomorphic to G acting on H and containing (H, H) as subgroup. Thus WLOG, (H, G) is a permutation group containing the (regular) subgroup (H, H). It will be necessary to distinguish between the product hg of elements $h \in H$ and $g \in G$, and the result h^g of applying the permutation $g \in G$ to the letter h. (The notation x^y for $y^{-1}xy$ will not be used in the next few sections.) Note that if $h \in H$ and $x \in H$, then $h^x = hx$, since we are dealing with the regular representation of H. The content of these remarks is summarized in the next theorem.

13.7.1.   *If G is a primitive overgroup of a regular group H, then, with change of notation, (H, G) is a primitive overgroup of the regular group (H, H), where $h^x = hx$ for $h \in H$ and $x \in H$.* ‖

If G is a finite group, let $J(G)$ denote the group ring of G over the integers, i.e., (see following Theorem 12.2.18) $J(G)$ is the set of elements $\Sigma \{c_g g \mid g \in G\}$, $c_g \in J$, where addition, multiplication, and multiplication by a scalar are defined as in the group algebra. If A is a nonempty subset of G, then $\bar{A}$ is defined by the equation

$$\bar{A} = \Sigma \{1x \mid x \in A\} = \Sigma \{x \mid x \in A\}.$$

If $(H, H) \subset (H, G)$, $x \in G$, and $\Sigma c_h h \in J(H)$, then we define

$$(\Sigma c_h h)^x = \Sigma c_h h^x.$$

13.7.2. *If* $(H, G) \supset (H, H)$, $h \in H$, *and* $g \in G$, *then* h^g *is the unique* $x \in H$ *such that (in* $J(G)$) $\bar{G}_e x = \bar{G}_e hg$. *Therefore,* $\bar{G}_e h^g = \bar{G}_e hg$.

Proof. $e^{G_e h^g} = e^{h^g} = h^g$ and $e^{G_e hg} = e^{hg} = h^g$. Since (Theorem 10.1.2) distinct right cosets of G_e have distinct effects on e, $G_e h^g = G_e hg$. Since $G_e \cap H = E$, h^g is the unique element x of H such that $G_e x = G_e hg$. The statements in the conclusion of the theorem follow.

13.7.3. *If* $(H, G) \supset (H, H)$, $x \in J(H)$, *and* $g \in G$, *then* x^g *is the unique* $y \in J(H)$ *such that* $\bar{G}_e y = \bar{G}_e xg$. *Hence,* $\bar{G}_e x^g = \bar{G}_e xg$.

Proof. Let $x = \Sigma c_h h$. By Theorem 13.7.2,

$$\bar{G}_e x^g = \bar{G}_e \Sigma c_h h^g = \Sigma c_h \bar{G}_e h^g = \Sigma c_h \bar{G}_e hg = \bar{G}_e xg.$$

The cosets $G_e h$ are distinct, so that the elements $\bar{G}_e h$ of $J(G)$ are linearly independent. Let $y = \Sigma d_h h^g \in J(H)$ and $\bar{G}_e y = \bar{G}_e xg$. Then

$$\Sigma d_h \bar{G}_e h^g = \bar{G}_e \Sigma d_h h^g = \bar{G}_e y = \Sigma c_h (\bar{G}_e h^g).$$

Hence $c_h = d_h$ for all $h \in H$, and $y = x^g$. ‖

If $(H, H) \subset (G, H)$, then define $J(H, G_e)$ to be the (free Abelian) additive subgroup of $J(H)$ generated by $\bar{T}_0, \ldots, \bar{T}_n$, where $T_0 = \{e\}$, $T_1, \ldots, T_n$ are the orbits of G_e. It will soon be shown that $J(H, G_e)$ is actually a subring of $J(H)$.

13.7.4. *If* $(H, H) \subset (G, H)$ *and* $x \in J(H)$, *then* $x \in J(H, G_e)$ *iff* $x^g = x$ *for all* $g \in G_e$.

Proof. Let $x = \Sigma c_h h$. Now $x \in J(H, G_e)$ iff, whenever h_1 and h_2 are in the same orbit of G_e, they have the same coefficient in x. But this is true iff $x^g = x$ for all $g \in G_e$.

13.7.5. *If* $(H, H) \subset (H, G)$ *and* $x \in J(H)$, *then the following statements are equivalent:*

(1)　$x \in J(H, G_e)$,

(2)　*if* $u \in G_e$, *then* $\bar{G}_e xu = \bar{G}_e x$.

(3)　$\bar{G}_e x \bar{G}_e = o(G_e)\bar{G}_e x$.

Proof. (1)$\Leftrightarrow$(2). By Theorem 13.7.4, $x \in J(H, G_e)$ iff $x^u = x$ for all $u \in G_e$. By Theorem 13.7.3, this is true iff $\bar{G}_e xu = \bar{G}_e x$.

(2) $\Rightarrow$ (3). $\bar{G}_e x \bar{G}_e = \Sigma \{\bar{G}_e xu \mid u \in G_e\}$
　　　　　　　$= \Sigma \{\bar{G}_e x \mid u \in G_e\}$
　　　　　　　$= o(G_e)\bar{G}_e x.$

(3) $\Rightarrow$ (2). If $u \in G_e$, then

$$o(G_e)\bar{G}_e xu = \bar{G}_e x \bar{G}_e u = \bar{G}_e x \bar{G}_e = o(G_e)\bar{G}_e x.$$

Statement (2) follows.

13.7.6. *If $(H, H) \subseteq (H, G)$ and $x \in J(H)$, then $x \in J(H, G_e)$ iff $\bar{G}_e x = x \bar{G}_e$.*

Proof. Let $x \in J(H, G_e)$. By linearity, it may be assumed that $x = \bar{T}$, where T is an orbit of G_e. By Theorem 13.7.5,

$$\bar{G}_e \bar{T} \bar{G}_e = o(G_e) \bar{G}_e \bar{T}.$$

Hence the set identity $G_e T G_e = G_e T$ holds, and $T G_e \subseteq G_e T$. Since T is a subset of H and $H \cap G_e = E$,

$$o(T G_e) = o(T) o(G_e) = o(G_e T), \qquad T G_e = G_e T.$$
$$\bar{G}_e x = \bar{G}_e \bar{T} = \bar{T} \bar{G}_e = x \bar{G}_e.$$

Now suppose that $\bar{G}_e x = x \bar{G}_e$. Then $\bar{G}_e x \bar{G}_e = \bar{G}_e^2 x = o(G_e) \bar{G}_e x$. By Theorem 13.7.5, $x \in J(H, G_e)$.

13.7.7. *If $(H, H) \subseteq (H, G)$, $x \in J(H, G_e)$, $g \in G$, and $u \in J(H)$, then $(xu)^g = xu^g$.*

Proof. By Theorems 13.7.6 and 13.7.3,

$$\bar{G}_e x u^g = x \bar{G}_e u^g = x \bar{G}_e ug = \bar{G}_e xug.$$

Since $xu^g \in J(H)$, by Theorem 13.7.3 again, $xu^g = (xu)^g$. ∥

Let H be a finite group. An *S-ring* over H is a subring R of $J(H)$ such that there is a partition $H = \cup\, T_i$, $T_i \neq \varnothing$, for which

(i) $T_0 = \{e\}$,

(ii) $T_i^* = T_j$ for some j, where $T_i^* = \{x \mid x^{-1} \in T_i\}$,

(iii) $\{\bar{T}_0, \ldots, \bar{T}_n\}$ is a basis of the free Abelian group $(R, +)$.

13.7.8. *If $(H, H) \subseteq (H, G)$, then $J(H, G_e)$ is an S-ring over H.*

Proof. Let x and y be elements of $J(H, G_e)$ and $u \in G_e$. Then by Theorems 13.7.7 and 13.7.4, $(xy)^u = xy^u = xy$. By Theorem 13.7.4 again, $xy \in J(H, G_e)$. Therefore, $J(H, G_e)$ is a ring. Only (ii) remains to be verified. Let $x \in T_i$, $y \in T_i$, and $x^{-1} \in T_j$. Then $\exists\, g \in G_e$ such that $x^g = y$. Therefore,

$$e^{xgy^{-1}} = x^{gy^{-1}} = y^{y^{-1}} = e, \qquad xgy^{-1} \in G_e,$$
$$(x^{-1})^{xgy^{-1}} = e^{gy^{-1}} = e^{y^{-1}} = y^{-1}.$$

Therefore, $y^{-1} \in T_j$ also. By symmetry, $T_i^* = T_j$. ∥

The decomposition $H = \cup\, T_i$ occurring in the definition of S-ring is determined by the S-ring R itself. An S-ring is *primitive* iff $i > 0$ implies $\langle T_i \rangle = H$. There is a trivial primitive S-ring R over any group H order >1, namely the one with basis $T_0 = \{e\}$ and $T_1 = H^\#$. In fact,

$$T_1^2 = (\bar{H} - e)(\bar{H} - e) = o(H)\bar{H} - 2\bar{H} + e$$
$$= (o(H) - 1)T_0 + (o(H) - 2)T_1,$$

so that R is a ring. The other requirements are immediately verified. This ring will be called the *trivial* primitive S-ring over H, and all others *nontrivial*.

The following theorem is clear.

13.7.9. *If* $(H, H) \subset (H, G)$, *then* G *is 2-transitive iff* $J(H, G_e)$ *is the trivial primitive S-ring.*

13.7.10. *If* $(H, H) \subset (H, G)$, *then* G *is primitive iff* $J(H, G_e)$ *is a primitive S-ring.*

Proof. If G is not primitive, then $\exists L$ such that $G_e < L < G$. Since $G = G_e H$ and $G_e \cap H = E$, $L = G_e K = K G_e$ with $E < K < H$ and $G_e \cap K = E$. Now $\bar{G}_e \bar{K} = \bar{L} = \bar{K} \bar{G}_e$. By Theorem 13.7.6, $\bar{K} \in J(H, G_e)$. Therefore $\bar{K}$ is a sum of some $\bar{T}_j$, hence K contains T_i for some $i > 0$. But then $\langle T_i \rangle \subset K < H$, and $J(H, G_e)$ is not primitive.

Now suppose that $J(H, G_e)$ is not primitive. In any case, it is an S-ring (Theorem 13.7.8). There is a subgroup K with $E < K < H$ and an orbit $T \neq \{e\}$ of G_e such that $\langle T \rangle = K$. If K is not a union of orbits of G_e, then there are an orbit U, an element $a \in U \cap K$, and $b \in U \backslash K$. Since T generates K, some $\bar{T}^m$ contains a in its expansion, hence contains b also (since $J(H, G_e)$ is an S-ring), so $b \in K$, a contradiction. Therefore, K is a set union of orbits, and $\bar{K} \in J(H, G_e)$. By Theorem 13.7.6, $\bar{G}_e \bar{K} = \bar{K} \bar{G}_e$. It follows that $G_e K = K G_e$, and therefore (Theorem 1.6.8) $G_e K$ is a subgroup with $G_e < G_e K < G$. Hence G_e is not a maximal proper subgroup, and therefore (Theorem 10.5.7) G is not primitive. ‖

A Burnside group, or *B-group*, is a finite group H such that any primitive group $(H, G) \supset (H, H)$ is 2-transitive.

13.7.11. *If* H *is a finite group and there is no nontrivial primitive S-ring over H, then H is a B-group.*

Proof. Let (H, G) be a primitive overgroup of (H, H). By Theorem 13.7.10, $J(H, G_e)$ is a primitive S-ring. By assumption, $J(H, G_e)$ is the trivial primitive S-ring. By Theorem 13.7.9, G is 2-transitive. Hence, H is a B-group. ‖

In connection with the last theorem there are two related open questions.

Question 1. Is it true that if H is a B-group, then there are no nontrivial primitive S-rings over H?

Question 2. Is it true that if R is a nontrivial primitive S-ring over a group H, then $R = J(H, G_e)$ for some (primitive) overgroup G of H?

If the answer to Question 2 is yes (which seems doubtful), then the answer to Question 1 is also yes by Theorems 13.7.9 and 13.7.10. Wielandt [6] gives an example of a nonprimitive S-ring R which is not a $J(H, G_e)$.

EXERCISES

13.7.12. If H is a finite simple group of order > 2, then there is a nontrivial primitive S-ring over H whose basis consists of all conjugate classes of H.

13.7.13. Show that $G = \mathrm{Alt}(6)$ furnishes an example of a primitive group which contains no regular subgroup.

13.7.14. The preceding exercise can be improved to read: there is a primitive but not 2-transitive group G which contains no regular subgroup. Let $H = \mathrm{Alt}(5)$, $P \in \mathrm{Syl}_3(H)$, and G the permutation group resulting from the representation of H on $N(P)$.

13.7.15. (Wielandt.) Let $p \in \mathscr{P}$ be such that $2^p - 1 \notin \mathscr{P}$, and $q \in \mathscr{P}$, $q \mid 2^p - 1$. Let F be a field of order 2^p, and let K be the subgroup of $F^{\#}$ of order q. Let G be the group of affine transformations T of the form: $xT = ax + b$, $a \in K$, $b \in F$, $x \in F$.

    (a)   The subgroup H of translations, $xT = x + b$, is regular and normal in G.

    (b) G is a Frobenius group with kernel H and complement (isomorphic to) K.

    (c) $K = G_0$.

    (d)   In $J_q^{\#}$, $o(2) = p$.

    (e) H is a minimal normal non-E subgroup of G [use (d)].

    (f) K is a maximal proper subgroup of G [use (e)].

    (g) G is a primitive but not 2-transitive group.

    (h)   An elementary Abelian group of order 2^p (where $2^p - 1$ is composite) is not a B-group.

Note that $2^{11} - 1 = 23 \cdot 89$, for example, so such primes p exist.

13.8   Primitive S-rings over groups of small order

In this section the nonexistence of nontrivial primitive S-rings over groups of certain small orders will be proved. In fact, if p is a prime such that $2 < p \leq 37$ and $o(H) = p + 1$, then there are no nontrivial primitive S-rings over H. Techniques sufficient to prove this statement will be exhibited (except for the case $o(H) = 38$, whose proof depends on facts given in later sections). However, the proof itself will not be carried out completely (see Exercise 13.8.12). The methods given here are too weak to handle groups of much larger order than those actually treated.

If R is a primitive S-ring over H, and $0 \neq x = \Sigma\, c_h h \in R$ with $c_h \geq 0$, define the *length* of x to be $L(x) = \Sigma\, c_h$, and $S(x)$ to be the subgroup of H

generated by the set of $h \in H$ such that $c_h > 0$. It is true, of course, that $S(x) = H$ or E always, but many of the proofs are by contradiction, and it is convenient to have the symbol available.

13.8.1. *If R is a nontrivial primitive S-ring over a finite group H of nonprime order, and $t \neq e$ is a basis element of R, then $L(t) \geq 3$.*

Proof. Let $o(H) = mn$ with $m > 1$ and $n > 1$. If $L(t) = 1$, then $t = x \in H$, and $\langle x \rangle = S(t) = H$. Therefore, $t^m = x^m \neq e$, and $S(t^m) = \langle x^m \rangle \neq H$, contradicting the primitivity of R.

Suppose that $L(t) = 2$, so that $t = x + y$ with x and y in H. If $y \neq x^{-1}$, then

$$tt^* = 2e + (xy^{-1} + yx^{-1}) = 2e + u,$$

where $u = a + a^{-1}$ is again a basis element. Hence *WLOG*, $y = x^{-1}$. It follows that $H = S(t) = \langle x \rangle$. Assume inductively that $x^i + x^{-i}$ is a basis element of R for $1 \leq i \leq j \leq m - 1$. Then

$$
\begin{aligned}
(x^j + x^{-j})t &= (x^j + x^{-j})(x + x^{-1}) \\
&= (x^{j+1} + x^{-(j+1)}) + (x^{j-1} + x^{-(j-1)}) \in R,
\end{aligned}
$$

so that, by the inductive assumption and the fact that there are no basis elements of length 1 (except e), $x^{j+1} + x^{-(j+1)}$ is a basis element. Therefore, $x^m + x^{-m}$ is a basis element such that $S(x^m + x^{-m}) < H$, which is impossible. ‖

If one knows one coefficient in the multiplication table of the basis elements of R, then two others (in general) may be determined by use of the next lemma.

13.8.2. *If t, u, and v are basis elements of a primitive S-ring R over a group H, with products*

$$tu = iv + \ldots, \quad uv^* = jt^* + \ldots, \quad v^*t = ku^* + \ldots,$$

where i, j, and k are integers, then $iL(v) = jL(t) = kL(u)$.

Proof. We have

$$(tu)v^* = iL(v)e + \ldots, \quad t(uv^*) = jL(t)e + \ldots,$$

hence, $iL(v) = jL(t)$. Similarly, the coefficient of e in v^*tu is $kL(u) = iL(v)$. ‖

If two basis elements (not e) have relatively prime lengths, more can be said. If s and t are elements of $J(H)$, call $s \geq t$ iff $s - t = \Sigma c_h h$ with all $c_h \geq 0$, and $s > t$ if $s \geq t$ but $s \neq t$.

13.8.3. *If* $t \neq e$ *and* $u \neq e$ *are basis elements of a nontrivial primitive S-ring R over a group H of nonprime order, and* $(L(t), L(u)) = 1$, *then*

(1) $tu = iv$, $i \in J$, *and* v *a basis element with* $L(v) > L(t)$ *and* $L(v) > L(u)$.

(2) *If* w *is a basis element, then* $t^*w > u$ *iff* $w = v$, *and* $t^*v = L(t)u + \ldots$.

(3) *If* w *is a basis element, then* $wu^* > t$ *iff* $w = v$, *and* $vu^* = L(u)t + \ldots$.

Proof. Let w be any basis element of T, and let

$$tu = iw + \ldots, \quad t^*w = ku + \ldots, \quad wu^* = jt + \ldots.$$

Then $w^*t = ku^* + \ldots$, and $uw^* = jt^* + \ldots$. By the preceding lemma, $iL(w) = jL(t) = kL(u)$. If any one of i, j, k is 0, so are the other two. If all are nonzero, then, since $L(u) \mid j$ while $j \leq L(u)$, $j = L(u)$ and $k = L(t)$. Let v be a basis element whose coefficient in the expansion of tu is positive: $tu = iv + \ldots$. Then, letting $w = v$ in the above equations,

$$L(iv) = iL(v) = kL(u) = L(t)L(u) = L(tu),$$

and $tu = iv$. The conclusions of the theorem are now evident except for the inequalities in (1).

By (2), $L(t)L(v) \geq L(t)L(u)$, so that $L(v) \geq L(u)$. Similarly by (3), $L(v) \geq L(t)$. Suppose that $L(t) < L(u) = L(v)$, for example. Then by (2), $t^*v = L(t)u$, so that $tt^*v = iL(t)v$ contrary to Exercise 13.8.14. Hence $L(v) > L(u)$ and $L(v) > L(t)$. ‖

This theorem has several corollaries.

13.8.4. *If* u *is the longest basis element and* $t \neq e$ *is a basis element of a nontrivial primitive S-ring over H, and* $o(H) \notin \mathscr{P}$, *then* $(L(t), L(u)) > 1$.

Proof. If $(L(t), L(u)) = 1$, then by Theorem 13.8.3, $tu = iv$ where v is basis and longer then u, an impossibility.

13.8.5. *If* $o(H) = p + 1$, $2 < p \in \mathscr{P}$, *and R is a nontrivial primitive S-ring over G with longest basis element* u, *then* $L(u)$ *is not a power of a prime.*

Proof. Suppose that $L(u) = q^i$, $q \in \mathscr{P}$. If $t (\neq e)$ is any basis element, then by the preceding theorem, $q \mid L(t)$. Hence $q \mid o(H) - 1$, so $q = p$, and R is trivial.

13.8.6. *If R is a nontrivial primitive S-ring over a group H of order* $p + 1$, $2 < p \in \mathscr{P}$, *then there are basis elements* $(\neq e)$ *of at least three different lengths.*

Proof. If all basis elements have the same length r, then $p = o(H) - 1 = kr$ and $r \neq 1$ by Theorem 13.8.1, so $k = 1$ and R is trivial. If there are just two distinct lengths, m and n, then by Theorem 13.8.4, $(m, n) = d > 1$. Hence $d \mid p$, a contradiction. Therefore there are at least three distinct lengths.

13.8.7. *If R is a nontrivial primitive S-ring over H with basis $t_0 = e$, $t_1, \ldots, t_n$, where $L(t_i) \le L(t_{i+1})$ for all i, then $L(t_1)L(t_i) \ge L(t_{i+1})$ for all i.*

Proof. Suppose that $L(t_1)L(T_j) < L(t_{j+1})$ for some j. Then if $0 \le i \le j$, $t_1 t_i$ is a linear combination of $t_0, \ldots, t_j$. But then $S(t_1) < H$, and R is not primitive (see Exercise 13.8.14). ‖

The nonexistence of nontrivial primitive S-rings over groups of order $p + 1$, where p is a small prime, will now be shown. If $p = 2$, then H is cyclic of prime order, and has a nontrivial primitive S-ring (Exercise 13.7.12). In all other cases, $p + 1$ is not a prime, so that by Theorems 13.8.1 and 13.8.6, $p \ge 3 + 4 + 5 = 12$. Therefore:

13.8.8. *If H is a group of order 4, 6, 8, or 12, then there are no nontrivial primitives S-rings over H.*

13.8.9. *There are no nontrivial primitive S-rings over a group of order 14.*

Proof. A primitive S-ring R over H must have a basis (e, t_1, t_2, t_3) with $L(t_1) = 3$, $L(t_2) = 4$, and $L(t_3) = 6$ by the theorems of this section. This will be described in the future by saying that R has *length pattern* 3–4–6. By Theorem 13.8.3, $t_1 t_2 = 2t_3$. Since $S(t_1) = H$, we cannot have $t_1^2 = 3e + 2t_1$. The only other possibility is $t_1^2 = 3e + t_3$. By Theorem 13.8.2 and the fact that each $t_i^* = t_i$ (or from the equation $t_1\bar{H} = 3\bar{H}$), $t_1 t_3 = 2t_1 + 3t_2$. Hence, $t_1^2(e + t_3) = ie + jt_3$, i and j integers. This contradicts Exercise 13.8.14.

13.8.10. *There are no nontrivial primitive S-rings over a group H of order 18.*

Proof. Suppose that R is such a ring over H. The only length pattern permitted by the theorems of this section is 3–4–4–6. By Theorem 13.8.3, $t_1 t_2 = 2t_4$ and $t_1 t_3 = 2t_4$, so that $t_1\bar{H} > 4t_4$, a contradiction.

13.8.11. *There are no nontrivial primitive S-rings over a group of order 20.*

Proof. Suppose that R is such a ring over H. The permissible length patterns are successively ruled out as follows.

3–4–12 $t_1^2 = 3e + 2t_1$, so that $S(t_1) < H$.

4–5–10 $S(t_1) < H$.

3–3–3–4–6 $t_1 t_4 = t_2 t_4 = t_3 t_4 = 2t_5$, hence $\bar{H}t_4 > 6t_5$.

3–4–6–6   If $P \in \text{Syl}_5(H)$, then $P \lhd H$. Hence, $\exists K \subset H$, with $[H:K] = 2$.

If the basis element t of R contains r elements of K and s elements of $H\backslash K$, then t will be said to be of type r-s. No basis element t is of type 0-s for this would imply that $S(t^2) \subseteq K$. Nor is $t \neq e$ of type r-0, for this implies $S(t) \subseteq K$. Now $WLOG$, $t_1^2 = 3e + t_3$. Since t_1 is of type 1–2 or 2–1, t_3 is of type 2–4. Also,

$$t_1^2 = (t_1^*)^2 = 3e + t_3^*,$$

hence $t_3 = t_3^*$. Therefore, $t_4 = t_4^*$. By Theorem 13.8.3, $t_1 t_2 = 2t_3$ or $2t_4$. One checks that this forces t_2 to be of type 2–2, $t_1 t_2$ of type 6–6; hence $t_1 t_2 = 2t_4$ and t_4 is of type 3–3. Hence, counting elements in K, t_1 is of type 2–1. By Theorem 13.8.2, $t_1 t_3 = 2t_1 + u$ where u is of type 4–8. Therefore, $t_1 t_3 = 2t_1 + 2t_3$, and

$$t_1(e + t_1 + t_3) = ie + jt_1 + kt_3$$

for some integers $i, j,$ and k, contradicting Exercise 13.8.14. $\parallel$

The fact that there are no nontrivial primitive S-rings over groups of order 24, 30, 32, is left as an exercise (13.8.12). For order 24, the proof is not much harder than for order 20, in spite of the fact that the technique used in the final case above is not available since there is a group of order 24 which has no subgroup of index 2 (the relative holomorph of the quaternion group by an automorphism of order 3). The orders 30 and 32 are considerably harder to handle and require a number of cases each. The fact that there are no nontrivial primitive S-rings over a group of order 38 follows from the fact that any such group is cyclic or dihedral and later theorems, 13.9.1 and 13.11.7.

EXERCISES

13.8.12. There are no nontrivial primitive S-rings over a group of order 24, 30, or 32.

13.8.13. If a primitive S-ring has exactly four basis elements e, t, u, v, and if $(L(t), L(u)) = 1$, then $tu = iv$ with $i > 1$.

13.8.14. If R is a primitive S-ring over H with basis
$B = (t_0 = e, t_1, \ldots, t_n),\quad y \in R, y > 0$,
U a subset of B which is neither B, $\{t_0\}$, nor empty, and for each $t_i \in U$, yt_i is a J-linear combination of the elements of U, then $y = ce$ for some integer c. (Show that otherwise $S(y) < H$.)

13.8.15. If R is a nontrivial primitive S-ring over H and H/K is an elementary Abelian p-group of order p^n, then any basis element $t \neq e$ of R has $L(t) \geq n + 1$.

13.9   S-rings over Abelian groups

The theorem in this section is due to Wielandt [6]. It is a generalization of earlier theorems of Burnside and Schur, and a special case of a theorem of Bercov [1].

13.9.1. *If H is a finite Abelian group not of prime order, and $P \neq E$ is a cyclic Sylow p-subgroup, then there is no nontrivial, primitive S-ring R over H. Hence, H is a B-group.*

Proof. Let K be the unique subgroup of order p. Let T be a nontrivial, primitive S-ring over H, and let $t \neq e$ be a basis element of R. If $t = \Sigma x_i$, $x_i \in H$, then, by the binomial theorem, $t^p \equiv \Sigma x_i^p \pmod{p}$. If $t^p \not\equiv je \pmod{p}$ for some integer j, then the set Q of elements of H whose coefficients in the expansion of t^p are not divisible by p generates a nontrivial subgroup of H (at most the subgroup of all pth powers), yet $\bar{Q} \in R$, a contradiction. Therefore, $t^p \equiv \Sigma x_i^p \equiv je \pmod{p}$. Now each element $h \in H$ has either no pth roots in H, or precisely an entire coset xK of them. Hence

$$t = a + b\bar{K}, \qquad a \in J(K), b \in J(H).$$

Since $S(t) = H$ and $H \neq K$, $b \neq 0$.

We assert that there is a basis element of the form $t = b\bar{K}$. If not, then there is one basis element $t_1 = a_1 + b_1\bar{K}$ with $L(b_1)/L(a_1)$ a maximum, and a basis element $t_2 = a_2 + b_2\bar{K} \neq t_1^*$ (possibly $t_2 = t_1$). Then

$$t_1 t_2 = a_1 a_2 + (a_1 b_2 + b_1 a_2 + b_1 b_2 o(K))\bar{K},$$

where e does not occur in the expansion of the right member, and

$$\frac{L(a_1 b_2 + b_1 a_2 + b_1 b_2 o(K))}{L(a_1 a_2)} > \frac{L(b_1 a_2)}{L(a_1 a_2)} = \frac{L(b_1)}{L(a_1)}.$$

This is a contradiction, since $t_1 t_2$ is a linear combination of $t_j \neq e$. Hence $t = b\bar{K}$ actually occurs as basis element.

The set M of $h \in H$ such that $th = t$ is a proper subgroup of H containing K. M is also the set of elements of H occurring with coefficient $L(t)$ in t^*t. Hence $\bar{M} \in R$. This contradicts the primitivity of R.

EXERCISES

13.9.2. If R is a primitive S-ring over H, $t \in R$, $t \neq c\bar{H}$, $h \in H$, and $th = t$, then $h = e$. (Reduce to the case where all coefficients in t are 0 or 1, and apply the argument at the end of Theorem 13.9.1.)

13.10   Nilpotent times dihedral or dicyclic

The main theorem of this section, 13.10.1, makes use of a number of earlier theorems, including the theorem of the preceding section. It is due to Huppert and Itô [1].

13.10.1. *If a finite group* $G = AB$, *A is nilpotent*, $[B:H] = 2$, *and H is cyclic, then G is solvable.*

Proof. Let G be a smallest counterexample. Since all subgroups of H are normal in B, $B \subset \text{Qua}(H)$. By Theorem 13.6.1, $o(A)$ is even, so that $P \neq E$ where $P \in \text{Syl}_2(A)$. Any subgroup or factor group of B is either cyclic or of the same type as B. Therefore, by induction and Kegel's theorem (13.2.9), G has no normal solvable subgroup except E. Note also that if $A \subset K < G$ or $B \subset K < G$, then K is solvable by induction or Theorem 13.2.9.

Let M be a maximal proper subgroup of G such that $M \supset A$. If $M \cap H \neq E$, then $(M \cap H)^G = (M \cap H)^{BA} \subset M$, so there is a normal solvable subgroup. Hence $M \cap H = E$. If $M = A$, then A is maximal, hence $N(P) = A$, and $P \in \text{Syl}(G)$. Since also $PQ \in \text{Syl}(G)$ for some $Q \in \text{Syl}(B)$ (Theorem 13.2.5), $Q \subset P$. Since $M \cap H = E$, $o(A \cap B) = 2$. But then

$$o(AH) = o(A)o(H) = \frac{o(A)o(B)}{2} = o(AB) = o(G),$$

$AH = G$, and G is solvable (nilpotent times nilpotent). Therefore, $A < M$, and there is a subgroup R of B such that

$$M = AR, \quad o(R) = 2, \quad A \cap R = E, \quad R \cap H = E,$$
$$R \subset B, \quad G = MH, \quad M \cap H = E.$$

Also $PR \in \text{Syl}_2(M)$.

Suppose that $2 \mid o(H)$. Let $Q \in \text{Syl}_2(H)$. Then $SQ \in \text{Syl}_2(G)$ for some $S \in \text{Syl}_2(M)$. Since the only Sylow 2-subgroup P of A is normal in M, it is contained in S. There is a subgroup U of SQ such that $[U:S] = 2$. Since $N(P) = M$, there are just two conjugates, P and P^* of P by elements of U. Then $[U:P \cap P^*] = 8$, and $N(P \cap P^*) \supset UA > M$. Since M is maximal, $P \cap P^* = E$ and $o(P) = 2$. Since $[SQ:Q]$ is now 4, it follows as above that Q has just one or two conjugates in SQ. Hence, Q or $Q \cap Q^y$ ($y \in SQ$) is a normal subgroup of SQ. But if Q contains a normal subgroup (not E) of SQ, it contains a central element x, and $N(P) \supset \langle M, x \rangle$, a contradiction. Therefore, $Q \cap Q^y = E$, and $o(Q) = 2$. Thus SQ is non-Abelian of order 8. Since SQ has distinct subgroups of order 2, it is not the quaternion group. It follows that SQ is dihedral.

In the representation of G on M, the representation is faithful and H is regularly represented. Since $o(H)$ is twice an odd number, an element y of order 2 in H is represented by the product of an odd number of 2-cycles, hence by an odd permutation. Therefore there is a subgroup K of index 2 in G. K cannot contain A or B (or it is solvable). Hence if $P = \langle x \rangle$, then $x \notin K$ and $y \notin K$, so that $xy \in K$. Since SQ is dihedral of order 8 while x and y are distinct noncentral elements of SQ of order 2, $o(xy) = 4$. Thus, K has a cyclic Sylow 2-subgroup. By a corollary of Burnside's theorem (6.2.11), K has a normal (hence characteristic) 2-complement L. Then $L \lhd G$, so L contains the 2-complements A_1 of A and B_1 of B. By orders, $L = A_1 B_1$, hence L is solvable, a contradiction.

Therefore $2 \nmid o(H)$. The representation of G on M is faithful and primitive, and H is regularly represented. If $o(H)$ is a prime, then by Theorem 12.9.2, G is 2-transitive. If $o(H)$ is not a prime, then by Theorem 13.9.1, G is 2-transitive. Thus G is 2-transitive in any case. Now M is the subgroup fixing one letter, so M is transitive on the rest of the letters. If $E < L \lhd M$, then $Ch(L) = 1$ (Theorem 10.1.7). The representation of G on $M = N(M)$ is similar to its representation on the conjugate class of M (Theorem 10.2.5). Hence

If $E \neq L \lhd M$ and $x \notin M$, then $L \not\subseteq M \cap M^x$. \hfill (1)

Now $PR \in \mathrm{Syl}_2(G)$. If $Z(PR) \subseteq A$, then $Z(PR) \lhd M$. If $x \in Z(PR) \backslash A$ and $y \in Z(P)$, then $C(y) \supset \langle P, x \rangle = PR$, so $y \in Z(PR)$. Hence,

Either $Z(PR) \lhd M$, or $Z(P) \lhd M$ and $Z(P) \subseteq Z(PR)$. \hfill (2)

By (1) and (2),

If $Z(PR) \subseteq M \cap M^x$, then $x \in M$. \hfill (3)

If $N(Z(PR)) \subseteq A$, then $Z(PR) \subseteq A$, and $Z(PR) \lhd M$, a contradiction. Thus, by (3),

$N(Z(PR)) \subseteq M$ but $N(Z(PR)) \not\subseteq A$. \hfill (4)

Suppose that G is not 2-normal. Then $\exists\, x \notin N(Z(PR))$ such that $Z(PR) \subseteq (PR)^x$. By (3), $x \in M$. The 2-complement A_1 of A is normal in M, so that $M = A_1(PR)$. Thus, $x = ab$ with $a \in PR$ and $b \in A_1$. Therefore, $(PR)^x = (PR)^b$ and $b \notin N(Z(PR))$. But if $c \in Z(PR)$, then $[c, b] \in A_1$ by the normality of A_1 in M, and

$$[c, b] = c^{-1} b^{-1} c b \in (Z(PR))(PR)^b = (PR)^b.$$

Since $(o(A_1), o(PR)) = 1$, it follows that $[c, b] = e$, hence $b \in N(Z(PR))$, a contradiction.

Therefore, G is 2-normal. By Grün's theorem,

$$G/G^1(2) \cong L/L^1(2), \qquad L = N(Z(PR)).$$

By (4), $[L:L \cap A] = 2$, so $L^1(2) \subseteq A$ and $L^1(2) \neq L$. It follows readily from Grün's theorem and Theorem 13.5.2 that

$$L \cap G^1(2) = L^1(2) \subseteq A$$

(this is Exercise 13.5.20). Since $G^1(2) \supset A_1$,

$$G^1(2) \cap M = A_1(G^1(2) \cap PR) \subseteq A_1(G^1(2) \cap L) \subseteq A.$$

Also $G^1(2) \supset H$ since H is of odd order. Therefore, since $G = MH$,

$$G^1(2) = (G^1(2) \cap M)H, \qquad G^1(2) \cap M \subseteq A.$$

Thus, $G^1(2)$ is solvable by Kegel's theorem, and we have a normal solvable subgroup again.

13.10.2. *If a finite group G is the product of a nilpotent subgroup and a dihedral or dicyclic subgroup, then G is solvable.*

Proof. Since dihedral and dicyclic groups have a cyclic subgroup of index 2, this follows from Theorem 13.10.1.

13.10.3. *If a finite group $G = AB$, where A is cyclic, $[B:H] = p \in \mathscr{P}$, $p = 2^n + 1$, $n \in J$, and $B \subset \mathrm{Qua}(H)$, then G is solvable.*

Proof. Let G be a counterexample of smallest order. It can be verified that any subgroup or factor group of B is either of the same type or nilpotent (in fact Dedekind). Therefore, G has no normal solvable subgroup except E. Any subgroup S such that $A \subset S < G$ or $B \subset S < G$ is solvable by induction or the nilpotent times nilpotent theorem. If $B \subset S < G$ and $A \cap S \neq E$, then

$$(A \cap S)^G = (A \cap S)^{AB} = (A \cap S)^B \subseteq S,$$

so $(A \cap S)^G$ is a normal solvable subgroup. Therefore (i) B is a maximal proper subgroup of G, and (ii) $A \cap B = E$.

If $2 \mid o(A)$, $2 \mid o(B)$, and $P \in \mathrm{Syl}_2(B)$, then $P \subset H$, so $N(P) > B$, a contradiction. Hence either $2 \nmid o(A)$ or $2 \nmid o(B)$. If $2 \mid o(A)$, then a Sylow 2-subgroup of A is one for G also, hence G has a cyclic Sylow 2-subgroup. By Theorem 6.2.11, G has a normal 2-complement S. But then, since B is of odd order, $B \subset S$, so S is solvable, a contradiction.

Therefore, A is of odd order. Suppose that A is a maximal proper subgroup. Then $N(A) = A$. If $A \cap A^x \neq E$ for some $x \notin A$, then $N(A \cap A^x) = G$. If $A \cap A^x = E$ for all $x \notin A$, then G is a Frobenius group with kernel B, hence B is a normal solvable subgroup. Therefore, A is not maximal.

Let $A < M < G$. If $M \cap H \neq E$, then

$$(M \cap H)^G = (M \cap H)^{BA} = (M \cap H)^A \subseteq M,$$

and $(M \cap H)^G$ is a normal solvable subgroup. Therefore

$$M = AR, o(R) = p, R \subseteq B, MH = G, M \cap H = E.$$

By Theorem 12.9.3, $M/\text{Core}_M(A)$ is either a 2-transitive group of degree p, or is between J_p and $\text{Hol}(J_p)$. The first case is impossible since M is of odd order. Since $o(\text{Hol}(J_p)) = p(p - 1) = 2^n p$, $M/\text{Core}_M(A) \cong J_p$, i.e., $A \lhd M$.

The representation of G on B is faithful and primitive, and A is regularly represented. By Theorems 12.9.2 and 13.9.1, G is 2-transitive. If the subgroup V of B fixing 2 letters is E, then G is a Frobenius group with kernel A, a contradiction. Hence $V \neq E$. By Theorem 10.1.7, V contains no normal subgroups of the singly transitive group B (except E). Therefore $o(V) = p$. Thus,

$$o(H) = [B:V] = \text{Deg}(G) - 1 = o(A) - 1. \tag{5}$$

If all $M \cap M^x$ have order p^i for some $i \geq 0$ when $x \notin M$, then each

$$[MxM:M] = [M:M^x \cap M] = c(o(A))/p^j$$

where c and j are nonnegative integers depending on x (Theorem 3.3.10). Therefore by (5),

$$o(A) - 1 = o(H) = [G:M]$$
$$= 1 + \sum c_i o(A)/p^k, \quad c_i \in J. \tag{6}$$

Thus $d(o(A)) = 2p^k$ for some positive integers d and k. Hence, $o(A) = p^r$ for some r, so M is a p-group. Therefore, $G = MH$ is the product of two nilpotent subgroups, hence is solvable.

Hence some $M \cap M^x$ has order divisible by a prime $q \neq p$. Then $q \mid o(A)$, and the unique subgroup L of order q of A is also the unique subgroup of order q of M and M^x. Therefore, $N(Q) \supset \langle M, M^x \rangle = G$, a contradiction. ‖

It seems likely that the preceding theorem is true without any restriction on the prime p. In fact, most of the proof goes through for this more general case. However, one cannot prove that $A \lhd M$, so additional complications arise.

13.11   Primitive S-rings over dihedral and dicyclic groups

It will be shown in this section that there are no nontrivial, primitive S-rings over (ordinary) dihedral groups, or over a certain type of generalized dicyclic group. The latter case, due to the author [4], will be considered first.

13.11.1. *If H is a finite Abelian group with just one element x of order 2, and $G = \text{Dic}(H, x)$, then there is no nontrivial, primitive S-ring over G, and G is a B-group.*

Proof. The second assertion follows from the first by Theorem 13.7.11. Suppose that R is a nontrivial, primitive S-ring over G. Recall that all elements of $G\backslash H$ have square equal to x. Let t be the basis element of R containing x. Then $t = t^*$. If $y \in G\backslash H$ occurs in t^2 as a product zh with $z \in G\backslash H$ and $h \in H$, then it also occurs as $h^{-1}z$, and conversely. Therefore,

$$t^2 = L(t)e + 2u + v$$

where $u \in J(G)$ and $v \in J(H)$. Now the sum of the elements of G with odd coefficient in t^2 is an element w of R such that $S(w) \subset H < G$. Since R is primitive,

$$t^2 = L(t)e + 2u, \qquad u \in R. \tag{1}$$

In the expansion of t^2 there are four types of products in $H^\#$. These are:

(i) y^2, $y \in G\backslash H$,

(ii) $y_1 y_2$, $y_1 \neq y_2$, $y_1 \neq y_2^{-1}$, $y_i \in G\backslash H$,

(iii) $h_1 h_2$, $h_1 \neq h_2$, $h_1 \neq h_2^{-1}$, $h_i \in H$,

(iv) h^2, $h \neq x$, $h \in H$.

Those of type (i) occur in equal pairs, since $y^2 = (y^{-1})^2 = x$. Again, $y_1 y_2 = y_1^{-1} y_2^{-1}$, so that the products of type (ii) occur in pairs. Since $h_1 h_2 = h_2 h_1$, the products of type (iii) occur in equal pairs. Therefore, from (1), the products of type (iv) occur in equal pairs also. But if $h_1^2 = h_2^2$, then $(h_1 h_2^{-1})^2 = e$, so that $h_1 h_2^{-1} = x$, and $h_1 = h_2 x$. Thus if $h \in H^\#$, $h \neq x$, and h occurs in t, so does hx, hence also $(hx)^{-1} = h^{-1}x$. Therefore, x occurs $L(t) - 1$ times in t^2 (as y^2 for $y \notin H$, and as $h(h^{-1}x)$ for $h \in H$ and $h \neq x$). Thus,

$$t^2 \geq L(t)e + (L(t) - 1)t.$$

But the right side has length

$$L(t) + (L(t) - 1)L(t) = (L(t))^2 = L(t^2).$$

Therefore,

$$t^2 = L(t)e + (L(t) - 1)t.$$

Hence $S(t) < G$, a contradiction. ‖

It is an open question whether the preceding theorem is true if the Sylow 2-subgroup of H is not cyclic.

The dihedral case is due to Wielandt [2]. As the proof is rather long, it will be broken up into several pieces. The first lemma is an amusing special case of Dirichlet's theorem about the infinitude of primes in an arithmetic progression.

13.11.2. *If G is the multiplicative group of the ring J_n ($n > 1$) and $H < G$, then there are an infinite number of primes whose residue class (mod n) is in G but not in H.*

Proof. Deny the theorem, and let $p_1, \ldots, p_r$ be the primes whose residue classes lie in $G \backslash H$ (possibly $r = 0$). If $r \neq 0$, then one of the integers $p_1 \ldots p_r$ and $p_1^2 p_2 \ldots p_r$ has residue class outside H. Call that integer m. If $r = 0$, let m be a positive integer whose residue class is in $G \backslash H$. Now $m + n$ is not divisible by any p_i or any prime dividing n. Therefore $m + n = \pi q_j$, where each q_j is a prime whose residue class is in H. But then the residue class of $m + n$ lies in H, which is not the case.

13.11.3. *If H is a finite Abelian group, $G = \mathrm{Dih}(H)$, R is a primitive S-ring over G, $t = t_1 + t_2 \in R$, and $u = u_1 + u_2 \in R$, where t_1 and u_1 are the parts of t and u in H and t_2 and u_2 the parts outside H, then R is commutative and $t_1 u_2 = u_2 t_1$.*

Proof. All elements of $G \backslash H$ have order 2. If $z \in R$, then either $z = ce$ or $z_2 \neq 0$, hence $z = z^*$. Therefore,

$$tu = t^* u^* = (ut)^* = ut,$$

and R is commutative. If $h \in H$ and $g \in G \backslash H$, then $hg = gh^{-1}$. It follows that $t_1 u_2 = u_2 t_1$.

13.11.4. *If $H = \langle h \rangle$ is a finite cyclic group, $G = \mathrm{Dih}(H)$, $o(G) = n$, and R is a nontrivial, primitive S-ring over G, then there is a nontrivial, primitive S-ring T over G such that if $u \in T$ and u contains ah^i ($a \in J$), then u also contains ah^{im} for all m such that $(m, n) = 1$.*

Proof. Let A be the multiplicative group of J_n. Let B be the set of $[m] \in A$ for which there is an endomorphism $u \longrightarrow u^{(m)}$ of R such that if $u_1 = \Sigma\, a_i h^i$, then $(u^{(m)})_1 = \Sigma\, a_i h^{im}$. Since no element v of R except 0 can have $v_1 = 0$, it follows that if such an endomorphism exists, it is unique. It is clear that $[1] \in B$, and that B is closed under multiplication, so that $B \subset A$.

Suppose that $B < A$. By Theorem 13.11.2, there are an infinite number of primes whose residue classes lie in $A \backslash B$. Hence $\exists\, [m] \in A \backslash B$ with an infinite number of primes $p \equiv m \pmod{n}$. Let

$$t = \Sigma\, a_i h^i + t_2 = t_1 + t_2 \in R.$$

Then $\exists\, p \equiv m \pmod{n}$ such that $p > 2\,|a_i|$ for all i. By Theorem 13.11.3 and the commutativity of H,

$$t^p = (t_1 + t_2)^p \equiv t_1^p + t_2^p \equiv \Sigma\, a_i h^{ip} + t_2^p \pmod{p},$$

and, since p is odd,

$$(t^p)_1 \equiv \Sigma\, a_i h^{ip}.$$

Let $t^{(m)}$ be the element of R formed by replacing all coefficients of t^p by their absolutely smallest residues (mod p). Then

$$t^{(m)} \equiv t^p \pmod{p}, \tag{2}$$

and, since $p > 2|a_i|$, $(t^{(m)})_1 = \Sigma a_i h^{ip} = \Sigma a_i h^{im}$. By an earlier argument, the definition of $t^{(m)}$ is independent of p. The map $t \longrightarrow t^{(m)}$ preserves addition trivially. If $t \in R$ and $u \in R$, then for large $p \equiv m \pmod{n}$, by (2) and the commutativity of R (Theorem 13.11.3)

$$(tu)^{(m)} \equiv (tu)^p = t^p u^p \equiv t^{(m)} u^{(m)} \pmod{p}. \tag{3}$$

Since the coefficients in the extreme members of (3) are finite while p is arbitrarily large, $(tu)^{(m)} = t^{(m)} u^{(m)}$. Therefore, $t \longrightarrow t^{(m)}$ is an endomorphism of R. Hence $[m] \in A$, a contradiction. Therefore, $B = A$. Now $u \longrightarrow u^{(1)}$ is the identity automorphism. If $[m] \in A$ and $[r] \in A$ are such that $[m][r] = [1]$, then it follows readily that $u^{(m)(r)} = u^{(1)} = u$. Therefore the map $u \longrightarrow u^{(m)}$ is an automorphism of R.

If t is a basis element, then $(t^{(m)})_1$ has all coefficients 0 or 1. It follows that $(t^{(m)})_1$ is the sum of first components of distinct basis elements and therefore, since the first component determines the element of R, that $t^{(m)}$ is the sum of distinct basis elements of R. Applying the inverse automorphism, we see that $t^{(m)}$ is a single basis element of R. Again, $t^2 = L(t)e + u$, where $u \in R$ and u does not involve e. Hence

$$(t^{(m)})^2 = (t^2)^{(m)} = L(t)e + u^{(m)},$$

where $u^{(m)}$ does not involve e either. Since also

$$(t^{(m)})^2 = L(t^{(m)})e + u^{(m)},$$

$L(t) = L(t^{(m)})$. From its very definition, $L(t_1) = L((t^{(m)})_1)$. Subtracting, we get $L(t_2) = L((t^{(m)})_2)$.

Let T be the set of all elements of R fixed under all automorphisms $u \longrightarrow u^{(m)}$. Then T is a subring of R. Also T has a basis such that each basis element u of T is the sum of the distinct images of a basis element t of R under these automorphisms. Since $t = t^*$, also $u = u^*$, and T is an S-ring. Since T is a subring of R, T is primitive.

Suppose that T is trivial. Then it has a single basis element u other than e. Then u is the sum of all basis elements ($\neq e$) of R, say r of them, and by earlier remarks, if $t \neq e$ is a basis element of R, then

$$r(o(L(t_1)) + 1 = o(H) = o(G \backslash H)$$
$$= r(o(L(t_2)).$$

But this implies that $r = 1$, hence R is trivial, a contradiction. Hence T is nontrivial. The final statement of the lemma follows from the definition of T. $\parallel$

An S-ring T over a dihedral group G of order n such that if $u \in T$ and u contains ah^i, then u contains ah^{im} when $(m, n) = 1$, will be called *rational*. A *simple* element of an S-ring is one all of whose coefficients are 0 or 1.

13.11.5. *If G is a finite dihedral group over H, $E < U < H$, T is a nontrivial, primitive, rational S-ring over G, $t \in T$, and $\bar{U} \mid t_2$ (in $J(G)$), then t is of the form $ae + b\bar{G}$.*

Proof. Any subgroup of U also divides t_2, hence $WLOG$, $o(U) = p \in \mathscr{P}$. Since cosets are equal or disjoint, $WLOG$, t is simple.

Since $\bar{U}^2 = p\bar{U} \equiv 0 (\bmod\, p)$,

$$t^p \equiv t_1^p + t_2^p \equiv t_1^p + (\bar{U}y)^p$$
$$\equiv t_1^p + \bar{U}^p y^p \equiv t_1^p (\bmod\, p),$$

where $y \in J(G)$, and $\bar{U}y = y\bar{U}$ since $U \lhd G$. In the expansion of t^p, the sum of the terms whose coefficients are not divisible by p is in T but contains no elements outside H. Hence $t_1^p \equiv ce (\bmod\, p)$ for some integer c. It follows that $t_1 = u + v\bar{U}$, where u is a sum of elements of U and $v \in J(H)$. Let $x \in U^{\#}$ occur in u, and let $1 < i < p$. Let $p^r \mid n = o(G)$ but $p^{r+1} \nmid n$. Then $[p]$ is an element of the multiplicative group of the ring J_{n/p^r}, so $\exists\, d$ such that $[d][p] = [1]$. Thus,

$$d(1 - i)p + i \equiv 1(\bmod\, n/p^r),$$
$$(d(1 - i)p + i, n) = 1.$$

By the rationality of T, $x^i = x^{d(1-i)p+i}$ occurs in u also. Hence $u = 0$, e, $\bar{U}$, or $\bar{U} - e$. Therefore one of t_1, $t_1 + e$, and $t_1 - e$ is divisible by $\bar{U}$. Since $\bar{U} \mid t_2$, one of t, $t + e$, and $t - e$ is divisible by $\bar{U}$. If $y \in U^{\#}$, then $ty = t$ (or $(t \pm e)y = t \pm e$). By Exercise 13.9.2, one of t, $t + e$, and $t - e$ equals $b\bar{G}$. In any case, the theorem holds. ‖

13.11.6. *If H is a finite cyclic group, $G = \mathrm{Dih}(H)$, $E < U \subset H$, T is a nontrivial, primitive, rational S-ring over G, t a simple element of G, $t \neq \bar{G}$, $\bar{U} \mid t_1$, and $L(t_1) = c$, then (i) $o(U) = 2$, (ii) H is not a 2-group, (iii) $c > o(H)/3$.*

Proof. Let $L(t_2) = d$. By Theorem 13.11.3, $(t^2)_2 = 2t_1t_2$, hence $\bar{U} \mid (t^2)_2$. Therefore, by Theorem 13.11.5, $t^2 = ae + b\bar{G}$. Since $(t^2)_1 = t_1^2 + t_2^2$, we have

$$\begin{cases} t_1^2 + t_2^2 = ae + b\bar{H}, \\ 2t_1t_2 = b(\bar{G} - \bar{H}). \end{cases} \tag{4}$$

Equating lengths gives $c^2 + d^2 = a + bo(H)$, $2cd = bo(H)$. Hence,

$$\begin{cases} a = (c - d)^2 \\ b = \dfrac{2cd}{o(H)}. \end{cases} \tag{5}$$

Let $x \in U^{\#}$. Since $\bar{U} \mid t_1$, $t_1 = x t_1$. Hence, $t_1^2 = x t_1^2 = x t_1 t_1^*$, which contains x exactly $L(t_1) = c$ times. Equating the coefficients of x in the first equation of (4), one obtains

$$\begin{cases} b - c = \text{coefficient of } x \text{ in } t_2^2, & (x \in U^{\#}), \\ b \geq c. \end{cases} \tag{6}$$

Now (5) and (6) yield $d \geq o(H)/2$. Since $\bar{G} - t$ is also a simple element of T such that $(\bar{G} - t)_1$ is divisible by $\bar{U}$, it follows from what has just been proved that $o(H) - d \geq o(H)/2$. Therefore, by (5),

$$d = o(H)/2, \qquad b = c. \tag{7}$$

If $o(U) > 2$, then since $L(t_2) = d = o(H)/2$, there are distinct elements x and y of U such that $x t_2$ and $y t_2$ have a $g \in G \backslash H$ in common, so that $x^{-1}g$ and $y^{-1}g$ are in t_2. Then

$$(x^{-1}g)(y^{-1}g) = (x^{-1}y)g^2 = x^{-1}y$$

occurs in t_2^2. However, by (6) and (7), no element of $U^{\#}$ occurs in t_2^2. Hence (i) is true.

Computing the coefficient of e in the first equation of (4), one obtains $c + d = a + b$. By (7), $a = d$. Hence by (5) and (7),

$$d = (c - d)^2, \qquad o(H) = 2(c - d)^2. \tag{8}$$

Another application of (7) gives

$$c = d + (c - d) = d \pm \sqrt{d} = \frac{o(H)}{2} \pm \sqrt{\frac{o(H)}{2}} \tag{9}$$

Suppose that H is a 2-group. By (8), d is an even power of 2:

$$d = 2^{2r}, \qquad c = 2^{2r} \pm 2^r. \tag{10}$$

Replacing t by $\bar{G} - t$ if necessary, it may be assumed that e does not occur in t. Since T is rational, t_1 is the sum of various s_q, where s_q is the sum of all elements of order 2^q in H. Since $L(s_q) = 2^{q-1}$, c is the sum of numbers of the form 2^{q-1}. Since by (10), $2^r \mid c$, summands less than 2^r do not occur, hence always $q > r$. Since all of the s_q are divisible by $\bar{U}_1$, where U_1 is the subgroup of H of elements of order at most 2^r, it must be that $\bar{U}_1 \mid t_1$. Hence by (i), $o(U_1) = 1$ or 2, so $r = 0$ or 1. If $r = 0$, then $d = 1$, $o(H) = 2$, $o(G) = 4$, and the case cannot arise by Theorem 13.8.8. Therefore, $r = 1$, and by (10), $d = 4$, $o(H) = 8$, and $c = 2$ or 6. By (5) and (7), $a = 4$, $b = c = 2$ or 6. Therefore by (4), all the coefficients of elements of G in $t_1^2 + t_2^2$ are even. Since $\bar{U} \mid t_1$, $t_1 = v\bar{U}$ where $v \in J(H)$. Hence, $t_1^2 = v^2 \bar{U}^2 = 2v^2 \bar{U}$, so all the coefficients of elements of H in t_1^2 are even. Hence all coefficients in t_2^2 are even. Also, by (6) and (i), t_2^2 does not contain the element y of order 2 in U. Since $d = 4$, t_2 is the sum of four elements of the form $h^i g$, where g is a fixed element of $G \backslash H$, and $H = \langle h \rangle$. The rule for multiplication in a dihedral group shows that the four numbers i are such that among their sixteen differences (mod 8), each occurs an even number of times and 4 does not occur.

We assert that such a set $S = \{i_1, i_2, i_3, i_4\}$ of four numbers (mod 8) does not exist. Since the set of differences is invariant under translation of S, it may be assumed that $i_1 = 0$. If the other numbers are 2, 4, and 6, then the forbidden difference of 4 occurs. Hence two of the i's differ by 1. By a further translation, $i_1 = 0$ and $i_2 = 1$. Hence the difference of 1 must occur again, but 4 and 5 are forbidden values of the i's. Thus, after another translation, $i_1 = 0$, $i_2 = 1$, and $i_3 = 2$, and therefore 4, 5, 6 are forbidden. It follows that the four values are consecutive and that the difference of 1 occurs exactly three times. This contradiction proves (ii).

If $o(H) > 18$, then $o(H)^2/36 > o(H)/2$, $o(H)/6 > \sqrt{o(H)/2}$. Equation (9) then gives $c > o(H)/2 - o(H)/6 = o(H)/3$. If $o(H) \leq 18$, then by (8), $o(H) =$ 2, 8, or 18. Now suppose that (iii) is false. By (ii), $o(H) = 18$, and by (9), $c = 6$. If the six elements in t_1 are all in the subgroup U_1 of order 6, then $\bar{U}_1 \mid t_1$, contradicting (i). Hence, t_1 contains an element y of order 9 or 18. Since T is rational, t_1 consists either of all six elements of order 9 in H or of all six elements of order 18 in H. Let u, a, and b be elements of H of orders 2, 9, and 18, respectively. Then $o(au) = 18$ and $o(bu) = 9$, so that $\bar{U} \nmid t_1$, a contradiction. Therefore, (iii) is true. $\parallel$

The main theorem can now be proved.

13.11.7. *If G is a finite dihedral group, then there is no nontrivial, primitive S-ring over G, and G is a B-group.*

Proof. Deny the theorem. By Theorem 13.11.4, there is a nontrivial, primitive, rational S-ring T over G. Let $H = \langle h \rangle$ be the cyclic subgroup of index 2 of G.

Suppose that $o(H) = 2^k$. There is a basis element t of T which contains neither e nor the element u of order 2 in H. By the rationality of T, if h^i occurs in t_1, so does $h^{2^{k-1}+i}$, so that t_1 is divisible by the subgroup U of H of order 2. This contradicts the preceding theorem.

Hence H is not a 2-group. Let t be a basis element such that $V = S(t_1) \neq E$ is of minimum order. If V is a 2-group, then $o(V) \leq o(H)/3$. Since T is rational, it follows as in the proof of Theorem 13.11.5 that either t_1 or $t_1 + e = (t + e)_1$ is divisible by $\bar{U}$, where $o(U) = 2$, $U \subset H$. By the preceding theorem, $o(V) > o(H)/3$, a contradiction.

Therefore, V is not a 2-group, and there is an odd prime divisor p of $o(V)$. By Theorem 13.11.3, $t^p \equiv t_1^p + t_2^p \pmod{p}$. Since p is odd, $(t^p)_1 \equiv t_1^p \pmod{p}$. Therefore the sum of elements occurring in t^p with coefficient not divisible by p is an element $u \in T$, and $S(u_1) \subset pV < V$. By the minimality of V, $u_1 = ae$, hence $t_1^p \equiv ae \pmod{p}$. Therefore, $t_1 = v + w\bar{U}_1$, where U_1 is the subgroup of H of order p, $v \in J(U_1)$, and $w \in J(H)$. Since T is rational t_1 or $t_1 + e$ is divisible by $\bar{U}_1$. By the preceding theorem, t_1 or $t_1 + e$ equals $\bar{G}$, contrary to the primitivity of T.

13.12   Dihedral or dicyclic times dihedral or dicyclic

The principal theorem here is that a group of the type described in the section title is solvable. Actually, a mild generalization is proved. The results are due to Huppert [3] and Scott [4].

Call a group H a *descendant* of G if H is a homomorphic image of a subgroup of G. Thus, H is a descendant of G iff H is a member of the smallest hereditary class of groups containing G as a member.

13.12.1. *If K is a descendant of H and H is a descendant of G, then K is a descendant of G.*

Proof. $K \cong A/B$ where $A \subset H$, and $H \cong U/V$ where $U \subset G$. Hence $K \cong R/S$ where $R \subset U/V$. Therefore, there are subgroups X and Y of U such that

$$K \cong \frac{X/V}{Y/V} \cong \frac{X}{Y}.$$

Hence, K is a descendant of G.

13.12.2. *If $G = AB$, where A and B each contain a cyclic subgroup of index 2 and are such that all noncyclic descendants of A and B are B-groups, then G is solvable.*

Proof. Let H and K be cyclic subgroups of index 2 in A and B, respectively. Let G be a counterexample of least order.

If there is a normal solvable subgroup $M \neq E$, then

$$G/M = (AM/M)(BM/M)$$

is such that each factor is either cyclic or satisfies the hypotheses of the theorem (using Theorem 13.12.1). Hence, by induction, the nilpotent by nilpotent theorem, or Theorem 13.10.1 (briefly "by induction"), G is solvable. Therefore, G has no normal solvable subgroups except E. Also, by induction, any M such that $A \subset M < G$ or $B \subset M < G$ is solvable.

Let M be a maximal proper subgroup of G containing A. If $M \cap K \neq E$, then

$$(M \cap K)^G = (M \cap K)^{BA} = (M \cap K)^A \subset M,$$

and there is a normal solvable subgroup. Hence $M \cap K = E$. It follows that either $M = A$ or

$$M = AR, \quad o(R) = 2, \quad R = M \cap B, \quad MK = G, \quad M \cap K = E.$$

The representation of G on M is faithful and primitive, and B or K is regularly represented (according as $M \cap B = E$ or not). By hypothesis, Theorems 12.9.2 and 13.9.1, G is 2-transitive. The subgroup V of M fixing two letters contains no normal subgroup of M, since M is transitive of degree $\mathrm{Deg}(G) - 1$ (Theorem 10.1.7). Now all Sylow p-subgroups of H are characteristic in A for odd p. If $P \in \mathrm{Syl}_2(A)$ is normal in A, then A is nilpotent, and G is solvable (13.10.1). In the other case, a Sylow 2-subgroup Q of H is the intersection of all Sylow 2-subgroups of A, hence is characteristic in A. Therefore, $H \in \mathrm{Char}(A)$, so $H \lhd M$. It follows that all subgroups of H are normal in M since H is cyclic. Thus, $V \cap H = E$. Now,

$$[M:V] = [M:HV][HV:V] = [M:HV][H:H \cap V]$$
$$= [M:HV]o(H)$$

is divisible by $o(H)$. But

$$[M:V] = \mathrm{Deg}(G) - 1 = o(K) - 1 \quad \text{or} \quad 2o(K) - 1.$$

Therefore, using symmetry,

$$i(o(K)) - 1 = r(o(H)), \qquad i = 1 \text{ or } 2, r > 0,$$
$$j(o(H)) - 1 = s(o(K)), \qquad j = 1 \text{ or } 2, s > 0.$$

Solving, one gets

$$o(H) = \frac{i + s}{ij - rs}, \qquad o(K) = \frac{j + r}{ij - rs}.$$

If $i = 1$, then $j = 2$, $r = s = 1$, $o(H) = 2$, $o(K) = 3$, $o(G) = 24$, and G is solvable. Hence, $i = 2$, and by symmetry, $j = 2$. If $rs = 1$, then $o(H) = o(K) = 1$, $o(G) = 4$, and G is solvable. If $rs = 2$, then $WLOG$, $s = 1$, and $o(H) = 3/2$, an impossibility. Since $ij - rs > 0$, it follows that $rs = 3$ and $WLOG$,

$$r = 1, \quad s = 3, \quad o(H) = 5, \quad o(K) = 3, \quad o(G) = 60.$$

Since G is not solvable, it follows easily that it has no normal 2-complement. By Burnside's theorem, if $S \in \mathrm{Syl}_2(G)$, then $N(S) \neq C(S)$, hence all the elements of S of order 2 are conjugate. Therefore all elements of G of order 2 are conjugate. In particular, this means that $\exists x \in G$ such that $A \cap B^x \neq E$. But $G = AB^x$ (Theorem 13.2.4), hence $A \cap B^x = E$, a contradiction.

13.12.3. *If $G = AB$ is finite, A is dihedral or dicyclic, and B is dihedral or dicyclic, then G is solvable.*

Proof. Any descendant of a dihedral or dicyclic group is dihedral, dicyclic, or cyclic. By 13.11.1, and 13.11.7, any noncyclic descendant group of A or B is a B-group. The theorem is now a corollary of the preceding theorem. ‖

There is at least one group which fulfills the hypotheses on A in Theorem 13.12.2 without falling into any of the categories: cyclic, dihedral, dicyclic, or nilpotent.

13.12.4. *If G is the direct sum of J_4 and* Sym(3), *then every descendant noncyclic group of G is a B-group, and G is not nilpotent, dihedral, or dicyclic.*

Proof. A noncyclic descendant H of G has order 24, 12, 8, 6, or 4. By Theorem 13.8.8 and Exercise 13.8.12, H is a B-group. G is not nilpotent since it contains the nonnilpotent subgroup Sym(3). It is not dihedral or dicyclic since its Sylow 2-subgroup (of order 8) is Abelian.

REFERENCES FOR CHAPTER 13

For Theorem 13.2.2, D. Higman [1]; Theorems 13.2.3 to 13.2.8, Wielandt [3] and [5]; Theorems 13.2.9 and 13.2.10, Kegel [1]; Theorem 13.3.1, Huppert [1] and Douglas [1]; Theorems 13.3.2 and 13.3.3, Itô [3]; Exercise 13.3.5, Scott [5]; Section 13.4, Wielandt [1]; Section 13.5, Zassenhaus [4]; Theorem 13.6.1, Scott [4]; Theorem 13.6.2, Ladner [1]; Section 13.7, Wielandt [6]; Section 13.8, Hanes [1], in part.

FOURTEEN

THE MULTIPLICATIVE GROUP OF A DIVISION RING

14.1 Wedderburn and Cartan-Brauer-Hua theorems

In this section two celebrated theorems about division rings will be proved.

Let D be a division ring. $D^{\#}$ will denote the multiplicative group of D. If $S \neq \varnothing$ is a subset of D, then the *centralizer* $C(S)$, *normalizer* $N(S)$, and *center* $Z(D)$ are defined as for groups, with the proviso that $0 \in C(S)$, $0 \in Z(D)$, but $0 \notin N(S)$. Then $C(S)$ is a subdivision ring and $Z(D)$ a subfield of D (some verifications, such as these, are left to the exercises). If K is a subdivision ring and $a \in D$, then $K(a)$ is the smallest subdivision ring of D containing K and a.

14.1.1. *If D is a division ring, K a subdivision ring, $x \in N(K)$, $x \notin K$, $x \notin C(K)$, and $y \in C(K) \cap K^{\#}$, then $x + y \notin N(K)$.*

Proof. Since $x \in N(K) \backslash C(K)$, $\exists\, u \in K^{\#}$ and $v \in K^{\#}$ such that $ux = xv$, $u \neq v$. If $x + y \in N(K)$, then $\exists\, w \in K^{\#}$ such that $u(x + y) = (x + y)w$. Therefore, $uy = yw + x(w - v)$. If $v \neq w$, then

$$x = y(u - w)(w - v)^{-1} \in K^{\#},$$

a contradiction. Hence, $v = w$ and $uy = yw$. Since $y \in C(K)$, $u = w = v$, a contradiction.

14.1.2. *If H and K are subdivision rings of a division ring D, $H \not\subset K$, and $H^{\#} \subset N(K)$, then $H \subset C(K)$.*

Proof. Deny. Then $\exists \ x \in H \backslash C(K)$ and $\exists \ y \in H \backslash K$. By Exercise 1.6.16, $\exists \ u \in H^{\#} \subset N(K)$ such that $u \notin C(K)$ and $u \notin K$. By Theorem 14.1.1, $u + 1 \notin N(K)$. Since $u + 1 \in H$ and $u \neq -1$, we have $u + 1 \in H^{\#} \subset N(K)$, a contradiction.

14.1.3. (*Cartan-Brauer-Hua.*) *If K is a subdivision ring of a division ring D, then $K^{\#} \lhd D^{\#}$ iff $K = D$ or $K \subset Z(D)$.*

Proof. If $K = D$ or $K \subset Z(D)$, then it is clear that $K^{\#} \lhd D^{\#}$. The converse follows by letting $H = D$ in Theorem 14.1.2.

14.1.4. (*Wedderburn.*) *A finite division ring is a field.*

Proof. Deny, and let D be a minimal counterexample. If there is a noncyclic subgroup H of $D^{\#}$ of order p^2, $p \in \mathscr{P}$, then (Exercise 14.1.8) there is a subfield F containing H. However $F^{\#}$ is not cyclic, a contradiction (Theorem 5.7.8). Therefore every subgroup of order p^2 is cyclic. It follows that every Sylow p-subgroup has only one subgroup of order p. But then (Theorem 9.7.3) the Sylow p-subgroups for odd p are cyclic, while the Sylow 2-subgroups are cyclic or generalized quaternion.

Suppose that a Sylow 2-subgroup of $D^{\#}$ is cyclic. Then all Sylow subgroups are cyclic. By Theorem 12.6.17, $D^{\#} = \langle x, y \rangle$ where $\langle x \rangle \lhd D^{\#}$ but $x \notin Z(D)$. Let F be a maximal subfield containing x. Then $x \in F \cap F^y$. If $F = F^y$, then $F^{\#} \lhd D^{\#}$, contradicting the Cartan-Brauer-Hua theorem. If $F \neq F^y$, then $C(x) \supset \langle F, F^y \rangle$, so $C(x) = D$ and $x \in Z(D)$, a contradiction.

Suppose that a Sylow 2-subgroup of $D^{\#}$ is generalized quaternion. Then there are elements x and y such that $x^2 = y^2$, $o(x) = 4$, and $xy = yx^{-1}$. Since $C(x^2)$ is a subdivision ring containing x and y, it is not a field, hence $C(x^2) = D$. Therefore, $x^2 \in Z(D)$. Since $x^4 = 1$, the characteristic of D is an odd prime p, and $x^2 = -1$. If P is the prime subfield, then the group ring $P(x, y)$ of the quaternion group $\langle x, y \rangle$ over P is closed under addition and multiplication, hence is a subdivision ring since D is finite. Since $P(x, y)$ is not commutative, $P(x, y) = D$. Now $P(x, y)$ is a vector space of dimension 4 over P (basis: 1, x, y, xy). Therefore, $o(D) = p^4$. Since $4 \nmid p^2 + 1$, there is an odd prime q dividing $p^2 + 1$. Then $q \mid o(D^{\#})$, so $\exists \ z \in D^{\#}$ with $o(z) = q$. Let F be a maximal subfield of D containing z. Since D is a vector space over F which is, in turn, a vector space over P, $o(F) = p^2$. By Lagrange's theorem, $q \mid p^2 - 1$. Since q is odd and divides $p^2 + 1$, this is a contradiction. ‖

There are a number of generalizations of Wedderburn's theorem. Just one of these (not best possible) will be given.

14.1.5. If D is a division ring of finite characteristic p, $a \in D\backslash Z(D)$, and $o(a)$ is finite, then $\exists\ x \in D$ such that $x^{-1}ax = a^i \neq a$.

Proof. Let P be the prime subfield. Then $P(a)$ is a finite field of order p^m, say. If $y \in P(a)$, let $y_R \in \text{End}(D, +)$ be given by $xy_R = xy$, and y_L by $xy_L = yx$. The subring of $\text{End}(D)$ generated by the y_R and y_L is commutative. The polynomials (over $P(a)$) $t^{p^m} - t$ and $\pi\{t - y \mid y \in P(a)\}$ are equal since they have the same roots and same leading coefficient. The map $y \longrightarrow y_R$ is an isomorphism of $P(a)$ into $\text{End}(D)$. Again,

$$(a_L - a_R)^p = a_L^p - a_R^p$$

since the characteristic of D is p. By induction, therefore,

$$(a_L - a_R)^{p^m} = a_L^{p^m} - a_R^{p^m} = a_L - a_R.$$

From all these remarks, it follows that

$$\pi\{a_L - a_R - y_R \mid y \in P(a)\} = (a_L - a_R)^{p^m} - (a_L - a_R) = 0.$$

Now $a \notin Z(D)$, hence $a_L - a_R \neq 0$. Therefore $\exists\ y \in P(a)^{\#}$ such that

$$\text{Ker}(a_L - a_R - y_R) \neq 0.$$

Thus $\exists\ x \in D$ such that

$$x(a_L - a_R - y_R) = 0, \qquad ax - xa - xy = 0,$$

$$x^{-1}ax = a + y \in P(a).$$

Since $P(a)$ is a finite field, $P(a)^{\#}$ is cyclic. Therefore, since $o(x^{-1}ax) = o(a)$, $x^{-1}ax = a^i \neq a$ for some i.

14.1.6. If D is a division ring and $D^{\#}$ is periodic, then D is a field.

Proof. If D is of characteristic 0, then $2 = 1 + 1$ has infinite order, a contradiction. Hence D is of prime characteristic p. Suppose $a \in D\backslash Z(D)$. By the previous theorem, $\exists\ x \in D$ such that $x^{-1}ax = a^i \neq a$. Therefore, $\langle a, x \rangle = H$ is a finite non-Abelian group. The group ring of H over the prime subfield P is then a finite noncommutative subdivision ring of D, contradicting Wedderburn's theorem. Therefore, $Z(D) = D$, and D is a field.

EXERCISES

14.1.7. Let D be a division ring and $S \neq \varnothing$ a subset of D.
  (a) $C(S)$ is a subdivision ring of D.
  (b) $Z(D)$ is a subfield of D.

14.1.8. (a) The intersection of a set of subdivision rings of a division ring is again a subdivision ring.

(b) If S is a subset of a division ring D, then there is a minimum subdivision ring of D containing S.

(c) If in (b) the elements of S commute, then there is a subfield F of D containing S.

(d) If K is a subfield of $Z(D)$ and $a \in D$, then $K(a)$ is a subfield of D.

14.1.9. A noncommutative division ring has an infinite subfield.

14.1.10. If K is a division ring of prime characteristic p and G is a finite subgroup of $K^{\#}$, then G is cyclic. (Amitsur [1] has determined all possible finite subgroups of division rings.)

14.2   Conjugates

This section contains a rather miscellaneous collection of theorems about the number of conjugates of elements and subdivision rings of a division ring.

14.2.1. *If K is a proper subdivision ring of an infinite division ring D, then* $[D^{\#}:K^{\#}] = o(D)$.

Proof. If $o(K) < o(D)$, then by Lagrange's theorem,

$$o(D) = o(D^{\#}) = [D^{\#}:K^{\#}]o(K^{\#}), \qquad [D^{\#}:K^{\#}] = o(D).$$

Suppose that $o(K) = o(D)$ and let $x \in D \backslash K$. Let

$$x + a \in K^{\#}(x + b), \qquad a \in K^{\#}, b \in K^{\#}.$$

Then $x + a = c(x + b)$ with $c \in K^{\#}$. By the linear independence of 1 and x over K, $c = 1$ and $a = b$. Hence the cosets $K^{\#}(x + a)$, $a \in K^{\#}$, are distinct. Therefore, $[D^{\#}:K^{\#}] \geq o(K^{\#}) = o(D)$. Hence again, $[D^{\#}:K^{\#}] = o(D)$.

14.2.2. *If D is a division ring and $x \in D \backslash Z(D)$, then $o(\mathrm{Cl}(x)) = o(D)$.*

Proof. D is infinite by Wedderburn's theorem. $C(x)$ is a proper subdivision ring of D. Hence, by Theorem 14.2.1,

$$o(\mathrm{Cl}(x)) = [D^{\#}:C(x)^{\#}] = o(D). \parallel$$

Since the normalizer of a subdivision ring is not necessarily a subdivision ring, the same reasoning cannot be used to show that if K is a noncentral proper subdivision ring of D, then $o(\mathrm{Cl}(K)) = o(D)$. In fact, it is an unsolved problem whether this is true. The following theorem of Faith [1] is a step in this direction.

14.2.3. *If D is a division ring and K a proper noncentral subdivision ring, then $o(\mathrm{Cl}(K))$ is infinite.*

Proof. First suppose that $C(x) \cap K$ is finite for all $x \in K \backslash Z(D)$. Then $o(x)$ is finite for all $x \in K^{\#} \backslash Z(D)$. If $K^{\#}$ is not periodic, then $\exists\, y \in K^{\#} \cap Z(D) \subset Z(K)$ and $x \in K^{\#} \backslash Z(D)$ such that y has infinite order. But then, xy is an element of $K^{\#} \backslash Z(D)$ of infinite order, contrary to an earlier remark. Therefore, $K^{\#}$ is periodic. By Theorem 14.1.6, K is a field. Since $C(x) \cap K = K$ for $x \in K$, this implies that K is finite. Hence, $K^{\#}$ is cyclic, say $K^{\#} = \langle u \rangle$. By Wedderburn, D is infinite. Hence, by Theorem 14.2.2, $o(\mathrm{Cl}(u)) = o(D)$. Since K is finite, it follows that $o(\mathrm{Cl}(K^{\#})) = o(D)$ also.

Now suppose that $x \in K \backslash Z(D)$ is such that $C(x) \cap K$ is infinite. Since $D^{\#} \backslash K^{\#}$ generates $D^{\#}$, $\exists\, y \in D \backslash K$ such that $y \notin C(x)$. Suppose that the theorem is false. Then there are distinct elements s_1, s_2, s_3 of $C(x) \cap K$ such that all $y + s_i$ are in the same right coset of $N(K)$. We have

$$y + s_1 = h(y + s_2), \qquad h \in N(K), h \neq 1,$$
$$(1 - h)y = hs_2 - s_1 = (s_2 - s_1) - (1 - h)s_2.$$
$$y = (1 - h)^{-1}(s_2 - s_1) - s_2. \tag{1}$$

Similarly,

$$y = (1 - h_1)^{-1}(s_2 - s_3) - s_2, \qquad h_1 \in N(K).$$

It follows that

$$(1 - h)(1 - h_1)^{-1} = (s_2 - s_1)(s_2 - s_3)^{-1}$$
$$= t \in C(x) \cap K,$$
$$1 - h = t - th_1,$$
$$h + (t - 1) = th_1 \in N(K),$$
$$h(t - 1)^{-1} + 1 = th_1(t - 1)^{-1} \in N(K),$$

where $t \neq 1$ since $s_1 \neq s_3$. By Theorem 14.1.1, $h(t - 1)^{-1} \in K$ or $C(K)$. If $h(t - 1)^{-1} \in K$, then $h \in K$, and by (1), $y \in K$, a contradiction. Hence $h(t - 1)^{-1} \in C(K)$. It follows that $h \in C(x)$. By (1), $y \in C(x)$, a contradiction. $\|$

The next theorem will be improved in a later section, but it is included since its proof is short and it is useful here.

14.2.4. *If D is a noncommutative division ring, then $Z_2(D^{\#}) = Z(D^{\#})$.*

Proof. Deny, and let $x \in Z_2(D^{\#}) \backslash Z(D^{\#})$. Let $y \in D \backslash Z(D)$. Then there are u and v in $Z(D^{\#})$ such that

$$xy = uyx, \qquad x(1 + y) = v(1 + y)x.$$

Subtraction gives

$$x = vx + (v - u)yx, \qquad 1 - v = (v - u)y.$$

If $v \neq 1$, then $(v - u)y \neq 0$, so $v - u \neq 0$. Therefore, $y \in Z(D)$, a contradiction. Hence $v = 1$, and $x(1 + y) = (1 + y)x$ so $xy = yx$. Thus x commutes with all $y \notin Z(D)$, therefore with all $y \in D$. Hence $x \in Z(D)$, a contradiction. ‖

Faith has generalized Theorem 14.2.3 to show that if K and L are subdivision rings of D and $N(L) \cap K < K^{\#}$, then, under various additional assumptions, $[K^{\#} : N(L) \cap K]$ is infinite. This theorem will be omitted, but a related theorem, due to Schenkman and Scott [1], is given below.

14.2.5. *If K and L are subdivision rings of a division ring D, and K is not a field, then $[K^{\#} : K \cap N(L)] \neq 2$.*

Proof. Deny. $K \cap L$ and $K \cap C(L)$ are proper subdivision rings of K. Now $N_K(K \cap L) \supset K \cap N(L)$, hence $o(\mathrm{Cl}_K(K \cap L))$ is finite. By Theorem 14.2.3, $K \cap L \subseteq Z(K)$. Similar remarks show that $K \cap C(L) \subseteq Z(K)$.

By Theorem 14.2.1, $\exists\, x \in (K \cap N(L)) \backslash Z(K)$. By Theorem 14.1.1, $x + 1$ and $x - 1$ are not in $N(L)$. But both are in K. Since $[K^{\#} : K \cap N(L)] = 2$,

$$x^2 - 1 = (x - 1)(x + 1) \in K \cap N(L).$$

Also $x^2 \in K \cap N(L)$. By Theorem 14.1.1 again, $x^2 \in L$ or $C(L)$, hence $x^2 \in Z(K)$. Thus $S = (K \cap N(L))/(Z(K) \cap N(L))$ is an elementary Abelian 2-group. The group $R = K^{\#}/Z(K)$ is either isomorphic to S or contains S as a normal subgroup of index 2. But R is centerless by Theorem 14.2.4. Hence $[R:S] = 2$. Since R is infinite, $S \neq E$. Let $u \in R \backslash S$. Since R is not Abelian, u induces an automorphism of order 2 on S. If $v \in S^{\#}$ is moved by u, then $vv^u \in S^{\#}$ is not. Hence $vv^u \in Z(R)$, a contradiction.

14.3 Subnormal subgroups

The main theorem (14.3.8) of this section, due to Stuth [1], contains as a special case the following generalization of the Cartan-Brauer-Hua theorem: If K is a proper noncentral subdivision ring of a division ring D, then $K^{\#}$ is not subnormal in $D^{\#}$. The discussion is not quite self-contained, and some outside references are required.

A subgroup H of a group G will be called *transfinitely subnormal* in G iff there is a well-ordered sequence $\{A_i\}$ of subgroups starting with H and ending with G such that $A_i \lhd A_{i+1}$ and, if j is a limit ordinal, then $A_j = \cup \{A_i \mid i < j\}$.

14.3.1. *If F is a subfield of a division ring D and $F^{\#}$ is transfinitely subnormal in $D^{\#}$, then $F = D$ or $F \subset Z(D)$.* *

Proof. Deny the theorem. D is not commutative. There is a well-ordered sequence $\{G_i\}$ of subgroups of $D^{\#}$ with first term $G_0 = F^{\#}$, last term $G_n = D^{\#}$, $G_i \lhd G_{i+1}$, and $G_j = \cup \{G_i \mid i < j\}$ for limit ordinals j.

Let F_i be the division ring generated by all F^x, $x \in G_i$. Then F_n is a normal subdivision ring of D containing F, hence not contained in $Z(D)$. By the Cartan-Brauer-Hua theorem, $F_n = D$, hence F_n is not commutative. There is a smallest ordinal j such that F_j is not a field. Since $F_0 = F$ is a field, $j \neq 0$. If j is a limit ordinal, then $F_j = \cup \{F_i \mid i < j\}$ is a union of an increasing sequence of fields, hence a field, a contradiction. Therefore, $j - 1$ exists, and F_{j-1} is a field. Let $x \in G_j$. Then $(F^x)^{\#} \subset G_{j-1}^x = G_{j-1}$. By the definition of F_{j-1}, $G_{j-1} \subset N(F_{j-1})$, hence $(F^x)^{\#} \subset N(F_{j-1})$. If $F^x \subset F_{j-1}$, then $F^x \subset C(F)$ since F and F^x are subfields of the field F_{j-1}. If $F^x \not\subset F_{j-1}$, then by Theorem 14.1.2, $F^x \subset C(F_{j-1})$. Hence $F^x \subset C(F)$ in any case. If also $y \in G_j$, then $F^{xy^{-1}} \subset C(F)$, so that $F^x \subset C(F)^y = C(F^y)$. Since the elements of the fields generating F_j all commute, F_j is commutative, a contradiction.

14.3.2. *If D is a division ring with prime subfield P, $1 \neq b \in Z(D)^{\#}$, and $n \in \mathcal{N}$, then there is a nonzero polynomial g over $P(b)$ such that if $x \in D^{\#}$, $y \in D^{\#}$,*

$$[x, y] = b, \quad x_1 = [1 + x, y], \quad x_{i+1} = [x_i, y], \quad x_n = 1,$$

then $g(x) = 0$.

Proof. Suppose that $[x, y] = b$. It follows inductively that $x^r y = y(bx)^r$. Hence if h is any polynomial over $Z(D)$, then $h(x)y = yh(bx)$. It then follows that if f is any rational function over $Z(D)$ such that $f(x)$ is defined, then $f(x)y = yf(bx)$. If also $f(x) \neq 0$, then

$$[f(x), y] = f(x)^{-1}y^{-1}f(x)y = f(x)^{-1}f(bx).$$

Hence, by the commutativity of $Z(D)(x)$,

$$x_1 = [1 + x, y] = (1 + x)^{-1}(1 + bx),$$
$$x_2 = [x_1, y] = (1 + bx)^{-1}(1 + x)(1 + bx)^{-1}(1 + b^2x)$$
$$= (1 + b^2x)(1 + bx)^{-2}(1 + x).$$

In general (Exercise 14.3.9).

$$x_k = [x_{k-1}, y] = \prod_{i=0}^{k} (1 + b^i x)^{\binom{k}{i}(-1)^{k-i}}$$

By hypothesis, $x_n = 1$, so (Exercise 14.3.10) $\exists \ a_i \in P(b)$ such that

$$(b - 1)^n + a_1 x + \ldots + a_r x^r = 0.$$

---

* Schenkman has a generalization of this theorem, unpublished at the time of writing.

Moreover, it is clear that the coefficients a_i depend only on b and n, and not on x or y.

14.3.3. *If D is a division ring, G a non-Abelian subgroup of $D^\#$, and $G^1 \subset Z(D)$, then G is not subnormal in $D^\#$.*

Proof. Deny the theorem, and let $G \lhd G_1 \lhd \ldots \lhd G_n = D^\#$. By hypothesis, $\exists\, x \in G$ and $y \in G$ such that $[x, y] = b \neq 1$, $b \in Z(D)^\#$. Then, using the notation of Theorem 14.3.2,

$$x_1 = [1 + x, y] \in [G_n, G] \subset G_{n-1}.$$

By induction, $x_n \in G$, hence $x_{n+1} = [x_n, y] \in Z(D)^\#$, $x_{n+2} = 1$. For all integers m such that $mx \neq 0$, $[mx, y] = b$ also, and $(mx)_{n+2} = 1$. If g is the polynomial guaranteed by the preceding theorem, then $g(mx) = 0$. Since $mx \in Z(D)(x)$ for all integers m, while g can have only a finite number of roots in a field, the characteristic of D is finite, say p.

Let P be the prime subfield and f the minimal polynomial of x over $P(b)$. Then,

$$f(x) = a_0 + \ldots + a_r x^r = 0, \quad a_i \in P(b), a_0 \neq 0, a_r \neq 0,$$

$$y^{-1}f(x)y = a_0 + \ldots + a_r b^r x^r = 0.$$

Hence,

$$a_1(b - 1) + \ldots + a_r(b^r - 1)x^{r-1} = 0,$$

and all coefficients are in $P(b)$. Since f was minimal, $b^r = 1$. Therefore, $P(b)$ is a finite field. Thus, x and, by symmetry, y are algebraic over the finite field $P(b)$. One then verifies (Exercise 14.3.11) that the set K of all finite sums $\Sigma\, a_{ij}x^i y^j$, $a_{ij} \in P(b)$ is finite and is closed under both addition and multiplication. Hence, K is a finite division ring, therefore, by Wedderburn's theorem, a field. Since $[x, y] \neq 1$, this is a contradiction. ‖

For any positive integer n, consider the following statement.

P_{nD}: If K is a division ring, H a subdivision ring,

$$G_1 \lhd G_2 \lhd \ldots \lhd G_n = K^\#, \qquad G_1 \subset N(H),$$

and $G_1 \not\subset Z(K)$, then $H \subset Z(K)$ or $H = K$.

Let P_{nF} denote the corresponding statement when H is a subfield (of course in this case $H \neq K$, so the conclusion will read $H \subset Z(K)$). The statement P_{nD} will be proved by induction by proving separately that $P_{n-1,D}$ implies P_{nF} and that P_{nF} implies P_{nD}. The Cartan-Brauer-Hua theorem, 14.1.3, is P_{1D}.

14.3.4. P_{nD} *implies* $P_{n+1,F}$.

Proof. Deny the theorem. Then there are K, H, $G_1, \ldots, G_{n+1}$ such that K is a division ring, H a subfield invariant under G_1,

$$G_1 \lhd \ldots \lhd G_{n+1} = K^\#,$$

$G_1 \nsubseteq Z(K)$, and $H \nsubseteq Z(K)$.

If $G_1 \subset C(H)$, then the subdivision ring G_1^* generated by G_1 is also in $C(H)$. But G_1^* is invariant under G_2. By P_{nD}, $G_1^* = K$. Hence, $C(H) = K$ and $H \subset Z(K)$, a contradiction. Therefore, $G_1 \nsubseteq C(H)$.

CASE 1. $\exists\, x \in H \backslash Z(K)$ such that x is algebraic over $Z(K)$. Then $f(x) = 0$ for some nonzero polynomial f over $Z(K)$. If $y \in G_1$, then x^y is also a root of f in the field H. Hence there is only a finite set $x_1, \ldots, x_m$ of such conjugates of x in H. Then $Z(K)(x_1, \ldots, x_m)$ is a field of finite dimension ($\neq 1$) over $Z(K)$ which is invariant under G_1. Therefore, K, H, $G_1, \ldots, G_n$ may be assumed to be such that H is of least possible finite dimension over $Z(K) \cap H$.

Suppose that $\exists\, y \in G_1 \backslash C(H)$ and $a \in H \backslash Z(K)$ such that $[y, a] = 1$. Then the minimality of $\mathrm{Dim}(H)$ is contradicted. For $C(a)$ is a division ring, H a subfield invariant under $G_1 \cap C(a)$, $G_i \cap C(a) \lhd G_{i+1} \cap C(a)$, $G_1 \cap C(a) \nsubseteq Z(C(a))$ since y is in the former group but not the latter, $H \nsubseteq Z(C(a))$ since $y \in C(a) \backslash C(H)$, and the dimension of H over $H \cap Z(C(a))$ is less than the dimension of H over $H \cap Z(K)$ since $a \in (H \cap Z(C(a))) \backslash Z(K)$.

It follows from the preceding paragraph that the natural homomorphism from G_1 into $\mathrm{Aut}(H)$ induces an isomorphism of $G_1/(G_1 \cap C(H))$ onto a nontrivial group of automorphisms of H such that the field of elements fixed by any automorphism (except I) is $H \cap Z(K)$. By Artin [3, pages 36 and 43], $G_1/(G_1 \cap C(H))$ is of prime order. Hence $G_1^1 \subset C(H)$. But $G_1^1 \lhd G_2$, so the subdivision ring $(G_1^1)^*$ is invariant under G_2. By P_{nD}, either $(G_1^1)^* = K$ or $G_1^1 \subset Z(K)$. If $(G_1^1)^* = K$, then $C(H) = K$, which is impossible. Hence $G_1^1 \subset Z(K)$. By Theorem 14.3.3, G_1 is Abelian. Therefore, G_1^* is a field invariant under G_2. Since $G_1^* \neq K$, this contradicts P_{nD}.

CASE 2. If $x \in H \backslash Z(K)$, then x is transcendental over $Z(K)$.

First suppose that $H \cap G_1 \subset Z(K)$. Since $G_1 \nsubseteq C(H)$, there are $y \in G_1$ and $x \in H$ such that $[x, y] = a \neq 1$. Using the notation of Theorem 14.3.2,

$$x_1 = [1 + x, y] \in [G_{n+1}, G_1] \subset [G_{n+1}, G_n] \subset G_n.$$

Hence, $x_1 \in H \cap G_n$. An easy induction gives $x_n \in H \cap G_1 \subset Z(K)$, and $x_{n+1} = 1$. Therefore, $\exists\, u \in H$ ($1 + x$ or an appropriate x_i) such that $[u, y] = b \neq 1$, $[b, y] = 1$, and since u and b are in H, $[u, b] = 1$. Let D be the subdivision ring generated by u and y. Thus $b \in Z(D)$. As above, one gets that $u_{n+1} = 1$. By Theorem 14.3.2, u is algebraic over $Z(D)$. One checks that $H \cap D$ is invariant under $G_1 \cap D$, $G_i \cap D \lhd G_{i+1} \cap D$, $H \cap D \nsubseteq Z(D)$

(since $[u, y] = b \neq 1$), and $G_1 \cap D \nsubseteq Z(D)$ (since $y \in (G_1 \cap D)\backslash Z(D)$). By
Case 1, $H \cap D = D$. Therefore, $y \in H$, so that $b = [u, y] = 1$, a contra-
diction.

Hence $H \cap G_1 \nsubseteq Z(K)$. Now $((H \cap G_1)^{G_2})^*$ is invariant under G_2, so
that it equals K by P_{nD}. If $(H \cap G_1)^{G_2} \subset C(H)$, then $C(H) = K$ and $H \subset$
$Z(K)$, a contradiction. Hence $\exists u \in G_2$ such that $(H \cap G_1)^u \nsubseteq C(H)$. Let
$y \in (H \cap G_1)^u\backslash C(H)$. $\exists v \in H$ such that $[v, y] \neq 1$. Since $y \in G_1^u = G_1$, it
follows as before that $v_n \in H \cap G_1$. Therefore,

$$v_{n+1} = [v_n, y] \in H \cap G_1 \cap H^u$$

since H^u is invariant under $G_1^u = G_1$. Since H^u is commutative, $v_{n+2} = 1$. As
in the preceding paragraph, this leads to a contradiction. $\|$

Before proving the other half of the inductive step, two lemmas will be
proved.

14.3.5. *If G is a group, $T \in \mathrm{Aut}(G)$, $h \in G$, $hT = h$, $L_0 = G$, and $L_{i+1} =$
h^{L_i}, then $L_n T = L_n$ for all n.*

Proof. $L_0 T = L_0$. Inductively suppose that $L_{n-1} T = L_{n-1}$. L_n is generated by $\{h^g \mid g \in L_{n-1}\}$, so $L_n T$ is generated by

$$\{(hT)^{gT} \mid g \in L_{n-1}\} = \{h^g \mid g \in L_{n-1}\}$$

since $L_{n-1} T = L_{n-1}$ by hypothesis. Hence $L_n T = L_n$.

14.3.6. *If K is a division ring, H a subdivision ring, and $g \in N(H)\backslash H$, then
$H \cap H^{1+g} = C(g) \cap H$.*

Proof. If $x \in C(g) \cap H$, then $x \in C(1 + g)$, hence

$$x = x^{1+g} \in H^{1+g} \cap H.$$

Conversely, let $h \in H \cap H^{1+g}$. Then $\exists h_1 \in H$ such that $h = h_1^{1+g}$. Since
$g \in N(H)$, $\exists h_2 \in H$ such that $h = h_2^g$. Thus

$$gh = h_2 g, \qquad (1 + g)h = h_1(1 + g).$$

Subtraction gives $h - h_1 = (h_1 - h_2)g$. Since $g \notin H$, $h_1 = h_2 = h$. Thus,
$gh = hg$ and $h \in C(g)$. Hence $h \in C(g) \cap H$. The theorem follows.

14.3.7. P_{nD} *implies* $P_{n+1,D}$.

Proof. Deny the theorem. Then $\exists K, H, G_1, \ldots, G_{n+1}$ such that K is a
division ring, H a subdivision ring invariant under G_1,

$$G_1 \lhd G_2 \lhd \ldots \lhd G_{n+1} = K^\#,$$

$G_1 \nsubseteq Z(K)$, $H \nsubseteq Z(K)$, and $H \neq K$.

We assert

$$\text{If } k \in K\backslash Z(K), \text{ then } \exists\, g \in G_1\backslash Z(K) \text{ such that } [g, k] \notin Z(K). \tag{1}$$

In fact, if (1) is false, then for all $g \in G_1$, $k^g = k[k, g] = kz$, $z \in Z(K)$, so $C(k) = C(k^g)$. In particular, $k \in C(k^g)$. A further conjugation shows that if $g_1 \in G_1$ and $g_2 \in G_1$, then $k^{g_1} \in C(k^{g_2})$. Therefore the subdivision ring $(k^{G_1})^*$ is actually a subfield invariant under G_1. Since $k \notin Z(K)$, this contradicts Theorem 14.3.4.

$$H \cap G_1 \nsubseteq Z(K). \tag{2}$$

Certainly $H \cap G_{n+1} \nsubseteq Z(K)$. Inductively suppose that

$$\exists\, h \in (H \cap G_{i+1})\backslash Z(K).$$

By (1), $\exists\, g \in G_1$ such that $[g, h] \notin Z(K)$. Now

$$[g, h] \in [G_1, H] \subset H, [g, h] \in [G_1, G_{i+1}] \subset [G_i, G_{i+1}] \subset G_i.$$

By finite induction, (2) is true.

$$\text{If } a \in (H \cap G_1)\backslash Z(K), \text{ then } C(a) \subset H. \tag{3}$$

Deny the statement, and let $y \in C(a)\backslash H$. Let $M_{n+1} = K^{\#}$, and inductively, let $M_i = a^{M_{i+1}}$. Since $a \in G_1$, an induction shows that $M_i \subset G_i$ for all i, and in particular, $M_1 \subset G_1$. Let $g \in M_1$. Then $a^g \in H$ since $g \in G_1$, and $g^y \in M_1$ by Theorem 14.3.5 with T the automorphism induced by y. Hence,

$$y^{-1}a^g y = (g^{-1})^y a^y g^y = (g^y)^{-1} a g^y = h \in H.$$

Thus, $a^g y = yh$. Since $1 + y \in C(a)\backslash H$ also, $a^g(1 + y) = (1 + y)h'$, with $h' \in H$. Subtracting, we get

$$a^g - h' = y(h' - h).$$

Since $y \notin H$, whereas all other terms in the preceding equation are in H, $h' - h = 0$, $a^g = h' = h$, $a^g y = yh = ya^g$, and $y \in C(a^g)$. Hence $(C(a)\backslash H) \subset C(a^g)$.

If $y' \in C(a) \cap H$, then both y and $y + y'$ are in $(C(a)\backslash H) \subset C(a^g)$. Hence $y' \in C(a^g)$. Thus, $C(a) \subset C(a^g)$ for all $g \in M_1$. Therefore, $C(a^{g^{-1}}) \subset C(a)$ for all $g \in M_1$. It follows that $C(a) = C(a^g) = C(a)^g$ for all $g \in M_1$, i.e., that $C(a)$ is invariant under M_1. Hence the field $Z(C(a))$ is invariant under M_1. Now $a \in M_1\backslash Z(K)$ and $a \in Z(C(a))\backslash Z(K)$. Since $M_1 \vartriangleleft \ldots \vartriangleleft M_{n+1}$, Theorem 14.3.4 yields a contradiction. Thus (3) holds.

$$C(H) = Z(K) = Z(H). \tag{4}$$

From (2) and (3), it follows that $C(H) \subset H$. Hence, $C(H) = Z(H)$. But $Z(K) \subset C(H)$, and since $Z(H)$ is invariant under G_1, it follows from Theorem 14.3.4 that $Z(H) \subset Z(K)$. Thus (4) holds.

$$\text{If } g \in N(H)\backslash H, \text{ then } G_1 \cap H \cap H^{1+g} \subset Z(K). \tag{5}$$

By Theorem 14.3.6, if (5) is false, then $\exists\, h \in (G_1 \cap H \cap C(g))\backslash Z(K)$. Hence, $g \in C(h) \subseteq H$ by (3). This contradiction proves (5).

If $Z(K) \subseteq G_1$ and $g \in N(H)\backslash H$, then $G_1 \cap H \cap H^{1+g} = Z(K)$. $\qquad$ (6)

By (4), $H \supset Z(K)$, hence $G_1 \cap H \cap H^{1+g} \supset Z(K)$. The statement now follows from (5).

There is no element of $(H \cap G_1)\backslash Z(K)$ which is algebraic over $Z(K)$. (7)

Deny, and let h be such an element. If $G_1 \subseteq H$, then by P_{nD}, $G_1^* = K$ and $K = H$, a contradiction. Therefore, $\exists\, g \in G_1\backslash H$. Then g induces an isomorphism of the subfields $Z(K)(h)$ and $Z(K)(h^g)$ of H. By Jacobson [2, p. 162], $\exists\, a \in H$ inducing the same isomorphism. Thus $h^a = h^g$ and $ag^{-1} \in C(h)$. By (3), $C(h) \subseteq H$. Hence, $ag^{-1} \in H$, so that $g \in H$, a contradiction.

If $h \in (H \cap G_1)\backslash Z(K)$, then $\exists\, g \in G_1\backslash H$ such that $[g, h] \in (H \cap G_1)\backslash Z(K)$.
$\qquad$ (8)

Deny the statement. Then $[g, h] \in Z(K)$ for all $g \in G_1\backslash H$. It was noted earlier that $G_1 \nsubseteq H$. If $g \in H \cap G_1$ and $g' \in G_1\backslash H$, then $gg' \in G_1\backslash H$, and $Z(K)$ contains

$$[gg', h] = [g, h]^{g'}[g', h] = [g, h]^{g'}z, \qquad z \in Z(K).$$

Hence, $[g, h] \in Z(K)$. Therefore, $[G_1, h] \subseteq Z(K)$. This contradicts (1).

$\exists\, g \in G_1\backslash H$ and $b \in (H \cap G_1)\backslash Z(K)$ such that $\qquad$ (9)

$$b^{1+g} \in G_1\backslash Z(K) \text{ and } b^{(1+g)^{-1}} \in G_1\backslash Z(K).$$

By (2), $\exists\, h \in (H \cap G_1)\backslash Z(K)$. By (8), $\exists\, g \in G_1\backslash H$ such that $[g, h] \in (H \cap G_1)\backslash Z(K)$. Let

$$a_2 = [g, h^{1+g}], \, a_{i+1} = [g, a_i],$$

$$b_2 = [g, h], \, b_{i+1} = [g, b_i], \, i > 2.$$

Since $a_2 = [g, h]^{1+g}$, $a_2 \in H^{1+g} \cap G_n$. Clearly, $b_2 \in H \cap G_1$. Also, $a_2 = b_2^{1+g}$. We assert that

$$a_i = b_i^{1+g} \in H^{1+g} \cap G_{n-i+2}, \qquad b_i \in H \cap G_1. \qquad (*)$$

Assume $(*)$ for $i - 1$. Then $b_i = [g, b_{i-1}] \in H \cap G_1$. Also,

$$a_i = [g, a_{i-1}] \in [G_1, G_{n-i+3}] \subseteq [G_{n-i+2}, G_{n-i+3}] \subseteq G_{n-i+2},$$

$$a_i = [g, a_{i-1}] = [g, b_{i-1}^{1+g}] = [g, b_{i-1}]^{1+g}$$

$$= b_i^{1+g} \in H^{1+g} \cap G_{n-i+2}.$$

Hence, $(*)$ is true by induction. In particular,

$$a_{n+1} = b_{n+1}^{1+g} \in H^{1+g} \cap G_1.$$

If $b_{n+1} \notin Z(K)$, then $a_{n+1} \in (H^{1+g} \cap G_1) \backslash Z(K)$, i.e., $g \in G_1 \backslash H$, $b_{n+1} \in (H \cap G_1) \backslash Z(K)$,

$$b_{n+1}^{1+g} \in (H^{1+g} \cap G_1) \backslash Z(K).$$

The same argument with $1 + g$ replaced by $(1 + g)^{-1}$ will show that

$$b_{n+1}^{(1+g)^{-1}} \in (H^{(1+g)^{-1}} \cap G_1) \backslash Z(K),$$

and (9) is proved in this case.

Now assume that $b_{n+1} \in Z(K)$. Letting $b_1 = h$, $\exists i$, $1 < i \leq n + 1$, such that $b_{i-1} \in (H \cap G_1) \backslash Z(K)$ and $b_i \in Z(K)$. If $b_i = 1$, then $1 = b_i = [g, b_{i-1}]$, so that $g \in C(b_{i-1}) \subseteq H$ by (3), whereas $g \notin H$. Hence $b_i \neq 1$, and by (*), $a_i = b_i \in Z(K)$. Let

$$c_1 = [a_{i-1} + 1, g], \qquad c_j = [c_{j-1}, g],$$
$$d_1 = [b_{i-1} + 1, g], \qquad d_j = [d_{j-1}, g], \quad j > 1.$$

Then, by (*),

$$c_1 = [(b_{i-1} + 1)^{1+g}, g] = d_1^{1+g} \in H^{1+g} \cap G_n.$$

By induction, $d_j \in H \cap G_{n-j+1}$. It follows by another induction that $c_j = d_j^{1+g} \in H^{1+g} \cap G_{n-j+1}$. If some $c_j \in Z(K)$, then $c_{j+1} = 1$, and, by Theorem 14.3.2, $a_{i-1} = b_{i-1}^{1+g}$ is algebraic over $Z(K)$, hence so is b_{i-1}, contrary to (7). Therefore, $c_j \notin Z(K)$ for all j. Let $b = d_{n+1}$. Then

$$b^{1+g} = d_{n+1}^{1+g} = c_{n+1} \in (H^{1+g} \cap G_1) \backslash Z(K).$$

Since $b_i \in Z(K)$, $b_i^{(1+g)^{-1}} \in Z(K)$. The argument above will then show that

$$b^{(1+g)^{-1}} \in (H^{(1+g)^{-1}} \cap G_1) \backslash Z(K).$$

This proves (9).

We now proceed to the proof of the theorem. P_{1D} is just the Cartan-Brauer-Hua theorem, 14.1.3. Let g and b be as in (9). Let $x = 1 + g$. Then $b^{x^{-1}} = a \in G_1 \backslash Z(K)$, and

$$c = [b, b^x] = [a^x, b^x] = [a, b]^x \in H^x,$$

since $[a, b] \in H$. Also by (9),

$$c = [b, b^x] \in [H \cap G_1, G_1] \subseteq H \cap G_1.$$

Therefore, by (5), $c \in G_1 \cap H \cap H^x \subseteq Z(K)$.

If $c = 1$, then $[a, b] = 1$ and by (3), $a \in C(b) \subseteq H$. Hence by (5),

$$b = a^x \in G_1 \cap H \cap H^x \subseteq Z(K),$$

a contradiction. Hence $c \neq 1$. Now,

$$d_1 = [b^x + 1, b] = [b + 1, a]^x \in H^x \cap G_n.$$

If $d_i = [d_{i-1}, b]$, then, after an induction, $d_n \in H^x \cap G_1$, hence,

$$d_{n+1} \in G_1 \cap H \cap H^x \subset Z(K),$$

so that $d_{n+2} = 1$. By Theorem 14.3.2, b^x is algebraic over $Z(K)$. Hence, b is also algebraic over $Z(K)$, contradicting (7).

Therefore, P_{nD} implies $P_{n+1,D}$.

14.3.8. *P_{nD} is true for all n, that is, if K is a division ring, H a subdivision ring invariant under a subgroup G which is subnormal in $K^{\#}$, $G \not\subset Z(K)$, and $H \not\subset Z(K)$, then $H = K$.*

Proof. This follows from Theorem 14.3.7 and the Cartan-Brauer-Hua theorem, 14.1.3.

EXERCISES

14.3.9. Prove the formula for x_k given in the proof of Theorem 14.3.2.

14.3.10. Prove the last equation asserted in the proof of Theorem 14.3.2.

14.3.11. Prove that if K is a division ring, F a finite subfield of $Z(K)$, x and y elements of K algebraic over F and such that $[x, y] \in F$, then the set R of all finite sums $\Sigma a_{ij} x^i y^j$, $a_{ij} \in F$, $i \in J$, $j \in J$, is a finite ring.

14.4 Subgroups of division rings

The principal theorem of the last section has a number of corollaries which will now be stated.

14.4.1. *If D is a division ring, then $D^{\#}/Z(D)^{\#}$ has no proper subnormal Abelian subgroups.*

Proof. Suppose that H is such a subgroup, and G its inverse image in $D^{\#}$. Then G is a subnormal subgroup of $D^{\#}$ such that $G^1 \subset Z(D)$. By Theorem 14.3.3, G is Abelian. Therefore, G^* is a subfield of D invariant under a subgroup G which is subnormal in $D^{\#}$ and $G \not\subset Z(D)$. This contradicts Theorem 14.3.8.

14.4.2. *If D is a division ring, G a subnormal subgroup of $D^{\#}$, and $G \not\subset Z(D)$, then $C(G) = Z(K)$.*

Proof. $C(G)$ is a subdivision ring invariant under G, and $C(G) \neq D$. The assertion now follows from Theorem 14.3.8.

14.4.3. *If D is a division ring, G is a subnormal subgroup, $G \not\subset Z(D)$, and $x \in D \backslash Z(D)$, then D is generated by x^G.*

Proof. $(x^G)^*$ is a subdivision ring of D invariant under G and not contained in $Z(D)$. The theorem follows from Theorem 14.3.8.

14.4.4. *If D is a division ring, G a subnormal subgroup of $D^{\#}$, and $G \not\subseteq Z(D)$, then G is not solvable.*

Proof. By Theorem 14.4.2, G is not Abelian. If $G^n = E$, then $\exists\, r$ such that $G^r \not\subseteq Z(D)$ and $G^{r+1} \subseteq Z(D)$. Since G^r is subnormal in D, this contradicts either Theorem 14.3.3 or the first sentence of this proof. Hence, G is not solvable.

14.4.5. *If K and M are subnormal subgroups of a division ring D which are not in $Z(D)$, then $K \cap M \not\subseteq Z(D)$.*

Proof. Deny, and let $K \lhd K_1 \lhd \ldots \lhd K_r = D^{\#}$, $M \lhd M_1 \lhd \ldots \lhd M_s = D^{\#}$, $K \not\subseteq Z(D)$, $M \not\subseteq Z(D)$, $K \cap M \subseteq Z(D)$, with $r + s$ a minimum. It is clear that $r \geq 1$ and $s \geq 1$. Thus $\exists\, x \in (K_1 \cap M) \backslash Z(D)$. Now (Exercise 15.2.9) $K \cap M_1$ is subnormal in $D^{\#}$ and is not in $Z(D)$. Hence, by Exercise 14.4.7, $\exists\, y \in K \cap M_1$ such that $[y, x] \notin Z(D)$. But

$$[y, x] \in [K, K_1] \cap [M_1, M] \subseteq K \cap M.$$

Hence $K \cap M \not\subseteq Z(D)$, a contradiction.

14.4.6. *If D is a division ring, G a subnormal subgroup of $D^{\#}$, $G \not\subseteq Z(D)$, H a subgroup of $D^{\#}$ invariant under G, and $H \not\subseteq Z(D)$, then* (i) $H \cap G \not\subseteq Z(D)$, (ii) $H^* = D$, (iii) $C(H) = Z(D)$.

Proof. Since H^* is invariant under G, (ii) follows from Theorem 14.3.8. Similarly $C(H)$ is a subdivision ring invariant under G and is not D, hence $C(H) = Z(D)$ and (iii) holds.

Let $G = G_0 \lhd G_1 \lhd \ldots \lhd G_n = D^{\#}$. Suppose inductively that $\exists\, x \in (H \cap G_{i+1}) \backslash Z(D)$. By Exercise 14.4.7, $\exists\, y \in G$ such that $[x, y] \notin Z(D)$. Then $[x, y] \in (H \cap G_i) \backslash Z(D)$. Therefore, $(H \cap G_i) \not\subseteq Z(D)$. By induction, $H \cap G \not\subseteq Z(D)$.

EXERCISE

14.4.7. If K is a division ring, G a subnormal subgroup not in $Z(K)$, and $x \in K \backslash Z(K)$, then $\exists\, y \in G$ such that $[x, y] \notin Z(K)$.

REFERENCES FOR CHAPTER 14

For Theorems 14.1.1 to 14.1.3, Schenkman [1]; Theorems 14.1.5 and 14.1.6, Herstein [1]; Theorems 14.2.1 and 14.2.2, Scott [3]; Sections 14.3 and 14.4, Stuth [1]; Theorem 14.3.3, Huzurbazar [1]; Theorem 14.3.4, Herstein and Scott [1]. See also Suprunenko [1].

FIFTEEN

TOPICS IN INFINITE GROUPS

15.1 *FC* groups

A group is an *FC group* (*finite conjugate group*) iff each $x \in G$ has only a finite number of conjugates. *FC* groups are similar to finite groups in many ways, as we shall see.

15.1.1. *A finite group is an FC group.*

15.1.2. *An Abelian group is an FC group.*

More generally:

15.1.3. *If $[G:Z(G)]$ is finite, then G is an FC group.*

Proof. If $x \in G$, then $C(x) \supset Z(G)$. Hence

$$o(Cl(x)) = [G:C(x)] \leq [G:Z(G)].$$

15.1.4. *A subgroup or factor group of an FC group is an FC group.* ‖

The converse of Theorem 15.1.3 is false (see Exercise 15.1.21), but a partial converse is given in the following theorem.

441

15.1.5. *If G is a finitely generated FC group, then $[G:Z(G)]$ is finite.*

Proof. Let S be a finite generating subset of G. Then

$$Z(G) = \cap \{C(x) \mid x \in S\}.$$

Since G is FC, $[G:C(x)]$ is finite for each $x \in S$. By Poincaré's theorem, 1.7.10, $[G:Z(G)]$ is finite.

15.1.6. *If a group $G = \langle S \rangle$, then G is an FC group iff $o(\mathrm{Cl}(x))$ is finite for each $x \in S$.*

Proof. If G is an FC group, then $o(\mathrm{Cl}(x))$ is finite for each $x \in S$ by definition of FC group. Conversely, suppose that $o(\mathrm{Cl}(x))$ is finite for each $x \in S$. Then if $x \in S$,

$$o(\mathrm{Cl}(x^{-1})) = [G:C(x^{-1})] = [G:C(x)] = o(\mathrm{Cl}(x)),$$

which is finite. Let $y \in G$. Then there are $x_1, \ldots, x_n$ in $S \cup S^{-1}$ such that $y = x_1 \ldots x_n$. If $g \in G$, then

$$g^{-1}yg = \pi(g^{-1}x_i g) \in \mathrm{Cl}(x_1) \cdots \mathrm{Cl}(x_n),$$

which is finite. Hence, $o(\mathrm{Cl}(y))$ is finite. Therefore, G is an FC group.

15.1.7. *If G is an FC group, then G^1 is periodic.*

Proof. Let $x \in G^1$. Then there are $a_1, \ldots, a_n, b_1, \ldots, b_n$ in G such that $x = \pi[a_i, b_i]$. Let $H = \langle a_1, \ldots a_n, b_1, \ldots, b_n \rangle$. Then (Theorem 15.1.4) H is a finitely generated FC group, and $x \in H^1$. By Theorem 15.1.5, $[H:Z(H)] = m$ is finite. Let T be the transfer of H into $Z(H)$. Now $\mathrm{Ker}(T) \supset H^1$ since HT is Abelian. By Exercise 3.5.8, $e = xT = x^m$. Hence, G^1 is periodic.

15.1.8. *If G^1 is periodic and T is the set of elements of finite order in G, then $T \subset G$.*

Proof. If $x \in T$, then, since $o(x^{-1}) = o(x)$, also $x^{-1} \in T$. Let $a \in T$ and $b \in T$. Then $\exists n \in \mathcal{N}$ such that $a^n = b^n = e$. Since G/G^1 is Abelian,

$$(ab)^n G^1 = a^n b^n G^1 = G^1.$$

Hence $\exists y \in G^1$ such that $(ab)^n = y$. Since G^1 is periodic, $\exists r \in \mathcal{N}$ such that $y^r = e$. Therefore, $(ab)^{nr} = e$, and $ab \in T$.

15.1.9. *If G is an FC group, and T is as in Theorem 15.1.8, then $T \subset G$.*

Proof. By Theorems 15.1.7 and 15.1.8.

15.1.10. *If G is a group, $x \in G$, and $o(x)$ and $o(\mathrm{Cl}(x))$ are finite, then x^G is a finite normal subgroup of G containing x.*

Proof. Certainly x^G is a normal subgroup of G containing x. Let $Cl(x) = \{x_1, \ldots, x_n\}$. Any $y \in x^G$ is of the form

$$y = x_{i_1} \ldots x_{i_r}. \tag{1}$$

We assert that y is also of the form

$$y = x_{j_1} \ldots x_{j_s}, \quad j_1 \leq j_2 \leq \ldots \leq j_s. \tag{2}$$

Assume that this is false, and choose a counterexample (1) with minimum r which is earliest in the lexicographic ordering of the words (1). Thus,

$$y = x_{i_1} \ldots x_{i_r}, \quad i_1 \leq i_2 \leq \ldots \leq i_t > i_{t+1}$$

(possibly $t = 1$). Now,

$$x_{i_t} x_{i_{t+1}} = x_{i_{t+1}} x_{i_{t+1}}^{-1} x_{i_t} x_{i_{t+1}} = x_{i_{t+1}} x_j$$

for some j. Replacement in (1) of $x_{i_t} x_{i_{t+1}}$ by $x_{i_{t+1}} x_j$ thus leads to an earlier word of length r which is not equal to any word of the form (2), a contradiction. Hence every $y \in x^G$ is of the form (2). Now $o(x_j) = o(x)$ for all j. It follows that if $y \in x^G$, then

$$y = x_1^{u_1} \ldots x_n^{u_n}, \quad 0 \leq u_k < o(x).$$

Hence $o(x^G) \leq o(x)^n$.

15.1.11. *If S is a finite subset of a group G, and $x \in S$ implies that both $o(x)$ and $o(Cl(x))$ are finite, then there is a finite, normal subgroup M of G such that S is contained in M.*

Proof. By Theorem 15.1.10, if $x \in S$, then there is a finite normal subgroup M_x of G containing x. By Theorem 2.3.2, $\langle M_x \mid x \in S \rangle$ is a finite, normal subgroup of G containing S. ‖

A group G is *locally normal* iff every finite subset S of G is contained in a finite, normal subgroup of G.

15.1.12. *A periodic FC group is locally normal.*

Proof. By Theorem 15.1.11.

15.1.13. *If $[G:Z(G)]$ is finite, then G^1 is finite.*

Proof. Let $G = \cup \{Z(G)x \mid x \in S\}$. If $z_i \in Z(G)$ and $x_i \in S$ for $i = 1, 2$, then $[z_1 x_1, z_2 x_2] = [x_1, x_2]$. Hence, G^1 is generated by the finite set U of all $[x, y]$ for $x \in S$ and $y \in S$. By Theorem 15.1.3, G is an *FC* group, hence G^1 is an *FC* group. By Theorem 15.1.7, G^1 is periodic. By Theorem 15.1.12, G^1 is locally normal. Since G^1 is finitely generated, it is finite.

15.1.14. *If G is a finitely generated FC group, then G^1 is finite.*

Proof. By Theorems 15.1.5 and 15.1.13.

15.1.15. *If a group G is such that $H \subset G$ implies that $o(\text{Cl}(H))$ is finite, then $H \subset G$ implies that $[H^G:H]$ is finite (hence every subgroup is of finite index in some normal subgroup of G).*

Proof. Let $x \in G$. Then $o(\text{Cl}(\langle x \rangle))$ is finite by assumption. Since a cyclic (perhaps infinite) group is generated by only a finite number of one element sets, $o(\text{Cl}(x))$ is also finite. Thus, G is an *FC* group.

Let $H \subset G$, and let $H_1, \ldots, H_n$ be the conjugates of H. $\exists\, x_i \in G$ such that $H^{x_i} = H_i$ for all i. If $h \in H \cap C(x_i)$, then $h = h^{x_i} \in H \cap H_i$, hence $H \cap C(x_i) \subset H \cap H_i$. It follows that

$$[H:H \cap H_i] \leqq [H:H \cap C(x_i)] \leqq [G:C(x_i)],$$

which is finite. Therefore,

$$[H:\text{Core}(H)] = [H: \cap (H \cap H_i)]$$

is finite. Thus, if $M = \text{Core}(H)$, then H/M is a finite subgroup of the *FC* group G/M (Theorem 15.1.4). By Theorem 15.1.11, there is a finite normal subgroup K/M of G/M containing H/M. Hence $K \lhd G$, $H \subset K$, and $[K:H]$ is finite. By the minimality property of H^G, it follows that $[H^G:H]$ is finite.

15.1.16. *If G is an FC group, then $G/Z(G)$ is locally normal.*

Proof. Let $x \in G$. Since G is an *FC* group, $[G:C(x)]$ is finite. Therefore, there is a finite subset S of G such that $G = \cup\, \{C(x)y \mid y \in S\}$. Let

$$M = C(x) \cap (\cap \{C(y) \mid y \in S\}).$$

Then $[G:M]$ is finite. Therefore, $G/\text{Core}(M)$ is finite. Hence, $\exists\, m > 0$ such that if $g \in G$, then $g^m \in \text{Core}(M) \subset M$. In particular, $x^m \in M$. If $g \in G$, then $g = cy$ for some $c \in C(x)$ and $y \in S$. Thus, $x^m g = g x^m$. Therefore $x^m \in Z(G)$. Hence, $G/Z(G)$ is periodic. Since $G/Z(G)$ is an *FC* group (Theorem 15.1.4), it is locally normal (Theorem 15.1.12).

15.1.17. *A group G is a finitely generated FC group iff it has a free Abelian subgroup of finite rank and finite index (in G) in its center.*

Proof. Suppose that A is a free Abelian subgroup of finite rank, $[G:A]$ is finite, and $A \subset Z(G)$. Since A is finitely generated and G/A is finite, G is also finitely generated. Since $[G:Z(G)]$ is finite, G is an *FC* group (Theorem 15.1.3).

Conversely, suppose that G is a finitely generated *FC* group. By Theorem 15.1.5, $[G:Z(G)]$ is finite. Therefore, $Z(G)$ is finitely generated (Exercise 8.4.33). Since $Z(G)$ is Abelian, $Z(G) = A + F$ where A is free Abelian of finite rank and F is finite (Theorem 5.4.2). Then A is a free Abelian central subgroup of finite rank and index.

15.1.18. (*Baer* [10].) *An infinite group G has an infinite number of endomorphisms.*

Proof. If $G/Z(G) \cong \text{Inn}(G)$ is infinite, we are done. Suppose $[G:Z(G)] = n < \infty$, and let T be the transfer of G into $Z(G)$. By Exercise 3.5.8, $gT = g^n$ for $g \in G$. Now, $U_r \colon yU_r = y^r$ is an endomorphism of $Z(G)$ for $r \in \mathcal{N}$. If $\text{Exp}(G)$ is not finite, then all of the endomorphisms TU_r of G are distinct.

Now suppose that $\text{Exp}(G)$ is finite. Since G^1 is finite (Theorem 15.1.13), there is a homomorphism of G onto the infinite Abelian group G/G^1. Both G/G^1 and $Z(G)$ are direct sums of an infinite number of cyclic groups of bounded prime power orders (Theorem 5.1.12). Hence there is a prime p such that $Z(G)$ has an infinite subset S of elements of order p. Since G^1 is finite, $\exists\ x \in S\backslash G^1$. Therefore one of the summands H of G/G^1 has order p^m. There is a homomorphism V of G onto a cyclic group P of order p (map G onto G/G^1, project onto H, then map H onto P). By following V with isomorphisms of P onto cyclic groups of order p in $Z(G)$, one gets an infinite number of endomorphisms of G. ‖

A related theorem is due to Alperin [1] (who also gave another proof of Theorem 15.1.18).

15.1.19. *If G is finitely generated, then* $\text{Aut}(G)$ *is finite iff G has a central cyclic subgroup H of finite index.*

Proof. (i) Suppose that $\text{Aut}(G)$ is finite. Then $n = [G:Z(G)]$ is finite. Therefore (Exercise 8.4.33), $Z(G)$ is finitely generated. If G has no central cyclic subgroup of finite index, then (Theorem 5.4.2) there is a free Abelian subgroup F of rank 2 such that $Z(G) = F + K$ for some K. The transfer T of G into $Z(G)$ maps each x onto x^n. Let $F = \langle x \rangle + \langle y \rangle = A + B$. Then following T by the projection onto A and the map $x \longrightarrow y$, one obtains a homomorphism U of G into F such that $xU = y^n$. More generally, for all $r \in \mathcal{N}$, there is a homomorphism U_r of G into F such that $xU = y^{nr}$. The map $I + U_r$ is an endomorphism of G (since $F \subseteq Z(G)$) which induces the identity on G/F. On F it is described by the matrix $\begin{bmatrix} 1 & 0 \\ rn & 1 \end{bmatrix}$ with determinant 1, hence $(I + U_r) \mid F$ is an automorphism. Therefore, $I + U_r$ is an automorphism of G, and distinct r yield distinct automorphisms. Thus $\text{Aut}(G)$ is infinite.

(ii) Suppose that G has a central cyclic subgroup H of finite index n. *WLOG*, H is infinite and $n > 1$. There is a characteristic subgroup K of G,

$K \subset H$, with $[G:K] = m$ finite (Theorem 7.1.7). The subgroup B of $\text{Aut}(G)$ of all T fixing K and G/K elementwise is of finite index in $\text{Aut}(G)$. Let $u \in G\backslash K$, $T \in B$, and $K = \langle x \rangle$. Then $uT = ux^i$ with $i \in J$. If $o(u) = r$ is finite, then $(ux^i)^r = x^{ir}$, hence $o(uT)$ is infinite unless $i = 0$, so $uT = u$. Suppose $o(u) = \infty$. Then $u^s = x^t$ for some $s > 0$ and t. Then

$$x^t = (x^t)T = (u^s)T = (ux^i)^s = u^s x^{is} = x^{is+t},$$

so $i = 0$, and again $uT = u$. Any $y \in K$ is $(yu^{-1})u$, hence $yT = y$ and $T = I$. Therefore, $B = E$, so $\text{Aut}(G)$ is finite.

15.1.20. (*Fedorov* [1].) *If an infinite group G is such that all subgroups $H \neq E$ are of finite index, then G is cyclic.*

Proof. Let $x \in G^{\#}$. Then $[G:\langle x \rangle]$ is finite, hence G is finitely generated, say $G = \langle S \rangle$ where S is finite. Now $Z(G) = \cap \{C(y) \mid y \in S\}$ is of finite index by Poincaré's theorem. By Theorem 15.1.13, G^1 is finite. Since G^1 is of infinite index, $G^1 = E$. Therefore, G is an infinite, finitely generated Abelian group, hence $G = \Sigma\, H_i$, where H_i is cyclic for each i, and H_1 is infinite. Since H_i is of infinite index for $i > 1$, it is E. Therefore $G = H_1$ is infinite cyclic.

EXERCISES

15.1.21. (a)   A direct sum of FC groups is an FC group.

(b)   Give necessary and sufficient conditions for a direct product of FC groups to be an FC group.

(c)   Give an example of an FC group G such that $[G:Z(G)]$ is infinite (or even larger than a preassigned cardinal).

(d)   Give an example of an FC group G with G^1 larger than a given cardinal A.

15.1.22. If H and G/H are FC groups, it does not follow that G is an FC group. Let K be the regular representation of a p^{∞}-group M. Let

$$H = \Sigma_E \{C_i \mid i \in M\}, \qquad o(C_i) = 2.$$

(a) K may be considered as a group of automorphisms of H.

(b)   Let $G = \text{Hol}(H, K)$. Then H and $G/H \cong K$ are Abelian, hence FC.

(c) G is not FC.

15.1.23. A torsion-free FC group is Abelian.

15.1.24. If $A > \aleph_0$ is a cardinal number and $[G:Z(G)] < A$, then $o(G^1) < A$.

15.1.25. If G^1 is finite, then G is FC.

15.1.26. Improve Theorem 15.1.18 to read: an infinite group G has an infinite number of endomorphisms U such that GU is infinite.

15.2   Composition subgroups and subnormal subgroups

A subgroup H of a group is a *composition* subgroup of *type n* iff there are $H_0 = G, H_1, \ldots, H_n = H$ such that H_i/H_{i+1} is simple and not E (and G is a composition subgroup of type 0). Clearly a composition subgroup is always subnormal, but the converse is false.

15.2.1. *If H is a composition subgroup of G of type n, $K \vartriangleleft \vartriangleleft G$, and $H < K < G$, then*

(i) K *is a composition subgroup of G of type $m < n$.*

(ii) H *is a composition subgroup of K of type less than n.*

Proof. There are normal series

$$G = H_0, \ldots, H_n = H, \tag{1}$$

and

$$G = K_0, \ldots, K_r = K$$

with $E \neq H_i/H_{i+1}$ simple. Then

$$G = K_0, \ldots, K_r = K, H_1 \cap K, \ldots, H_n \cap K = H, \tag{2}$$

is a normal series containing H and K. By the refinement theorem, 2.10.1, there are equivalent refinements of (1) and (2). Both conclusions follow immediately. ‖

In particular, if K is a composition subgroup of G, then its type is less than n.

15.2.2. *If H and K are composition subgroups of G of types m and n, then $H \cap K$ is a composition subgroup of G of type $\leq m + n$.*

Proof. The theorem is obvious if $m = 0$ or $n = 0$. Induct on the ordered pair $(\min(m, n), \max(m, n))$. Let $G = H_0, \ldots, H_m = H$ and $G = K_0, \ldots, K_n = K$ be such that H_i/H_{i+1} and K_j/K_{j+1} are simple. If $m = n = 1$, then either $(G, H, H \cap K)$ or $(G, H \cap K)$ has simple factors, hence $H \cap K$ is a composition subgroup of type 1 or 2. Therefore $WLOG, m > 1$. By induction, $H_1 \cap K$ is a composition subgroup of G of type $\leq n + 1$. By Theorem 15.2.1, $H_1 \cap K$ is a composition subgroup of H_1 of type $\leq n$. By induction, $H \cap K = H \cap (H_1 \cap K)$ is a composition subgroup of H_1 of type $\leq (m - 1) + n$, hence a composition subgroup of G of type $\leq m + n$.

15.2.3. *If A and B are composition subgroups of G, then $L = \langle A, B \rangle$ is a composition subgroup of G.*

Proof. Deny, and choose a counterexample (G, A, B) with the composition types m and n of A and B such that the ordered pair $(\max(m, n), \min(m, n))$ is as small as possible (lexicographically). Let $G = A_0, \ldots, A_m = A$ be such that A_i/A_{i+1} is simple.

If $A < A^L$, then $\exists\ b \in B$ such that $A^b \nsubseteq A$. Now A and A^b are composition subgroups of A_1 of type $m - 1$, hence by induction, $\langle A, A^b \rangle \subset L$ is a composition subgroup of A^L, hence of G, whose type in G is $< m$ by Theorem 15.2.1. This contradicts the minimality properties of m and n since $L = \langle A, A^b, B \rangle$. Hence $A = A^L$ and $A \lhd L$. Similarly, $B \lhd L$. By induction, $M = \langle A_{m-1}, B \rangle$ is a composition subgroup of G. Hence, L is not a composition subgroup of M. If $M < G$, then by Theorem 15.2.1, A and B are composition subgroups of M of types (m', n') with $m' < m$ and $n' < n$, a contradiction. Thus, $M = G$, and

$$N(A) \supset \langle A_{m-1},\ L \rangle \supset \langle A_{m-1},\ B \rangle = M = G.$$

Therefore, $A \lhd G$ and, similarly, $B \lhd G$. Hence, $L = AB \lhd G$. By Theorem 15.2.1, L is a composition subgroup of G.

15.2.4. *If A and B are subnormal subgroups of a finite group G, then $A \cap B$ and $\langle A, B \rangle$ are subnormal subgroups of G.*

Proof. In the finite case, a subgroup is a composition subgroup iff it is subnormal. The theorem therefore follows from Theorems 15.2.2 and 15.2.3. ∥

In order to avoid complex wording, we shall refer to isomorphic composition factors of a group (or groups) as equal in the following discussion. If a group G has a composition series, then its set of composition factors is the set of factors arising from any one composition series (by the refinement theorem). Note that the number of occurrences of a given factor in a composition series is ignored, once it is known to occur.

15.2.5. *If a group G has a composition series, then any subnormal subgroup A is a composition subgroup.*

Proof. For the composition series and a normal series through A have equivalent refinements, making A a composition subgroup.

15.2.6. *If G has a composition series, and A and B are subnormal subgroups, then the set of composition factors of $L = \langle A, B \rangle$ is the union of the sets for A and for B.*

Proof. By Theorems 15.2.3 and 15.2.5, A, B, and L occur in composition series, hence the sets of their composition factors exist. By taking refinements of various series, it is clear that there is a composition series of G passing through A and L. Therefore all composition factors of A are composition factors of L, and, similarly, all composition factors of B are composition factors of L.

Suppose that there is a composition factor of L which is not one for A or for B. Choose this counterexample so that the type m of A is a minimum. Since the type of A in L is less than m if $L < G$, $L = G$. If $A \lhd G$, then

$$\frac{G}{A} = \frac{AB}{A} \cong \frac{B}{A \cap B},$$

so the composition factors of G are those of A and B. Hence $A \not\lhd G$. Thus $\exists\, D \lhd G$ such that $A < D < G$. Also $\exists\, b \in B$ such that $A^b \not\subseteq A$. Now (Theorem 15.2.3) $S = \langle A, A^b \rangle$ is a composition subgroup of G, $S \subseteq D < G$, and $A < S$. The type of A in S is smaller than m, hence the composition factors of S are those of A and A^b, i.e., those of A. But $G = \langle S, B \rangle$, and the type of S is less than m, hence the composition factors of G are those of S and B, hence those of A and B, a contradiction. $\|$

In $G = J_{12}$ the composition factors (with multiplicity) J_2, J_2, J_3 appear in all possible orders as the composition factors of composition series. In fact, the list of composition series of G and ordered composition factors is as follows:

Composition Series	Ordered Composition Factors
(J_{12}, J_6, J_3, J_1)	$J_2, J_2, J_3,$
(J_{12}, J_6, J_2, J_1)	$J_2, J_3, J_2,$
(J_{12}, J_4, J_2, J_1)	$J_3, J_2, J_2.$

The question naturally arises, which groups have this property? This question is answered in the next theorem.

15.2.7. (*Berman and Lyubimov* [1].) *If a group G has a composition series, then its composition factors (with multiplicity) occur in all possible orders in composition series iff* $G = H_1 + \ldots + H_n$, *where, for each* i, *all the composition factors of* H_i *are the same.*

Proof. If $G = H_1 + \ldots + H_n$ and all the composition factors of H_i are the same, then it is clear that, given an ordering of the composition factors of G (with multiplicity), then there is a composition series of G with the factors occurring in this order.

Conversely, suppose that the composition factors of G occur in all possible orders. Let $A_1, \ldots, A_n$ be the distinct composition factors, where A_i occurs r_i times as a factor in each composition series. For each i, there is a composition series where all the factors A_i occur at the end. Hence there is a composition subgroup H_i of G which has as composition factors A_i r_i times and no others. If H_i is not normal in G, then $\exists\, x \in G$ such that $H_i^x \nsubseteq H_i$. By Theorems 15.2.3 and 15.2.6, $L = \langle H_i, H_i^x \rangle$ is a composition subgroup of G, all of whose composition factors are A_i. Since $L > H_i$, A_i occurs more than r_i times as a composition factor of L, a contradiction (to the refinement theorem). Hence $H_i \lhd G$. Now if $M = H_1 + \ldots + H_{k-1}$ exists inductively, then M and H_k have no composition factors in common, hence $M \cap H_k = E$. Therefore, $H_1 + \ldots + H_k$ exists. Thus, $H_1 + \ldots + H_n$ exists and is a composition subgroup of G. Since it has each A_i as a composition factor r_i times, $G = H_1 + \ldots + H_n$.

EXERCISES

15.2.8. It is false that if A is a composition subgroup of G and $A < H < G$, then H is a composition subgroup of G.

15.2.9. If A and B are subnormal subgroups of G, so is $A \cap B$.

15.2.10. If A and B are subnormal in G, AB is not necessarily a subgroup of G.

15.2.11. The product AB of a normal subgroup A and a subnormal subgroup B need not be normal.

15.2.12. Give an example of a group G and subnormal subgroup H such that $N(H)$ is not subnormal.

15.3 Complete groups

A group G is *complete* iff $Z(G) = E$ and $\mathrm{Aut}(G) = \mathrm{Inn}(G)$. If $Z(G) = E$, then G is naturally isomorphic to $\mathrm{Inn}(G)$, and if G is complete, then G is naturally isomorphic to $\mathrm{Aut}(G)$.

15.3.1. *If a complete group H is a normal subgroup of G, then H is a direct summand of G, in fact, $G = H + C(H)$.*

Proof. Since $H \lhd G$, $C(H) \lhd G$. Since $Z(H) = E$, $H \cap C(H) = E$. If $g \in G$ and T_g is the induced automorphism, then $T_g \mid H = T_{h'} \mid H$ for some $h \in H$, since $\mathrm{Aut}(H) = \mathrm{Inn}(H)$. Therefore, $gh^{-1} \in C(H)$ and $g \in C(H)H$. Hence, $G = C(H)H$ and $G = H + C(H)$.

15.3.2. *If H is a complete group and $H^1 < H$, then there is no group G such that $G^1 = H$.*

Proof. Deny. By the preceding theorem, $G = C(H) + H$. Hence, $G^1 = C(H)^1 + H^1$. Since $G^1 = H$, this implies that $C(H)^1 = E$ and $H^1 = H$, contrary to assumption.

15.3.3. *If* $Z(G) = E$, *then* $C_{\text{Aut}(G)}(\text{Inn}(G)) = E$.

Proof. Let $T \in C_{\text{Aut}(G)}(\text{Inn}(G))$. Then for all $x \in G$, $T_x T = T T_x$. Therefore, if $y \in G$, then

$$(xT)^{-1}(yT)(xT) = yT_x T = yTT_x = x^{-1}(yT)x.$$

Hence, $x(xT)^{-1} \in C(yT)$ for all $y \in G$, so that $x(xT)^{-1} \in Z(G) = E$. Therefore $x = xT$ for all $x \in G$, and $T = I_G$. This proves the result.

15.3.4. *If* $Z(G) = E$, *then there is a natural isomorphism* f *of* Aut(G) *onto* Aut(Inn(G)) *given by*

$$T_x(Tf) = T^{-1}T_x T = T_{xT}, \qquad T_x \in \text{Inn}(G), \ T \in \text{Aut}(G).$$

Proof. The fact that $T^{-1}T_x T = T_{xT}$ was noted in Theorem 2.11.4. The proof may now be completed directly or by recalling or checking that (i) $x \longrightarrow T_x$ is an isomorphism of G onto Inn(G), and (ii) if s is an isomorphism of G onto H, then d is an isomorphism of Aut(G) onto Aut(H), where

$$(xs)(Td) = (xT)s, \qquad x \in G, \ T \in \text{Aut}(G).$$

In our case, the above equation is just $T_x(Tf) = T_{xT}$, so that f has the asserted properties.

15.3.5. *If* $Z(G) = E$ *and* Inn(G) *is characteristic in* Aut(G), *then* Aut(G) *is complete.*

Proof. Let $A = \text{Aut}(G)$. By Theorem 15.3.3, $Z(A) = E$. It remains only to show that Aut(A) = Inn(A). Let $U \in \text{Aut}(A)$. Since Inn(G) $\in$ Char(A), $U^* = U \mid \text{Inn}(G)$ is an automorphism of Inn(G). Using the notation and result of Theorem 15.3.4, $\exists \ T \in A$ such that $Tf = U^*$. The equation for Tf shows that this means that $\exists \ S \in \text{Inn}(A)$ such that $S \mid \text{Inn}(G) = U^*$. Hence $SU^{-1} \in \text{Aut}(A)$ and $SU^{-1} \mid \text{Inn}(G) = I$. If $SU^{-1} = I$, then $U = S \in \text{Inn}(A)$. Suppose $V = SU^{-1} \neq I$. Then $\exists \ R \in A$ such that $RV \neq R$. Let $x \in G$. Then, since $V \mid \text{Inn}(G) = I$,

$$R^{-1}T_x R = (R^{-1}T_x R)V = (RV)^{-1}T_x(RV),$$
$$R(RV)^{-1} \in C_A(T_x).$$

Hence $R(RV)^{-1} \in C_A(\text{Inn}(G)) = E$ by Theorem 15.3.3. Hence, $RV = R$, a contradiction.

15.3.6. *If G is a simple non-Abelian group, then* Inn(G) *is strictly charac-teristic in* Aut(G).

Proof. Inn(G) $\triangleleft A =$ Aut(G). Now deny the theorem. Then there is an endomorphism T of A onto A such that Inn(G)$T \nsubseteq$ Inn(G). Since Inn(G) $\cong$ G, Inn(G) is simple. Therefore, since Ker(T) $\nsupseteq$ Inn(G), $T \mid$ Inn(G) is an isomorphism. Hence Inn(G)T is simple, so that Inn(G) $\nsubseteq$ Inn(G)T. Also Inn(G)$T \triangleleft A$. By simplicity,

$$\text{Inn}(G) \cap \text{Inn}(G)T = E, \qquad \text{Inn}(G)T \subset C_A(\text{Inn}(G)),$$

contradicting Theorem 15.3.3.

15.3.7. *If G is a simple non-Abelian group, then* Aut(G) *is complete.*

Proof. $Z(G) = E$. By Theorem 15.3.6, Inn(G) is characteristic in Aut(G). By Theorem 15.3.5, Aut(G) is complete.

EXERCISE

15.3.8. (a)   Sym(n) is complete for $3 \leq n$, $n \neq 6$ (n may be infinite).

   (b)   If $n \geq 3$ is finite and not 6, then there is no group G such that $G^1 = $ Sym(n).

15.4   Existence of infinite Abelian subgroups

It has been conjectured that any infinite group G contains an infinite Abelian subgroup. This is obviously the case if G is not periodic. It will be proved here in the case where G is locally finite, where a group is locally finite iff every finite subset is contained in a finite subgroup. The first two results are due to Brauer and Fowler [1], and the main theorem to P. Hall.*

15.4.1. *If G is a finite centerless group,* $h = \max o(C(u))$ *for* $u \in G^\#$, $x \in G$, $o(x) = 2$, $K_0 = 1 \cdot e$, $K_1 = \overline{\text{Cl}(x)}, \ldots, K_r$ *are such that* K_i *is the sum of the elements of the i'th conjugate class of G* (*in the group algebra*), *and* $K_1^2 = \Sigma \, a_i K_i$, *then* $a_i \leq h$ *for all* $i \neq 0$.

Proof. Let $a_i > 0$, and let $u \in G$ appear in K_i. Then $u = y_1 y_2$, where $y_i \in \text{Cl}(x)$. This implies that $y_1^{-1} u y_1 = y_2 y_1 = u^{-1}$. But the set of $z \in G$ such

---

   * As of this writing, Professor Hall's proof, which also depends on the Feit-Thompson theorem, is not yet published. The author wishes to thank Professor Hall for permission to include this theorem. Another proof of the theorem is given in Kargopolov [1].

that $z^{-1}uz = u^{-1}$ is a right coset of $C(u)$. Hence the number of $y \in \text{Cl}(x)$ such that $\exists \, y' \in \text{Cl}(x)$ for which $yy' = u$ is at most h.

15.4.2. *If L is a finite group, then there are only a finite number of isomorphism classes S of finite simple groups such that $\exists \, x \in G \in S$ with $o(x) = 2$ and $C(x) \cong L$.*

Proof. Suppose that x, G, and S are such. Let $o(G) = g$, and $o(C(x)) = o(L) = m$. Using the notation of the previous lemma, $m \leq h$, and the coefficient of e in K_1^2 is $g/m < g$. By the lemma, $(g/m)^2 < g + hg < 2hg$, so that $g/h < 2m^2$. But g/h is the index of a subgroup $C(u)$, so that (Theorem 10.2.6), $o(G) < (2m^2)!$ ‖

In order to prove the main result, we assume
(i)   (Feit and Thompson [1]) Any group of odd order is solvable.

15.4.3. *If G is a locally finite, infinite group, then G has an infinite Abelian subgroup.*

Proof. Deny. There is a subgroup A which is maximal with respect to the properties: (1) A is Abelian, (2) $C(A)$ is infinite. *WLOG*, $G = C(A)$. Then A is finite, hence G/A is locally finite and infinite. If there is a nontrivial, finite subgroup B/A of G/A with infinite normalizer, and $x \in B\backslash A$, then $\langle A, x \rangle$ is Abelian and has an infinite centralizer, contradicting the maximality of A. In particular, G/A has no infinite Abelian subgroup. Hence *WLOG*,
(∗)   No finite, nontrivial subgroup of G has infinite normalizer.
By local finiteness, *WLOG*, $G = \cup H_n$ where H_n is finite and $H_n < H_{n+1}$ for all n. Let M_n be a minimal normal non-E subgroup of H_n. Taking subsequences, it is easy to reduce the discussion to two cases.

CASE 1. $M_i \cap M_j = E$ if $i \neq j$.

By (∗), every element of $G^{\#}$ has finite centralizer. Then *WLOG*, $M_j \cap C(x) = E$ for all $x \in H_i^{\#}$ and $i < j$. We assert that $M_1 M_2 M_3$ is a Frobenius group with kernel M_3 and complement $M_1 M_2$. In fact, $M_2 \lhd H_2$ so $M_1 M_2 \subseteq H_2$. Since $M_3 \lhd H_3$, $M_1 M_2 M_3 \subseteq H_3$. If $y \in M_3^{\#}$, $x \in (M_1 M_2)^{\#}$ and $y^{-1}xy = x' \in M_1 M_2$, then $yx'x^{-1} = xyx^{-1} \in M_3$, so $x'x^{-1} \in M_3 \cap H_2 = E$. Thus $x' = x$ and $y \in C(x)$, a contradiction. Therefore (Theorem 12.6.2), $M_1 M_2 M_3$ is a Frobenius group with kernel M_3 and complement $M_1 M_2$. In a similar but easier manner, one sees that $M_1 M_2$ is a Frobenius group. This contradicts Theorem 12.6.11.

CASE 2. $M_n \subseteq M_{n+1}$ for all n.

Since there is no finite, normal subgroup of G, *WLOG*, $M_n < M_{n+1}$ for all n. Now replace H_n by $H_n^* = M_n$. The minimal normal non-E subgroups

M_n^* of H_n^* are all simple (Theorem 4.4.3). By taking subsequences, one either reduces to Case 1 again, or to the case $M_n^* < M_{n+1}^*$. Hence $WLOG$, $G = \cup \, M_n^*$ where, by the Feit-Thompson theorem, the simple groups M_n^* are all of even order. Let $x \in M_1^*$, $o(x) = 2$. $WLOG$, $C(x) \subset M_1^*$. This contradicts Theorem 15.4.2.

EXERCISES

15.4.4. Prove that if x and y are elements of order 2 in a finite group G, then $\langle x, y \rangle$ is a 4-group or a dihedral group. What happens if G is infinite?

15.4.5. (Brauer and Fowler.) If G is a group of even order not 2, then there is a proper subgroup of G of order $n > \sqrt[3]{o(G)/2}$. (See the proof of Theorem 15.4.2.)

15.4.6. Justify the division into cases in the proof of Theorem 15.4.3.

15.4.7. (Brauer and Fowler.) Let G be a finite group having more than one conjugate class of involutions (elements of order 2). Let x and y be involutions in G. Prove

    (a)   If x is not conjugate to y, then $\exists\, u \neq e$ such that $u \in C(x)$ and $y \in C(u)$. (Use Exercise 15.4.4.)

    (b)   There is an involution z not conjugate to x such that $z \in C(x)$ [use part (a)].

    (c)   In any case, $\exists\, u \neq e$ and $v \neq e$ such that $u \in C(x)$, $v \in C(u)$, and $y \in C(v)$ (use parts (a) and (b)).

15.4.8. (Compare with Theorem 15.4.3.) It is false that if G is an infinite locally finite group and $x \in G$, then there is an infinite Abelian subgroup of G containing x. Let $H = \Sigma\, H_n$ where $o(H_n) = 2$. Show that there is an automorphism x of H of order 3 whose only fixed point is e. Let $G = \mathrm{Hol}(H, x)$, and show that $C(x) = \langle x \rangle$.

15.5 Miscellaneous exercises

15.5.1. If a group G has a subgroup H of finite index such that $Z(H)$ is non-denumerable, then there is an infinite, descending, invariant series of subgroups.

15.5.2. If $Z(G)$ is torsion-free, then each factor $Z_{i+1}(G)/Z_i(G)$ of the transfinite upper-central series of G is torsion-free.

15.5.3. (Unsolved problem.) (a) If $G = AB$ and A and B satisfy the minimal condition for subgroups, then G does also.

    (b)   Likewise for the maximal condition.

15.5.4. If H is finite and $o(\text{Lat}(G/H)) = A$ is infinite, then $o(\text{Lat}(G)) = A$.

15.5.5. If $o(\text{Lat}(G)) = A$ is infinite and $[H:G]$ is finite, then $o(\text{Lat}(H)) = A$.

15.5.6. If $\text{Lat}(G)$ is dual to $\text{Lat}(H)$, then G is periodic.

15.5.7. If G is a finite 2-group and G/G^1 a noncyclic group of order 4, then G^1 is cyclic. (Hint: a minimal counterexample would have order 16.)

15.5.8. If $H \in \text{Hall}(G)$, G is finite, and G^1 is nilpotent, then H^1 is characteristic in G.

15.5.9. Let G be a finite group.

    (a)   There is a maximum normal, solvable subgroup R_G of G (the *radical* of G).

    (b)   If $R_G = E$ and $H \lhd G$, then $R_H = E$.

    (c)   The following four statements are equivalent:

        (1) $R_G \neq E$.
        (2) G has an Abelian normal subgroup $A \neq E$.
        (3) G has a solvable, subnormal subgroup $S \neq E$.
        (4) G has an Abelian subnormal subgroup $H \neq E$.

    (d) G/R_G has radical E.

15.5.10. (Automorphism tower of finite Abelian groups.)

    (a)   If $G = A + B$ where A and B are cyclic of orders p^i and p^j ($i > 0$, $j > 0$, $p \in \mathscr{P}$), then $\text{Aut}(G)$ is not Abelian.

    (b)   If G is finite Abelian but not cyclic, then $\text{Aut}(G)$ is not Abelian.

    (c)   If $n = 1, 2, 4, 5, 10, 11, 22, 23, 46, 47, 94, 3^i, 2 \cdot 3^i$, or, if $2 \cdot 3^i + 1$ is prime, $2 \cdot 3^i + 1$ or $4 \cdot 3^i + 2$ ($i > 0$), then all $\text{Aut}^k(J_n)$ are cyclic, and some $\text{Aut}^k(J_n) = E$. Determine the smallest such k.

    (d)   If $n \in \mathscr{N}$ is not one of the numbers listed in (c), then some $\text{Aut}^k(J_n)$ is not Abelian.

15.5.11. Example of a finite group G such that G^1 is not the set of commutators. Let $S = \{0, a, b, c, d, f, g, h, i, 1\}$ with multiplication: $ab = f$, $ac = g$, $ad = h$, $bc = i$, $1x = x1 = x$ for all $x \in S$, and all other products 0.

    (a) S is a semi-group (i.e., multiplication is associative).

Let R be the semi-group ring of S over the field J_2. Let G be the set of matrices over R of the form $I + A$, where

$$A = \begin{bmatrix} 0 & x & y \\ 0 & 0 & x \\ 0 & 0 & 0 \end{bmatrix}.$$

    (b) G is a finite group.

(c)   If

$$B = \begin{bmatrix} 0 & u & v \\ 0 & 0 & u \\ 0 & 0 & 0 \end{bmatrix},$$

then $[I + A, I + B] = I + \begin{bmatrix} 0 & 0 & ux + xu \\ 0 & 0 & 0 \\ 0 & 0 & 0 \end{bmatrix}.$

(d)   A *ring commutator* is an element of the form $yz - zy$. Show that $h + i$ is a sum of two ring commutators, but is not a ring commutator in R.

(e)   Use (c) and (d) to show that there is an element of G^1 which is the product of two commutators but is not a commutator.

15.5.12.   Give an example of increasing sequences $\{A_n\}$ and $\{B_n\}$ of groups such that $A_n \cong B_n$ for all n, but $\cup A_n \ncong \cup B_n$.

15.6   A final word

Needless to say, we have merely scratched the surface of group theory. It seems appropriate to mention some areas untouched or largely untouched in the book.

The theory of representations is treated by several authors, Curtis and Reiner [1], Robinson [1], and Boerner [1], in a largely nonoverlapping fashion. Applications to physics are given in several recent books. Curtis and Reiner give a brief guide to the application of character theory to the theory of groups at the end of their book. Kurosh [1] and Specht [1] give much fuller treatments of infinite groups. Specht also extends much of the theory of groups to operator groups. The most important line of research in groups today is probably that which is headed toward the classification of all finite simple groups. The previously mentioned guide in Curtis and Reiner is relevant here. In addition, there are a number of papers on simple groups discovered in the last few years, for example, Suzuki [3], Ree [1], Chevalley [1], and Steinberg [1]. The connection of groups with geometry is studied in Dieudonné [2], Artin [2], and M. Hall [1, Chapter 20]. The important theory of p-solvable groups, as well as application of linear methods in group theory, will be covered in a forthcoming book by Higman and Blackburn [1] (see also M. Hall [1, Chapter 18]). The lattice of subgroups is studied in Suzuki [2]. Permutation groups are handled by Wielandt [6].

In addition to general theories, there are a number of beautiful single theorems which we have not mentioned. For example, the theorem of Wielandt (see Zassenhaus [4, Appendix G]) that if G is finite and $Z(G) = E$, then the sequence of automorphism groups comes to a standstill eventually. In this connection, two problems seem to be open:

(1) Is it true that if G is a finite group, then the automorphism sequence is eventually periodic (up to isomorphism)?

(2) If G is infinite and $Z(G) = E$, does the automorphism sequence come to a standstill (perhaps transfinitely)?

REFERENCES FOR CHAPTER 15

For Section 15.1, J. Erdös [1] and Neumann [2]; Theorem 15.1.15, Neumann [4] and Scott [2]; Section 15.2, M. Hall [1]; Section 15.3, Zassenhaus [4]; Exercise 15.5.6, Baer [3]; Exercise 15.5.7, Taussky [1]; Exercise 15.5.8, see Huppert [4]; Exercise 15.5.10, Dubisch [1].

BIBLIOGRAPHY

The bibliography includes only items referred to in the text.

Alperin, J.

[1] "Groups with Finitely Many Automorphisms," *Pac. J. Math.*, 12 (1962) 1–5.

Amitsur, N.

[1] "Finite Subgroups of Division Rings," *Trans. Amer. Math. Soc.*, 80 (1955) 361–386.

Artin, E.

[1] "The Orders of the Linear Groups," *Commun. Pure Appl. Math.*, 8 (1955) 355–365.

[2] *Geometric Algebra*, New York: Interscience, 1957.

[3] *Galois Theory*, Notre Dame Mathematical Lectures, No. 2 (2nd ed.), 1946.

Baer, R.

[1] "Situation der Untergruppen und Struktur der Gruppe," *S.-B. Heidelberg. Akad. Math.-Nat. Klasse* 2 (1933) 12–17.

[2] "Die Kompositionsreihe der Gruppe aller eineindeutigen Abbildungen einer unendlichen Menge auf sich," *Studia Math.* 5 (1935) 15–17.

[3] "Duality and Commutativity of Groups," *Duke Math. J.*, 5 (1939) 824–838.

[4] "Nilpotent Groups and Their Generalizations," *Trans. Amer. Math. Soc.*, 47 (1940) 393–434.

[5] "Sylow Theorems for Infinite Groups," *Duke Math. J.*, 6 (1940) 598–614.

[6] *Linear Algebra and Projective Geometry*, New York: Academic, 1952.

[7] "Group Elements of Prime Power Index," *Trans. Amer. Math. Soc.*, 75 (1953) 20–47.

[8] "Das Hyperzentrum einer Gruppe III," *Math. Zeit.*, 59 (1953) 299–338.

[9] "Auflösbare Gruppen mit Maximalbedingungen," *Math. Ann.*, 129 (1955) 139–173.

[10] "Finite Extensions of Abelian Groups with Minimum Conditions," *Trans. Amer. Math. Soc.*, 79 (1955) 521–540.

Bercov, R.

[1] The Double Transitivity of a Class of Permutation Groups (unpublished).

Berman, S. and Lyubimov, V.

[1] "Groups Allowing Arbitrary Permutation of the Factors of their Composition Series" (Russian), *Uspekhi Mat. Nauk*, 12 No. 5, (1957) 181–183.

Birkhoff, G. and MacLane, S.

[1] *A Survey of Modern Algebra* (rev. ed.), New York: Macmillan, 1953.

Boerner, H.

[1] *Darstellungen von Gruppen*, Grundlehren der Math. Wiss., LXXIV, Berlin: Springer-Verlag, 1955.

Brauer, R. and Fowler, K.

[1] "On Groups of Even Order," *Ann. Math.*, 62 (1955) 565–583.

Brodkey, J.

[1] "A Note on Finite Groups with an Abelian Sylow Group," *Proc. Amer. Math. Soc.*, 14 (1963) 132–133.

Bruijn, N. de

[1] "Embedding Theorems for Infinite Groups," *Indag. Math.*, 19 (1957) 560–569.

Burnside, W.

[1] *Theory of Groups* (2nd ed.) 1911, New York: Dover, 1955.

Cartan, H. and Eilenberg, S.

[1] *Homological Algebra*, Princeton, N.J.: Princeton U. P., 1956.

Chevalley, C.

[1] "Sur Certains Groupes Simples," *Tôhoku Math. J.*, 7 (1955) 14–66.

Cohn, P.

[1] "A Remark on the General Product of Two Infinite Cyclic Groups," *Arch. Math.*, 7 (1956) 94–99.

Cole, F.

[1] "Simple Groups from Order 201 to Order 500," *Amer. J. Math.*, 14 (1892) 378–388.

[2] "Simple Groups as Far as Order 660," *Amer. J. Math.*, 15 (1893) 303–315.

Coxeter, H. and Moser, W.

[1] *Generators and Relations for Discrete Groups*, Berlin: Springer-Verlag, 1957.

Curtis, C. and Reiner, I.

[1] *Representation Theory of Finite Groups and Associative Algebras*, Pure and Applied Math., Vol. XI, New York: Interscience, 1962.

Dickson, L. E.

[1] "Definitions of a Group and a Field by Independent Postulates," *Trans. Amer. Math. Soc.*, 6 (1905) 198–204.

Dieudonné, J.

[1] "Les déterminants sur un corps non commutatif," *Bull. Soc. Math. France*, 71 (1943) 27–45.

[2] *La Géométrie des Groupes Classiques*, Ergebnisse der Mathematik und Ihrer Grenzgebiete, Springer-Verlag, 1955.

Dinkines, F.

[1] "Semi-automorphisms of Symmetric and Alternating Groups," *Proc. Amer. Math. Soc.*, 2 (1951) 478–486.

Dlab, V.

[1] "The Frattini Subgroups of Abelian Groups," *Czech. Math. J.*, 10 (1960) 1–16.

Dlab, V. and Kořínek, V.

[1] "The Frattini Subgroup of a Direct Product of Groups," *Czech. Math. J.*, 10 (1960) 350–358.

Douglas, J.

[1] "On the Supersolvability of Bicyclic Groups," *Proc. Natl. Acad. Sci. U.S.*, 47 (1961) 1493–1495.

Dubisch, R.

[1] "A Chain of Cyclic Groups," *Amer. Math. Monthly*, 66 (1959) 384–386.

Erdös, J.

[1] "The Theory of Groups with Finite Classes of Conjugate Elements," *Acta Math. Acad. Sci. Hungar.*, 5 (1954) 45–58.

Faith, C.

[1] "On Conjugates in Division Rings," *Canad. J. Math.*, 10 (1958) 374–380.

Fedorov, Yu.

[1] "On Infinite Groups of Which All Nontrivial Subgroups Have a Finite Index" (Russian), *Uspekhi Mat. Nauk*, 6 No. 1 (1951) 187–189.

Feit, W. and Thompson, J.

[1] "Solvability of Groups of Odd Order," *Pac. J. Math.*, 13 (1963) 775–1029.

Fitting, H.

[1] "Beiträge zur Theorie der Gruppen endlicher Ordnung," *Jahr. Deutsch. Math. Ver.*, 48 (1938) 77–141.

Fox, R.

[1] "Discrete Groups and Their Presentations," Lecture notes, Princeton U., 1955.

Frobenius, G. and Stickelberger, L.

[1] "Über Gruppen von vertauschbaren Elementen," *J. Reine Angew. Math.*, 86 (1878) 217–262.

Fuchs, L.

[1] *Abelian groups*, Budapest: Publishing house of Hung. Acad. Sci., 1958.

Gaschütz, W.

[1] "Zur Erweiterungstheorie der endlichen Gruppen," *J. Reine Angew. Math.*, 190 (1952) 93–107.

[2] "Über die Φ-Untergruppe endlicher Gruppen," *Math. Zeit.*, 58 (1953) 160–170.

Gilbert, J.

[1] "A Note on a Theorem of Fuchs," *Proc. Amer. Math. Soc.*, 12 (1961) 433–435.

Green, J.

[1] "On the Number of Automorphisms of a Finite Group," *Proc. Roy. Soc. (London)*, Ser. A 237 (1956) 574–581.

Hall, M.

[1] *The Theory of Groups*, New York: Macmillan, 1959.

Hall, P.

[1] "A Note on Soluble Groups," *J. London Math. Soc.*, 3 (1928) 98–105.

[2] "A Characteristic Property of Soluble Groups," *J. London Math. Soc.*, 12 (1937) 198–200.

[3] "On the Sylow Systems for a Soluble Group," *Proc. London Math. Soc.*, 43 (1937) 316–323.

[4] "Theorems Like Sylow's," *Proc. London Math. Soc.*, 6 (1956) 286–304.

[5] "Nilpotent Groups," Lecture notes, Fourth Canad. Math Cong., 1957.

[6] "Some Sufficient Conditions for a Group to be Nilpotent," *Illinois J. Math.*, 2 (1958) 787–801.

[7] "The Frattini Subgroups of Finitely Generated Groups," *Proc. London Math. Soc.*, 11 (1961) 327–352.

Hanes, H.

[1] "Primitive S-rings," Master's thesis, U. of Kansas (1959).

Herstein, I.

[1] "Wedderburn's Theorem and a Theorem of Jacobson," *Amer. Math. Monthly*, 68 (1961) 249–251.

Herstein, I. and Adney, J.

[1] "A Note on the Automorphism Group of a Finite Group," *Amer. Math. Monthly*, 59 (1952) 309–310.

Herstein, I. and Scott, W.

[1] "Subnormal Subgroups of Division Rings," *Canad. J. Math.*, 15 (1963) 80–83.

Higman, D.

[1] "Focal Series in Finite Groups," *Canad. J. Math.*, 5 (1953) 477–497.

Higman, G.

[1] "On Infinite Simple Permutation Groups," *Publ. Math. Debrecen*, 3 (1954) 221–226.

Higman, G. and Blackburn, N.

[1] *Linear Methods in Finite Groups*, Englewood Cliffs, N.J.: Prentice-Hall (to appear).

Higman, G., Neumann, B., and Neumann, H.

[1] "Embedding Theorems for Groups," *J. London Math. Soc.*, 24 (1949) 247–254.

Hirsch, K.

[1] "On Infinite Soluble Groups I," *Proc. London Math. Soc.*, 44 (1938) 53–60.

[2] "On Infinite Soluble Groups II," *Proc. London Math. Soc.*, 44 (1938) 336–344.

[3] "On Infinite Soluble Groups III," *Proc. Lond. Math. Soc.*, 49 (1946) 184–194.

[4] "On Infinite Soluble Groups V," *J. London Math. Soc.*, 29 (1954) 250–251.

Hobby, C.

[1] "The Frattini Subgroup of a p-group," *Pac. J. Math.*, 10 (1960) 209–212.

Hölder, O.

[1] "Die Gruppen der Ordnungen p^3, pq^2, pqr, p^4," *Math. Ann.*, 43 (1893) 301–412.

Howarth, J.

[1] "On the Power of a Prime Dividing the Order of the Automorphism Group of a Finite Group," *Proc. Glasgow Math. Assoc.*, 4 (1960) 163–170.

Huppert, B.

[1] "Über das Produkt von paarweise vertauschbaren zyklischen Gruppen," *Math. Zeit.*, 58 (1953) 243–264.

[2] "Monomiale Darstellung endlicher Gruppen," *Nagoya Math. J.*, 6 (1953) 93–94.

[3] "Über die Auflösbarkeit faktorisierbarer Gruppen," *Math. Zeit.*, 59 (1953) 1–7.

[4] "Normalteiler und maximale Untergruppen endlicher Gruppen," *Math. Zeit.*, 60 (1954) 409–434.

[5] "Scharf dreifach transitive Permutationsgruppen," *Arch. Math.*, 13 (1962) 61–72.

Huppert, B. and Itô, N.

[1] "Über die Auflösbarkeit faktorisierbarer Gruppen II," *Math. Zeit.*, 61 (1954) 94–99.

Huzurbazar, M.

[1] "The Multiplicative Group of a Division Ring" (Russian), *Dokl. Akad. Nauk SSSR*, 131 (1960) 1268–1271.

Itô, N.

[1] "Note on A-groups," *Nagoya Math. J.*, 4 (1952) 79–81.

[2] "On the Factorizations of the Linear Fractional Group $LF(2, p^n)$," *Acta Sci. Math. Szeged*, 15 (1953) 79–84.

[3] "Über das Produkt von zwei abelschen Gruppen," *Math. Zeit.*, 62 (1955) 400–401.

[4] "Normalteiler mehrfach transitiver Permutationsgruppen," *Math. Zeit.*, 70 (1958) 165–173.

[5] "Über den kleinsten p-Durchschnitt auflösbarer Gruppen," *Arch. Math.*, 9 (1958) 27–32.

Iwasawa, K.

[1] "Über die Einfachheit der speziellen projektiven Gruppen," *Proc. Imp. Acad. Tokyo*, 17 (1941) 57–59.

[2] "Über die struktur der endlichen Gruppen deren echte Untergruppen sämtlich nilpotent sind," *Proc. Phys. Math. Soc. Japan*, 23 (1941) 1–4.

[3] "Über die endliche Gruppen und die Verbände ihrer Untergruppen," *J. Fac. Sci. Tokyo*, 4 (1941) 171–199.

Jacobson, N.

[1] *Lectures in Abstract Algebra*, Vol. I, Princeton, N.J.: Van Nostrand, 1951.

[2] *Structure of Rings*, Colloquium Publications, Amer. Math. Soc., Vol. XXXVII, 1956.

Janko, Z.

[1] "Verallgemeinerung eines Satzes von B. Huppert und J. G. Thompson," *Arch. Math.*, 12 (1961) 280–281.

Kaplansky, I.

[1] *Infinite Abelian Groups*, Ann Arbor: U. of Mich. Press, 1954.

Kargopolov, M.

[1] "On a problem of O. Yu. Schmidt" (Russian), *Sib. Mat. Zhurnal*, 4 (1963) 232–235.

Karrass, A. and Solitar, D.

[1] "Some Remarks on the Infinite Symmetric Groups," *Math. Zeit.*, 66 (1956) 64–69.

Kegel, O.

[1] "Produkte nilpotenter Gruppen," *Arch. Math.*, 12 (1961) 90–93.

Kemhadze, S.

[1] "On the Determination of Regular p-groups" (Russian), *Uspekhi Mat. Nauk*, 7 No. 6 (1952) 193–196.

[2] "On Regularity of p-groups for $p = 2$" (Russian), *Soob. Akad. Nauk Gruzin. SSR*, 11 (1950) 607–611.

Kneser, M. and Swierczkowski, S.

[1] "Embeddings in Groups of Countable Permutations," *Colloq. Math.*, 7 (1959–1960) 177–179.

Kulikov, L.

[1] "On the Theory of Abelian Groups of Arbitrary Power" (Russian), *Mat. Sbornik*, 9 (1941) 165–182.

[2] "On the Theory of Abelian Groups of Arbitrary Power" (Russian), *Mat. Sbornik*, 16 (1945) 129–162.

Kurosh, A.

[1] *Theory of Groups*, two volumes, K. A. Hirsch, trans., New York: Chelsea, 1955.

Ladner, G.

[1] "On Solvability of Factorizable Groups," Ph.D. dissertation, U. of Kansas (1957).

Landau, E.

[1] "Über die Klassenzahl der binären quadratischen Formen von negativer Discriminante," *Math. Ann.*, 56 (1903) 671–676.

Ledermann, W. and Neumann, B.

[1] "On the Order of the Automorphism Group of a Finite Group II," *Proc. Roy. Soc. (London)*, Ser. A 235 (1956) 235–246.

Lyndon, R.

[1] "The Cohomology Theory of Group Extensions," *Duke Math. J.*, 15 (1948) 271–292.

Miller, G.

[1] "Note on a Group of Isomorphisms," *Bull. Amer. Math. Soc.*, 6 (1900) 337–339.

Miller, G. and Moreno, H.

[1] "Non-Abelian Groups in Which Every Subgroup is Abelian," *Trans. Amer. Math. Soc.*, 4 (1903) 398–404.

Nagao, H.

[1] "A Note on Extensions of Groups," *Proc. Jap. Acad.*, 25 No. 10 (1949) 11–14.

Neumann, B.

[1] "Adjunction of Elements to Groups," *J. London Math. Soc.*, 18 (1943) 4–11.

[2] "Groups with Finite Classes of Subgroups," *Proc. London Math. Soc.*, 1 (1951) 178–187.

[3] "Appendix" to Kurosh, A., *Gruppentheorie*, Berlin: Akademie-Verlag, 1953.

[4] "Groups with Finite Classes of Conjugate Subgroups," *Math. Zeit.*, 63 (1955) 76–96.

Parker, E.

[1] "On a Question Raised by Garrett Birkhoff," *Proc. Amer. Math. Soc.*, 2 (1951) 901.

Pazderski, G.

[1] "Die Ordnungen zu denen nur Gruppen mit gegebener Eigenschaft gehören," *Arch. Math.*, 10 (1959) 331–343.

Redei, L.

[1] "Zur Theorie der faktorisierbaren Gruppen I, "*Acta Math. Acad. Sci. Hungar.*, 1 (1950) 74–98.

Ree, R.

[1] "A Family of Simple Groups Associated with the Simple Lie Algebra of Type F_4," *Amer. J. Math.*, 83 (1961) 401–420.

Remak, R.

[1] "Über Untergruppen direkter Produkte von drei Faktoren," *J. f. Math.*, 166 (1931) 65–100.

Robinson, G.

[1] *Representation Theory of the Symmetric Groups*, Toronto: U. of Toronto, 1961.

Rosenberg, A.

[1] "The Structure of the Infinite General Linear Group," *Ann. of Math.*, 68 (1958) 278–294.

Schenkman, E.

[1] "Some Remarks on the Multiplicative Group of a Sfield," *Proc. Amer. Math. Soc.*, 9 (1958) 231–235.

Schenkman, E. and Scott, W.

[1] "A Generalization of the Cartan-Brauer-Hua Theorem," *Proc. Amer. Math. Soc.*, 11 (1960) 396–398.

Schottenfels, I.

[1] "Two Non-isomorphic Simple Groups of the Same Order 20160," *Ann. of Math.*, 1 (1900) 147–152.

Schreier, J. and Ulam, S.

[1] "Über die Automorphismen der Permutationsgruppe der natürlichen Zahlenfolge," *Fund. Math.*, 28 (1936) 258–260.

Schreier, O.

[1] "Über die Erweiterung von Gruppen I," *Monatsh. Math. Phys.*, 34 (1926) 165–180.

[2] "Die Untergruppen der freien Gruppen," *Hamburg Abh.*, 5 (1927) 161–183.

Schur, I.

[1] "Zur Theorie der einfach transitiven Permutationsgruppen," *Sitz. Preuss. Akad. Wiss. Berlin, Phys-math. Kl.* (1933) 598–623.

Scorza, G.

[1] *Gruppi Astratti*, Edizioni Cremonese, Perella, Rome, 1942.

Scott, W.

[1] "Groups and Cardinal Numbers," *Amer. J. Math.*, 74 (1952) 187–197.

[2] "On a Result of B. H. Neumann," *Math. Zeit.*, 66 (1956) 240.

[3] "On the Multiplicative Group of a Division Ring," *Proc. Amer. Math. Soc.*, 8 (1957) 303–305.

[4] "Solvable Factorizable Groups," *Illinois J. Math.*, 1 (1957) 389–394.

[5] "Solvable Factorizable Groups II," *Illinois J. Math.*, 4 (1960) 652–655.

Scott, W., Holmes, C., and Walker, E.

[1] "Contributions to the Theory of Groups," Report No. 5 NSF-G 1126 U. of Kansas, 1956.

Shaw, R.

[1] "Remark on a Theorem of Frobenius," *Proc. Amer. Math. Soc.*, 3 (1952) 970–972.

Specht, W.

[1] *Gruppentheorie*, Grundlehren der Math. Wiss., Band 82, Berlin: Springer-Verlag, 1956.

Speiser, A.

[1] *Die Theorie der Gruppen von endlicher Ordnung*, Dritte Auflage 1937, New York: Dover, 1945.

Steinberg, R.

[1] "Variations on a Theme of Chevalley," *Pac. J. Math.*, 9 (1959) 875–891.

Stuth, C.

[1] "A Generalization of the Cartan-Brauer-Hua Theorem," *Proc. Amer. Math. Soc.*, 15 (1964) 211–217.

Suprunenko, D.

[1] "On Solvable Subgroups of the Multiplicative Group of a Skew Field" (Russian), *Izv. Akad. Nauk SSSR Ser. Mat.*, 26 (1962) 631–638.

Suzuki, M.

[1] "On the Lattice of Subgroups of Finite Groups, *Trans. Amer. Math. Soc.*, 70 (1951) 345–371.

[2] *Structure of a Group and the Structure of Its Lattice of Subgroups*, Ergebnisse der Math. und ihrer Grenzgebiete, Heft 10, Berlin: Springer-Verlag, 1956.

[3] "A New Type of Simple Groups of Finite Order," *Proc. Natl. Acad. Sci. U.S.*, 46 (1960) 868–870.

Sylow, L.

[1] "Théorèmes sur les groupes de substitutions," *Math. Ann.*, 5 (1872) 584–594.

Szép, J. and Itô, N.

[1] "Über die Faktorisation von Gruppen," *Acta Sci. Math. Szeged*, 16 (1955) 229–231.

Taussky, O.

[1] "A Remark on the Class Field Tower," *J. London Math. Soc.*, 12 (1937) 82–85.

Thompson, J.

[1] "Finite groups with fixed-point-free automorphisms of prime order," *Proc. Natl. Acad. Sci. U.S.* 45 (1959) 578–581.

Waerden, B. van der

[1] *Modern Algebra*, two volumes, New York: Ungar, 1949.

Weir, A.

[1] "The Reidemeister-Schreier and Kurosh Subgroup Theorems," *Mathematika*, 3 (1956) 47–55.

Wiegold, J.

[1] "On Direct Factors in Groups," *J. London Math. Soc.*, 35 (1960) 310–319.

Wielandt, H.

[1] "*p*-Sylowgruppen und *p*-Faktorgruppen," *J. Reine Angew. Math.*, 182 (1940) 180–193.

[2] "Zur Theorie der einfach transitiven Permutationsgruppen II," *Math. Zeit.*, 52 (1949) 384–393.

[3] "Über das Produkt paarweise vertauschbarer nilpotenter Gruppen," *Math. Zeit.*, 55 (1951) 1–7.

[4] "Zum Satz von Sylow," *Math. Zeit.*, 60 (1954) 407–408.

[5] "Über Produkte von nilpotenten Gruppen," *Illinois J. Math.*, 2 (1958) 611–618.

[6] *Permutation Groups*, New York: Academic Press, 1964.

Wielandt, H. and Huppert, B.

[1] "Normalteiler mehrfach transitiver Permutationsgruppen," *Arch. Math.*, 9 (1958) 18–26.

Witt, E.

[1] "Die 5-fach transitiven Gruppen von Mathieu," *Hamburg Abh.*, 12 (1938) 256–264.

Yonaha, M.

[1] "Factorization of Metabelian Groups," Ph.D. dissertation, U. of Kansas, 1962.

Zassenhaus, H.

[1] "Kennzeichnung endlicher linearer Gruppen als Permutationsgruppen," *Hamburg Abh.*, 11 (1936) 17–40.

[2] "Über endliche Fastkörper," *Hamburg Abh.*, 11 (1936) 187–220.

[3] "A Group Theoretic Proof of a Theorem of Maclagan-Wedderburn," *Proc. Glasgow Math. Assoc.*, 1 (1952) 53–63.

[4] *The Theory of Groups* (2nd ed.), New York: Chelsea, 1958.

Zorn, M.

[1] "A Remark on Method in Transfinite Algebra," *Bull. Amer. Math. Soc.*, 41 (1935) 667–670.

Zuravskii, V.

[1] "On the Group of Abelian Extensions of Abelian Groups," *Soviet Math. Dokl.*, 1 (1961) 1007–1010.

INDEX OF NOTATION

471

$G \wr H$	Wreath product, 215	
$[G:H]$	Index of H in G, 20	
$[x, y]$	$x^{-1}y^{-1}xy$, 56	
$[x, y, z]$	$[[x, y], z]$, 56	
$[H, K]$	Subgroup generated by all $[h, k]$, $h \in H$, $k \in K$, 58	
$[i]_n$	Class of i modulo n, 34	
$(1, 2, 3, 4)$	Cyclic permutation, 9	

Infinitary operations

$\{,\ldots,\}$	Set whose members are, 1
$\{x \mid P\}$	The set of all x such that P is true, 1
Σ	Direct sum, 70; also usual meaning
Σ_E	External direct sum, 15 (Example 12)
π	Direct product, 14 (Example 11); also usual meaning
$\times$	Cartesian product, 14 (Example 10)
$\cup$	Set union, 1
$\overset{.}{\cup}$	Disjoint set union, 2
$\langle\ \rangle$	Subgroup generated by, 17
$\overset{*}{\pi}$	Free product, 175
$(\overset{*}{\pi})_A$	Free product with amalgamated subgroup A, 178

Other

$\exists$	There is
$\exists\,\vert$	There is a unique
$\aleph_0$	Aleph null, the smallest infinite cardinal
$\varnothing$	Empty set, 1

ALPHABETIC

Script

$\mathscr{C}$	Field of complex numbers
$\mathscr{N}$	Set of natural numbers
$\mathscr{P}$	Set of primes
$\mathscr{P}\infty$	Set of prime powers ($\neq 1$)
$\mathscr{R}$	Rationals (set, group, or field)

Non-script

$\mathrm{Aff}(V)$	Group of affine transformations of a vector space V, 278
$\mathrm{Alt}(M)$	Alternating group on set M, 267

$o(S)$	Number of elements in set S, 1
$o(x)$	Order of x (smallest positive n such that $x^n = e$), 34
O_{SG}	Zero function from S into group G: $sO_{SG} = e$, 15 (Example 13)
PGL	Projective general linear group, 278
PSL	Projective special linear group, 278
$P\Gamma L$	Projective semi-linear group, 281
$\mathrm{Qua}(H)$	Subgroup of all elements of G normalizing all subgroups of H, 396
Rng	Range, 2
$SL(n, F)$	Group of $n \times n$ matrices of determinant 1 over field F, 125
Soc	Sockel of. Subgroup generated by all minimal normal subgroups, 168
Syl	Set of all Sylow subgroups of, 132
Syl_p	Set of all Sylow p-subgroups of, 132
$\mathrm{Sym}(A, B)$	Group of permutations of A letters moving $< B$ of them (A and B infinite), 301
$\mathrm{Sym}(M)$	Symmetric group on set M, 9
$\mathrm{Sym}(n)$	Symmetric group of degree n, 28
Tr	Trace, 322
Trans	Group of factor systems which are translations, 244
WLOG	Without loss of generality
X_T	Character of representation T, 322
X_j^i	Value of ith irreducible character on jth conjugate class, 327
Z	Center of, 50
$Z_n(G)$	nth term of upper central series of G, 140
$Z^n(G)$	nth term of lower central series of G, 142
ΓL	Group of semi-linear transformations of, 280

INDEX

A CATALOG OF SELECTED
DOVER BOOKS
IN ALL FIELDS OF INTEREST

A CATALOG OF SELECTED DOVER
BOOKS IN ALL FIELDS OF INTEREST

LASERS AND HOLOGRAPHY, Winston E. Kock. Sound introduction to burgeoning field, expanded (1981) for second edition. 84 illustrations. 160pp. 5⅜ × 8¼. (EUK)   24041-X Pa. $3.50

FLORAL STAINED GLASS PATTERN BOOK, Ed Sibbett, Jr. 96 exquisite floral patterns—irises, poppie, lilies, tulips, geometrics, abstracts, etc.—adaptable to innumerable stained glass projects. 64pp. 8¼ × 11.   24259-5 Pa. $3.50

THE HISTORY OF THE LEWIS AND CLARK EXPEDITION, Meriwether Lewis and William Clark. Edited by Eliott Coues. Great classic edition of Lewis and Clark's day-by-day journals. Complete 1893 edition, edited by Eliott Coues from Biddle's authorized 1814 history. 1508pp. 5⅜ × 8½.
21268-8, 21269-6, 21270-X Pa. Three-vol. set $22.50

ORLEY FARM, Anthony Trollope. Three-dimensional tale of great criminal case. Original Millais illustrations illuminate marvelous panorama of Victorian society. Plot was author's favorite. 736pp. 5⅜ × 8½.   24181-5 Pa. $10.95

THE CLAVERINGS, Anthony Trollope. Major novel, chronicling aspects of British Victorian society, personalities. 16 plates by M. Edwards; first reprint of full text. 412pp. 5⅜ × 8½.   23464-9 Pa. $6.00

EINSTEIN'S THEORY OF RELATIVITY, Max Born. Finest semi-technical account; much explanation of ideas and math not readily available elsewhere on this level. 376pp. 5⅜ × 8½.   60769-0 Pa. $5.00

COMPUTABILITY AND UNSOLVABILITY, Martin Davis. Classic graduate-level introduction th theory of computability, usually referred to as theory of recurrent functions. New preface and appendix. 288pp. 5⅜ × 8½.   61471-9 Pa. $6.50

THE GODS OF THE EGYPTIANS, E.A. Wallis Budge. Never excelled for richness, fullness: all gods, goddesses, demons, mythical figures of Ancient Egypt; their legends, rites, incarnations, etc. Over 225 illustrations, plus 6 color plates. 988pp. 6⅛ × 9¼. (EBE)   22055-9, 22056-7 Pa., Two-vol. set $20.00

THE I CHING (THE BOOK OF CHANGES), translated by James Legge. Most penetrating divination manual ever prepared. Indispensable to study of early Oriental civilizations, to modern inquiring reader. 448pp. 5⅜ × 8½.
21062-6 Pa. $6.50

THE CRAFTSMAN'S HANDBOOK, Cennino Cennini. 15th-century handbook, school of Giotto, explains applying gold, silver leaf; gesso; fresco painting, grinding pigments, etc. 142pp. 6⅛ × 9¼.   20054-X Pa. $3.50

AN ATLAS OF ANATOMY FOR ARTISTS, Fritz Schider. Finest text, working book. Full text, plus anatomical illustrations; plates by great artists showing anatomy. 593 illustrations. 192pp. 7⅞ × 10¼.   20241-0 Pa. $6.50

EASY-TO-MAKE STAINED GLASS LIGHTCATCHERS, Ed Sibbett, Jr. 67 designs for most enjoyable ornaments: fruits, birds, teddy bears, trumpet, etc. Full size templates. 64pp. 8¼ × 11.   24081-9 Pa. $3.95

TRIAD OPTICAL ILLUSIONS AND HOW TO DESIGN THEM, Harry Turner. Triad explained in 32 pages of text, with 32 pages of Escher-like patterns on coloring stock. 92 figures. 32 plates. 64pp. 8¼ × 11.   23549-1 Pa. $2.95

THE BOOK OF BEASTS: Being a Translation from a Latin Bestiary of the Twelfth Century, T. H. White. Wonderful catalog real and fanciful beasts: manticore, griffin, phoenix, amphivius, jaculus, many more. White's witty erudite commentary on scientific, historical aspects. Fascinating glimpse of medieval mind. Illustrated. 296pp. 5⅜ × 8¼. (Available in U.S. only) 24609-4 Pa. $5.95

FRANK LLOYD WRIGHT: ARCHITECTURE AND NATURE With 160 Illustrations, Donald Hoffmann. Profusely illustrated study of influence of nature—especially prairie—on Wright's designs for Fallingwater, Robie House, Guggenheim Museum, other masterpieces. 96pp. 9¼ × 10¾. 25098-9 Pa. $7.95

FRANK LLOYD WRIGHT'S FALLINGWATER, Donald Hoffmann. Wright's famous waterfall house: planning and construction of organic idea. History of site, owners, Wright's personal involvement. Photographs of various stages of building. Preface by Edgar Kaufmann, Jr. 100 illustrations. 112pp. 9¼ × 10.
23671-4 Pa. $7.95

YEARS WITH FRANK LLOYD WRIGHT: Apprentice to Genius, Edgar Tafel. Insightful memoir by a former apprentice presents a revealing portrait of Wright the man, the inspired teacher, the greatest American architect. 372 black-and-white illustrations. Preface. Index. vi + 228pp. 8¼ × 11. 24801-1 Pa. $9.95

THE STORY OF KING ARTHUR AND HIS KNIGHTS, Howard Pyle. Enchanting version of King Arthur fable has delighted generations with imaginative narratives of exciting adventures and unforgettable illustrations by the author. 41 illustrations. xviii + 313pp. 6⅛ × 9¼. 21445-1 Pa. $5.95

THE GODS OF THE EGYPTIANS, E. A. Wallis Budge. Thorough coverage of numerous gods of ancient Egypt by foremost Egyptologist. Information on evolution of cults, rites and gods; the cult of Osiris; the Book of the Dead and its rites; the sacred animals and birds; Heaven and Hell; and more. 956pp. 6⅛ × 9¼.
22055-9, 22056-7 Pa., Two-vol. set $20.00

A THEOLOGICO-POLITICAL TREATISE, Benedict Spinoza. Also contains unfinished *Political Treatise*. Great classic on religious liberty, theory of government on common consent. R. Elwes translation. Total of 421pp. 5⅜ × 8½.
20249-6 Pa. $6.95

INCIDENTS OF TRAVEL IN CENTRAL AMERICA, CHIAPAS, AND YUCATAN, John L. Stephens. Almost single-handed discovery of Maya culture; exploration of ruined cities, monuments, temples; customs of Indians. 115 drawings. 892pp. 5⅜ × 8½. 22404-X, 22405-8 Pa., Two-vol. set $15.90

LOS CAPRICHOS, Francisco Goya. 80 plates of wild, grotesque monsters and caricatures. Prado manuscript included. 183pp. 6⅜ × 9⅜. 22384-1 Pa. $4.95

AUTOBIOGRAPHY: The Story of My Experiments with Truth, Mohandas K. Gandhi. Not hagiography, but Gandhi in his own words. Boyhood, legal studies, purification, the growth of the Satyagraha (nonviolent protest) movement. Critical, inspiring work of the man who freed India. 480pp. 5⅜ × 8½. (Available in U.S. only)
24593-4 Pa. $6.95

ILLUSTRATED DICTIONARY OF HISTORIC ARCHITECTURE, edited by Cyril M. Harris. Extraordinary compendium of clear, concise definitions for over 5,000 important architectural terms complemented by over 2,000 line drawings. Covers full spectrum of architecture from ancient ruins to 20th-century Modernism. Preface. 592pp. 7½ × 9⅝. 24444-X Pa. $14.95

THE NIGHT BEFORE CHRISTMAS, Clement Moore. Full text, and woodcuts from original 1848 book. Also critical, historical material. 19 illustrations. 40pp. 4⅝ × 6. 22797-9 Pa. $2.25

THE LESSON OF JAPANESE ARCHITECTURE: 165 Photographs, Jiro Harada. Memorable gallery of 165 photographs taken in the 1930's of exquisite Japanese homes of the well-to-do and historic buildings. 13 line diagrams. 192pp. 8⅜ × 11¼. 24778-3 Pa. $8.95

THE AUTOBIOGRAPHY OF CHARLES DARWIN AND SELECTED LETTERS, edited by Francis Darwin. The fascinating life of eccentric genius composed of an intimate memoir by Darwin (intended for his children); commentary by his son, Francis; hundreds of fragments from notebooks, journals, papers; and letters to and from Lyell, Hooker, Huxley, Wallace and Henslow. xi + 365pp. 5⅜ × 8. 20479-0 Pa. $5.95

WONDERS OF THE SKY: Observing Rainbows, Comets, Eclipses, the Stars and Other Phenomena, Fred Schaaf. Charming, easy-to-read poetic guide to all manner of celestial events visible to the naked eye. Mock suns, glories, Belt of Venus, more. Illustrated. 299pp. 5¼ × 8¼. 24402-4 Pa. $7.95

BURNHAM'S CELESTIAL HANDBOOK, Robert Burnham, Jr. Thorough guide to the stars beyond our solar system. Exhaustive treatment. Alphabetical by constellation: Andromeda to Cetus in Vol. 1; Chamaeleon to Orion in Vol. 2; and Pavo to Vulpecula in Vol. 3. Hundreds of illustrations. Index in Vol. 3. 2,000pp. 6⅛ × 9¼. 23567-X, 23568-8, 23673-0 Pa., Three-vol. set $36.85

STAR NAMES: Their Lore and Meaning, Richard Hinckley Allen. Fascinating history of names various cultures have given to constellations and literary and folkloristic uses that have been made of stars. Indexes to subjects. Arabic and Greek names. Biblical references. Bibliography. 563pp. 5⅜ × 8½. 21079-0 Pa. $7.95

THIRTY YEARS THAT SHOOK PHYSICS: The Story of Quantum Theory, George Gamow. Lucid, accessible introduction to influential theory of energy and matter. Careful explanations of Dirac's anti-particles, Bohr's model of the atom, much more. 12 plates. Numerous drawings. 240pp. 5⅜ × 8½. 24895-X Pa. $4.95

CHINESE DOMESTIC FURNITURE IN PHOTOGRAPHS AND MEASURED DRAWINGS, Gustav Ecke. A rare volume, now affordably priced for antique collectors, furniture buffs and art historians. Detailed review of styles ranging from early Shang to late Ming. Unabridged republication. 161 black-and-white drawings, photos. Total of 224pp. 8⅜ × 11¼. (Available in U.S. only) 25171-3 Pa. $12.95

VINCENT VAN GOGH: A Biography, Julius Meier-Graefe. Dynamic, penetrating study of artist's life, relationship with brother, Theo, painting techniques, travels, more. Readable, engrossing. 160pp. 5⅜ × 8½. (Available in U.S. only) 25253-1 Pa. $3.95

CHILDREN'S BOOKPLATES AND LABELS, Ed Sibbett, Jr. 6 each of 12 types based on *Wizard of Oz, Alice,* nursery rhymes, fairy tales. Perforated; full color. 24pp. 8¼ × 11. 23538-6 Pa. $3.50

READY-TO-USE VICTORIAN COLOR STICKERS: 96 Pressure-Sensitive Seals, Carol Belanger Grafton. Drawn from authentic period sources. Motifs include heads of men, women, children, plus florals, animals, birds, more. Will adhere to any clean surface. 8pp. 8½ × 11. 24551-9 Pa. $2.95

CUT AND FOLD PAPER SPACESHIPS THAT FLY, Michael Grater. 16 colorful, easy-to-build spaceships that really fly. Star Shuttle, Lunar Freighter, Star Probe, 13 others. 32pp. 8¼ × 11. 23978-0 Pa. $2.50

CUT AND ASSEMBLE PAPER AIRPLANES THAT FLY, Arthur Baker. 8 aerodynamically sound, ready-to-build paper airplanes, designed with latest techniques. Fly *Pegasus, Daedalus, Songbird,* 5 other aircraft. Instructions. 32pp. 9¼ × 11¼. 24302-8 Pa. $3.95

SIDELIGHTS ON RELATIVITY, Albert Einstein. Two lectures delivered in 1920-21: *Ether and Relativity* and *Geometry and Experience.* Elegant ideas in non-mathematical form. 56pp. 5⅜ × 8½. 24511-X Pa. $2.25

FADS AND FALLACIES IN THE NAME OF SCIENCE, Martin Gardner. Fair, witty appraisal of cranks and quacks of science: Velikovsky, orgone energy, Bridey Murphy, medical fads, etc. 373pp. 5⅜ × 8½. 20394-8 Pa. $5.95

VACATION HOMES AND CABINS, U.S. Dept. of Agriculture. Complete plans for 16 cabins, vacation homes and other shelters. 105pp. 9 × 12. 23631-5 Pa. $4.95

HOW TO BUILD A WOOD-FRAME HOUSE, L.O. Anderson. Placement, foundations, framing, sheathing, roof, insulation, plaster, finishing—almost everything else. 179 illustrations. 223pp. 7⅞ × 10¾. 22954-8 Pa. $5.50

THE MYSTERY OF A HANSOM CAB, Fergus W. Hume. Bizarre murder in a hansom cab leads to engrossing investigation. Memorable characters, rich atmosphere. 19th-century bestseller, still enjoyable, exciting. 256pp. 5⅜ × 8. 21956-9 Pa. $4.00

MANUAL OF TRADITIONAL WOOD CARVING, edited by Paul N. Hasluck. Possibly the best book in English on the craft of wood carving. Practical instructions, along with 1,146 working drawings and photographic illustrations. 576pp. 6½ × 9¼. 23489-4 Pa. $8.95

WHITTLING AND WOODCARVING, E.J Tangerman. Best book on market; clear, full. If you can cut a potato, you can carve toys, puzzles, chains, etc. Over 464 illustrations. 293pp. 5⅜ × 8½. 20965-2 Pa. $4.95

AMERICAN TRADEMARK DESIGNS, Barbara Baer Capitman. 732 marks, logos and corporate-identity symbols. Categories include entertainment, heavy industry, food and beverage. All black-and-white in standard forms. 160pp. 8⅛ × 11. 23259-X Pa. $6.95

DECORATIVE FRAMES AND BORDERS, edited by Edmund V. Gillon, Jr. Largest collection of borders and frames ever compiled for use of artists and designers. Renaissance, neo-Greek, Art Nouveau, Art Deco, to mention only a few styles. 396 illustrations. 192pp. 8⅜ × 11¼. 22928-9 Pa. $6.00

ILLUSTRATED GUIDE TO SHAKER FURNITURE, Robert Meader. All furniture and appurtenances, with much on unknown local styles. 235 photos. 146pp. 9 × 12. 22819-3 Pa. $7.95

WHALE SHIPS AND WHALING: A Pictorial Survey, George Francis Dow. Over 200 vintage engravings, drawings, photographs of barks, brigs, cutters, other vessels. Also harpoons, lances, whaling guns, many other artifacts. Comprehensive text by foremost authority. 207 black-and-white illustrations. 288pp. 6 × 9.
24808-9 Pa. $8.95

THE BERTRAMS, Anthony Trollope. Powerful portrayal of blind self-will and thwarted ambition includes one of Trollope's most heartrending love stories. 497pp. 5⅜ × 8½. 25119-5 Pa. $8.95

ADVENTURES WITH A HAND LENS, Richard Headstrom. Clearly written guide to observing and studying flowers and grasses, fish scales, moth and insect wings, egg cases, buds, feathers, seeds, leaf scars, moss, molds, ferns, common crystals, etc.—all with an ordinary, inexpensive magnifying glass. 209 exact line drawings aid in your discoveries. 220pp. 5⅜ × 8½. 23330-8 Pa. $3.95

RODIN ON ART AND ARTISTS, Auguste Rodin. Great sculptor's candid, wide-ranging comments on meaning of art; great artists; relation of sculpture to poetry, painting, music; philosophy of life, more. 76 superb black-and-white illustrations of Rodin's sculpture, drawings and prints. 119pp. 8⅜ × 11¼. 24487-3 Pa. $6.95

FIFTY CLASSIC FRENCH FILMS, 1912–1982: A Pictorial Record, Anthony Slide. Memorable stills from Grand Illusion, Beauty and the Beast, Hiroshima, Mon Amour, many more. Credits, plot synopses, reviews, etc. 160pp. 8¼ × 11.
25256-6 Pa. $11.95

THE PRINCIPLES OF PSYCHOLOGY, William James. Famous long course complete, unabridged. Stream of thought, time perception, memory, experimental methods; great work decades ahead of its time. 94 figures. 1,391pp. 5⅜ × 8½.
20381-6, 20382-4 Pa., Two-vol. set $19.90

BODIES IN A BOOKSHOP, R. T. Campbell. Challenging mystery of blackmail and murder with ingenious plot and superbly drawn characters. In the best tradition of British suspense fiction. 192pp. 5⅜ × 8½. 24720-1 Pa. $3.95

CALLAS: PORTRAIT OF A PRIMA DONNA, George Jellinek. Renowned commentator on the musical scene chronicles incredible career and life of the most controversial, fascinating, influential operatic personality of our time. 64 black-and-white photographs. 416pp. 5⅜ × 8¼. 25047-4 Pa. $7.95

GEOMETRY, RELATIVITY AND THE FOURTH DIMENSION, Rudolph Rucker. Exposition of fourth dimension, concepts of relativity as Flatland characters continue adventures. Popular, easily followed yet accurate, profound. 141 illustrations. 133pp. 5⅜ × 8½. 23400-2 Pa. $3.50

HOUSEHOLD STORIES BY THE BROTHERS GRIMM, with pictures by Walter Crane. 53 classic stories—Rumpelstiltskin, Rapunzel, Hansel and Gretel, the Fisherman and his Wife, Snow White, Tom Thumb, Sleeping Beauty, Cinderella, and so much more—lavishly illustrated with original 19th century drawings. 114 illustrations. x + 269pp. 5⅜ × 8½. 21080-4 Pa. $4.50

SUNDIALS, Albert Waugh. Far and away the best, most thorough coverage of ideas, mathematics concerned, types, construction, adjusting anywhere. Over 100 illustrations. 230pp. 5⅜ × 8½. 22947-5 Pa. $4.00

PICTURE HISTORY OF THE NORMANDIE: With 190 Illustrations, Frank O. Braynard. Full story of legendary French ocean liner: Art Deco interiors, design innovations, furnishings, celebrities, maiden voyage, tragic fire, much more. Extensive text. 144pp. 8⅜ × 11¾. 25257-4 Pa. $9.95

THE FIRST AMERICAN COOKBOOK: A Facsimile of "American Cookery," 1796, Amelia Simmons. Facsimile of the first American-written cookbook published in the United States contains authentic recipes for colonial favorites—pumpkin pudding, winter squash pudding, spruce beer, Indian slapjacks, and more. Introductory Essay and Glossary of colonial cooking terms. 80pp. 5⅜ × 8½. 24710-4 Pa. $3.50

101 PUZZLES IN THOUGHT AND LOGIC, C. R. Wylie, Jr. Solve murders and robberies, find out which fishermen are liars, how a blind man could possibly identify a color—purely by your own reasoning! 107pp. 5⅜ × 8½. 20367-0 Pa. $2.00

THE BOOK OF WORLD-FAMOUS MUSIC—CLASSICAL, POPULAR AND FOLK, James J. Fuld. Revised and enlarged republication of landmark work in musico-bibliography. Full information about nearly 1,000 songs and compositions including first lines of music and lyrics. New supplement. Index. 800pp. 5⅜ × 8¼. 24857-7 Pa. $14.95

ANTHROPOLOGY AND MODERN LIFE, Franz Boas. Great anthropologist's classic treatise on race and culture. Introduction by Ruth Bunzel. Only inexpensive paperback edition. 255pp. 5⅜ × 8½. 25245-0 Pa. $5.95

THE TALE OF PETER RABBIT, Beatrix Potter. The inimitable Peter's terrifying adventure in Mr. McGregor's garden, with all 27 wonderful, full-color Potter illustrations. 55pp. 4¼ × 5½. (Available in U.S. only) 22827-4 Pa. $1.75

THREE PROPHETIC SCIENCE FICTION NOVELS, H. G. Wells. *When the Sleeper Wakes, A Story of the Days to Come* and *The Time Machine* (full version). 335pp. 5⅜ × 8½. (Available in U.S. only) 20605-X Pa. $5.95

APICIUS COOKERY AND DINING IN IMPERIAL ROME, edited and translated by Joseph Dommers Vehling. Oldest known cookbook in existence offers readers a clear picture of what foods Romans ate, how they prepared them, etc. 49 illustrations. 301pp. 6⅛ × 9¼. 23563-7 Pa. $6.00

SHAKESPEARE LEXICON AND QUOTATION DICTIONARY, Alexander Schmidt. Full definitions, locations, shades of meaning of every word in plays and poems. More than 50,000 exact quotations. 1,485pp. 6½ × 9¼. 22726-X, 22727-8 Pa., Two-vol. set $27.90

THE WORLD'S GREAT SPEECHES, edited by Lewis Copeland and Lawrence W. Lamm. Vast collection of 278 speeches from Greeks to 1970. Powerful and effective models; unique look at history. 842pp. 5⅜ × 8½. 20468-5 Pa. $10.95

THE BLUE FAIRY BOOK, Andrew Lang. The first, most famous collection, with many familiar tales: Little Red Riding Hood, Aladdin and the Wonderful Lamp, Puss in Boots, Sleeping Beauty, Hansel and Gretel, Rumpelstiltskin; 37 in all. 138 illustrations. 390pp. 5⅜ × 8½. 21437-0 Pa. $5.95

THE STORY OF THE CHAMPIONS OF THE ROUND TABLE, Howard Pyle. Sir Launcelot, Sir Tristram and Sir Percival in spirited adventures of love and triumph retold in Pyle's inimitable style. 50 drawings, 31 full-page. xviii + 329pp. 6½ × 9¼. 21883-X Pa. $6.95

AUDUBON AND HIS JOURNALS, Maria Audubon. Unmatched two-volume portrait of the great artist, naturalist and author contains his journals, an excellent biography by his granddaughter, expert annotations by the noted ornithologist, Dr. Elliott Coues, and 37 superb illustrations. Total of 1,200pp. 5⅜ × 8.

Vol. I 25143-8 Pa. $8.95
Vol. II 25144-6 Pa. $8.95

GREAT DINOSAUR HUNTERS AND THEIR DISCOVERIES, Edwin H. Colbert. Fascinating, lavishly illustrated chronicle of dinosaur research, 1820's to 1960. Achievements of Cope, Marsh, Brown, Buckland, Mantell, Huxley, many others. 384pp. 5¼ × 8¼. 24701-5 Pa. $6.95

THE TASTEMAKERS, Russell Lynes. Informal, illustrated social history of American taste 1850's-1950's. First popularized categories Highbrow, Lowbrow, Middlebrow. 129 illustrations. New (1979) afterword. 384pp. 6 × 9.

23993-4 Pa. $6.95

DOUBLE CROSS PURPOSES, Ronald A. Knox. A treasure hunt in the Scottish Highlands, an old map, unidentified corpse, surprise discoveries keep reader guessing in this cleverly intricate tale of financial skullduggery. 2 black-and-white maps. 320pp. 5⅜ × 8½. (Available in U.S. only) 25032-6 Pa. $5.95

AUTHENTIC VICTORIAN DECORATION AND ORNAMENTATION IN FULL COLOR: 46 Plates from "Studies in Design," Christopher Dresser. Superb full-color lithographs reproduced from rare original portfolio of a major Victorian designer. 48pp. 9¼ × 12¼. 25083-0 Pa. $7.95

PRIMITIVE ART, Franz Boas. Remains the best text ever prepared on subject, thoroughly discussing Indian, African, Asian, Australian, and, especially, Northern American primitive art. Over 950 illustrations show ceramics, masks, totem poles, weapons, textiles, paintings, much more. 376pp. 5⅜ × 8. 20025-6 Pa. $6.95

SIDELIGHTS ON RELATIVITY, Albert Einstein. Unabridged republication of two lectures delivered by the great physicist in 1920–21. *Ether and Relativity* and *Geometry and Experience*. Elegant ideas in non-mathematical form, accessible to intelligent layman. vi + 56pp. 5⅜ × 8½. 24511-X Pa. $2.95

THE WIT AND HUMOR OF OSCAR WILDE, edited by Alvin Redman. More than 1,000 ripostes, paradoxes, wisecracks: Work is the curse of the drinking classes, I can resist everything except temptation, etc. 258pp. 5⅜ × 8½. 20602-5 Pa. $3.95

ADVENTURES WITH A MICROSCOPE, Richard Headstrom. 59 adventures with clothing fibers, protozoa, ferns and lichens, roots and leaves, much more. 142 illustrations. 232pp. 5⅜ × 8½. 23471-1 Pa. $3.95

CATALOG OF DOVER BOOKS

PLANTS OF THE BIBLE, Harold N. Moldenke and Alma L. Moldenke. Standard reference to all 230 plants mentioned in Scriptures. Latin name, biblical reference, uses, modern identity, much more. Unsurpassed encyclopedic resource for scholars, botanists, nature lovers, students of Bible. Bibliography. Indexes. 123 black-and-white illustrations. 384pp. 6 × 9. 25069-5 Pa. $8.95

FAMOUS AMERICAN WOMEN: A Biographical Dictionary from Colonial Times to the Present, Robert McHenry, ed. From Pocahontas to Rosa Parks, 1,035 distinguished American women documented in separate biographical entries. Accurate, up-to-date data, numerous categories, spans 400 years. Indices. 493pp. 6½ × 9¼. 24523-3 Pa. $9.95

THE FABULOUS INTERIORS OF THE GREAT OCEAN LINERS IN HISTORIC PHOTOGRAPHS, William H. Miller, Jr. Some 200 superb photographs capture exquisite interiors of world's great "floating palaces"—1890's to 1980's: Titanic, Ile de France, Queen Elizabeth, United States, Europa, more. Approx. 200 black-and-white photographs. Captions. Text. Introduction. 160pp. 8⅜ × 11¼. 24756-2 Pa. $9.95

THE GREAT LUXURY LINERS, 1927–1954: A Photographic Record, William H. Miller, Jr. Nostalgic tribute to heyday of ocean liners. 186 photos of Ile de France, Normandie, Leviathan, Queen Elizabeth, United States, many others. Interior and exterior views. Introduction. Captions. 160pp. 9 × 12. 24056-8 Pa. $9.95

A NATURAL HISTORY OF THE DUCKS, John Charles Phillips. Great landmark of ornithology offers complete detailed coverage of nearly 200 species and subspecies of ducks: gadwall, sheldrake, merganser, pintail, many more. 74 full-color plates, 102 black-and-white. Bibliography. Total of 1,920pp. 8⅜ × 11¼. 25141-1, 25142-X Cloth. Two-vol. set $100.00

THE SEAWEED HANDBOOK: An Illustrated Guide to Seaweeds from North Carolina to Canada, Thomas F. Lee. Concise reference covers 78 species. Scientific and common names, habitat, distribution, more. Finding keys for easy identification. 224pp. 5⅜ × 8½. 25215-9 Pa. $5.95

THE TEN BOOKS OF ARCHITECTURE: The 1755 Leoni Edition, Leon Battista Alberti. Rare classic helped introduce the glories of ancient architecture to the Renaissance. 68 black-and-white plates. 336pp. 8⅜ × 11¼. 25239-6 Pa. $14.95

MISS MACKENZIE, Anthony Trollope. Minor masterpieces by Victorian master unmasks many truths about life in 19th-century England. First inexpensive edition in years. 392pp. 5⅜ × 8½. 25201-9 Pa. $7.95

THE RIME OF THE ANCIENT MARINER, Gustave Doré, Samuel Taylor Coleridge. Dramatic engravings considered by many to be his greatest work. The terrifying space of the open sea, the storms and whirlpools of an unknown ocean, the ice of Antarctica, more—all rendered in a powerful, chilling manner. Full text. 38 plates. 77pp. 9¼ × 12. 22305-1 Pa. $4.95

THE EXPEDITIONS OF ZEBULON MONTGOMERY PIKE, Zebulon Montgomery Pike. Fascinating first-hand accounts (1805-6) of exploration of Mississippi River, Indian wars, capture by Spanish dragoons, much more. 1,088pp. 5⅜ × 8½. 25254-X, 25255-8 Pa. Two-vol. set $23.90

CATALOG OF DOVER BOOKS

A CONCISE HISTORY OF PHOTOGRAPHY: Third Revised Edition, Helmut Gernsheim. Best one-volume history—camera obscura, photochemistry, daguerreotypes, evolution of cameras, film, more. Also artistic aspects—landscape, portraits, fine art, etc. 281 black-and-white photographs. 26 in color. 176pp. 8⅜ × 11¼. 25128-4 Pa. $12.95

THE DORÉ BIBLE ILLUSTRATIONS, Gustave Doré. 241 detailed plates from the Bible: the Creation scenes, Adam and Eve, Flood, Babylon, battle sequences, life of Jesus, etc. Each plate is accompanied by the verses from the King James version of the Bible. 241pp. 9 × 12. 23004-X Pa. $8.95

HUGGER-MUGGER IN THE LOUVRE, Elliot Paul. Second Homer Evans mystery-comedy. Theft at the Louvre involves sleuth in hilarious, madcap caper. "A knockout."—Books. 336pp. 5⅜ × 8½. 25185-3 Pa. $5.95

FLATLAND, E. A. Abbott. Intriguing and enormously popular science-fiction classic explores the complexities of trying to survive as a two-dimensional being in a three-dimensional world. Amusingly illustrated by the author. 16 illustrations. 103pp. 5⅜ × 8½. 20001-9 Pa. $2.00

THE HISTORY OF THE LEWIS AND CLARK EXPEDITION, Meriwether Lewis and William Clark, edited by Elliott Coues. Classic edition of Lewis and Clark's day-by-day journals that later became the basis for U.S. claims to Oregon and the West. Accurate and invaluable geographical, botanical, biological, meteorological and anthropological material. Total of 1,508pp. 5⅜ × 8½. 21268-8, 21269-6, 21270-X Pa. Three-vol. set $25.50

LANGUAGE, TRUTH AND LOGIC, Alfred J. Ayer. Famous, clear introduction to Vienna, Cambridge schools of Logical Positivism. Role of philosophy, elimination of metaphysics, nature of analysis, etc. 160pp. 5⅜ × 8½. (Available in U.S. and Canada only) 20010-8 Pa. $2.95

MATHEMATICS FOR THE NONMATHEMATICIAN, Morris Kline. Detailed, college-level treatment of mathematics in cultural and historical context, with numerous exercises. For liberal arts students. Preface. Recommended Reading Lists. Tables. Index. Numerous black-and-white figures. xvi + 641pp. 5⅜ × 8½. 24823-2 Pa. $11.95

28 SCIENCE FICTION STORIES, H. G. Wells. Novels, *Star Begotten* and *Men Like Gods,* plus 26 short stories: "Empire of the Ants," "A Story of the Stone Age," "The Stolen Bacillus," "In the Abyss," etc. 915pp. 5⅜ × 8½. (Available in U.S. only) 20265-8 Cloth. $10.95

HANDBOOK OF PICTORIAL SYMBOLS, Rudolph Modley. 3,250 signs and symbols, many systems in full; official or heavy commercial use. Arranged by subject. Most in Pictorial Archive series. 143pp. 8⅛ × 11. 23357-X Pa. $5.95

INCIDENTS OF TRAVEL IN YUCATAN, John L. Stephens. Classic (1843) exploration of jungles of Yucatan, looking for evidences of Maya civilization. Travel adventures, Mexican and Indian culture, etc. Total of 669pp. 5⅜ × 8½. 20926-1, 20927-X Pa., Two-vol. set $9.90

CHRISTMAS CUSTOMS AND TRADITIONS, Clement A. Miles. Origin, evolution, significance of religious, secular practices. Caroling, gifts, yule logs, much more. Full, scholarly yet fascinating; non-sectarian. 400pp. 5⅜ × 8½.
23354-5 Pa. $6.50

THE HUMAN FIGURE IN MOTION, Eadweard Muybridge. More than 4,500 stopped-action photos, in action series, showing undraped men, women, children jumping, lying down, throwing, sitting, wrestling, carrying, etc. 390pp. 7⅞ × 10⅝.
20204-6 Cloth. $19.95

THE MAN WHO WAS THURSDAY, Gilbert Keith Chesterton. Witty, fast-paced novel about a club of anarchists in turn-of-the-century London. Brilliant social, religious, philosophical speculations. 128pp. 5⅜ × 8½.
25121-7 Pa. $3.95

A CEZANNE SKETCHBOOK: Figures, Portraits, Landscapes and Still Lifes, Paul Cezanne. Great artist experiments with tonal effects, light, mass, other qualities in over 100 drawings. A revealing view of developing master painter, precursor of Cubism. 102 black-and-white illustrations. 144pp. 8¾ × 6⅜.
24790-2 Pa. $5.95

AN ENCYCLOPEDIA OF BATTLES: Accounts of Over 1,560 Battles from 1479 B.C. to the Present, David Eggenberger. Presents essential details of every major battle in recorded history, from the first battle of Megiddo in 1479 B.C. to Grenada in 1984. List of Battle Maps. New Appendix covering the years 1967–1984. Index. 99 illustrations. 544pp. 6½ × 9¼.
24913-1 Pa. $14.95

AN ETYMOLOGICAL DICTIONARY OF MODERN ENGLISH, Ernest Weekley. Richest, fullest work, by foremost British lexicographer. Detailed word histories. Inexhaustible. Total of 856pp. 6½ × 9¼.
21873-2, 21874-0 Pa., Two-vol. set $17.00

WEBSTER'S AMERICAN MILITARY BIOGRAPHIES, edited by Robert McHenry. Over 1,000 figures who shaped 3 centuries of American military history. Detailed biographies of Nathan Hale, Douglas MacArthur, Mary Hallaren, others. Chronologies of engagements, more. Introduction. Addenda. 1,033 entries in alphabetical order. xi + 548pp. 6½ × 9¼. (Available in U.S. only)
24758-9 Pa. $11.95

LIFE IN ANCIENT EGYPT, Adolf Erman. Detailed older account, with much not in more recent books: domestic life, religion, magic, medicine, commerce, and whatever else needed for complete picture. Many illustrations. 597pp. 5⅜ × 8½.
22632-8 Pa. $8.50

HISTORIC COSTUME IN PICTURES, Braun & Schneider. Over 1,450 costumed figures shown, covering a wide variety of peoples: kings, emperors, nobles, priests, servants, soldiers, scholars, townsfolk, peasants, merchants, courtiers, cavaliers, and more. 256pp. 8⅜ × 11¼.
23150-X Pa. $7.95

THE NOTEBOOKS OF LEONARDO DA VINCI, edited by J. P. Richter. Extracts from manuscripts reveal great genius; on painting, sculpture, anatomy, sciences, geography, etc. Both Italian and English. 186 ms. pages reproduced, plus 500 additional drawings, including studies for *Last Supper, Sforza* monument, etc. 860pp. 7⅞ × 10¾. (Available in U.S. only) 22572-0, 22573-9 Pa., Two-vol. set $25.90

HOW THE OTHER HALF LIVES, Jacob A. Riis. Journalistic record of filth, degradation, upward drive in New York immigrant slums, shops, around 1900. New edition includes 100 original Riis photos, monuments of early photography. 233pp. 10 × 7⅞. 22012-5 Pa. $7.95

CHINA AND ITS PEOPLE IN EARLY PHOTOGRAPHS, John Thomson. In 200 black-and-white photographs of exceptional quality photographic pioneer Thomson captures the mountains, dwellings, monuments and people of 19th-century China. 272pp. 9⅜ × 12¼. 24393-1 Pa. $12.95

GODEY COSTUME PLATES IN COLOR FOR DECOUPAGE AND FRAMING, edited by Eleanor Hasbrouk Rawlings. 24 full-color engravings depicting 19th-century Parisian haute couture. Printed on one side only. 56pp. 8¼ × 11. 23879-2 Pa. $3.95

ART NOUVEAU STAINED GLASS PATTERN BOOK, Ed Sibbett, Jr. 104 projects using well-known themes of Art Nouveau: swirling forms, florals, peacocks, and sensuous women. 60pp. 8¼ × 11. 23577-7 Pa. $3.50

QUICK AND EASY PATCHWORK ON THE SEWING MACHINE: Susan Aylsworth Murwin and Suzzy Payne. Instructions, diagrams show exactly how to machine sew 12 quilts. 48pp. of templates. 50 figures. 80pp. 8¼ × 11. 23770-2 Pa. $3.50

THE STANDARD BOOK OF QUILT MAKING AND COLLECTING, Marguerite Ickis. Full information, full-sized patterns for making 46 traditional quilts, also 150 other patterns. 483 illustrations. 273pp. 6⅞ × 9⅜. 20582-7 Pa. $5.95

LETTERING AND ALPHABETS, J. Albert Cavanagh. 85 complete alphabets lettered in various styles; instructions for spacing, roughs, brushwork. 121pp. 8¾ × 8. 20053-1 Pa. $3.95

LETTER FORMS: 110 COMPLETE ALPHABETS, Frederick Lambert. 110 sets of capital letters; 16 lower case alphabets; 70 sets of numbers and other symbols. 110pp. 8⅛ × 11. 22872-X Pa. $4.50

ORCHIDS AS HOUSE PLANTS, Rebecca Tyson Northen. Grow cattleyas and many other kinds of orchids—in a window, in a case, or under artificial light. 63 illustrations. 148pp. 5⅜ × 8½. 23261-1 Pa. $2.95

THE MUSHROOM HANDBOOK, Louis C.C. Krieger. Still the best popular handbook. Full descriptions of 259 species, extremely thorough text, poisons, folklore, etc. 32 color plates; 126 other illustrations. 560pp. 5⅜ × 8½. 21861-9 Pa. $8.50

THE DORÉ BIBLE ILLUSTRATIONS, Gustave Doré. All wonderful, detailed plates: Adam and Eve, Flood, Babylon, life of Jesus, etc. Brief King James text with each plate. 241 plates. 241pp. 9 × 12. 23004-X Pa. $8.95

THE BOOK OF KELLS: Selected Plates in Full Color, edited by Blanche Cirker. 32 full-page plates from greatest manuscript-icon of early Middle Ages. Fantastic, mysterious. Publisher's Note. Captions. 32pp. 9¾ × 12¼. 24345-1 Pa. $4.50

THE PERFECT WAGNERITE, George Bernard Shaw. Brilliant criticism of the Ring Cycle, with provocative interpretation of politics, economic theories behind the Ring. 136pp. 5⅜ × 8½. (Available in U.S. only) 21707-8 Pa. $3.00

THE RIME OF THE ANCIENT MARINER, Gustave Doré, S.T. Coleridge. Doré's finest work, 34 plates capture moods, subtleties of poem. Full text. 77pp. 9¼ × 12. 22305-1 Pa. $4.95

SONGS OF INNOCENCE, William Blake. The first and most popular of Blake's famous "Illuminated Books," in a facsimile edition reproducing all 31 brightly colored plates. Additional printed text of each poem. 64pp. 5¼ × 7. 22764-2 Pa. $3.50

AN INTRODUCTION TO INFORMATION THEORY, J.R. Pierce. Second (1980) edition of most impressive non-technical account available. Encoding, entropy, noisy channel, related areas, etc. 320pp. 5⅜ × 8½. 24061-4 Pa. $4.95

THE DIVINE PROPORTION: A STUDY IN MATHEMATICAL BEAUTY, H.E. Huntley. "Divine proportion" or "golden ratio" in poetry, Pascal's triangle, philosophy, psychology, music, mathematical figures, etc. Excellent bridge between science and art. 58 figures. 185pp. 5⅜ × 8½. 22254-3 Pa. $3.95

THE DOVER NEW YORK WALKING GUIDE: From the Battery to Wall Street, Mary J. Shapiro. Superb inexpensive guide to historic buildings and locales in lower Manhattan: Trinity Church, Bowling Green, more. Complete Text; maps. 36 illustrations. 48pp. 3⅞ × 9¼. 24225-0 Pa. $2.50

NEW YORK THEN AND NOW, Edward B. Watson, Edmund V. Gillon, Jr. 83 important Manhattan sites: on facing pages early photographs (1875-1925) and 1976 photos by Gillon. 172 illustrations. 171pp. 9¼ × 10. 23361-8 Pa. $7.95

HISTORIC COSTUME IN PICTURES, Braun & Schneider. Over 1450 costumed figures from dawn of civilization to end of 19th century. English captions. 125 plates. 256pp. 8⅜ × 11¼. 23150-X Pa. $7.50

VICTORIAN AND EDWARDIAN FASHION: A Photographic Survey, Alison Gernsheim. First fashion history completely illustrated by contemporary photographs. Full text plus 235 photos, 1840-1914, in which many celebrities appear. 240pp. 6½ × 9¼. 24205-6 Pa. $6.00

CHARTED CHRISTMAS DESIGNS FOR COUNTED CROSS-STITCH AND OTHER NEEDLECRAFTS, Lindberg Press. Charted designs for 45 beautiful needlecraft projects with many yuletide and wintertime motifs. 48pp. 8¼ × 11. 24356-7 Pa. $2.50

101 FOLK DESIGNS FOR COUNTED CROSS-STITCH AND OTHER NEEDLE-CRAFTS, Carter Houck. 101 authentic charted folk designs in a wide array of lovely representations with many suggestions for effective use. 48pp. 8¼ × 11. 24369-9 Pa. $2.25

FIVE ACRES AND INDEPENDENCE, Maurice G. Kains. Great back-to-the-land classic explains basics of self-sufficient farming. The one book to get. 95 illustrations. 397pp. 5⅜ × 8½. 20974-1 Pa. $4.95

A MODERN HERBAL, Margaret Grieve. Much the fullest, most exact, most useful compilation of herbal material. Gigantic alphabetical encyclopedia, from aconite to zedoary, gives botanical information, medical properties, folklore, economic uses, and much else. Indispensable to serious reader. 161 illustrations. 888pp. 6½ × 9¼. (Available in U.S. only) 22798-7, 22799-5 Pa., Two-vol. set $16.45

DECORATIVE NAPKIN FOLDING FOR BEGINNERS, Lillian Oppenheimer and Natalie Epstein. 22 different napkin folds in the shape of a heart, clown's hat, love knot, etc. 63 drawings. 48pp. 8¼ × 11. 23797-4 Pa. $1.95

DECORATIVE LABELS FOR HOME CANNING, PRESERVING, AND OTHER HOUSEHOLD AND GIFT USES, Theodore Menten. 128 gummed, perforated labels, beautifully printed in 2 colors. 12 versions. Adhere to metal, glass, wood, ceramics. 24pp. 8¼ × 11. 23219-0 Pa. $2.95

EARLY AMERICAN STENCILS ON WALLS AND FURNITURE, Janet Waring. Thorough coverage of 19th-century folk art: techniques, artifacts, surviving specimens. 166 illustrations, 7 in color. 147pp. of text. 7⅞ × 10¾. 21906-2 Pa. $9.95

AMERICAN ANTIQUE WEATHERVANES, A.B. & W.T. Westervelt. Extensively illustrated 1883 catalog exhibiting over 550 copper weathervanes and finials. Excellent primary source by one of the principal manufacturers. 104pp. 6⅛ × 9¼. 24396-6 Pa. $3.95

ART STUDENTS' ANATOMY, Edmond J. Farris. Long favorite in art schools. Basic elements, common positions, actions. Full text, 158 illustrations. 159pp. 5⅜ × 8½. 20744-7 Pa. $3.95

BRIDGMAN'S LIFE DRAWING, George B. Bridgman. More than 500 drawings and text teach you to abstract the body into its major masses. Also specific areas of anatomy. 192pp. 6½ × 9¼. (EA) 22710-3 Pa. $4.50

COMPLETE PRELUDES AND ETUDES FOR SOLO PIANO, Frederic Chopin. All 26 Preludes, all 27 Etudes by greatest composer of piano music. Authoritative Paderewski edition. 224pp. 9 × 12. (Available in U.S. only) 24052-5 Pa. $7.50

PIANO MUSIC 1888-1905, Claude Debussy. Deux Arabesques, Suite Bergamesque, Masques, 1st series of Images, etc. 9 others, in corrected editions. 175pp. 9⅜ × 12¼. (ECE) 22771-5 Pa. $5.95

TEDDY BEAR IRON-ON TRANSFER PATTERNS, Ted Menten. 80 iron-on transfer patterns of male and female Teddys in a wide variety of activities, poses, sizes. 48pp. 8¼ × 11. 24596-9 Pa. $2.25

A PICTURE HISTORY OF THE BROOKLYN BRIDGE, M.J. Shapiro. Profusely illustrated account of greatest engineering achievement of 19th century. 167 rare photos & engravings recall construction, human drama. Extensive, detailed text. 122pp. 8¼ × 11. 24403-2 Pa. $7.95

NEW YORK IN THE THIRTIES, Berenice Abbott. Noted photographer's fascinating study shows new buildings that have become famous and old sights that have disappeared forever. 97 photographs. 97pp. 11⅜ × 10. 22967-X Pa. $7.50

MATHEMATICAL TABLES AND FORMULAS, Robert D. Carmichael and Edwin R. Smith. Logarithms, sines, tangents, trig functions, powers, roots, reciprocals, exponential and hyperbolic functions, formulas and theorems. 269pp. 5⅜ × 8½. 60111-0 Pa. $4.95

HANDBOOK OF MATHEMATICAL FUNCTIONS WITH FORMULAS, GRAPHS, AND MATHEMATICAL TABLES, edited by Milton Abramowitz and Irene A. Stegun. Vast compendium: 29 sets of tables, some to as high as 20 places. 1,046pp. 8 × 10½. 61272-4 Pa. $19.95

REASON IN ART, George Santayana. Renowned philosopher's provocative, seminal treatment of basis of art in instinct and experience. Volume Four of *The Life of Reason*. 230pp. 5⅜ × 8. 24358-3 Pa. $4.50

LANGUAGE, TRUTH AND LOGIC, Alfred J. Ayer. Famous, clear introduction to Vienna, Cambridge schools of Logical Positivism. Role of philosophy, elimination of metaphysics, nature of analysis, etc. 160pp. 5⅜ × 8½. (USCO)
20010-8 Pa. $2.75

BASIC ELECTRONICS, U.S. Bureau of Naval Personnel. Electron tubes, circuits, antennas, AM, FM, and CW transmission and receiving, etc. 560 illustrations. 567pp. 6½ × 9¼. 21076-6 Pa. $8.95

THE ART DECO STYLE, edited by Theodore Menten. Furniture, jewelry, metalwork, ceramics, fabrics, lighting fixtures, interior decors, exteriors, graphics from pure French sources. Over 400 photographs. 183pp. 8⅜ × 11¼.
22824-X Pa. $6.95

THE FOUR BOOKS OF ARCHITECTURE, Andrea Palladio. 16th-century classic covers classical architectural remains, Renaissance revivals, classical orders, etc. 1738 Ware English edition. 216 plates. 110pp. of text. 9½ × 12¾.
21308-0 Pa. $11.50

THE WIT AND HUMOR OF OSCAR WILDE, edited by Alvin Redman. More than 1000 ripostes, paradoxes, wisecracks: Work is the curse of the drinking classes, I can resist everything except temptations, etc. 258pp. 5⅜ × 8½. (USCO)
20602-5 Pa. $3.95

THE DEVIL'S DICTIONARY, Ambrose Bierce. Barbed, bitter, brilliant witticisms in the form of a dictionary. Best, most ferocious satire America has produced. 145pp. 5⅜ × 8½. 20487-1 Pa. $2.50

ERTÉ'S FASHION DESIGNS, Erté. 210 black-and-white inventions from *Harper's Bazar*, 1918-32, plus 8pp. full-color covers. Captions. 88pp. 9 × 12.
24203-X Pa. $6.50

ERTÉ GRAPHICS, Erté. Collection of striking color graphics: *Seasons, Alphabet, Numerals, Aces* and *Precious Stones*. 50 plates, including 4 on covers. 48pp. 9⅜ × 12¼. 23580-7 Pa. $6.95

PAPER FOLDING FOR BEGINNERS, William D. Murray and Francis J. Rigney. Clearest book for making origami sail boats, roosters, frogs that move legs, etc. 40 projects. More than 275 illustrations. 94pp. 5⅜ × 8½. 20713-7 Pa. $2.25

ORIGAMI FOR THE ENTHUSIAST, John Montroll. Fish, ostrich, peacock, squirrel, rhinoceros, Pegasus, 19 other intricate subjects. Instructions. Diagrams. 128pp. 9 × 12. 23799-0 Pa. $4.95

CROCHETING NOVELTY POT HOLDERS, edited by Linda Macho. 64 useful, whimsical pot holders feature kitchen themes, animals, flowers, other novelties. Surprisingly easy to crochet. Complete instructions. 48pp. 8¼ × 11.
24296-X Pa. $1.95

CROCHETING DOILIES, edited by Rita Weiss. Irish Crochet, Jewel, Star Wheel, Vanity Fair and more. Also luncheon and console sets, runners and centerpieces. 51 illustrations. 48pp. 8¼ × 11. 23424-X Pa. $2.50